SURVEYING NATURAL POPULATIONS

Surveying
Natural Populations

QUANTITATIVE TOOLS FOR ASSESSING BIODIVERSITY

SECOND EDITION

LEE-ANN C. HAYEK & MARTIN A. BUZAS

COLUMBIA UNIVERSITY PRESS

NEW YORK

COLUMBIA UNIVERSITY PRESS
PUBLISHERS SINCE 1893
NEW YORK   CHICHESTER, WEST SUSSEX
COPYRIGHT © 2010 THE SMITHSONIAN INSTITUTION
ALL RIGHTS RESERVED

LIBRARY OF CONGRESS CATALOGING-IN-PUBLICATION DATA
HAYEK, LEE-ANN C.
SURVEYING NATURAL POPULATIONS : QUANTITATIVE TOOLS FOR ASSESSING
BIODIVERSITY / LEE-ANN C. HAYEK AND MARTIN A. BUZAS. — 2ND ED.
P. CM.
ISBN 978-0-231-14620-3 (CLOTH : ALK. PAPER)
1. POPULATION BIOLOGY–STATISTICAL METHODS.   2. PALEOECOLOGY–
STATISTICAL METHODS.   I. BUZAS, MARTIN A.   II. TITLE.

QH532.H39 2010
577.8'80727–DC22

2010009475

COLUMBIA UNIVERSITY PRESS BOOKS ARE PRINTED ON PERMANENT AND
DURABLE ACID-FREE PAPER.
THIS BOOK IS PRINTED ON PAPER WITH RECYCLED CONTENT.
PRINTED IN THE UNITED STATES OF AMERICA

C 10 9 8 7 6 5 4 3 2 1

To Christine, Matthew, Katherine, Aron, and my new daughter Kate

and

To Barbara, Pamela, Jeffrey, and Thomas

CONTENTS

PREFACE

WE BEGAN THIS WORK as a revision of our text *Surveying Natural Populations* published in 1997. Fortunately, our colleagues at many universities and colleges kindly identified textual errors or misleading writing that we incorporated. Those who used the book as a classroom text and/or reference gave especially helpful advice. However, as we began this updating and especially when writing about the exciting new and newly developed material on SHE analysis, we realized that this book would require not just a major revision, but also new chapters. In particular, we incorporated entire developmental chapters on quantitative biodiversity. This book now contains 7 chapters on the analysis and interpretation of the quantitative results that are obtained from a study of biodiversity. Chapters 12 through 18 provide a synthesis of the entire field of biodiversity assessment. Most importantly, in this book all of the major ecological and biodiversity indices have been related mathematically under the umbrella of information measures. We show how the use of a single measure is not a solitary numerical calculation but a purposeful selection of a member of a mathematical family, each selection having repercussions for the interpretation of the resultant biodiversity data. We also demonstrate how many suggested diversity indices must be rejected because they lack vital properties and they do not follow an information–theoretic scheme wherein all measures are mathematically linked. Use of such a scheme not only relates all measures within a study but also provides the standardization necessary for individual studies to be related over time or space allowing for more insight and stronger generalizations from biodiversity analyses.

In the intervening decade since the first book was published, the plight of both humans and other species has been brought more clearly to the attention of both scientists and the public. Our declining global conditions of anthropogenic origin are sending imperatives to those who study natural populations; we can no longer utilize qualitative assessments of biodiversity. We must use the latest and most mathematically useful methods for the evaluation of affected and potentially affected species assemblages and related environmental conditions. We believe that this book is needed now more than ever.

As we stated in the Preface to our original book, the degradation of environments and the fate of natural populations that inhabit them has become a topic of paramount concern throughout the world. Obtaining collections from the field along with pertinent observations, as well as the identification and enumeration of organisms, falls within the domain of field biologists. Because of the enormous quantities involved, all organisms cannot be directly counted and population parameters must be estimated. Estimation from sample to population falls within the domain of mathematical statisticians. Unfortunately, the two groups of researchers often do not communicate well. This book attempts to bridge the gap.

One author (LCH) is a mathematical statistician with extensive experience in statistical theory and practice for developing and interpreting statistical science for solving ecological, natural history, and biodiversity problems; the other (MAB) is a researcher in quantitative ecology–paleoecology who is familiar with statistics usage. Their knowledge and experience form overlapping sets. Together, they integrate the intuition of the experienced field researcher with the statistical rationale required to survey natural populations.

This is not a book for statisticians. The authors do not offer mathematical proofs of their results. Statistical knowledge is extrapolated without the type of documentation that mathematical statisticians expect. The concepts and complexities contained in the area of mathematical application are, however, requisite knowledge for field biologists and conservation biologists with marine or terrestrial specialties as well as for site managers, paleontologists, and archaeologists. This book is for you.

No prior statistical knowledge is assumed, and only a familiarity with, or, at least not an aversion to algebra is required. This is a versatile book that can be used at many levels: (1) serious students can learn biological and statistical principles fundamental to sound quantitative surveys; (2) experienced researchers can, for the first time in a single readable source, find the underlying statistical theory for their commonly used field approaches; and (3) graduate students and researchers can discover many new relationships, derivations, and measures that can open new research areas suggested throughout the book. For field researchers in natural systems, this book is a reference, a text, a tool, and a guide.

ACKNOWLEDGMENTS

W E THANK F. DALLMEIER, Director of the Smithsonian Man and the Biosphere Program at the Smithsonian's National Zoological Park, for graciously providing the primary data sets used in this book. Data sets and information on the Indian River field project were provided by S. Reed of the Smithsonian's Marine Station at Ft. Pierce, Florida, and for the amphibian work by R. Heyer of the Smithsonian's National Museum of Natural History.

Without J. Jett this book could never have been completed; we thank her for her invaluable assistance, organization, checking, for her exceptional effort of compiling the index and for putting up with us and our hectic schedule. A. Hayek entered data for some of the problems and ran some analyses for the solutions. We thank M. Parrish from the NMNH Department of Paleobiology for the cover and interior artwork.

SURVEYING NATURAL POPULATIONS

1

INTRODUCTION

SURVEYING NATURAL POPULATIONS is not a new endeavor for *Homo sapiens*. Hunters and gatherers needed to classify, find, and catch organisms in order to survive. Communication of what was edible and where it could be found were the essential bits of information necessary for the survival of primitive populations.

The problem of *what* still requires careful observation of animals and plants so that one group can be discriminated from another. After all, survival could depend on it. For example, strangers to the forest gather and eat mushrooms with considerable risk. Local inhabitants with taxonomic skills recognize the subtle differences required to classify the fungi into edible and nonedible. Their taxonomic skills provide them with reliable sources of food and medicine.

Once we can identify *what,* the problem of *where* requires knowledge about the organism's habitat. Clearly, we do not expect to find fish on dry land. However, the identification of what organism lives where demands a little more sophistication. We may now ask how many of the "whats" live where. We may also consider the best way to catch or observe them. If we know something about an organism's behavior and how many of these organisms we are likely to find, the likelihood of catching or observing greatly increases. We do not rake an aquatic environment to catch fish, but may find such a technique quite effective for acquiring clams.

Even though we may have figured out what and where, most likely we also need to know *when*. Organisms are distributed not only in space but also in time. *When* along with *where*, for example, is important if we wish to study the migration of birds.

Once *Homo sapiens* became efficient at hunting, gathering, and domesticating (farming) organisms, "natural" curiosity led us to expand our observations in an attempt to understand or explain the world about us. Understanding, in some cases, can of course also allow for more efficient exploitation. We are still trying to understand more and more about the natural world; our curiosity demands it. Such knowledge is now crucial for conservation as well as exploitation.

In this book we show why it is necessary to quantify observations of natural populations. We then present the basic building blocks of descriptive and inferential statistics. We aim to help readers—for example, students, conservation administrators, conservationists, and experienced researchers and fieldworkers—understand the mathematical concepts necessary for undertaking a survey of any natural population.

We believe that biodiversity studies should be well organized, cost-efficient, and detailed. These standards apply to inventories for producing species lists as well as more comprehensive monitoring efforts to acquire data on abundance, environmental relationships, and time trends. This book reflects those concerns. In many chapters we explain the usual field or analysis procedures but then offer variations to strengthen the design of the work and the inferences from it. We also present new ways to conceptualize or analyze the study data in order to make the most efficient use of time, personnel, and funds.

Key Concepts

Systematics. Simple identification of the "what" essential for survival has become much more sophisticated with the science of *systematics*. Systematists not only classify organisms but arrange them in a hierarchical fashion, postulating evolutionary pathways in an attempt to explain ancestral relationships. Such systematic studies have produced large collections of specimens, along with data about where and when each was captured. The specimens are lodged in natural history museums, such as the Museum of Natural History at the Smithsonian Institution in Washington, DC. Although these collections are only a small sample of the species that inhabit(ed) the world, they nevertheless provide a baseline for designing rigorous surveys of natural populations. Museum collections are often overlooked, however, when surveys are initiated and pilot data are sought, resulting in recollection of information that is already available.

Biogeography and ecology. All organisms have the potential to reproduce at a geometric rate, and hence to populate the world. The *where* and *when* observations show that they do not. The fields of *biogeography* and *ecology* attempt to explain why not. The study of the distribution of organisms (what and how many) in space (where) and time (when), combined with the study of biotic and abiotic factors that may explain the distribution, form the kernel of these two fields. Using all the pieces of information together in an evolutionary context, biogeographers and ecologists try to understand the natural world and our place in it. What we see has been aptly described as the "ecological theater and the evolutionary play" (Hutchinson 1965). Considering the extent of the biosphere and the vastness of geologic time, this production is ambitious indeed.

Surveying natural populations. To chart the distribution of organisms over vast amounts of space and time is a monumental task. Even a modest approach is quite difficult. Fundamentally, for each taxon of interest we search for data that will reveal:

1. *How many of what*—that is, the number of individuals, the weight, or size or biomass
2. *Where*—that is, the space they occupy in some selected units
3. *When*—that is, the time of observation, which can be measured by some type of clock

These tasks are inherently quantitative. How shall we begin? Imagine a single locality where we observe the comings and goings of organisms over time. Because of the limitations of our own senses and memory capacity, we could imagine making a motion picture or videotape of life's proceedings at this locality. At the end of the observation period we would perhaps begin by examining the contents of a single frame. The still picture captures one moment in time at a single locality. The observation is thus fixed in space and time. We can search the contents of this *frame* for whatever organism or organisms may be the object of our study. Fortunately, nature sometimes provides us with a frame frozen in time, that is, fossils.

If we are concerned with the number of individuals of a particular type, all we need to do is count them. If we are focusing on plants or a colonial animal, we might wish to measure the amount of space covered. Direct measurement may be difficult, however; so we may need to rely on a secondary measure, such as the number of chirps or croaks heard at a site in a given time. We call the number enumerated in a designated space (frame) the *density*. Whatever "what" we choose to observe (population, species, other category or grouping), this category becomes the *target population*.

Scale. The size of the locality or frame depends on a chosen viewpoint because, unlike a camera's lens, human imagination is not restricted. We may be interested in the North Atlantic or part of a pond in New England. Likewise, what we are interested in may vary from a local population of a single species to several phyla; there is no single correct scale.

Patterns exist at all scales, even though the mechanisms we invoke to explain them may be restricted to one or several scales (see the review by Levin 1992; Aronson 1994). On a large enough scale (for example, all land) even the broadest categories of organisms will be patchy, whereas on smaller scales (1 hectare of land), the patchiness may disappear. The importance of scale is also apparent temporally, and the same phenomenon can reappear in time and space.

Figure 1.1 shows a distinct periodicity of some measure over a period of a year. (The wave is generated by $\sin \frac{\pi}{3} \cdot$ weeks.) Many species exhibit a seasonal periodicity. A researcher suspecting such a periodicity might choose a seasonal sampling plan, with sampling times selected to coincide with the middle of each of the distinctive seasons.

If a researcher decided that sampling once in the middle of each of the, say, four seasons (4 samples) were sufficient to describe the suspected seasonal periodicity, the weeks 3, 16, 29, and 42 (the middle of each season) might be chosen. The pattern observed with this sampling scheme (Figure 1.1(B)) would be one of decrease from winter to fall. However, if the sampling were to detect the pattern shown in Figure 1.1(A), the scale of the observations would have been wrong. The researcher would have completely missed the truth and would draw erroneous conclusions. The scale of weeks selected for Figure 1.1 was arbitrary, and could, of course, have been months, years, or millions of years; cm or km. The point is that we must design studies such that spatial and temporal scales can reveal the patterns we wish to discern. Unfortunately, without some prior, pilot information, such a design may be impossible. In the example, we would need to know the required scale is weeks and not seasons.

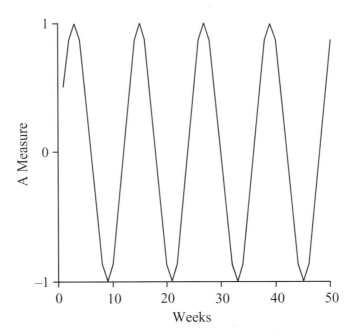

FIGURE I.I (A) A periodicity of some measure over a period of 1 year by weeks.

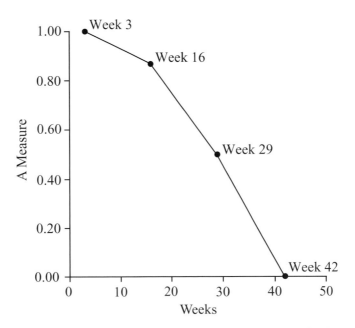

FIGURE I.I (B) Results of sampling from Figure 1.1(A) in the middle of each season or period: at weeks 3, 16, 29, and 42.

Statistical field surveys. Counting all the individuals in a target population that are within a chosen frame is ideal. Sometimes this is possible. Often, however, individuals are too numerous to count. When we are interested in a large area, this is nearly always the case. For example, we could count the grains in a milliliter or so of sand, but we cannot count all the sand grains on a beach. If a target population is large, we often examine just a segment—the *sample population*, or *sample*. On the basis of this sample, we then estimate some characteristic(s) of the target population. Generalization about a populationwide characteristic drawn from a sample(s) of that population is termed *statistical inference*. Such generalizations cannot be made with perfect certainty. By using suitable statistical methods, however, we can measure the degree of uncertainty in terms of probability.

The degree of uncertainty in our estimate is often stated as a degree of confidence in a particular statement concerning the characteristic. The various polls used to estimate public opinion on a wide variety of subjects use such statistical methods. All such polls have a frame, a target population, and a sample population. For example, we might be interested in the number of adult females in Boston who prefer a particular political candidate. The frame is Boston; the target population is females over age 18, from which a suitable sample population must be chosen. Providing that a suitable sample has been drawn from the target population, we can estimate the number of females preferring a particular candidate with some degree of confidence (a probability statement). The frame is, however, frozen at the time the survey was conducted—which is why the pollsters say, "If the election were held today. . . ." Estimating population characteristics from a sample not only saves time and money but also may be the only reasonable alternative when the target population is large.

The trees in a forest, like the sand grains on a beach, are usually too numerous to count. Consequently, we may select a representative area for our sample population and enumerate the trees within it. This representative area containing the sample population is called a *quadrat*. This term is now in such common usage that it is employed even when the sampling unit is not strictly a square area. Thus the term *quadrat* may refer to the volume from a core, a catch of a net, or some other contents of a sampling device. This book is not about how to sample for one particular kind of organism (manuals exist for this purpose—for example, the guide published by Heyer and colleagues in 1994) or about the characteristics of any particular organism. Instead, it is about characteristics of populations and samples that are applicable to any group we may choose.

If the target population is the total number of trees in a particular forest (the frame), we can count the number of trees in a known subarea of the forest (the quadrat, or sampled population, or sample). We can then simply multiply the number of trees in the chosen quadrat by the total number of possible quadrats in the forest to estimate the total number of trees in the forest. If we choose a second quadrat of the same size within the forest and count the trees within it, we usually find that the number of trees in the 2 quadrats is not exactly the same. Results may differ again in a third and fourth quadrat. Obviously, estimation is a little more complicated than simple multiplication. The simple procedure of making a count in a single quadrat and multiplying by the number of possible quadrats gives us an estimate, but we have no idea how reliable or precise that estimate is. Variation among the quadrat counts must be taken into account, and any estimate of the total trees in a forest will have to include some notice of the discrepancies.

A large body of statistical methodology exists to solve such problems. *Statistical surveys* are concerned with measuring in a probabilistic way so that statements can be made with some known amount of confidence. In addition to taxonomic expertise, sampling gear, field and laboratory methodology, and so on, the surveyor or sampler of natural populations needs these statistical tools also. The purpose of this book is to provide the mathematical statistical tools necessary for conducting samples or surveys of natural populations.

How to Use This Book

The Illustrative Example

For the purposes of illustrating statistical procedures in this book, we chose the results from an inventory of trees in a single hectare in the Beni Biosphere Reserve in Bolivia. This inventory is one of many conducted by the Smithsonian's Man and the Biosphere Biodiversity Program. If the entire forest of the reserve (about 90,000 hectares) is our frame, then this single hectare could be considered a quadrat within the frame. Such an approach, however, would not serve the purposes of this book because we wanted a single target population that was known. Instead, we chose the hectare itself as the frame.

Dallmeier and colleagues (1991) have identified each tree to species and have mapped the position of each individual tree within this hectare. From their exhaustive survey we know the location of each individual of each species. Our frame of 1 hectare contains a population consisting of 663 trees distributed among 52 species. This is the target population, and we know it exactly. Appendixes 1 and 2 contain the full data set.

The distribution of the 663 individual trees is represented by the dots shown in Figure 1.2. This figure is a graphical representation of our frame frozen at the time of mapping (1991). The figure is an incomplete representation because it does not contain 52 symbols to represent the different species. Depending on what we wish to illustrate at any particular point in this book, we will divide this hectare either into 100 quadrats or into 25 quadrats. We will then sample these quadrats as a basis for estimating various population characteristics of the target population. Usually, the estimates for these population characteristics will be accompanied by a confidence statement. Because we always know the true values of the target population, the reader will always know how well the statistical methods estimated the true values. No faith is required.

Although we occasionally use other data sets to illustrate a point or as the target populations for the problems at the ends of the chapters, and although those data sets may contain individuals from a wide variety of groups (not just trees), we use the Bolivian example again and again throughout this book. This approach has advantages and disadvantages. One advantage is that readers do not have to reorient themselves constantly to a new set of data with different organisms and different scales. Another advantage, previously mentioned, is that the true values are known, and so the student becomes comfortable with the definitions and explanations and is reassured about how well statistical methodology really works, even when all the assumptions are not met. An obvious

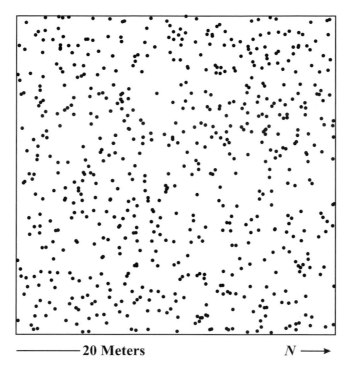

20 Meters $N \longrightarrow$

FIGURE I.2   The distribution of all 663 trees in plot 01 within the Beni Biosphere Reserve in Bolivia.

disadvantage is that the data set becomes so familiar that the reader longs for another set of numbers (we certainly did). We also worried that the reader might think that we have written a book about trees or might assume that we are using a single example and generalizing from it. This is not the case. Instead, general principles—valid across all natural populations—are illustrated through a single example.

An Overview of Terminology and Statistical Methods

You may have noticed that when we introduce terms we want you to remember they are italicized on first use. In this overview section, however, we will introduce technical terms without italics and without definition. Our intent is to provide an overview of what this book contains and what you will learn. Rest assured that each term will be adequately described in the chapters that follow.

We begin by using the data to produce a distribution of all trees in a single hectare of the 90,000 hectare Beni Biosphere Reserve in Bolivia (Figure 1.2). We could prepare a graphical representation that also distinguishes each of the 52 species represented in the plot. Whether we are examining individual species or total trees, the basic, or primary, data consist of dots representing individual trees. If we wish to estimate how many trees are present, without counting them all, we could divide the plot into quadrats and count the trees in some of them. This procedure allows us to estimate the average number per quadrat, or the mean density. By also examining the variability (the variance) and the statistical sampling distributions, we are able to make confidence statements about the mean density. This, in turn, allows us to estimate how many quadrats are required for any selected degree of confidence. We indicate the sort of precision that can be expected for estimates of the mean density of natural populations, given time, effort, and cost considerations.

The power curve is used to examine the relationship of the variance and the mean. This relationship is an effective way of examining spatial distributions by evaluating the spatial homogeneity of the space. The power curve is remarkably similar for a wide variety of organisms.

We also examine random, systematic, stratified, and cluster sampling schemes, individually and in combination. Consideration of these schemes and their interrelationships allows us to plan a sampling program efficiently; that is, to determine which quadrats should be sampled and to assess how the use of various sampling schemes will affect the estimates.

A secondary and very popular population characteristic that can be derived from the primary (original) data is the proportion that each species or taxon comprises of the total. These species proportions (expressed as percent) are often called percent species, relative abundance, or percentage abundance. We demonstrate in detail the method of calculation of confidence limits for species proportions and of the number of individuals required for any particular degree of confidence. We show that the methods for calculating confidence limits are quite different, depending on whether individuals or quadrats are sampled.

Surveyors of natural populations are interested not only in population characteristics of individual species or groups but also in the entire ensemble. In the Beni Biosphere Reserve, a few species are abundant, some are common, and most are rare. A similar pattern of species abundances is observed for most natural populations. We examine a number of mathematical distributions (Negative binomial, Log series, Broken Stick, Log normal, and others) to determine how well they describe or fit the observations. The strengths and weakness of each are explored.

When a homogeneous habitat is divided into a large number of quadrats, the more abundant species occur in many (sometimes all) of them, the common ones in fewer, and the rarest species, represented by 1 individual, in only 1 quadrat. There is, then, a distribution of species occurrences. Given certain stipulations, these occurrences can serve as a surrogate for the actual densities through the use of regression analysis. Species occurrences are also a way of looking at species abundances on a different scale, one in which much of the variability is reduced. Most of the statistical estimators used in this book can be applied to species occurrences. And yet, occurrence data have been greatly neglected as a source of information. The large number of museum collections, along with their accompanying data, are an untapped resource that may, by themselves, provide interesting insights to the naturalist. They may also serve as pilot studies for more detailed fieldwork.

The number of species (species richness) observed within some frame is the core measure for most biodiversity studies. The semilog and log–log relationships of the number of individuals to the number of species, as well as the relationship of area versus species, are examined using regression. We also explore several methods of estimating the number of species that would occur if samples were reduced to fewer individuals (rarefaction). Similarly, we investigate the problem of how many species would be observed if the number of individuals were increased (abundifaction).

Many researchers desire diversity indices that are independent of the number of individuals sampled and/or take into account the species abundances as well. The literature is replete with searches for indices that summarize the number of species and their abundances within a frame, regardless of the number of samples or individuals observed. In this book we explain and test a number of diversity indices. We show how many of the commonly used indices violate repeatability and decomposition. They are discarded. Particular attention is given to the Log series α and measures from information theory H, J, E, V. We show that if the species abundances conform to a particular distribution, a single measure such as α is adequate. However, if species abundances do not so conform, a more general solution must be sought. We abandon the concept of a single measure except in cases that are severely limited in space or time considerations and instead suggest an original methodology, using the number of species (S), a measure of information (H), and a measure of evenness (E) based on the information function—all of which are evaluated as a function of the number of individuals (N). We show how Simpson's and Shannon's indices are measures of information that incorporate evenness and can be described by a single generalized notation. Developed in this book, for the first time, our SHE analysis methodology is the most comprehensive method of quantitative biodiversity evaluation at the present time. It not only provides a succinct definition of the community structure but also allows for the unique identification of particular distributions that may have generated the field data. Fundamental to this approach, we present our new decomposition of diversity into its two integral parts: richness and evenness. A new graphic method called a biodiversity-gram plots the measures lnS, H, and lnE against lnN on a single diagram. Its evaluation is simplified by using a plot of the Log series on the same figure as a null model against which the observed data can be evaluated. If the reader is primarily interested in measurement of biodiversity, and is familiar with the material presented in the earlier chapters, then Chapters 12 though 18 provide not only a succinct review but new ideas and methods for the measurement and analysis of biodiversity.

Mathematical Equations and End-of-Chapter Problems

There are many equations in this book. For those who are uncomfortable with symbols and equations, we also explain each equation with words. All terms, notation, and equations are defined and explained in a manner similar to the translation of a foreign language. We give this translation in a simple way so that "mathematical sentences" make the going easier. Most people who are not comfortable with mathematics assume that those who are can read, with comprehension, a text containing complicated equations as quickly as they can read a verbal presentation. They usually cannot. Thus this first chapter should be the only chapter that you can or should read quickly. Although only knowledge of arithmetic and algebra is required, you should read slowly. Read less at a sitting; read for comprehension. When you come to an equation, pause. Look it over, follow our example through the arithmetic in detail to see where our values come from and then try inserting a few "dummy" numbers to see whether you can get the equation to work. You may find that when you come to the end of a chapter, you believe that you have understood it perfectly. Nevertheless, we suggest that you do some of the problems at the end of the chapter. What seemed to be perfectly clear may in reality not be so.

The problems at the end of each chapter use real data. Either the entire set is available in a table or that part of the data set we use for the problem is provided in the chapter (with a citation to the publication that contains the data set in full). Each data set consists of actual field data collected by one of the curators or research associates of the Smithsonian Natural History Museum. The wide range of research interests and the tremendously varied collections, plus the graciousness of the scientists in contributing their data, allowed us to incorporate problems across a wide spectrum of taxa. With these problems, and their detailed answers, we want to emphasize the applicability of the book's contents and formulas to all natural populations. The answers to each problem are calculated and described with specifics to enhance understanding.

2

DENSITY:
MEAN AND VARIANCE

A STATEMENT SUCH AS "I found 100 clams" may be important for someone preparing a clambake, but it contains little information about the size or distribution of a clam population in nature. All organisms are distributed in space and time. The particular time at which a survey is conducted constitutes a look at the distribution of organisms at a fixed moment. In this book, however, our main concern is with distributions in space, and that space must be circumscribed. A meaningful statement for our purposes would be "There were 100 clams in Tisbury Pond on 27 December, 1943."

Distribution in space is generally described in terms of *density*, which is the number of individuals per unit space. Some researchers prefer the term *absolute abundance* or *numerical abundance*. Still others simply use the number of individuals per some unit or amount of space. Density is a fundamental measure upon which depends a wide variety of indices describing various attributes of natural populations. Often a specimen consists of a discrete individual; but specimens of colonial organisms or of plant rhizomes may be composed of many individuals. The chosen unit of space by which we anchor the enumeration will depend on the size of the organism studied as well as on the available sampling gear. The area for a survey of large trees might cover km^2; of smaller vegetation, like herbs or grasses, m^2; or of minute seeds, cm^2. Similarly, for some organisms, such as plankton that occupy a portion of the water column, volume is a more suitable unit of space than area. In turn, the volume sampled may depend on the sizes of the available nets and the vessels to tow them. Sometimes a researcher can obtain only material that was collected for other purposes, with sampling gear and procedures less than ideal for collecting the particular organism of interest. In such cases the unit of space is often an estimate or guess.

If the organisms of interest are relatively large and the unit of space relatively small, then a *direct count*, or *census*, may be possible. For example, in the Beni Biosphere Reserve in Bolivia, Dallmeier and colleagues (1991) surveyed the exact position of trees in 1-hectare (10,000 m^2) plots. A map showing the position of each tree in the Bolivian plot 01 is shown in Figure 2.1. By direct count, this plot is known to contain exactly 663 trees. Such precision is, however, unusual. A researcher is usually unable to make a direct count of the entire area of interest. Instead, a small portion of the entire area (for example, a quadrat, core, transect,

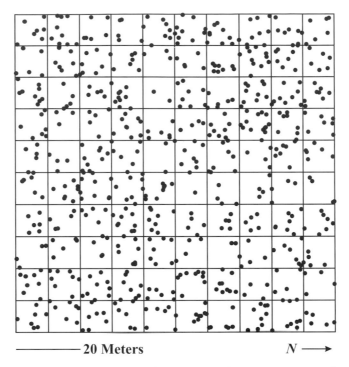

20 Meters $N \longrightarrow$

FIGURE 2.1   Individual trees in plot 01 of the Beni Biosphere Reserve. Here 1 hectare is divided into $N = 100$ quadrats.

subplot, net, or biological sample) is surveyed and an *estimate* of the total number is calculated.

For illustrative purposes, we have divided the Beni plot into 100 subplots, or quadrats, each with an area of 100 m². A simple estimate of the total number of trees can be obtained by choosing 1 quadrat or sample, counting the trees within it, and multiplying by 100. Arbitrarily, we have chosen a sample that is 2 quadrats from the left and 5 up from the bottom. This sample contains 8 trees. (When a quadrat bisects a tree, a rule must be made for including or excluding that tree. The rule must then be applied consistently over the entire plot.) Based on our sample, we would estimate that the entire area contains 8 trees. Of course, if we did not know, as is usually the case, how many trees were actually in the area, we would have no way of knowing how close the estimate is to the true value. A glance at Figure 2.1 and a little common sense would lead us to believe that if we counted more than 1 quadrat, our estimate should be closer to the true value. By sampling more than 1 quadrat, we can obtain an *average* or *mean* number of trees per quadrat.

MEAN DENSITY

The *mean density* is the average number of individuals per unit of space in the entire census (the *population*) or in the sample. In statistical notation the mean density of the *target population* is designated as μ (mu). In algebraic notation we write

$$\mu = \sum_{i=1}^{N} \frac{X_i}{N} = \frac{\sum_{i=1}^{N} X_i}{N}. \tag{2.1}$$

This mathematical equation is read like an English sentence from left to right. It translates to say that the parameter μ that we are looking for equals the sum (Σ) of all the *sampling units* (X_i) divided by the total N. We add these units one at a time, together. Another way to express this addition is to say that the sum is taken over the i sampling units, with i ranging (or going) from 1 to N. Then, after adding all units, we divide by the total number, N, to give the mean number per quadrat or per amount of area represented by the quadrat.

It is usual to call X_i the ith *observation* in statistics. For our density application, N is the total number of observations, or quadrats, in the *target population*—that is, in our area or grouping of current interest. In our example we have $i = 1, 2, \ldots, 100$ because our statistical observations (or sampling units) are not the individual trees in the 1-hectare plot, but the total number of trees within 1 biological sample or quadrat, and there are 100 quadrats.

In summary, for the Bolivian example the mean density of trees in the forest can be obtained by adding the total number of trees in each of the 100 quadrats and dividing this sum by 100. Then, $663/100 = 6.63$ is the mean density of trees per quadrat in the hectare. Because $\mu = 6.63$ is a value obtained from using the entire population, it is called a *population parameter*. This is the value that we wish to know or to estimate. Notice, again, that our N statistical observations in this case were not the trees but the 100 individual quadrats (biological samples).

One of the major purposes of statistics is to obtain a reasonable estimate of μ without the necessity of counting every individual, or all the trees in all 100 quadrats. What we wish to obtain is a *statistical estimate* $\hat{\mu}$ (mu hat) of this population parameter μ. A statistical estimate is a value obtained from incomplete data; that is, from a sample of the entire population. What we want is an estimate $\hat{\mu}$ of μ that is unbiased. An *unbiased estimate* of a population mean is one whose value would exactly equal the true mean if all possible samples were enumerated. An unbiased set of samples will not systematically either overestimate or underestimate the true mean. A biased set of samples may.

The reader may have noticed at this point that the word *sample* has become a problem. Earlier we equated the words *quadrat, subplot,* and *sample*. When biologists or geologists take a core, make a tow, examine the contents of a quadrat, or in some way make an enumeration or measurement at a locality in the field, they refer to each of these as a sample. Thus, each of the 100 quadrats or subplots in our example is a *biological sample*. For the statistician, however, each biological sample is an *observation*. In our example, one observation constitutes the entire contents of any 1 quadrat. In this case we say the quadrat itself, not the 100-quadrat hectare, is the *sampling unit,* or object of the sampling. A sample in the statistical sense consists of more than one observation. The number of observations, n (n is a *subset*, or small group, of N, the total number of quadrats), is our *statistical sample,* even though the researcher in the field refers to this as n *samples*.

It is, of course, possible to count the trees in only 1 quadrat or biological sample and to estimate the total from that sample alone. However, the concept of a mean density

TABLE 2.1   Distinctions Between the Terminology Used by Researchers and Statisticians

EXAMPLE	RESEARCHERS/BIOLOGISTS	STATISTICIANS
1 tree	1 individual	1 individual
1 quadrat	1 biological sample 1 subplot	1 observation The sampling unit
All 100 quadrats	Target area Target population All individuals	The statistical population
8 quadrats	8 samples	8 observations; the 1 statistical sample

NOTE: In this example, the hectare of Bolivian forest has been divided into 100 quadrats, with a sample conducted in just 8 of the quadrats.

would then be meaningless. Needless to say, this confusion often hinders communication between researchers and statisticians. The word *sample* is so ingrained in the biological and geological literature, however, that any adherence to a strict definition is unlikely to gain wide acceptance. A summary of the synonymous usages is provided in Table 2.1. We will try to keep our meaning clear throughout what follows, but the reader should keep this potential source of ambiguity in mind when perusing the literature and reading this book.

RANDOM SELECTION

The best way to ensure an unbiased estimate of the mean density is through *random sampling*. The word *random* does not describe the data in the observed sample, but the process by which it is obtained. That is, randomness in a quantitative survey is a property of the process by which you obtain your sample, not a characteristic of your individual sample. *Simple random sampling* is *unrestricted sampling* that gives each possible combination of *n* biological samples, or selections, an identical chance of becoming the observed (selected) statistical sample. This process ensures that each possible biological sample has an equal chance of being selected—not, as is often stated, that each individual in the population has an equal chance of being in the sample. This is so because our primary sampling unit is the quadrat and we must accept the individuals that we find within it. In some instances, the individual itself can function as the primary sampling unit, and in these cases, the individuals are selected at random. Any other method of sampling will have some attached pattern of variability; only samples generated by a random process will have a known pattern that can be used in statistical theory and methods.

For the Bolivian tree example, we wish to ensure that each of the 100 quadrats has an equal chance of being selected. For a statistical sample of just 1, the chance (probability) of any particular quadrat's being selected would be $\frac{1}{100}$. More generally, the probability is $\frac{1}{N}$ of selecting 1 from N. Now one might assume that selecting n samples at random from the N possibilities could be achieved by thinking up n numbers between 1 and 100. Unfortunately, many studies show that people cannot think randomly. In statistical jargon such a procedure is referred to as *haphazard selection*, and it almost never produces a situation in which each biological sample has an equal chance of selection. To ensure randomness, a table or a computer generation of random numbers is used. A small table of random numbers is provided in Appendix 3.

As an illustration, for the Beni study plot 01 let us assume that we have used a table of random numbers for choosing a statistical sample of size $n = 8$ biological samples (quadrats) from the possible $N = 100$. The chosen quadrats are those numbered 90, 15, 69, 16, 57, 03, 29, and 67. As Figure 2.2 shows, a simple random sample does not necessarily cover the entire area in an orderly manner. Our visual interpretation suggests that in two areas the samples are "clumped." Nevertheless, these are termed *random samples*, and their arrangement in the plot demonstrates why it is nearly impossible for individuals to choose what they perceive as random samples.

19	20	39	40	59	60	79	80	99	100
17	18	37	38	X	58	77	78	97	98
X	X	35	36	55	56	75	76	95	96
13	14	33	34	53	54	73	74	93	94
11	12	31	32	51	52	71	72	91	92
9	10	X	30	49	50	X	70	89	X
7	8	27	28	47	48	X	68	87	88
5	6	25	26	45	46	65	66	85	86
X	4	23	24	43	44	63	64	83	84
1	2	21	22	41	42	61	62	81	82

—————— **20 Meters** $N \longrightarrow$

FIGURE 2.2 Location of 8 randomly selected quadrats in plot 01 of the Beni Biosphere Reserve.

The numbers of trees in these 8 randomly selected quadrats are 4, 9, 4, 5, 4, 8, 9, and 5, with a sum of 48 individuals. The formula for the *unbiased estimate* of the *mean* is similar to Equation 2.1 for the true mean. Here,

$$\hat{\mu} = \sum_{i=1}^{n} \frac{x_i}{n} = \frac{\sum_{i=1}^{n} x_i}{n} \tag{2.2}$$

where x_i is the number of individuals in i^{th} sample quadrat, $i = 1, \ldots, 8$ (the *limits of the summation* are 1 and 8), and n is the number of observations (biological samples) selected at random—that is, the *sample size*. This is also called the *arithmetic mean* to distinguish its value, based upon addition of the observations, from other possible ways of computing a mean or average value.

$$\text{Hence,} \sum_i \frac{x_i}{n} = \frac{(x_1 + x_2 + x_3 + x_4 + x_5 + x_6 + x_7 + x_8)}{8} = \frac{(4 + 9 + 4 + 5 + 4 + 8 + 9 + 5)}{8} =$$

$6 = \hat{\mu}$. Notice that the symbol for the summation can be written, as it was in Equation 2.2, with its limits of 1 and n; or it can be written without those limits, as long as it remains clear that the limits are defined unambiguously. Often the summation sign Σ is used with no limits when the addition problem is clearly defined for the subscript used.

The calculated value of 6 is called the *statistical estimate*, the *parameter estimate*, or simply the *sample estimate*. The formula by which we define the estimate $\hat{\mu}$—that is, Equation 2.2—is called the *estimator*. If we wish to estimate the total number of trees, we would calculate $\hat{\mu}N = 6 \cdot 100 = 600$. Thus the formula $\hat{\mu}N$ is an estimator for the total number of individuals in the target population.

Although we now have an unbiased estimate of the true mean density, μ, we find ourselves in the same situation as when we calculated the total number of trees based on a single arbitrarily chosen quadrat. Unless the total number of trees is actually known (they were all counted in our example), there is no way to evaluate how close either estimate really is to the true value.

In most sample situations, of course, we do not know the true value. We have, however, estimated one statistical parameter, μ, and by estimating one more, the variance, we will be able to evaluate how close our estimate is to the true value, even though we do not know it. This is one reason why statistical science is so powerful.

VARIANCE AND STANDARD DEVIATION FOR DENSITY ESTIMATES

Variance is a quantitative statement used to describe how much the individual observations x_i (either single observations or totals from each sample, quadrat, core, or transect) are scattered about their mean, $\hat{\mu}$. The mean is called a *measure of central tendency*. Central tendency implies that when data are summarized into categories there is a tendency for more observations to be near the middle (or center) rather than at the extremes of the data set. A quantitative statement of this spread is the variance.

The *amount of deviation* of each count from the mean can be measured by $d_i = x_i - \hat{\mu}$. This is the difference between any observation or count x_i and the mean of all the counts $\hat{\mu}$ in the sample. Note that this difference could also be expressed as $d_i = \hat{\mu} - x_i$ but the first notation is the conventional one. The d_i values, or differences, are usually called *residuals*. In quantitative work, these residuals are not called "errors." An error is a deviation from a true population value. However, for a large number of observations, on the average, these differences do approach the true error.

Just as we summed the counts when obtaining the mean, here we wish to sum the deviations. Summing the deviations is the first step in obtaining a measure of all the scatter about the mean density. However, because the numbers are measured around the mean, some differences are positive and some are negative; so their sum always equals 0. This is true regardless of whether there is large or small dispersion of the observations. Consequently, it is convenient to use the squared quantity $(x_i - \hat{\mu})^2$ as the measure of deviation for each count. In this way we sum only positive (but squared) differences. The sum of these squared quantities—that is, the *sum of the squared residuals*—is then divided by the number of observations to obtain a measure of variability or spread called the *variance*. The symbol used for the true, or population, variance is small sigma squared, σ^2. Its estimator, obtained from the sample, is given the symbol $\hat{\sigma}^2$, or sometimes, s^2. In algebraic notation we write

$$\sigma^2 = \sum_{i=1}^{N} \frac{(X_i - \mu)^2}{N} = \frac{\sum_{i=1}^{N}(X_i - \mu)^2}{N} \tag{2.3}$$

where μ is the mean of the population consisting of N observations, quadrats, or counts, denoted by X_i, and where i can be any number between 1 and N observations, quadrats, or counts ($i = 1, 2, \ldots, N$).

Thus, this formula says that the population variance, called σ^2, is obtained by summing, or adding, these squared differences and dividing by N to make the quantity an average.

If a sample variance is used to estimate a population variance, we can still use Equation 2.3 and substitute sample values for the population values. The sample estimator of the variance, like that of the mean, uses n (the number of observations) rather than N. This sample estimator is given by

$$\hat{\sigma}^2 = \sum_{i=1}^{n} \frac{(x_i - \hat{\mu})^2}{n} = \frac{\sum_{i=1}^{n}(x_i - \hat{\mu})^2}{n}. \tag{2.4}$$

On average this formula underestimates σ^2. That is, the sample variance is a *biased estimator* because the residual usually underestimates the true error when there is only a small number of samples. Therefore, Equation 2.4 is *corrected*, or *adjusted*. The usual formula is

$$\hat{\sigma}^2 = \sum_{i=1}^{n} \frac{(x_i - \hat{\mu})^2}{n-1} = \frac{\sum_{i=1}^{n}(x_i - \hat{\mu})^2}{n-1}. \tag{2.5}$$

That is, by convention, to obtain a better estimate of the variance and to simplify results for more advanced statistical inferential methods we must multiply Equation 2.4 by $\dfrac{n}{(n-1)}$ to obtain Equation 2.5. The result is that the sum then is divided by $n-1$ rather than n. This correction is important only when n is small. Its effect becomes negligible when n gets beyond about 100. Even for 25 or 30 observations, it will often be of little consequence.

As long as a simple random sampling plan is followed, $\hat{\mu}$ (Equation 2.2) and $\hat{\sigma}^2$ (Equation 2.5) will be unbiased estimates of the true mean and variance. In some cases it is preferable to use the square root of the variance, σ, estimated by $\hat{\sigma}$, which is called the *standard deviation*, or, in some of the older literature, the *root mean squared deviation*. One important reason for using the standard deviation is that it is expressed in the same units of measure as are the original observations (x_i). For example, if we were to make measurements in numbers of individuals, both the mean and the standard deviation would be in individuals, but the variance would be a measure in (individuals)^2.

We will now illustrate the variance calculation with our Bolivian plot data. Table 2.2 shows the calculation of $\hat{\sigma}^2$ for the $n=8$ randomly chosen quadrats from Bolivia. Although calculators and computers make such laborious calculations unnecessary, the display in Table 2.2 allows us to better understand what is involved.

Because the last column involves squares, it should be apparent that one large deviation from the mean (an *outlier*) will greatly increase $\hat{\sigma}^2$. For example, if the second-to-last observation (x_7) of 9 trees were changed to 12, then ($x_7 - \hat{\mu})^2$ would equal 36, and the resulting sum of the squared residuals would be 63. These changes would result in a value for the variance of $63/7 = 9.00$, instead of 5.14. One outlier can thus make a considerable difference. It can represent a meaningful observation or a procedural

TABLE 2.2   Steps in Calculating Sample Variance, $\hat{\sigma}^2$

OBSERVATION NUMBER (i)	QUADRAT NUMBER	NUMBER OF TREES (x_i)	RESIDUALS FROM THE MEAN ($x_i - \hat{\mu}$)	SQUARED RESIDUALS ($x_i - \hat{\mu})^2$
1	90	4	−2	4
2	15	9	3	9
3	69	4	−2	4
4	16	5	−1	1
5	57	4	−2	4
6	03	8	2	4
7	29	9	3	9
8	67	5	−1	1

NOTE: $\Sigma x_i = 48$; $\Sigma(x_i - \mu) = 0$; $\Sigma(x_i - \mu)^2 = 36$; $\hat{\mu} = \dfrac{48}{8} = 6.00$; $\hat{\sigma}^2 = 36(8-1) = 5.14$

mistake. The researcher who not only relies on the number printed out by a machine but also understands how the number was obtained will be in a much better position to evaluate and interpret the original observations.

INTERPRETING VARIANCE OF SMALL AND LARGE SAMPLES: TCHEBYCHEV'S THEOREM

The standard deviation σ^2 is a measure of the spread or variability in a population. If the standard deviation of a data set is small, the values are concentrated near the mean; if the standard deviation is large, the values are spread considerable distances away from the mean.

The use of the words *small* and *large* is always disconcerting in statistical applications because they are undefined. Tchebychev (see Cramér 1946) provided a method for presenting statements on standard deviation in more concrete terms. He showed that for any kind or shape of data set (populations or samples), and regardless of the statistical distribution, the proportion (p) of the observations that will fall within any selected number of *units*, say k, of their mean will be at least

$$p_k \geq 1 - \frac{\sigma^2}{k^2}. \tag{2.6}$$

This relationship is known as *Tchebychev's theorem* or *Tchebychev's inequality*.

Let us assume that we wish to know what proportion of the samples is within 2 standard deviations of the mean. In this case, our unit is a standard deviation. That is, we want to know what proportion of the observations will fall within k units, or 2 standard deviations ($k = 2$) of the mean, for this example. Therefore, we let $k = 2 \cdot \sigma$ in Equation 2.6. Because the σ in both the numerator and the denominator cancel, we have $1 - \frac{1}{2^2} = 0.75$, or 75%. We would expect, then, that at least 75% of the observations are within 2 standard deviations of the mean. This gives us an idea of the sense in which the standard deviation or its square, the variance, is used as a measure of dispersion. We expect to find this statement to be true of at least 75% of the data, even though we have no idea of what the true value for the standard deviation may be. Tchebychev's theorem is also always true for any set of values for which we can calculate a mean and a variance.

As another example, returning to the Bolivian trees, recall from the data set with $N = 100$ quadrats that a sample of $n = 8$ yielded the estimates $\hat{\mu} = 6.00$ and $\hat{\sigma} = \sqrt{5.14} = 2.27$ (Table 2.2). Therefore, we expect at least 75% of the counts per quadrat to fall in the interval $6.00 \pm 2(2.27)$; that is, between 1.46 and 10.54. We do not need to know anything about the distribution of these counts in order to use this theorem; it will always be true for any set of numbers we obtain. In fact, for this data set, 98% actually are within these limits (Appendix 1). This example thus illustrates how the magnitude of the standard deviation is related to the concentration of the observed values about the mean. With the aid of a few more tools, which will be introduced a little later, we shall be able to make more precise statements.

Summary

1. The true mean density of a population is the average number of individuals per quadrat (biological sample). For density estimation, the sampling unit is the quadrat— not the individual. The true mean is a statistical parameter.

2. An unbiased estimate of the population mean is one whose value would equal the true mean if all possible samples were enumerated. An unbiased estimate will not systematically overestimate or underestimate the true mean. An unbiased estimate of the true mean is assured through random sampling.

3. For density estimation, the quadrats (biological samples), not the individuals, are chosen at random. Each individual does not have an equal probability of being chosen, but each quadrat does.

4. The variance and its square root, the standard deviation, are measures of the spread or scatter of the observations about the mean.

5. When the observations form a sample, then quantitative measures of that sample, such as the mean and the variance, are called estimates, and their formulas are called estimators. When the observations constitute the entire census or population count, then these quantities are called parameters. We use estimates based on samples to learn about, or estimate, population parameters.

6. A lower limit for the proportion of observations lying within any selected number of standard deviations from the true mean is fixed by Tchebychev's theorem. No information on either the field distribution of the organisms or the statistical distribution of the sample values is required to apply this theorem.

Problems

2.1 *Scheelea princeps* (species #1) is the most abundant species in the Beni Biosphere Reserve plot 01. *Calycophyllum spruceanum* (species #4) is a commonly occurring species. *Aracia loretensis* (species #18) is a rarely occurring species. For each of these 3 species, sample $n = 4$ quadrats selected randomly (also called *random quadrats*) from the data given in Appendix 1 (Appendix 3 is a random numbers table).

a. Find $\hat{\mu}$ and $\hat{\sigma}^2$ for each of these species.

b. Repeat the exercise for a sample of size $n = 8$.

c. Using Tchebychev's theorem, calculate the proportion of observations within 2 standard deviations of the mean when using first the statistical sample of size 4 and then the sample of size 8.

d. Comment on the differences you obtained for an abundant, a common, and a rare species.

2.2 In 1974, Heyer sampled anuran larvae with 3 different types of sweeps. The 3 sweep types sampled different microhabitats within one pond. The unequal distribution of larvae among sweep types reflects habitat partitioning (Heyer 1976, 1979). Larval samples were taken in 25 m sweeps for each sample once a week from 9 March to 29 June, 1974 (16 weeks). Here are the results:

| | Sweep Type | | | |
Species	Surface	Midwater	Bottom	Total
A. maculatum	11	2	4	17
A. opacum	6	3	3	12
H. chysoscelis	6	17	3	26
H. crucifer	530	46	58	634
R. clamitans	3	1	8	12
R. palustris	0	1	1	2
R. sylvatica	7	130	54	191
B. americaus	0	0	0	0
				884

a. Find the mean density per week over all sweeps of *A. maculatum*.

b. Find the mean density of *H. crucifer* for each microhabitat and for the total.

c. Put bounds on the microhabitat and total mean density of *H. crucifer*.

2.3 The sipunculid *Phascolion cryptum* was collected in the Indian River, near the Smithsonian Marine Station at Link Port, Florida (Rice, and colleagues 1983). Three different sieve meshes were used. The data for replicates from June 1994 are given below, where size 1: size > 2.83 mm; size 2: 2.00 < size < 2.83 mm; size 3: 1.0 < size < 2.00 mm.

Replicate	Size 1	Size 2	Size 3
1	1,511	178	2,889
2	1,156	222	978
3	889	489	3,156
4	667	178	89
Total	4,223	1,067	7,112

a. Each sample size was 0.025 m². What is the density per m² for the first replicate?

b. Estimate the mean and total density per 0.025 m² replicate during the month of June.

c. What is the deviation from the mean per replicate? What is the deviation from the mean for all replicates combined?

d. Use Tchebychev's theorem to calculate for each of the 4 samples the proportion of *P. cryptum* that is expected to be within 2 standard deviations of the mean. Calculate the intervals.

3
NORMAL AND SAMPLING DISTRIBUTIONS
FOR FIELDWORK

I N THE PREVIOUS CHAPTER we introduced 2 statistical parameters of population density: the mean, μ (Equation 2.1), and the variance, σ^2 (Equation 2.3). We also introduced $\hat{\mu}$ (Equation 2.2) and $\hat{\sigma}^2$ (Equation 2.5) as statistical estimators of these parameters. These estimates are once removed from the original field observations. The summarizations they provide, however, are powerful and well worth the price. These statistical tools are, moreover, essential for much biological fieldwork, as a succinct summary.

In statistics, a *random variable* is a quantity that may take on any of a specified set of values. For example, human age is a random variable because it can take on any value in the set of numbers from 0 to 120 years as far as we now know. The mean and variance of the counts from a sample are estimates, but these estimates are also random variables because they may take on any of the values determined by the data. In addition, the total count is a random variable, and of a particular kind. The total count, which varies from sample to sample and could take on any value from 0 to some large number (equal to or less than 663 in our example), is a *discrete random variable. Discrete* means that the count can only be a whole number (an *integer*). Counts are not the only examples of discrete variables, but they are the most common. The descriptor "random" for the variable is probably a poor choice, however, because it means something quite different from the idea of randomness common in ecological work. Use it we must, as this term is well established.

FIELD DATA SUMMARY

Field counts may vary widely from sample to sample. Therefore, it is more convenient to summarize them than to provide a list of each count. By categorizing the data from our Bolivian forest example into *intervals*, or groups, according to the quadrat counts obtained, we change the representation of the data from a list of all individual values to a summary of the values in tabular form (Table 3.1). A summarized representation of the data in tabular form is called a *frequency table*. In a frequency table we show the number of individuals counted and the *frequency*, or number of times a particular count was obtained. The tabular arrangement in Table 3.1 shows the distribution of the frequency of occurrence of the trees within the quadrats in the Beni study plot. The table can be read as follows: Row 1 shows that 3 of the quadrats in our example each

TABLE 3.1   Frequency Table for Total Trees in Bolivian Beni Plot 01

NUMBER OF TREES	FREQUENCY (NO. QUADRATS WHERE PRESENT)	TOTAL TREES IN A QUADRAT (COL. 1 · COL. 2)
2	3	6
3	6	18
4	8	32
5	14	70
6	17	102
7	13	91
8	20	160
9	11	99
10	5	50
11	2	22
12	0	0
13	1	13

NOTE: $N = 100$ quadrats, where each quadrat = 100 m². Total trees = 663.

contained 2 individuals, for a total number of individuals observed in those 2 quadrats of $3 \cdot 2 = 6$. The second row shows that 6 additional quadrats each contained 3 individuals. The total observed for these quadrats is 13, and so on. The sum of the information in the frequency column is the total number of quadrats; that is, 100. The frequency count times the number of quadrats gives the total number of individuals in each interval, and their sum provides the total number of individuals in the entire plot. The values for the intervals, also called *classes*, we selected (that appear in column 1) were the actual counts, but we could have selected other classes. We could, for example have selected *ranges* (or groupings) of counts, perhaps (0, 2), (3, 5), and so on. The data summary in this table, or its graphical representation, is called a *frequency distribution*.

In constructing a frequency distribution for our data set, we discard detailed information on the exact numbers of individuals found in each quadrat. In return, we obtain a summary of information that emphasizes the essential, overall characteristics of the data set that are obscured when we try to look at a mass of unorganized values. Notice from Table 3.1 that quadrats with counts between 5 and 8 have the highest frequency. Recall also that the true, or population, mean is $\mu = 6.63$. This example thus reaffirms the notion of the mean as a measure representative of the middle or center of the data, which is why the mean is called a measure of central tendency.

Histograms

One possible graph of the Bolivian data is presented in Figure 3.1. This figure is a *histogram* of the tabular layout shown in Table 3.1. The height of each column, the y value,

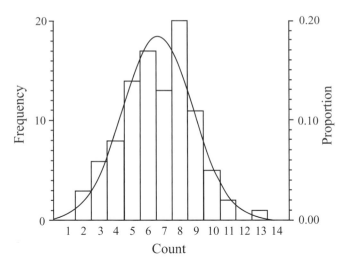

FIGURE 3.1   Histogram and Normal curve for Bolivian plot 01 data, with $N=100$ quadrats.

or the *ordinate*, is the frequency of the particular count. The value of the variable on the horizontal axis, or *abscissa*, is the x value; namely, the numerical value of the count. Connecting the midpoints of each of the intervals (which we did not do) yields an alternative graphical presentation called a *frequency polygon*, or *frequency curve*. In this case the frequency is the area under the curve between 2 points on the x axis. The total area under the entire curve equals N. In this example, N equals 100 quadrats, the total frequency. Sometimes it may be useful to mathematically manipulate to make the total area equal 1 instead of N or n. Under certain conditions we discuss, such a transformation allows us to equate area to probability, because probability always sums to 1.

When considering frequency distributions, either in tabular or in graphic form, it is important to distinguish between those observations that, by their very nature, can assume only certain values and those that may take on any value within the data range. The first type is called a *discrete distribution*. For example, our field counts cannot be negative, fractions, or 0; they can only be positive, real-valued, whole numbers (integers). Alternatively, we could have a *continuous distribution*, in the sense that any value for the random variable is possible, provided the measurement is made at a fine enough scale. Depending on what was measured, the values in a continuous distribution may range from positive to negative, or they may be all positive or all negative values. An example of continuous positive measures is available from the same Bolivian study plot 01. In addition to recording the position of each tree in the plot, Dallmeier and colleagues (1991) measured the breast height diameter of each tree, which is the diameter of the tree at the height of the researcher's chest. These data, unlike the counts, provide us with a continuous distribution. The random variable in this case is called a *continuous random variable*; its frequency polygon would be a continuous frequency curve.

A third type of frequency distribution may also be encountered when sampling natural populations. This type is representative of data sets in which the observations are "logically" discrete, but appear continuous. For example, say we formed a table for the total number of individual insects found in 500 quadrats placed along a 10 km transect in the Bolivian forest. The numbers of individuals in each quadrat might range from one to tens of thousands or more. Thus, the *unit distances* between the possible values, which are distances between each count value 1, 2, 3, . . .—that is, $(4 - 3) = 1$, $(3 - 2) = 1$, and so on—are extremely small when compared to the range of the data. A frequency table would have such a very large number of classes that the data would not actually be grouped in any appreciable way. In such cases it is advisable to use *class intervals* (such as 0–10, 11–20) and to treat the data as being continuous for all further statistical treatment, even though the data really are discrete.

The Normal Distribution

There are many discrete and continuous distributions used in statistical application to describe and compare data. By far the most important and well known is the *Normal distribution*. A Normal distribution is one in which the data set includes many observations near the center and a decreasing number toward each side. This distribution graphs into a symmetrical, bell-shaped frequency curve.

A Normal distribution underlies much of the theory for common statistical tests. More importantly, the Normal distribution is completely defined (in statistics it is said to be "specified" or "determined") by only two parameters: the mean and the standard deviation (or its square, the variance). In other words, there is a precise mathematical equation with which we can define the Normal distribution or Normal curve once we know the mean and the standard deviation. If we have the mean and the standard deviation, the area under the curve between any 2 points on the horizontal scale can be calculated. For example, consider Figure 3.1, the histogram of the data set from the Beni study plot 01. The mean of the data set is $\mu = 6.63$ and the standard deviation is $\sigma = 2.17$. Using these 2 values, the smooth curve represents a Normal distribution with this same mean and standard deviation. This Normal curve is thus *fitted* to the data. (Fitting a statistical distribution to data is a rather complex procedure, which is discussed, but not in depth, in Chapter 17).

For a long time it had been thought that any measurement of a natural phenomenon was distributed Normally because the Normal distribution was theorized to be a law of nature. We now know that this assumption is incorrect and that many natural phenomena follow other distributions. Nevertheless, the Normal distribution is ubiquitous and still extremely useful for practical fieldwork because many distributions of organisms or their characteristics can be approximated by a Normal. Many times in this text we shall assume that field data can be well approximated by this bell-shaped distribution. In some cases, like that illustrated in Figure 3.1, the data may not be an exact fit (especially for data sets that are small or in which the random variable is discrete), but the Normal distribution still provides a reasonable approximation that can be used when working with the observed information and that allows reasonable inferences to be made.

Fitting models to data involves approximations of the Normal shape. Often data ranges are *truncated*; that is, they do not have an infinite number of x values, and thus have some *cut-off point*. These observed distributions are, however, treated statistically as if they were the ideal (infinite). This is a legitimate statistical practice and usually has very little effect on the results obtained. Many field-oriented scientists shy away from use of the Normal distribution and dismiss it with a simple statement concerning the discrete nature of, and lack of, Normality in their data sets. In our opinion, this cavalier attitude severely limits their horizons. In fact, Normal distribution methods do indeed give the correct results on numerous examples from Bolivian plot 01, even though the data set is relatively small and not ideally Normal at all (Figure 3.1). Remember, we have specifically chosen the Beni Reserve example because we know the true values, so we can examine the correctness of various statistical procedures ourselves. We don't just estimate confidence intervals in this book; we can evaluate them against the ground truth.

If we are willing to assume a Normal distribution for the Bolivian data, we can make much more precise statements about sample variability than we could when we relied on Tchebychev's theorem (Equation 2.6). For example, using this theorem, which holds for any distribution, we can calculate that the proportions of observations that fall within, say, 2 or 3 standard deviations of the mean are 0.75 and 0.89, respectively. Using the procedures for a Normal distribution (see, for example, Feller 1957), the respective values are 0.95 and 0.99. Consider the Bolivian data with 100 quadrats and $\mu = 6.63$ and $\sigma = 2.17$. Despite the fact that this is a discrete frequency distribution and not ideally Normal (Figure 3.1), if we assume Normality, then, instead of stating that at least 75% of the observations are within $\mu \pm 2\sigma = 6.63 \pm 4.34$ (from Tchebychev's theorem), we can state that at least 95% of the observations should be between 6.63 ± 4.34. In other words, we would expect that 95% of the quadrats would contain between 2.29 and 10.97 individuals. Let us look for any values outside of these limits in the frequency table (Table 3.1). We can see that 3 quadrats have 2 individuals (that is, less than 2.29); 2 quadrats have 11 (that is, more than 10.97); and 1 quadrat has 13. This gives us $3 + 2 + 1 = 6$ quadrats lying outside of $\mu \pm 2\sigma$, instead of our expected 5 (95% of 100 quadrats is 95, so we expect 5 quadrats to be outside of this range) if this were really distributed according to a Normal. Considering the rather poor fit between the histogram and the Normal curve shown in Figure 3.1, this example illustrates why statisticians so often rely on a Normal approximation even when the fit of the data at first glance does not seem encouraging.

Standardized Normal Distribution

A Normal distribution is uniquely defined by, or known from, its mean and standard deviation. These quantities, of course, differ for each distinct data set, which results in a unique equation to describe the distribution of each data set. Thus, because each different set of one mean and one standard deviation can define one Normal curve, there can be an infinite number of Normal curves. In addition, because the frequency is the area under the curve, it can also differ.

For example, Figure 3.2 shows graphs of two Normal distributions. Inspection of this figure shows that the shaded area under each curve between 2 points (say, from 10 to 15) is not the same for the 2 curves. Unfortunately, the only way to find the areas

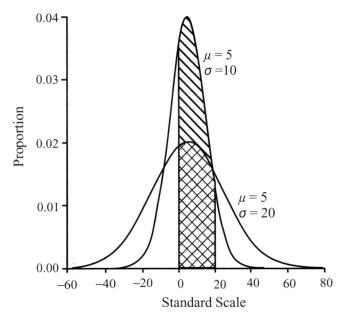

FIGURE 3.2   Two Normal distributions with $\mu = 5$, $\sigma = 10$ and $\mu = 5$, $\sigma = 20$. Cross-hatching shows the area under each curve between 0 and 20.

under each of these curves is by integration methods of calculus. However, for convenience, this area's mathematical integration can be done and all the values put in a table. But we would then require an infinite number of tables, one for each set of $\hat{\mu}$, $\hat{\sigma}$, and n.

Luckily, there is a mechanism for dealing with the problem that we can have an infinite number of Normal curves, each with a different amount of area underneath, so that we need not form an infinite number of tables of areal values. What is needed is simply to transform different data sets to a standardized format so that they become comparable, with equal areas. The transformation of a data set is accomplished by changing the scale on the x axis from a scale defined by the observed values to a scale of *standard* values. This process is no different from changing 6 apples, 1 watermelon, and a pint of raspberries into 0.5 kg of apples, 4 kg of watermelon, and 0.25 kg of berries, only in our case, instead of changing different types of fruits into kilograms, we transform each observed value (whatever its original unit of measure) into units of standard deviation, which indicate the distance of the original value from its mean (above or below). Such a standard scale is called the *z scale*, or *z axis*. It is obtained by the following formula:

$$z = \frac{(x - \hat{\mu})}{\hat{\sigma}} \tag{3.1}$$

Once the values of x (in x units) are converted into z values, z scores, or *standard scores* (in standardized z units), in a very simple way we can obtain the area under whatever Normal curve is required to fit the data. Note that this change of scale can be done for any distribution that has a mean and standard deviation, not just for the Normal.

Let us examine Equation 3.1 more closely. The mean of our sample of x values (that is, of $x_1, x_2, \ldots, x_n$) has been designated $\hat{\mu}$. The numerator of the equation indicates that we must subtract this mean value, $\hat{\mu}$, from the selected (any one you choose) observation. We could have written both x and z with a subscript i, but conventionally this is omitted because we always consider one selected value at a time. If we subtract $\hat{\mu}$ from all values of x and recalculate the mean using the new values, then the new mean value will be 0 (the *origin* of the x–y scale). Before we calculate the new mean, if we first divide each new x value ($x - \mu$) by the standard deviation, then when we calculate the new mean using these values, $\frac{(x - \mu)}{\sigma}$, it will still equal 0 but we will also have a standard deviation of 1.

Next, the area under this *standard Normal distribution* (a Normal with $\mu = 0$, $\sigma^2 = 1$, and $\sigma = 1$) is put in a table or found by computer program and can easily be related back to each curve representing a particular data set by use of the z value (z score). (See the example given in Table 3.2.) We can now turn any x value from any Normal distribution into a *standard score*. A standard score, or z score, indicates the distance that score is above or below the mean of 0 in standard deviation terms, or units. Because the z score is also a distance, it is called, alternatively, a *standard deviate*. The standard Normal distribution has been adjusted so that the area between the curve itself and the horizontal axis always equals 1. Consequently, the z score can help us find the *proportion* of x values whose equivalent z scores lie below that particular z value; that is, the *probability*, p, that any given x value will have a z score in that range. Some selected values of z and p are shown in Table 3.2.

In Chapter 2 we randomly selected 8 quadrats from the Bolivian data set, and from that calculated an estimated mean of $\hat{\mu} = 6.00$. The estimated standard deviation was

TABLE 3.2   Values of z (the Standard Normal Variable) Corresponding to p (Probability) for Any Normal Curve

z	p	z	p
−3.09	.001	0.25	.600
−2.58	.005	0.52	.700
−2.33	.010	0.84	.800
−1.96	.025	1.28	.900
−1.64	.050	1.64	.950
−1.28	.100	1.96	.975
−0.84	.200	2.33	.990
−0.52	.300	2.58	.995
−0.25	.400	3.09	.999
0.00	.500		

$\hat{\sigma} = 2.27$. Let us assume that we now wish to know the expected proportion of values equal to or less than 11 individuals—that is, how many of the 100 quadrats contain 11 or fewer individual trees? From Equation 3.1, we calculate $z = \dfrac{11-6}{2.27} = 2.20$. This value of 2.20 indicates that the value of 11 is 2.2 standard deviation units above the mean. Table 3.2 reveals that about 0.98, or 98%, of the observations in the plot should lie below this value. Looking at Table 3.1, we see that for the observed data in the Beni plot 01, only one observation is actually above 11. Because we have a total of 100 observations (quadrats), this means that 99%, or 99/100, of them have 11 or fewer trees.

Similarly, now using the 8-quadrat sample again, if we wish to know the expected number of observations of fewer than 3 trees, for example, we have $z = \dfrac{3-6}{2.27} = -1.32$. Table 3.2 indicates that roughly 10% of the observations should fall below this value. Going back to Table 3.1 we see that 3 quadrats have a value of 2, and none has 1 or 0. Consequently, we are well within our expectation, based on an assumption of a Normal distribution.

SAMPLING DISTRIBUTION OF THE MEAN

The Normal distribution greatly helps not only in describing observations but also for inferring the values of population parameters from *sample statistics*. Recall that the mean of the sample, $\hat{\mu}$, is used as an estimate of the population mean μ. Usually we calculate only 1 value of this estimate from our statistical sample. However, had we taken a different sample, in all likelihood we would have obtained a different numerical value for the estimate. This new outcome has nothing to do with the competence of the investigator, but is instead due to chance.

The realization that different samples will yield different values for the estimate of the mean density may, at first, appear to be cause for some alarm. Not so. Statistical inference and knowledge of the Normal distribution provide us with a method of estimating limits within which the true mean must lie. Those limits can be ascertained with any desired degree of closeness by use of just a single estimate of the mean. This remarkable bit of statistical inference is possible even though the values of the estimate vary, and even though we do not know the true value of the mean.

Because different values of $\hat{\mu}$ are obtained from sampling or choosing different sets of field samples, this sample statistic, which is also a random variable, must have its own distribution. The distribution of a sample statistic is called a *sampling distribution*. Note that this is not the distribution of the observations (the counts). This sampling distribution is, rather, the distribution of the mean of the counts.

We will illustrate by considering an example of manageable size consisting of the counts of the number of the trees in the 8 quadrats shown in Figure 2.2. For the purposes of Chapter 2, these 8 randomly chosen quadrats comprised a sample of $n = 8$ from the $N = 100$ quadrats of Bolivian trees. The calculations for this sampling are shown in Table 2.1. For our present purpose, these 8 quadrats now become our target population of $N = 8$ quadrats. We will choose samples of size $n = 2$ from the 8 quadrats. Remember, for statistical purposes any cohesive set of data we find of interest can be designated as our target population. The population mean for the $N = 8$ will remain the same as when these 8 quadrats were our sample. Therefore, we have $\mu = 6.00$, calculated from Equation 2.1. We calculate the population variance for this target population using Equation 2.3, rather than Equation 2.4 or 2.5, which are estimators for sample variance. We must divide by N (not $n - 1$), and Equation 2.3 becomes $\sigma^2 = \dfrac{36}{8} = 4.50$. The population standard deviation is $\sqrt{4.50} = 2.12$.

Let us consider samples of size $n = 2$ from this unrealistically small, but manageable, population. From combinatorial analysis (discussed next) we calculate that 28 different sample combinations of size $n = 2$ are possible from a group of 8 quadrats. The counts listed in Table 2.1 clearly show that various 2-quadrat combinations of the same 8 quadrats will yield different estimates of the mean.

Combinatorial Analysis

It is intuitively clear that there are many possible samples of size 2, but how many? The answer comes from *combinatorial analysis*. We wish to know how many ways we can choose n elements from N without regard to their order. The order is not important, because a sample consisting of quadrats #90 and #15, say, gives the same result as one in which we selected #15 and then chose #90. A theorem from a branch of statistics called combinatorics says that we can find how many different samples of size $n < N$ there are from a population of N elements. Alternatively, we can say we will find the number of different combinations of N things considered n at a time. This number is called the *binomial coefficient*, and the notation to express it is written as

$$_N C_n \text{ or } \binom{N}{n}. \tag{3.2a}$$

The binomial coefficient is given by

$$_N C_n = \frac{N!}{n!(N-n)!}. \tag{3.2b}$$

The exclamation point (!) after a number represents a *factorial*, so that $N!$ is read as N factorial. For any number N, $N!$ is the number of different orderings for N elements. N factorial is defined to be

$$N! = N \cdot (N-1) \cdot \ldots \cdot 2 \cdot 1. \tag{3.3}$$

Returning to our example in which $N = 8$ and $n = 2$, from Equation 3.2 we obtain

$${}_8C_2 = \frac{(8 \cdot 7 \cdot 6 \cdot 5 \cdot 4 \cdot 3 \cdot 2 \cdot 1)}{(2 \cdot 1 \cdot (6 \cdot 5 \cdot 4 \cdot 3 \cdot 2 \cdot 1))}.$$

We could simply multiply out the numerator and the denominator. However, they both contain $6 \cdot \ldots \cdot 1$, which cancel, so we are left with $\frac{(8.7)}{(2.1)} = 28$. That is, 28 different samples of size $n = 2$ can be obtained from a population of $N = 8$. It should be apparent that as N becomes large, the number of possible samples can become enormous, which is, of course, why we chose a target population of manageable size. Computation becomes difficult if we wish to calculate the number of possible samples for large N, but these formulas are very useful for statistical inference.

Mean of the Sampling Distribution

All possible samples of size $n = 2$ from the $N = 8$ quadrats are enumerated in Table 3.3. Simple random sampling requires that each of these 28 possible quadrat combinations has an equal chance (1/28) of being selected as our observed sample. However, many of the count combinations resulting from the quadrat pairs appear more than once, and a given mean can be calculated from more than one combination of counts. For example, 4, 9 and 5, 8 both give $\hat{\mu} = 6.5$.

Table 3.3 shows that a mean value of 6.50 occurs in 8 different combinations. Therefore, the actual probability of obtaining a mean of 6.50 is 8/28. This implies that in repeated sampling from this population, about 8 of the 28 samples would have 6.50 as an average number of trees.

In Table 3.4 we form a *distribution of the means*. This table is the *sampling distribution of the mean*. It is also called a *probability distribution of sample means*. Here we have $N = 28$ observations of the mean, so the arithmetic average of this distribution is $168/28 = 6.00$. This value is identical to the population mean because we have simply counted each sample value (1 mean) once and divided by the number of possible means. It is always true that the mean of the sampling distribution ($\hat{\mu}$) equals the population mean (μ) when simple random sampling is used. In addition, now we have another way to consider the meaning of an unbiased estimate. In general, a statistic is an unbiased estimator of a parameter (in this case the mean density, μ) when the sampling distribution of that statistic has a mean equal to the parameter being estimated. Note that this lack of bias does not imply that any one sample statistic or sample result exactly equals the true mean. Any sample value usually has a definite positive or negative error (or deviation from the mean), owing to sampling.

The variance of this sampling distribution (Table 3.4) is simply $\sigma^2 = \frac{54}{28} = 1.93$. The standard deviation is $\sqrt{1.93} = 1.39$. Note that the values for the variance and the standard deviation of the sampling distribution are not the same as those for the variance and the standard deviation of the population, which were $\sigma^2 = 4.50$ and $\sigma = 2.12$, respectively. For the sampling distribution, they are smaller because the use of the means dampens the variation observed in the original counts.

TABLE 3.3   All Possible Samples of Size 2 from 8 Quadrats of Bolivian Tree Plot 01

QUADRAT NUMBERS	COUNTS OF TREES	MEAN
90, 15	4, 9	6.50
90, 69	4, 4	4.00
90, 16	4, 5	4.50
90, 57	4, 4	4.00
90, 03	4, 8	6.00
90, 29	4, 9	6.50
90, 67	4, 5	4.50
15, 69	9, 4	6.50
15, 16	9, 5	7.00
15, 57	9, 4	6.50
15, 03	9, 8	8.50
15, 29	9, 9	9.00
15, 67	9, 5	7.00
69, 16	4, 5	4.50
69, 57	4, 4	4.00
69, 03	4, 8	6.00
69, 29	4, 9	6.50
69, 67	4, 5	4.50
16, 57	5, 4	4.50
16, 03	5, 8	6.50
16, 29	5, 9	7.00
16, 67	5, 5	5.00
57, 03	4, 8	6.00
57, 29	4, 9	6.50
57, 67	4, 5	4.50
03, 29	8, 9	8.50
03, 67	8, 5	6.50
29, 67	9, 5	7.00

THE STANDARD ERROR

To avoid confusion, the standard deviation of the sampling distribution of the mean is designated as σ_μ and is called the *standard error*. Whenever simple random sampling has been used, the formula for the standard error is

$$\sigma_\mu = \frac{\sigma}{\sqrt{n}} \cdot \sqrt{\frac{N-n}{N-1}}. \qquad (3.4)$$

TABLE 3.4   Distribution of the Means Shown in Table 3.3

MEAN $\hat{\sigma}$	FREQUENCY	TOTAL TREES	$(\hat{\mu} - \mu)$	$(\hat{\mu} - \mu)^2$	$(\hat{\mu} - \mu)^2 \cdot$ FREQ.
4.00	3	12	−2	4	12.00
4.50	6	27	−1.5	2.25	13.50
5.00	1	5	−1	1	1.00
6.00	3	18	0	0	0.00
6.50	8	52	0.5	0.25	2.00
7.00	4	28	1	1	4.00
8.50	2	17	2.5	6.25	12.50
9.00	1	9	3	9	9.00
	$\Sigma = 28$	$\Sigma = 168$			$\Sigma = 54$

$$\mu = \frac{168}{28} = 6.00 \qquad \sigma^2_\mu = \frac{54}{28} = 1.93 \qquad \sigma_\mu = 1.39$$

NOTE: The sample size is $n = 2$, from a population of $N = 8$ quadrats of Bolivian trees.

To obtain a value for σ_μ, we never need to find all possible values in the sampling distribution as we did above. In Equation 3.4 the standard deviation of the population, σ, is merely substituted directly. Thus, in practice the laborious presentation of enumerating all possible values in the sampling distribution, as illustrated in Tables 3.3 and 3.4, is unnecessary. The use of σ sometimes confuses researchers into thinking that the standard error, σ_μ, is simply another formula for the population standard deviation of the total population divided, or adjusted, by the square root of n. It is not. Although σ, the standard deviation of the counts in the population, is used for the calculation, σ_μ is the standard deviation of the sampling distribution of the mean. For our example (with $N = 8$, $n = 2$), we have

$$\sigma_\mu = \frac{2.12}{\sqrt{2}} \cdot \frac{\sqrt{(8-2)}}{(8-1)} = 1.50 \cdot 0.93 = 1.39.$$

This is the same answer, obtained more simply, that we obtained from calculating the variance and the standard deviation of the complete sampling distribution of means shown in Table 3.4.

The second part of Equation 3.4 under the square root symbol; that is, $\frac{N-n}{N-1}$, is called the *finite population correction* (fpc). The purpose of the fpc is to correct for the use of a small sample. In our particular example, which uses a very small population

$N = 8$, we obtain $\sqrt{\dfrac{N-n}{N-1}} = 0.93$. In almost any faunal or floral survey, however, the number of possible biological samples, N, is quite large, so the fpc is very close to 1 and can be safely ignored. If the ratio $\dfrac{n}{N}$ is less than about 5% $\left(\dfrac{n}{N} < 0.05\right)$, the fpc need not be considered. Consequently, the formula for the standard error is usually written as

$$\sigma_\mu = \frac{\sigma}{\sqrt{n}}. \qquad (3.5)$$

Although we do not need to consider the fpc when calculating standard errors for most surveys of natural populations, it nevertheless affords a remarkable statistical insight that may be contrary to intuition. Here is why: The standard error is a measure of the precision of the random variable's value as an estimate of the related parameter from the target population. Most of us would suppose that a sample of $n = 8$ from an area containing a total of $N = 100$ quadrats would give a more precise estimate of the population mean than would a sample of $n = 8$ from an area containing an $N = 1,000$. Or, more realistically and dramatically, we would expect that a sample of size 8 from an area containing 1,000 possible quadrats would be considerably better than taking this same size sample of 8 from an area with 10,000 possible quadrats. But Equation 3.4 shows that this is not so. For an $N = 1,000$ the fpc would be $\sqrt{\dfrac{1,000 - 8}{1,000 - 1}} = 0.9965$. For an $N = 10,000$ we have $\sqrt{\dfrac{10,000 - 8}{10,000 - 1}} = 0.9996$. The difference between the two resultant values of fpc is only 0.0031.

To demonstrate the insignificance of this difference, recall that $\sigma = 2.17$ for the total number of quadrats in the Bolivian plot ($N = 100$), so that from Equation 3.5, $\sigma_\mu = \dfrac{2.17}{\sqrt{8}} = 0.7672$. Now let us instead pretend we had a plot of 1,000 quadrats and yet obtained these same values. Then, using the fpc for $N = 1,000$, we would have $0.7672 \cdot 0.9965 = 0.7645$. For $N = 10,000$, under this same condition, we would get $0.7672 \cdot 0.9996 = 0.7669$. In other words, error in our estimate of the population mean is almost as small when we sample 8 from a possible 1,000 as it is when we sample 8 from a possible 10,000. This is why surveys or polls, providing the target population is chosen correctly, can so accurately predict outcomes from very small samples. Although contrary to their intuition, it should afford some comfort to those who sample natural populations.

To estimate the standard error, σ_μ from a sample, we simply use the estimate of the standard deviation, $\hat{\sigma}$, so that

$$\hat{\sigma}_\mu = \frac{\hat{\sigma}}{\sqrt{n}} \cdot \mathrm{fpc}. \qquad (3.6)$$

Or, without the fpc,

$$\hat{\sigma}_{\mu} = \frac{\hat{\sigma}}{\sqrt{n}}. \qquad (3.7)$$

For our sample of $n = 8$ random quadrats (Table 2.1), we had $\hat{\sigma}^2 = 5.14$ and $\hat{\sigma} = 2.27$. Using the fpc from Equation 3.4, we obtain $\sigma_{\mu} = \frac{2.27}{\sqrt{8}} \cdot \sqrt{0.93} = 0.80 \cdot 0.96 = 0.77$. If we ignore the fpc, then the result would be $\hat{\sigma}_{\mu} = \frac{2.27}{\sqrt{8}} = 0.80$, and the standard error would have been overestimated; an error on the "safe" side.

Thus, we conclude that the sampling distribution of μ has parameters μ and $\sigma_{\mu} = \frac{\sigma}{\sqrt{n}}$, and corresponding statistics $\hat{\mu}$ and $\hat{\sigma}_{\mu} = \frac{\hat{\sigma}}{n}$. Consequently, on average, the estimated values from a sample of size n, which has $(\hat{\mu}, \hat{\sigma})$, can be expected to equal the true population values (μ, σ). In addition, the standard deviation of the distribution of the means of all possible samples will equal the population standard deviation divided by the square root of the sample size. These estimated population characteristics can be used to great advantage in samples from natural populations because they approximate the true population parameters and can be used to make precise statements about the population values. Note that the standard error (the standard deviation divided by the square root of n) will always be smaller than the standard deviation of the population, thus enhancing our ability to make confident statements about the precision of the mean.

THE CENTRAL LIMIT THEOREM

As we have already stated, researchers often lament that the distributions of many natural populations are far from Normal; that is, they are *skewed*. In a skewed distribution the frequency distribution is asymmetrical, with some values being disproportionately more or less frequent. For example, the tendency of a type of organism to aggregate makes the more abundant individuals very abundant in very few quadrats. Realizing this, researchers sometimes shy away from methods of inference that impose Normal distributions.

The sampling distribution of the mean that we developed above, however, circumvents the apparent problem. Regardless of the population's distribution, the distribution of the means that pertain to that population will not be skewed. The *Central Limit theorem* states that the distribution of sample means from repeated samples of N independent observations approaches a Normal distribution, with a mean of μ and a variance $\frac{\sigma^2}{N}$. Notice that this theorem requires no assumption about the form of the original population. It works for any configuration. What makes this theorem so important is that it is true regardless of the shape or form of the population distribution. If the sample size is large enough, the sampling distribution of the means is always Normal.

The major importance of this result is for statistical inference (as explained in Chapter 1). The Central Limit theorem provides us with the ability to make even more exact statements. Instead of relying on Tchebychev's theorem, which ensures that at least 75% of the observations are within 2 standard deviations, we know from mathematical statistics that the proportion of observations falling within 2 or 3 standard deviations of the mean for a Normal distribution is 0.95 or 0.99, respectively. For example, the value of σ_μ shown in Table 3.4 is 1.39, and the mean is $\mu = 6.00$. Two standard deviations about the mean is then 6.00 ± 2.78, or $3.22 \leq \mu \leq 8.78$. Looking at all the means shown in Table 3.3, we find none below 4.00 and only one above 8.78. Recalling that there are 28 means in all, $\frac{1}{28} = 0.04$, so that in fact for our Bolivian data set $1 - 0.04 = 0.96$ of the means fall within 2 standard deviations.

Summary

1. Counts of organisms are integers (whole numbers) and are called *discrete random variables*.

2. Frequency tables are tabulations of data into ordered intervals, or classes, such as the number of organisms that are in 1, 2, . . . , *n* observations or samples. This tabular information is called a *frequency distribution*. A continuous curve for a frequency distribution drawn from either discrete or continuous variables is called a *frequency curve*.

3. The Normal distribution has a bell-shaped frequency curve that is completely specified by the mean and standard deviation (or its square, the variance).

4. Any observation can be standardized so that the distribution of all the observations in the data set will have a mean of 0 and a standard deviation of 1. This standardized observation is called a *z score*.

5. If a *z* score is compared to a Normal distribution, then the proportion of observations in standard deviation units is known, and the proportion of observations occurring below or above any specified standard score can be assessed.

6. The distribution of a sample statistic, such as the mean, is called a *sampling distribution*. When simple random sampling is used, the mean of the sampling distribution always equals the population mean. The standard deviation of the sampling distribution of the mean is called the *standard error*.

7. The Central Limit theorem states that if the sample size is large enough, the distribution of sample means is Normal. Even when the sample size is as small as 10 or so, this theorem can still give results that are useful for characterizing natural populations.

8. We can make improvements on the information afforded by Tchebychev's theorem.

   a. Tchebychev's theorem ensures that at least 75% of the observations are expected to be within 2 standard deviations of the mean at some chosen confidence level, using no assumptions on the distribution of the random variable.

   b. An assumption of Normality increases this amount so that 95% of the observations are expected to be within 2 standard deviations of the mean.

   c. Consideration of the sampling distribution of the mean and the Central Limit theorem tells us that 95% of mean values are expected to be within $\frac{2\sigma}{\sqrt{N}}$ of the mean.

In each of these three cases (a, b, c), successively more restricted inferences are being made.

9. The standard error is a measure of the precision of the random variable's value as an estimate of the related parameter of the target population.

PROBLEMS

3.1 Using Appendix 1, prepare a frequency table and a histogram for *Brosimum lactescens* (species no. 2). Comment on the central tendency.

3.2 Take a random sample (Appendix 3 has random numbers) of *Brosimum lactescens* from Appendix 2, using $n = 4$.

a. Find the $\hat{\mu}$ and $\hat{\sigma}$.

b. Based on these estimates, about what proportion of the observations would you expect to produce fewer than 10 trees? Does your expectation agree with the data?

3.3 Using the sample size of 4 that you took in Problem 3.2,

a. Calculate the standard error, using the fpc for the sample of $n = 4$ of *Brosimum lactescens* calculated in Problem 3.2.

b. Is the true mean within $2\hat{\sigma}_\mu$ of $\hat{\mu}$? Had you ignored the fpc, would you still have come to the same answer?

3.4 For the Rice and colleagues (1983) data for *P. cryptum* given in Problem 2.3 in Chapter 2,

a. Divide the values for each replicate by the appropriate standard deviation estimate. Calculate the new standard deviation of the $\left(\dfrac{x_i}{\hat{\sigma}}\right)$ divided values.

b. Standardize the 4 replicate observations on *P. cryptum*. Check that the standardized mean and variance are correct.

3.5 Your population consists of 6 quadrats, with counts of 4, 6, 2, 8, 0, and 4 organisms. You took a sample of $n = 2$ quadrats. We have said that the mean density is an estimate and also a random variable. What is the set of values that your sample mean density can take on?

4

CONFIDENCE LIMITS AND
INTERVALS FOR DENSITY

A N ARITHMETIC MEAN, such as the mean density, is a single number called a *point estimate* in statistics. The use of such a point estimate is common in ecology, but it has the disadvantage that we do not know how good it is. That is, a single value cannot tell us how close it is to the quantity we wish to estimate. This is why a point estimate must always be accompanied by some information upon which its usefulness as an estimate can be judged. The properties of the Normal distribution (Chapter 3) allow us to calculate such information, even though the true population value for the mean density is unknown.

CONFIDENCE LIMITS ON MEAN DENSITY

We saw in the previous chapter that the size of the chance (random) variation in the mean obtained in our sample depends on two things: (1) the sample size (n), and (2) the variability (the standard deviation σ) in the population. However, neither of these quantities provides information on the reliability of our sample results nor on the merit of $\hat{\mu}$ as an estimate of the true mean density, μ. We need a way to bracket μ with a quantitative statement. In other words, we need to provide a range of values (an interval) for the mean, accompanied by a statement with a probability (the level of confidence); that is, with high probability, the true mean falls in that range, or between the limits of that range. In this way we obtain an *interval estimate* of a population parameter. In statistics, *interval estimation* is estimation by specifying a range of values between some upper and lower limits, within which the true value lies. The true value, or population parameter, is a fixed number, and the limits of the range are random variables for purposes of natural population surveys. We use interval estimation when the primary concern is to obtain an estimate of some parameter, not to test a hypothesis.

By convention, the probability is usually set at 0.95, which is to say, we are confident that 95% of the time the range of values we select for $\hat{\mu}$ will bracket the true mean density, μ. There is nothing magical about 95%, and in many situations other levels of confidence may be more appropriate. This selected value is an arbitrary convention, and we will point out a little later, contrary to the opinion of some reviewers and editors of scientific journals, that 0.95 may not always be desirable.

Of the 5% of the values that may be expected to fall outside this range, some will be greater than the upper limit, and others will be smaller than the lower limit. If we do not know where a value may fall, we assume symmetry. *Symmetry* means that half of this 5.0%, or 2.5%, will be above the upper limit, and half, or the remaining 2.5%, will be below the lower limit. Such situations are termed *two-tailed*, because half of the error lies at each end, or tail. Sometimes, however, we are concerned only with whether the true value is at least as big. That is, interest is in whether values exceed the limit (or, conversely, whether they are below the limit). The situation is thus described as *one-tailed*. Use of a two-tailed or a one-tailed confidence situation is determined by the nature of the research question.

The information we have already generated on the sampling distribution of the mean, from Chapter 3, will now help us with this problem of setting confidence limits.

Limits and Intervals for Large Samples

Recall that the standard deviate, z, is the distance of a count from the mean value, μ, in standard deviation units. For any given value of z, the proportion of observations in the population having smaller values of z can be read from Table 3.2. If more detail is desirable, the proportion can be obtained from a more extensive table or statistical computer program for the Normal distribution. Inspection of a standard Normal table shows that 95% of the area under the curve falls between $z = -1.96$ and $z = +1.96$. These values of z are denoted $-z$ and $+z$, respectively. Note that in Table 3.2 the p value for $z = -1.96$ is 0.025; for +1.96 it is 0.975. Because we are interested in 95% of the area under the curve, and because the curve is symmetric and bell-shaped, there is 0.025 at each end, or tail. That is, the end area beyond the point +1.96 is $1.000 - 0.975 = 0.025$. Consequently, 1.96 is the *two-tailed* value for 95% (Figure 4.1). A *one-tailed* value for 95% (Table 3.2) would be 1.64 (or −1.64). Should we wish some confidence value other than 95%, we simply choose the appropriate value of z.

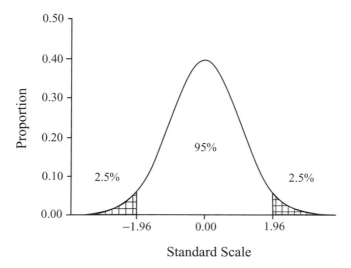

FIGURE 4.1   Standard scores at tails of the Normal distribution. Percent values refer to area under the curve.

We know from the Central Limit theorem (Chapter 3) that the sampling distribution of the mean density, μ, can be approximated quite well by a Normal distribution having this same population mean of μ $\left(\text{and standard deviation of } \sigma_\mu = \dfrac{\sigma}{\sqrt{n}}\right)$. We also know that about 95% of the time, any sample mean will differ from the population mean μ by less than $1.96 \cdot \dfrac{\sigma}{\sqrt{n}}$. Another way to say this is that the probability is 0.95 that a sample mean will vary in either direction (that is, above or below the true value) by less than $1.96 \cdot \dfrac{\sigma}{\sqrt{n}}$.

Summarizing these steps in order to make a statement about how good our estimate is, we find that the error made by using $\hat{\mu}$ instead of μ is given by the difference $\hat{\mu} - \mu$. And we know that the size of this error must be less than $1.96 \cdot \dfrac{\sigma}{\sqrt{n}}$ (95% of the time). We can write these two pieces of information in more succinct form using algebraic symbols as

$$-z\frac{\sigma}{\sqrt{n}} \leq \hat{\mu} - \mu \leq +z\frac{\sigma}{\sqrt{n}}. \tag{4.1}$$

Using algebraic manipulation to rearrange these terms, we obtain the commonly accepted statistical formulation

$$\hat{\mu} - z\frac{\sigma}{\sqrt{n}} \leq \mu \leq \hat{\mu} + z\frac{\sigma}{\sqrt{n}}. \tag{4.2}$$

For any selected sample, Equation 4.2 is true at the level of probability associated with the value of z that is selected. This statement is called an *interval estimate* of the mean density. When there is a large amount of *sampling error* (that is, when $\dfrac{\sigma}{\sqrt{n}}$ is large), the *confidence interval* calculated from any one of our statistical samples will be large. A confidence interval, or interval estimate, is an estimated range of numbers that has a given probability of including the true population value in repeated sampling.

The endpoints of confidence intervals are called *confidence limits* (CL). Confidence limits are given by

$$CL = \hat{\mu} \pm z\frac{\sigma}{\sqrt{n}}. \tag{4.3}$$

When using the CL, 0.95 or any other value we might choose is the probability value selected to be our *degree of confidence*, or *confidence coefficient*. The degree of confidence can be decreased or increased. For example, if we choose 0.99 by using $z = 2.58$ (Table 3.2), the resultant confidence interval becomes wider and will tell us less about the mean density. A probability value is always greater than or equal to 0, and at the same time, less than or equal to 1. However, when a probability value is used to describe the confidence in our computations, it is usually selected to be between 0.80 and 0.995. In ecological investigations, values of less than 0.95 are seldom used.

If we knew the true standard deviation in the target population, σ, this would be a solution to the problem of describing the merit of our sample estimate of the population mean obtained by using Equation 4.2. Such fortunate circumstances are not at all likely for biological samples. Luckily, we can merely substitute the sample estimate $\hat{\sigma}$ for σ if the sample size is large. Most texts define large as greater than 25 to 30. As our example will show, the sample size may be smaller than 25 and yet still allow for the substitution. By making this substitution, we obtain an alternative to Equation 4.2:

$$\hat{\mu} - z\frac{\hat{\sigma}}{\sqrt{n}} \leq \mu \leq \hat{\mu} + z\frac{\hat{\sigma}}{\sqrt{n}} \tag{4.4}$$

or, equivalently, remembering that,

$$\hat{\mu} - z\hat{\sigma}_{\mu} \leq \mu \leq \hat{\mu} + z\hat{\sigma}_{\mu}. \tag{4.5}$$

Our problem is solved once again by forming a *large-sample confidence interval*.

Limits and Intervals for Small Samples

In Equation 4.5 we can recognize three facts about sampling from natural populations. First, the true value for the standard deviation, σ, is never known. Second, for large samples, there is not much chance of making a sizable error by using the estimate, $\hat{\sigma}$, in place of σ in the formula for the standard error of the mean. Finally, the Central Limit theorem (Chapter 3) reveals that whatever the details of a large sample, the mean density's sampling distribution displays a Normal distribution curve.

Alternatively, for small samples, we cannot have as much faith in our estimate of σ_{μ}. For small samples we expect the value of this variability estimate to be considerably wider-ranging than an estimate based on a larger sample. Because this variability is large, the standardized Normal distribution is not as reasonable a *reference distribution*—that is, the distribution to which we refer when we make our inference. Instead we use what is called a *Student's t value*. The Student's t value is used in place of the standard deviate z. Then we use a *Student's t distribution* as the reference distribution, which can be used to set confidence intervals about the mean, independently of the population variance.

The reason for substituting a Student's t distribution for the Normal distribution when working with small samples is based on a very important fact in mathematical statistics. Whenever we take random observations from a natural population, the sample mean and standard deviation (either $\hat{\sigma}$ or σ) are independent when the population has a Normal distribution. For large sample sizes we have shown that the assumption of a Normal distribution is reasonable; for small samples this cannot be done. If such independence is not possible, our inferences are affected. Thus we must make an adjustment to our procedures for forming confidence intervals. We make this adjustment by changing the reference distribution.

Recall our Bolivian tree example in which we chose a statistical sample of $n = 8$ quadrats (observations) and calculated a mean density estimate of $\hat{\mu} = 6.00$, with $\hat{\sigma} = 2.27$. What would be the effect of using Equation 4.5 to form a confidence interval if we had only the knowledge of this small sample? Because we would have neither the true value for σ nor a large sample size, we should expect that we would have less confidence in the inequality statement. Because the sample standard deviation estimate, $\hat{\sigma}$, which is the basis for our estimate of the standard error, $\hat{\sigma}_{\mu}$, varies from sample to sample (whereas the population value does not), the widths of confidence intervals based on repeated samples from the same population will also vary. That is, the confidence intervals will not have the desired constant widths. If the sample standard deviation is smaller than the population quantity, the resulting confidence interval (Equation 4.5) will be narrower, and if larger, then it will be wider. The width of the interval is an indication of the *precision of the estimate*. Thus for our small-sample example, we will be unsure how good our estimate is.

To be able to construct a suitable confidence interval in a small-sample situation, we must assume that the population from which we are sampling has approximately a Normal distribution. Only then will the information in the variance be unaffected by the value of the mean (independence). We then can relate the degree of confidence to values from a Student's t distribution for various sample sizes. This distribution is continuous. It has a mean of 0, and it is symmetrical about that mean, thus making it similar to a Normal distribution. However, its shape depends on the sample size.

This dependence is expressed via a parameter called the *degrees of freedom* (df). The df equals the sample size minus 1 (df $= n - 1$) for our purposes here. The usual notation for a Student's t value is $t_{c,df}$, where c denotes our selected confidence level. For example, $t_{0.95,7}$ is the t value for a 95% level of confidence when the sample size is 8 (and thus df $= 8 - 1 = 7$). The topic of degrees of freedom is much more complex than that involved in this confidence interval construction. The interested reader is referred to Feller (1957). Simply described, the shape of the Student's t distribution is distinct for each different sample size, and the degrees of freedom help to discriminate among the many possibilities.

Table 4.1 lists the most common values of t for one-tailed and two-tailed limits. Note that the values of t for infinity (very large sample size) are exactly the same as those of z from the standard Normal table for the same confidence (Table 3.2). This is so because as the sample sizes enlarge, the t distribution becomes (or approaches the shape of) a Normal distribution.

Just as we did for the large samples, we now can form a confidence interval for small samples. We do this by using a value from the t distribution in place of a value from the Normal (z) distribution. There is one added point of importance. It is possible to enclose the mean with many other intervals that would also contain 95% of the

TABLE 4.1   Values of *t*, Student's *t* Distribution, Corresponding to p for One- and Two-Tailed Confidence Limits with Degrees of Freedom

df		CONFIDENCE LIMITS p			
	ONE-TAILED	0.90	0.95	0.975	0.995
	TWO-TAILED	0.80	0.90	0.95	0.99
1		3.08	6.31	12.71	63.66
2		1.89	2.92	4.30	9.92
3		1.64	2.35	3.18	5.84
4		1.53	2.13	2.78	4.60
5		1.48	2.01	2.57	4.03
6		1.44	1.94	2.45	3.71
7		1.41	1.89	2.36	3.50
8		1.40	1.86	2.31	3.36
9		1.38	1.83	2.26	3.25
10		1.37	1.81	2.23	3.17
20		1.32	1.72	2.09	2.84
30		1.31	1.70	2.04	2.75
50		1.30	1.68	2.01	2.68
100		1.29	1.66	1.98	2.62
(∞)		1.28	1.64	1.96	2.58

area under the curve. However, when we use the phrase "95% confidence interval" about the mean density, this particular *central and symmetric interval* is defined to give the shortest possible range of values about the true parameter μ. Confidence intervals for other parameters besides the mean (for example, proportions) do not always yield the shortest possible range.

Our *small-sample confidence interval* is

$$\hat{\mu} - t\frac{\hat{\sigma}}{\sqrt{n}} \leq \mu \leq \hat{\mu} + t\frac{\hat{\sigma}}{\sqrt{n}} \tag{4.6}$$

or, the alternative form,

$$\hat{\mu} - t\hat{\sigma}_\mu \leq \mu \leq \hat{\mu} + t\hat{\sigma}_\mu. \tag{4.7}$$

The formulas in Equations 4.6 and 4.7 simply replace the *z* in Equations 4.4 and 4.5 with *t* and accomplish the same result for small samples. Another way to look at this statement is that if we desire 95% confidence limits, we cut off 0.025 of the area under the curve at each end, or tail, of the distribution. Thus the interval includes the central

0.95 of the area under the entire curve. Selection of a two-tailed value from Table 4.1 would give us the appropriate number for this particular small-sample case. Just as we saw with the large-sample case, if we instead wanted to place all the *critical area* in one tail for this small-sample case, we would have selected a *one-tailed* value. In one sense, all the area under the curve is critical. However, we are selecting a particular value for making a decision, so that the critical area is defined as the area corresponding to noncompliance, or the area in which we agree to consider the observed value an error.

The reason for choosing a value to use in the confidence interval formula from either a one-tailed or two-tailed table (regardless of the reference distribution) depends on the statement of the research question. That is, for the usual type of natural population survey, a two-tailed value for the mean density may be all that is required. Alternatively, a value from one tail of the distribution might be of interest if a target population is thought to be above or below some predetermined critical limit.

Continuing our illustration of how to form a confidence interval for a small sample, in the Bolivian plot 01, recall once again our random sample of $n = 8$, $\hat{\mu} = 6$, and $\hat{\sigma} = 2.27$. The estimate of the standard error is $\hat{\sigma}_\mu = \dfrac{2.27}{\sqrt{8}} = 0.80$, without the fpc, or 0.77, from Equation 3.4, with it. Note here, as elsewhere in this book, that we provide the calculations both with and without the fpc for illustrative purposes only, to emphasize that an fpc is useful only when $\dfrac{n}{N}$ is not close to about 0.05. If we wish to be 95% confident that the true mean lies within the interval given by Equations 4.6 or 4.7, then from Table 4.1 (for $n = 8$, df = $n - 1 = 7$) we select $t = 2.36$ (two-tailed) at 0.95, or 95%.

Plugging this information into Equation 4.6, we can calculate the small-sample confidence interval as follows: $6.00 - 2.36 \cdot 0.77 \le \mu \le 6.00 + 2.36 \cdot 0.77$, or, $4.18 \le \mu \le 7.82$.

The result tells us that we can be 95% confident, or would expect that 95 times out of 100 in similar samples (based on our random sample), that the true mean of population density lies between 4.18 and 7.82 (roughly between 4 and 8 trees per quadrat). In our Bolivian case, because all the individuals in all 100 quadrats were actually counted by the researchers, we know that the true value μ is 6.63. So, it appears that an assumption of population Normality was indeed reasonable.

If we are willing to be less confident, the limits can be narrowed even further. For an 80% two-tailed confidence, $t = 1.41$ and the small-sample confidence interval becomes $6.00 - 1.41 \cdot 0.77 \le \mu \le 6.00 + 1.41 \cdot 0.77$, so we obtain the interval $4.91 \le \mu \le 7.09$. Here we can state with 80% confidence that the true mean density lies between about 5 and 7 individuals, which is, in this case, correct. In other words, if we are willing to accept less confidence (in this example, 80%), the confidence interval can usually be narrowed considerably.

Thus far we have been basing our estimate of tree density on only 8 quadrats ($n = 8$) of the 100 quadrats in the Bolivian forest plot. Sometimes the work involved in enumerating even a seemingly small number of quadrats (biological samples), such as 8, especially for microorganisms, may require more time or money than the researcher has available. Consequently, an even smaller statistical sample will have to suffice. Let us now consider the sizes of confidence intervals as we select different sample sizes.

Returning to the census of $N = 100$ quadrats on plot 01, we will calculate the confidence limits for a sample of $n = 4$. Using a table of random numbers, the 4 quadrats selected were numbers 63, 95, 06, and 76. The total numbers of individuals in these samples are 9, 9, 5, and 10, respectively (see Appendix 1). Using Equation 2.2, we calculate that the mean density for this 4-quadrat sample is 8.25. This figure then serves as the estimated mean density of the entire 100-quadrat population, $\hat{\mu} = 8.25$. From Equation 2.5, we obtain an estimate of the variance, $\hat{\sigma}^2$, and then an estimate of its square root, $\hat{\sigma}$, which is the standard deviation. In this 4-quadrat case, $\hat{\sigma} = 2.22$. From

Equation 3.6, we estimate the standard error, $\hat{\sigma}_\mu = \dfrac{2.22}{\sqrt{4}} \cdot \sqrt{\dfrac{100-4}{99}} = \dfrac{2.22 \cdot 0.98}{2} = 1.09.$

Note how little the fpc, 0.98, contributes to the estimate. Had we ignored the fpc, we would have obtained $\dfrac{2.22}{2} = 1.11$, rather than 1.09. This small difference occurs because the ratio $\dfrac{n}{N}$ is 0.04, or 4%, for $n = 4$, whereas it was 8% for $n = 8$. With this information for the sample of size 4, we can calculate $t_{0.95,\,3} \cdot \sigma_\mu$ to obtain the small-sample confidence limits from Equation 4.7. From Table 4.1 the value of t (two-tailed) required for a 95% confidence limit for a sample size of 4 is 3.18. The limits are $8.25 \pm 3.18 \cdot 1.09$, or, 8.25 ± 3.47. From Equation 4.7, we calculate that the desired interval is $4.78 \leq \mu \leq 11.72$. We know, of course, that for this plot the true mean density is $\mu = 6.63$, which can be seen to be within our calculated, though rather wide, confidence interval.

If a researcher is willing to accept 0.90 confidence instead of 0.95 (the values are arbitrary; 0.90 is still a grade of "A"), the limits are narrowed. For 0.90 and $n = 4$, df $= 3$, $t = 2.35$. This means that using Equation 4.6 and the fpc, then the confidence interval is calculated to be $8.25 \pm 2.35 \cdot 1.09$, or 8.25 ± 2.56, or $5.69 \leq \mu \leq 10.81$.

Ignoring the fpc, which is small, we can turn to Equation 3.7 to see that $\sigma = \dfrac{\hat{\sigma}}{\sqrt{n}}$. The confidence limits are calculated as CL $= \hat{\mu} \pm t\hat{\sigma}_\mu$. Then, if we round the value $t = 2.35$ to $t = 2$, we have $2 \cdot \dfrac{2.22}{2} = 2.22$. The interval is thus $6.03 \leq \mu \leq 10.47$. In other words, the confidence limits for 0.90 when $n = 4$ are approximately equal to $\hat{\mu} \pm \hat{\sigma}$ (because $t = 2$ cancels with $\sqrt{4} = 2$ in the formula). Because statistical packages routinely print the $\hat{\mu}$ and $\hat{\sigma}$, knowledge of this approximate result can be very useful when examining a printed output from numerous localities that were sampled with $n = 4$.

Limits and Intervals for Total Population

If we wish to place confidence limits on the total number of individuals in a population, we simply multiply each term in the *inequality* (Equation 4.6 or 4.7) by N. Thus,

$$N\hat{\mu} - Nt\hat{\sigma}_{\mu} \leq N_{\mu} \leq N\hat{\mu} + Nt\hat{\sigma}_{\mu}. \qquad (4.8)$$

Of course, this interval estimate only works if we are using a sampling plan that allows us to know the total number of quadrats that would cover the area. For example, Equation 4.8 would be useful if a grid or map system were used, but it would have little utility when taking cores from bathyl depths.

For our sample 8 quadrats ($n = 8$), we are, then, 95% confident that the total number of individuals in the 100-quadrat population ($N = 100$) is a number somewhere between 418 and 782. If we were content with 80% confidence, then we would predict the true number to be between 491 and 709.

Had we not used the fpc, then the 95% confidence limits for the true mean would be $6.00 - 2.36 \cdot 0.80 \leq \mu \leq 6.00 + 2.36 \cdot 0.80$. The result is $4.11 \leq \mu \leq 7.89$. This means that the number of trees in the total population would be between 411 and 789. If a table for Student t distribution were unavailable, but we remembered 1.96 as the usual two-tailed 95% confidence value, we would have the large-sample interval $6.00 - 1.96 \cdot 0.80 \leq \mu \leq 6.00 + 1.96 \cdot 0.80$. This yields $4.43 \leq \mu \leq 7.57$ for the mean, and $443 \leq N\mu \leq 757$ for the total. In the first instance the interval is wider and in the second, smaller. In either case, the difference is rather small and most researchers would be content—and correct.

PRECISION OF THE CONFIDENCE INTERVAL

Continuing with our comparison of sample sizes for the Bolivian forest plot 01, recall that when we took $n = 8$ quadrats at the 0.95 confidence level, the confidence limits about the true mean density were $\hat{\mu} \pm t\hat{\sigma}_{\mu} = \hat{\mu} \pm (2.36 \cdot 0.77) = \hat{\mu} \pm 1.82$. When $n = 4$ at this same 95% level, we obtain $\hat{\mu} \pm t\hat{\sigma}_{\mu} = \hat{\mu} \pm 3.18 \cdot 1.09$, or $\hat{\mu} \pm 3.47$. Although twice the amount of work was required to enumerate the individuals in 8 samples versus those in 4 samples, we did not receive twice the reward. Let us examine why not.

In addition to the sample mean density estimate, the confidence limit depends on two components: t and $\hat{\sigma}_{\mu}$. We will examine the contribution of each of these two parts.

First, recall that for the sampling distribution of the mean, we saw that the mean of that distribution was identical to the true mean μ, and the standard deviation was the standard error σ_{μ} (Chapter 3). It would help if we had a *relative measure*, also called a *dimensionless measure*, that could provide us with a way to compare distributions. This dimensionless measure must not be in the units of measure of either of the distributions being compared. A common such measure of variability is the *coefficient of variation* (CV), which is a measure of degree of variability (standard deviation) as a proportion of the mean. That is, a $CV = \dfrac{\sigma}{\mu}$, which is a standard deviation divided by its related mean. Often the CV is multiplied by 100 and expressed as a percentage.

In order to investigate the components making up the confidence interval, we are interested in the standard deviation of the sampling distribution. We therefore use

$$CV = \frac{\sigma_{\mu}}{\mu}. \qquad (4.9)$$

This Equation 4.9 represents the CV of the sampling distribution of the mean density. We can further examine this ratio using the true population parameters for the total individuals from the $N = 100$ quadrats; namely, $\mu = 6.63$ and $\sigma = 2.17$. Using these numbers with $n = 4$, the CV $= \dfrac{\sigma_\mu}{\mu} = \dfrac{1.09}{6.63} \cdot 100 = 16\%$. For $n = 8$, the CV $= \dfrac{0.77}{6.63} \cdot 100 = 12\%$. Thus, as we increase the sample size from $n = 4$ to $n = 8$, the CV does not decrease proportionally; 16% does not decrease to half of this amount, $\dfrac{16\%}{2} = 8\%$, but instead to $\dfrac{16\%}{\sqrt{2}} = 12\%$.

The second component of a confidence limit, the value for t, will vary with the sample size (degrees of freedom) and with the desired confidence level. A look at the t table (Table 4.1) shows that t decreases with increasing sample size and also decreases with the desired confidence level. At the same time, for any degree of confidence selected, the differences between the values for differing sample sizes are smaller at lower levels of confidence. For example, at $n = 4$ and df $= 3$, with a 0.95 two-tailed confidence, we get $t = 3.18$. At $n = 8$, we get $t = 2.36$ for the same 95% confidence level. This is a difference of 0.82. For the 0.90 level the difference between the t values for $n = 4$ and $n = 8$ is $2.35 - 1.89 = 0.46$. By the time we get to the 0.80 confidence level, the values of t for different sample sizes differ very little. The influence of t as a multiplier, then, decreases with increasing sample size and as the degree of desired confidence is lowered. At the higher sample sizes and/or lower levels of confidence, t values are quite similar.

The CV and t taken together can be used to define the *precision*, or *deviation* (*d*), of the confidence interval. This is simply the width of the confidence interval above or below the estimated mean, expressed as a proportion of the mean itself. Thus we have

$$d = \frac{t\hat{\sigma}_\mu}{\mu}. \tag{4.10}$$

The numerator, $t\hat{\sigma}_\mu$, is the estimated quantity from Equation 4.7 that is necessary to obtain the small-sample confidence interval. This numerator is the amount added to and subtracted from the estimate of the mean density to yield the confidence limits. As already pointed out, the first component of Equation 4.10, t, varies with sample size and level of confidence (Table 4.1). The second component in the formula, the standard error (σ_μ), varies with the square root of the sample size n. Even if t is constant, the value in the numerator will vary with the size of the square root of n, but not with n itself.

We can, as before, substitute the estimated values of μ and σ into Equation 4.10 to get an answer for our particular set of samples. In the example for $n = 8$, we have $d = \dfrac{2.36 \cdot 0.77}{6.00} = 0.30$. From this we see that the precision, or deviation, of the confidence limit about the mean is $\pm 30\%$. This means that 95% of the time, $\hat{\mu}$, will under- or overestimate the true population density by a factor of $(0.30 \cdot \hat{\mu})$, or less.

In Table 4.2 we list the precision, or deviation, for small-sample sizes using the true population parameters and confidence limits (two-tailed) of 0.80, 0.90, and 0.95. With only one biological sample, a single statistical observation, it is not possible to calculate confidence limits at all. The minimum number of biological samples is $n = 2$. The confidence limits for this small statistical sample will be very large because of the large values of t (Table 4.1). Notice, however, that as we double the sample size to $n = 4$, the rewards are enormous. At the 0.95 level, the confidence limits from $n = 2$ to $n = 4$ get about 6 times better; that is, more precise. At the 0.90 level the calculation of confidence limits is about 4 times better. Even so, many researchers are chagrined to learn that with $n = 4$, which often requires much more effort (i.e., work) than they wish to expend, the 0.95 confidence limits for the mean density have only about $d = \pm 50\%$ (Table 4.2). This means that a 95% confidence interval about the population mean density has only 50% precision, despite all the work that went into examining the 4 samples. We might wish to inquire, for example, what percentage of the quadrats could be expected to contain 10 or fewer individuals. Here we are interested in only one tail; we are interested in only the 10 or "fewer" section of the distribution. As shown in the last section, we can *standardize*; that is, we can obtain a standard deviate for any distribution, not just the Normal. In the small-sample case, instead of z, we wish to refer to the Student's t distribution. Therefore, $t = \dfrac{(x - \hat{\mu})}{\hat{\sigma}} = \dfrac{(10 - 6.00)}{2.27} = 1.76$.

Table 4.1 can then be consulted to find that for $n = 8$ and df $= 7$, with $t = 1.89$ (one-tail), we would expect 95% of the quadrats to contain 10 or fewer individuals. Observing the table of actual frequencies in the known population (Appendix 1), we see that 5 quadrats contain 10 individuals, and 92 contain fewer. There are, then, 97 quadrats with 10 or fewer individuals. Therefore, with a value of $t = 1.76$, we are well within our expectation. Our calculations turned out to be correct, and on the safe side, even though we counted the individuals in only 8 of the possible 100 quadrats. Had we ignored the number of observations (n) and simply used the figure for large samples (∞), then t would have had a value of 1.64 at a 95% confidence. Our interval would not have been too narrow.

TABLE 4.2   Coefficient of Variation of the Sampling Distribution for Some Small-Sample Sizes from Bolivia

n	df	$\dfrac{\sigma_\mu}{\mu}$	$t_{0.80}\,\dfrac{\sigma_\mu}{\mu}$	$t_{0.90}\,\dfrac{\sigma_\mu}{\mu}$	$t_{0.95}\,\dfrac{\sigma_\mu}{\mu}$
2	1	0.23	0.71	1.45	2.92
4	3	0.16	0.26	0.38	0.51
8	7	0.12	0.17	0.23	0.28
31	30	0.06	0.08	0.10	0.12

NOTE: Coefficient of variation of the sampling distribution, CV $= \dfrac{\sigma_\mu}{\mu}$. Deviation, $d = \dfrac{t\sigma_\mu}{\mu}$ for two-tailed confidence limits of 0.80, 0.90, and 0.95.

SOURCE: Data from Bolivian tree plot 01, $N = 100$, $\mu = 6.63$, $\sigma = 2.17$.

INTERPRETATION OF CONFIDENCE LIMITS AND INTERVALS

Overall, one of the reasons the t distribution was used for constructing interval estimates for small samples was because of the increased uncertainty in the estimate of the standard error when the sample size was not large. However, it is also true that the t distribution is theoretically appropriate whenever we must estimate the standard error, regardless of sample size. If, for example, our sample was of size $n = 1,000$, we actually should construct our estimate by using a t table with df = 999. Tables for t for all possible samples would be unwieldy. We can take advantage of the fact that whenever n is greater than about 30, the t values are approximately the same as the z values from the table of the Normal distribution (Table 4.1). So, our use of the Normal z values for large samples (with unknown population standard deviations) really are used as good approximations to the t values. For this reason we assumed that the sampling distribution was Normal in large samples, despite the fact that we had to estimate the standard errors.

The wording of statements in the research literature involving interval estimates presents many problems. To state that the inequality in, say, Equation 4.7 is made with 95% confidence, or has a confidence coefficient of 0.95, does not mean that there is a probability of 0.95 that the true mean density is somewhere between two confidence limit values. Nor does it say that the true mean is in that interval at least 95% of the time. The true mean density is some well-defined, although unknown, number that either is or is not between the limit values of the confidence interval. There is no chance involved at all ($p = 0.00$ or $p = 1.00$). However, whether the inequality is right or wrong depends on whether the sample mean we calculated actually falls within 1.96 standard errors from the population mean; this we cannot know from just one field sampling. We do know, however, that if we took a very large number of similar random samples from this same population, and if we then constructed a confidence interval with a 95% confidence coefficient on each one, or if we constructed this interval for a great many different problems, about 95% of all of these statements would be correct.

Thus the important point is to realize that any one such statement we make may be right or wrong. We will never really know, on the basis of only one statistical sample, whether we have made the infrequently occurring (about 5% of the time) single statement that is wrong. We simply proceed as if it were correct. Thus, our single-sample results sometimes will mislead us. But this risk, due to sampling error, cannot be eliminated as long as we are dealing with partial information.

We must therefore stress that the confidence coefficient signifies the probability that the procedure itself will lead to a correct interval estimate—not that our sample will do so. If this confidence coefficient is high (near 1.00), we proceed with our analysis as if the interval estimate derived from our sample were correct. We do so because the procedure is known to produce a high rate of correct estimates. Thus we must be careful in interpreting a confidence interval about the true density (mean or total) of the population we are studying. Because μ is a constant, a confidence interval is not a probability statement about μ. Instead, it is a probability statement about the *random interval*, having random limits or endpoints, and we are estimating the location of the true mean in that interval.

Statistical methods enable us to measure the magnitude of the sampling error for any simple random sample. However, the extraneous factors that necessitated obtaining the sample information are what determine whether the sampling error is sufficiently small to make the estimate a useful one. Because the width of the confidence interval is related to the risk that the interval estimate will be incorrect, this risk or error should be clearly stated in the research plan. The risk should not be set (say, to 0.95) simply to make a journal editor or reviewer happy.

In general, the more data obtained for investigating a density problem, the better off we are. This is because the biological information is more representative. Hence, any inferential statements will carry more weight of evidence. However, in many, if not most, field surveys a small amount of data is all that can be obtained. The proper use of statistical science in such cases is to draw a conclusion from the sample evidence in such a way as to take into account the limitations on the amount of data. We have shown that we have a procedure for taking account of the smallness of the sample and that we can draw conclusions from small data sets when we allow our estimate to be less specific (i.e., our confidence interval to be wider).

Summary

1. Variation in the point estimate of the mean density, $\hat{\mu}$, obtained from a field sample depends on sample size (n) and variability (σ) in the population.

2. The point estimate of the standard deviation, $\hat{\sigma}$, can be substituted for the true value of σ when calculating the standard error, $\hat{\sigma}_\mu = \dfrac{\hat{\sigma}}{\sqrt{n}}$.

3. A confidence interval is an estimated range of numbers that has a given probability of including the true population value in repeated sampling. Confidence limits (CL) are the end points of this interval.

4. For large sample sizes, $n > 30$, the standard Normal deviate, z, and the standard error, $\hat{\sigma}_\mu$, are used to calculate confidence intervals (Equations 4.2 and 4.5).

5. For small sample sizes, $n < 30$, values of Student's t distribution with degrees of freedom equal to ($n - 1$) and the estimate of the standard error are used to calculate confidence intervals (Equation 4.6).

6. For an $n = 4$ and 90% confidence probability, the confidence limits for mean density are approximately $\hat{\mu} \pm \hat{\sigma}$. This short formula provides a quick and easy way to get an approximate answer from a computer printout.

7. A dimensionless measure of variability is the coefficient of variation (CV), which is defined as a standard deviation divided by the related mean.

8. For the sampling distribution of the mean density, the CV equals the standard error divided by the mean density.

9. Precision, or deviation, of the confidence interval is the product of the appropriate value of t and the CV of the sampling distribution of the mean (Equation 4.10).

10. For one biological sample ($n = 1$), no confidence limits are possible. The increase in precision from a statistical sample size of 2 biological samples ($n = 2$) to 4 biological samples ($n = 4$) is very large.

11. For $n = 4$, the 95% confidence interval for the Bolivian tree example has a precision of only about $\pm 50\%$. Experience with other data sets indicates this is usual.

PROBLEMS

4.1 Take a random sample of $n = 4$ of *Scheelea princeps* from Appendix 1. Calculate the confidence limits and the confidence interval. Does the true mean lie within the interval?

4.2 Using the estimates from Problem 4.1, calculate the precision of the confidence limits. Use the true population parameters (Appendix 1) and calculate the precision. How close are the estimates to the parameters?

4.3 Take a random sample of $n = 4$ of *Scheelea princeps* from Appendix 2. Calculate the confidence limits and the confidence interval. Does the true mean lie within the interval?

4.4 Using the estimates from Problem 4.3, calculate the precision of the confidence limits. Use the true population parameters (Appendix 2) and calculate the precision. How close are the two precision values?

4.5 Calculate the total number of *Scheelea princeps* and its confidence interval from first the $N = 100$ and then the $N = 25$ data sets. How do the results from these two sets compare?

4.6 Take a sample of $n = 2$ of *Pouteria macrophylla* from Appendix 2. In what percentage of the quadrats and total area would you expect to find no *P. macrophylla*? How does your answer compare with the actual tally?

4.7 A study by Hart and colleagues (1985) examined the numbers of ostracods commensal on crayfish. Densities in relation to sex of the host crayfish were examined but found not to be of consequence. Table 8 in that study gave the following information for numbers of the ostracod *Uncinocythere occidentalis* that were found on 45 crayfish, based on seasonal sampling.

		OSTRACODS					
SEASON	CRAYFISH (NUMBER OF)	TOTAL	MEAN	ST. DEV.	CV	MAX.	MIN.
Fall	10	3,856	384.60	273.06	70.98	935	151
Winter	10	2,432	243.20	168.21	69.17	532	29
Spring	15	2,930*	195.33	180.12	92.21	575	17
Summer	10	955	95.50	68.87	73.16	260	12

* Proportionally correcting this total for 10 instead of 15 crayfish would give a total of 1,953.

The authors say that this table contains the total seasonal counts from samples that indicate possible seasonal differences. They say that their methods of testing show that the ostracod counts are similar for the first three seasons, but different for the summer.

a. Put 95% confidence intervals on the number of ostracods per season.

b. If the confidence intervals in part a do not overlap (do not contain some of the same range of values), then the means are said to be "significantly different." Do your confidence intervals show that the authors' contention of a significant difference in counts for the summer is correct?

4.8 Using the larval data of Heyer (1979) in Problem 2.2, Chapter 2, find the mean density and variance for *R. sylvatica* over all sweeps. Which type of sweep's count is the most deviant from the mean density in standard deviation units?

5
HOW MANY FIELD SAMPLES?

W E NOW KNOW how to estimate the mean density and its variability. By making an assumption of Normality and understanding sampling distributions, we were able to construct confidence limits and estimate precision even for small numbers of biological samples. This same knowledge, with a little manipulation, allows us to choose a sample size that will be sufficient to provide any desired degree of accuracy for surveying natural populations in the field. Because the larger the sample size, the smaller the error we can expect to make in using an estimate in place of the true value, we can select the sample size (n) required to make the measure of that error (σ_μ) any size (for example, as small as) we want. In fact, there will even be some value for n that will be a *sufficient sample size*. By *sufficient*, in this instance, a statistician means that the selected size will be just large enough to achieve the desired accuracy of estimation, but not so large as to exceed it. In this way, the minimum time, effort, and money will be expended in obtaining field samples.

MINIMALLY SIZED SAMPLES WITH NO PILOT DATA

Law of Large Numbers

We can identify a sufficient sample size by using what is called the *Law of Large Numbers*, which is related to Tchebychev's theorem. The Law of Large Numbers states that if a randomly selected sample is "large," then the probability (P) that the sample mean $\hat{\mu}$ is quite close to the population mean (μ) is related to the standard error (and therefore to the sample size). The law is expressed mathematically as follows:

$$P(\hat{\mu} - \mu) \geq 1 - \frac{\sigma_\mu^2}{k^2}. \tag{5.1}$$

Only a mathematical statistician would consider this algebraic statement equivalent to the word form of the Law of Large Numbers. However, with the insight gained from prior chapters, the algebraic form should become clear. Following is our explanation of how to achieve this understanding.

The Law of Large Numbers, and its mathematical expression as Equation 5.1, is a statement for large samples. The larger the sample size n, the closer the error gets to 0.

This is because the error is defined as $\sigma_\mu = \dfrac{\sigma}{\sqrt{n}}$, which has the large number (n) in the denominator. That is, regardless of what value we get for the standard deviation, σ, if we divide it by a large number, this error estimate is small, or near 0. Consequently, no matter how small a value we select for the constant, or criterion, (k), the probability statement gets closer and closer to its maximum of 1 because the ratio in Equation 5.1, $\dfrac{\sigma_\mu^2}{k^2}$, will be very close to 0. This is so because the numerator, σ_μ^2, is a value very close to 0; this says that the value in the denominator, k^2, becomes unimportant. This ratio quantity, then, essentially will equal 0, and the right side of the inequality will reduce to 1. This exercise should reinforce our faith in the representativeness of the sample mean density and also will provide us with help for determining a sample size.

In each of the following sections, we present a method for obtaining a reasonable estimate of sample size, based on some set of conditions or information. In each successive section we assume that the researcher has available some additional facts about the target population. The aim is to show that we can obtain a sample size estimate for fieldwork with minimal or no target population pilot data. But each additional increment of information can allow the field researcher, who is developing a sampling plan, to reduce the number of biological samples needed to obtain an equivalent amount of confidence in the results.

Sample Size with No Assumption of Normality

Let us assume that we know nothing about the distribution because the area of interest has never been sampled before (for example, we don't know the extent of the patchiness). We have no pilot data, and therefore we do not want to assume Normality. Nevertheless, we would like the probability to be quite high so that our mean density estimate $\hat{\mu}$ is close to the true μ. We can decide on what constitutes a sufficient sample size by employing the Law of Large Numbers. If, for example, we want $P(\hat{\mu} - \mu) = 0.95$ (that is, 95% probability), we can use Equation 5.1. We begin,

$$0.95 \geq 1 - \frac{\sigma_\mu^2}{k^2}$$

Substituting $\dfrac{\sigma^2}{n}$ for the standard error σ_μ^2, we get

$$0.95 \geq 1 - \frac{\sigma^2/n}{k^2}$$

$$1 - \frac{\sigma^2/n}{\sigma^2} \leq 0.95 \tag{5.2}$$

$$1 - \frac{1}{n} \leq 0.95.$$

Next we can solve Equation 5.2 for n, the sample size. We do this by entering any values of σ^2 and k that we choose. However, because we know nothing about our target population, we have the problem that any value we choose for σ^2 or for k will be a guess. To circumvent this problem we employ a mathematical "trick." For convenience we will let k be a multiple of σ. That is, the value that we substitute for k will be $k = \sigma$. By making this substitution, the mean density estimate will be estimated to within 1 standard deviation ($\sigma = 1.0 \cdot \sigma$) of the true mean density. Of course, we could use any such multiple, 2σ, 3σ, . . . , but our choice of 1σ is the simplest for explanatory purposes. We want the error ($\hat{\mu} - \mu$) to be less than our criterion of $k = \sigma$, with a confidence of 0.95. Thus,

$$1 - \frac{(\sigma^2/n)}{\sigma^2} \le 0.95.$$

Canceling the σ^2 from both the numerator and the denominator (which is why we chose k as a multiple of σ in the first place) yields

$$1 - \frac{1}{n} \le 0.95$$

so that $\frac{1}{n} \ge 0.05$, and solving for n, we have

$$n \le 20.$$

Twenty is thus the largest number of observations (biological samples) needed to meet our chosen criterion so that the error we make by estimating μ is less than one population standard deviation (1σ). This formula (Equation 5.2) is used when we don't know what assumptions to make about the population. The disadvantage of using this method to obtain a sample size is that regardless of the size of σ, we will get the same sample size estimate for each target population and confidence level. When σ is quite large, this criterion may be too weak and thus may not provide us with much usable information. When there is additional information on the sampling distribution of the estimate of interest, the sample size required usually can be lowered. That additional information is discussed next.

Sample Size with Assumption of Normality

Even though we may have no prior information on the target population's distribution, the Central Limit theorem (Chapter 3) can provide meaningful information on the sampling distribution for most natural populations from which field samples will be selected. Because the distribution of sample means is Normal (if n is large enough), an added consequence is that consideration of this theorem allows for the computation of a sufficient sample size for our sampling purposes.

Suppose, again for convenience, that we select the same criterion of $k = 1.0\sigma$, but we also assume that the sampling distribution of the mean is approximately Normal, which is reasonable (as explained in Chapter 3). Once again we wish to be very confident; say, 95% confident, that the difference between the estimate and the true value is within our criterion of k. We know from Equation 3.1 that $z = (x - \mu)/\sigma$, where x represents a value of any random variable that has a mean denoted by μ and a standard deviation denoted by σ. The sample mean density itself is a random variable (Chapter 3), so we substitute it for x in the equation. Recall that the standard deviation of the sampling distribution is σ_μ, so we substitute it for the symbol σ.

Putting all this together, the result is

$$z = \frac{(\hat{\mu} - \mu)}{\sigma_\mu}, \tag{5.3}$$

which is now a variable in *standard form*. Most importantly, recall that $\sigma_\mu = \frac{\sigma}{\sqrt{n}}$, so that now this equation can have n in its formula and we can solve for n to obtain our required sample size.

Now continuing with the determination of sample size, we want the error; that is, the difference between our estimate and the true value $(\hat{\mu} - \mu)$, to be within the criterion value k. Therefore, we input $k = \hat{\mu} - \mu$ into Equation 5.3 to obtain

$$z = \frac{k}{\sigma/\sqrt{n}}. \tag{5.4}$$

For $k = \sigma$ we thus have $z = \dfrac{\sigma}{\left(\dfrac{\sigma}{\sqrt{n}}\right)}$. It should be clear that we chose this value of k so that we could get rid of, or cancel, the population standard deviation σ from the equation. Next we enter the number for z, the standard deviate. Using the standard Normal table (Tables 3.2 or 4.1), we find that $z = 1.96$ for the two-tailed value at the 0.95 confidence level. Remember that the justification for selecting the standard Normal table as the reference is the Central Limit theorem. That theorem ensures that the sampling distribution of the mean is Normal, even if the true distribution of individuals within the entire population being sampled is not.

Solving Equation 5.4 for n we obtain $1.96 = \dfrac{\sigma}{\left(\dfrac{\sigma}{\sqrt{n}}\right)}$. Alternatively, we could reorganize the terms to get $1.96 = \dfrac{\left(\sigma\sqrt{n}\right)}{\sigma}$. Canceling σ from both the numerator and the denominator, we have $1.96 = \sqrt{n}$. Squaring both sides of the equation gives $n = (1.96)^2$. So $n = 3.84$. We conclude that about 4 biological samples are needed in this instance to estimate a mean population density at a 95% confidence level.

By contrasting the answer of $n = 4$ (the sample size we obtain when using one mild statistical assumption) with $n = 20$ (the size we get when we make no assumption on the distribution), we encounter another example of the advantages of statistical theory.

Notice that for these exercises we need no prior knowledge of the population standard deviation, σ. This is because we stated our criterion k in population standard deviation terms; because k was not a number but a value with σ in it, it canceled out. This cancellation is a great help not only for performing the arithmetic but also for obtaining a sample size when little or nothing is known about the population we wish to sample.

MINIMALLY SIZED SAMPLES WITH PRELIMINARY PILOT DATA

Sample Size with an Estimate of the Standard Deviation

Consider what we might do if we already have a preliminary estimate of population variance before we begin the population survey in the field. We can know this additional information perhaps because we examined a museum collection or perhaps because we performed some research in the relevant literature or took a preliminary field inventory. Whatever its source, prior knowledge of population variance can be a real help in determining the appropriate number of quadrats to sample.

If we have a preliminary estimate of the population variance, or standard deviation, we can refine our determination of how many quadrats we actually need to sample in order to achieve a chosen level of confidence. Let us say we wanted to perform a sampling of the Bolivian plot 01 and that we wanted to obtain a sufficient sample size to be 95% confident that the estimated mean density obtained would lie within a certain percentage of the true mean density, or the mean plus or minus a few individuals. In both these cases (when k is not a percentage of the standard deviation but a proportion of the total or the mean or a count), a preliminary estimate of the sample standard deviation (from a pilot field or museum study or a literature search) is required.

For our pilot study, we will use the $n = 8$ randomly selected quadrats (from $N = 100$) that we selected in Chapter 2. The preliminary estimates then will be $\hat{\mu} = 6.00$ and $\hat{\sigma} = 2.27$. We can make these preliminary estimates whenever we have data from a previous study of the target area. Remember, to the extent that our estimate is in error, our final results will be inaccurate also, but, statistically, we are willing to ignore this risk. Now if we want our final sample, with 95% confidence, to have the estimate correct within, say, 25% of the true mean density, we let $k = 0.25 \cdot 6.00 = 1.50$. That is, now that we have some representative information on our target population (in the form of estimates $\hat{\mu}$ and $\hat{\sigma}$), we can use an actual number for our criterion k in Equation 5.4.

We have from Equation 5.4, $z = \dfrac{k}{\left(\dfrac{\hat{\sigma}}{\sqrt{n}}\right)}$. Solving for n gives

$$n = \frac{z^2 \hat{\sigma}^2}{k^2}. \tag{5.5}$$

Equation 5.5 is a very common formula, often found in statistical texts for calculating sample size. It can be used when we have an estimated value for the standard deviation. Remember, however, it is a result for large samples. Then k can be any number we choose—a multiple of σ, a proportion, or a count. For our example, Equation 5.5 becomes $n = \dfrac{1.96^2 \cdot 2.27^2}{(1.50)^2} = 8.79 \approx 9$. Thus we would need 9 biological samples or observations to achieve the desired tolerance, accuracy, or precision of $\pm 25\%$ of the mean density, with 95% confidence.

This value of 9 should come as no surprise, because from the example in Chapter 4 on confidence limits we obtained comparable information for the same data. In Chapter 4 we estimated the standard error as $\dfrac{2.27}{\sqrt{8}} = 0.80$ if we ignored the fpc, which is the correction for small size. If we took account of the fpc, we obtained 0.77. So using Equation 5.4 with t instead of z, as we did for the confidence interval, we get $\dfrac{k}{\hat{\sigma}_\mu} = t$.

Rearranging the quantities, we get $k = t\hat{\sigma}_\mu$. For this example, when $n = 8$, $t_{0.95,7} = 2.36$ (Table 4.1), and we obtain $k = 2.36 \cdot 0.77 = 1.81$.

This value of $k = 1.81$ is slightly larger than the $k = 1.50$ we calculated above, when we wanted to estimate within 25% of the mean. In the present case, $\dfrac{k}{\mu} = t \cdot \hat{\sigma}_\mu = 0.30$ instead of 0.25. That is, we have 95% confidence that the true mean density will be within $\pm 30\%$ (not $\pm 25\%$) of the estimated value for this example if we use a sample of size 8. When $n = 9$, we would have $k = 2.31 \cdot 0.72 = 1.68$, so that $\dfrac{1.68}{6.00} = 0.28$. This difference is due mostly to the change in the value of t and not to the small change in n.

However, we arrived at Equation 5.5 when we were interested in large samples. What we don't know is whether a pilot study of size $n = 8$ is large enough to satisfy the conditions of the Law of Large Numbers. Alternatively, we could have used Equation 4.10 for our calculation, which does not require large numbers. Using the estimates instead of the true values, we recall that $d = \dfrac{t\hat{\sigma}_\mu}{\mu}$. This means that for $n = 8$ the value for d is obtained as, $d = \dfrac{2.36 \cdot 0.77}{6.00} = 0.30$.

It should be apparent that k in Equation 5.5 to obtain a sample size is the same as $d\mu$ in Equation 4.10. Consequently, we could solve Equation 4.10 using a t for obtaining the size instead of a z value for the calculation. Thus we would obtain

$$n = \frac{t^2\hat{\sigma}^2}{(d\mu)^2}. \tag{5.6}$$

Solving for n, using these same values, we find that $n = \dfrac{(2.36)^2(2.27)^2}{(1.50)^2} = 12.76 \approx 13$ biological samples. The difficulty with this answer of $n = 13$ is that we used a $t_{0.95,7} = 2.36$ to obtain it. Since $df = 7$, this t value actually refers to a sample size of 8. But for $n = 13$, $t_{0.95,12} = 2.18$ (df of 12 refers to an $n = 13$). Recalculation for the precision would then be $d = \dfrac{2.13 \cdot 2.27}{\dfrac{\sqrt{13}}{6.00}} = 0.23$, instead of 0.30. This discrepancy presents a prob-

lem: We cannot use an equation to obtain a value of n that requires us to know a value of n for the solution.

Overall, Equation 5.5 is left as our only choice, yet it holds true only for large samples. That is, it produces asymptotic results. Nevertheless, it might serve as a reasonable approximation even for rather small samples, because the approximation, or convergence of the Normal distribution, is quite good. Let us investigate further.

Large Versus Small Sample Results

Just as with confidence intervals, when we know the population standard deviation, or when the sample sizes will be very large, we use the standard Normal deviate z in the calculations. Alternatively, when sample sizes will be small, or any time we must use the estimate in the calculation, we might think that the deviate from the Student's t distribution is a better choice. Unfortunately, the value of t in Equation 5.6 depends on n, which is, of course, the quantity we are trying to find. The simplest and most practical solution to the problem of small amounts of data is to use a z of about 2 and accept a confidence of about 90% to 95%. For our Bolivian forest example we would then have $n = \dfrac{4.00 \cdot 5.15}{2.25} = 9$ biological samples. This is an approximation for the field sample size problem.

Let us return to an earlier example in which we had no population information; that is, we did not know the value of σ. Let us make an assumption of Normality and take a preliminary sample of $n = 4$ upon which to base our estimates, and then examine how well we can do with this simplest solution of using $z = 2$.

Using the Central Limit theorem as justification, we calculated the number of samples required for 95% confidence with a criterion of $k = \sigma$. Without any prior knowledge of the distribution in Bolivia, and setting $k = \sigma$ and $z = 1.96$ (or, about 2), we would use Equation 5.5 to calculate that $n = \dfrac{2^2 \cdot \sigma^2}{\sigma^2}$, or $n = 4$ biological samples.

In Chapter 3, we actually took a random sample of $n = 4$ from the $N = 100$ quadrats and obtained $\hat{\mu} = 8.25$. We found that 8.25 was a very poor estimate for the density per quadrat of the entire population (because we knew the true value was 6.63). In that calculation, the standard deviation, $\hat{\sigma}$, was 2.22. To calculate a confidence interval from these values, with $z = 2$, we use Equation 4.4 to obtain $\hat{\mu} \pm 2 \cdot \left(\dfrac{2.22}{\sqrt{4}} \right) = \hat{\mu} \pm 2.22$.

Plugging in the value of the mean estimate, the confidence interval would be $8.25 - 2.22 \le \mu \le 8.25 + 2.22$, which gives $6.03 \le \mu \le 10.47$. Despite the fact that our initial estimate was a poor one, and using only an approximate value of z, with this small sample the true mean, 6.63, does indeed lie within the large-sample confidence interval.

Thus when we deal with a small number of field samples and have some initial estimates of μ and σ, a simple solution is to use $z = 2$ to give us a reasonable idea of the number of samples (n) required for the fieldwork.

As a further example, recall the exercise in Chapter 3 in which a sample size of just 4 quadrats out of 100 could be used at a 95% confidence level. In that exercise we used $t_{0.95,3} = 3.18$ to find that $4.78 \leq \mu \leq 11.72$. For 90% confidence $t_{0.90,3} = 2.35$, and the confidence interval was $5.64 \leq \mu \leq 10.86$. These confidence intervals do indeed contain the true mean density (omnisciently known to be 6.63), based on only $n = 4$. It should be clear that although we arrived at this sample size of $n = 4$ by choosing $k = \sigma$ when we had no prior data on the target population, we could have chosen $k = 0.5 \cdot \sigma$, or $k = 1.5 \cdot \sigma$, or indeed any combination or multiple of σ we desired.

In practical terms what we have done is to demonstrate that without any pilot data whatsoever, our estimate from Equation 5.5 indicates that a sample of $n = 4$ quadrats of a total population of $N = 100$ quadrats will yield about a 90% confidence that the true mean will be within $\pm \sigma$. Actually, taking the $n = 4$ quadrats, which contain a total of 33 trees, we estimated with 90% confidence that the true mean was between 6 and 11. If this interval is acceptable, we have saved ourselves the trouble of counting 630 more trees to get the exact figure.

Sample Size with a Bound on Mean Density

As a final example of sample size determination for fieldwork, let us assume we wanted to know the true mean density within 2 individuals, more or less, with 95% confidence. Recall that for Equation 5.5 we stated that k, the criterion, could be any value we wanted. For our example here we will choose $k = 2$, and $z = 2$, so that we obtain $n = \dfrac{2^2 \cdot 2.27^2}{2^2} = 5.15 \approx 5$ biological samples.

To see how well we would do with $n = 5$, the quadrats numbered 85, 29, 76, 31, and 70 were chosen with a table of random numbers (Appendix 3). These quadrats contained 8, 9, 10, 10, and 6 individuals, respectively. The estimated mean $\hat{\mu}$ is 8.60. This result is within our desired $\mu \pm 2$ individuals ($6.63 + 2 = 8.63$).

Although the numerical value of k may be selected by the researcher, the resultant number for n may not be realistic for fieldwork. For example, because 0.95 (or $0.05 = 1 - 0.95$) is so ingrained as a "significant" value, when asked for the confidence or precision acceptable in estimates of density, a number of researchers have told us that they wish to be 95% confident that their estimate is within 5% of the true mean density. As an interesting exercise, let us calculate the number of samples required to achieve the magic 0.05 precision for Bolivian plot 01, which is, by nature's standards, quite homogeneous. In this case, using the $n = 8$ example for our pilot information, we would have $k = 6.00 \cdot 0.05 = 0.30$. The value for k^2 is therefore 0.09. Hence the number of samples required using Equation 5.5 to achieve 95% confidence that our estimate is within 5% of the true mean density works out to be $n = \dfrac{2^2 \cdot 2.27^2}{0.30^2} = 229.$

This answer of 229 is clearly absurd. We cannot sample 229 quadrats if the entire population we are sampling constitutes only 100 quadrats. The investigators must set their sights for precision and confidence much lower. Still at the 95% level of confidence, but with a precision of 10% (rather than the 5% that yielded a need for 229 quadrats), the researchers would need to include about 57 quadrats in their statistical sample. For a precision of 15% about 25 would be needed. For most researchers these close precisions require an intolerable amount of work and translate into impossible demands. Consequently, in most instances, a precision of somewhere between 25% and 50% is all that is reasonably attainable.

QUADRAT SIZE OR AREA

Homogeneous Case: Bolivian Trees

Until now we have been sampling the Bolivian tree census by dividing the Beni Reserve plot 01 into $N = 100$ quadrats, or subplots. In the original census, Dallmeier and colleagues (1991) divided the entire plot into $N = 25$ subplots in order to facilitate the measurements required to position each tree precisely in the plot. Their original counts for $N = 25$ are given in Appendix 2.

Earlier in this book we introduced an exercise in which we randomly selected 8 of the 100 quadrats as our own sample. The area of this example is exactly the same as if we had selected $n = 2$ of the original $N = 25$ quadrats. Presumably we would be counting and identifying about the same number of trees in $n = 8$ of $N = 100$ as we would in $n = 2$ of $N = 25$. A question immediately arises: Providing the same amount of work is required, which quadrat size would be the better strategy for obtaining confidence limits of the true mean?

Recalling that the total number of individuals is 663, we can use Equation 2.1 with $N = 25$ to obtain a value for the true mean density of $\mu = \dfrac{6.63}{25} = 26.52$. Using Equation 2.3, the value of the true variance for $N = 25$ is $\sigma^2 = 26.65$; its square root, the standard deviation, is then $\sigma = 5.16$. As before, the table of random numbers (Appendix 3) was consulted to select $n = 2$ from the possible $N = 25$ quadrats. The lucky selections were 09 and 20. Consulting Appendix 2, we find that these quadrats contained 29 and 37 individuals, respectively. The total number of individuals counted, then, is 66. Using Equation 2.2, the estimated mean is $\hat{\mu} = \dfrac{66}{2} = 33.00$. For the curious, the total number of individuals counted in the example of $n = 8$ (from $N = 100$) was 48. Using Equation 2.5, we estimate the variance for $N = 25$ as $\hat{\sigma}^2 = 32.00$ and the standard deviation as $\hat{\sigma} = 5.66$.

Having calculated $\hat{\sigma}$, the standard error is easily obtained from Equation 3.6. Here, $\hat{\sigma}_\mu = \dfrac{5.66}{\sqrt{2}} \cdot \sqrt{\dfrac{25-2}{25-1}} = \dfrac{5.66}{1.41} \cdot 0.98 = 3.93$. The 95% confidence level of t with $n = 2$, df = 1 is 12.71 (Table 4.1). Consequently, the term $t\hat{\sigma}_\mu$ is 12.71 · 3.93 = 49.95. The upper and lower confidence limits are then $\hat{\mu} \pm t\hat{\sigma}_\mu = 33.00 + 49.95$, which yields a confidence interval of $-16.95 \le \mu \le +82.95$. As an aside here, the negative value is, of course, unrealistic. We read this merely as 0 and proceed. Had we chosen a 90% confidence limit, then $t_{0.90,1} = 6.31$ and $t\hat{\sigma}_\mu = 6.31 \cdot 3.93 = 24.80$, resulting in a confidence interval of $+ 8.20 \le \mu \le + 57.80$. Because the true mean is 26.52, the confidence intervals do, as we would expect, bracket the true mean. However, the very large spread (interval width or range) is clearly undesirable. Although it is probably obvious that the very large values of t are causing the wide intervals, let us further examine the results.

We have previously defined precision (Equation 4.10) by dividing the quantity $t\hat{\sigma}_\mu$

by the mean. At the 95% level, we have for our $n = 2$ quadrat example, $\dfrac{t\hat{\sigma}_\mu}{\hat{\mu}} = \dfrac{49.95}{33.00} = 1.51$.

Another way to see the implication of this is to say that at the 95% level, the confidence bound on μ (the value $t\hat{\sigma}_\mu$) is within 151% of the true mean $33.00 \cdot 1.51 = 49.95 = t\hat{\sigma}_\mu$.

At the same 95% level for $n = 8$ (from $N = 100$), we obtained $\dfrac{t\hat{\sigma}_\mu}{\hat{\mu}} = \dfrac{1.82}{6.00} = 0.30$, or 30%.

Now, if we rewrite $\dfrac{t\hat{\sigma}_\mu}{\hat{\mu}}$ as $t \cdot$ CV, where CV (the coefficient of variation) $= \dfrac{\hat{\sigma}_\mu}{\hat{\mu}}$ from

Equation 4.9, we can see a further consequence. The t's in this two-component formula change greatly, but the CV's do not. For the $n = 2$ case the CV is $\dfrac{3.93}{33.00} = 0.12$,

and for the $n = 8$ example the CV is $\dfrac{0.77}{6.00} = 0.13$. Our problem with the very wide

interval for $n = 2$ lies entirely with the fact that we have only one degree of freedom—that is, we have only a very small sample size—which results in a large value for t (Table 4.1). Consequently, and providing that the amount of effort (work), time, and expense required to count the individuals is about the same, a sample of $n = 8$ biological samples affords much greater precision and ought to be our choice. If the differences in the area of the quadrats in the sampling scheme were such that larger sample sizes (n) were required in both the 25- and the 100-quadrat instances, then the differences between the required values for t would diminish (Table 5.1).

The decision as to what value of n is large enough for our needs, without being excessive, is somewhat subjective. In Table 5.1 we show the precision for various values of n with a constant CV of 0.13. Both the degree of confidence as well as the number of observations or samples (n) must enter into any decision about design. Clearly, however, a sample size of 2 does not give much precision. The gain from $n = 3$ to $n = 4$ is substantial. Once $n = 4$ is reached the gain is smaller, especially at 90% confidence. We conclude that, provided n is a large enough share of N to be biologically representative, quadrat size will be of little importance in a homogeneous target area, such as our Bolivian forest example. Homogeneity, however, is not always to be found in nature.

TABLE 5.1 Precision (Deviation) of Confidence Interval for Bolivian Trees Plot 01 with CV = 0.13

SAMPLE SIZE (n)	DEGREES OF FREEDOM (df)	PRECISION	
		$(t_{0.90}$ CV)	$(t_{0.95}$ CV)
2	1	0.82	1.65
3	2	0.40	0.56
4	3	0.30	0.41
5	4	0.28	0.36
6	5	0.26	0.33
7	6	0.25	0.32
8	7	0.24	0.31
9	8	0.24	0.30
10	9	0.24	0.29
31	30	0.22	0.26
∞		0.21	0.25

NOTE: The coefficient of variation of the sampling distribution of density, $CV = \dfrac{\hat{\sigma}_\mu}{\hat{\mu}} = 0.13$, from Equation 4.9. The value of precision (deviation), $d = t\dfrac{t\hat{\sigma}_\mu}{\hat{\mu}}$, from Equation 4.10. Confidence interval about mean density from Equation 4.7.

General Cases

Recall that in Chapter 1, Figure 1.2 demonstrates graphically that the Bolivian trees are fairly homogeneously dispersed throughout the study area. Where $N = 100$ and $\mu = 6.63$, the value of σ^2 is 4.69. Where $N = 25$ and $\mu = 26.52$, then $\sigma^2 = 26.65$. Notice the congruency in each of these examples of μ and σ^2. Indeed, for homogeneously dispersed organisms we would expect that the value of the population mean density per quadrat, μ, would approximate the square of the standard deviation, σ^2. We discuss this further in Chapter 6. In nature, however, many organisms are not so homogeneously distributed. Often the organisms exhibit some degree of aggregation. In such instances of aggregation, the density per quadrat (μ) will be less than σ^2. In cases of severe aggregation, μ will approach the value of σ.

We next examine two hypothetical cases, one in which $\mu = \sigma^2$ (spatially homogeneous) and the second for which $\mu = \sigma$ (severe spatial aggregation). For each example we take two sets of statistical samples. The quadrat size or area sampled for the first will be 25% of the size of the second, just as in our Bolivian tree example (in which we divided the total population into 100 quadrants in one instance, and 25 in another).

Where the mean equals the variance. First we examine the case in which $\mu = \sigma^2$. Let us pretend we have selected a quadrat size that yields $\mu = 10$, $\sigma^2 = 10$, and $\sigma = 3.16$ for the first field sampling. Then let us make each quadrat 4 times larger for the second sampling, thus yielding $\mu = 40$, $\sigma^2 = 40$, and $\sigma = 6.32$ (because of the homogeneity). Of course, we will have fewer total quadrats for the second selection, since the area will remain the same. The numbers of individuals counted are the same in the two series of samples.

Table 5.2 shows a comparison for $N = 100$ quadrats in the top block and $N = 25$ quadrats in the bottom block. The corresponding sample sizes are $n = 8, 16, 24, 32$ (when $\mu = 10$, $\sigma = 3.16$) and 2, 4, 6, 8 (when $\mu = 40$, $\sigma = 6.32$). The results show that when $\mu = \sigma^2$, regardless of quadrat size, the coefficient of variation of the sampling distribution of the mean, CV $= \dfrac{\sigma_\mu}{\mu}$, used as our measure of precision, will depend on the number of individuals counted. Let us see why.

By using the definition of the standard error, we know that $\dfrac{\sigma_\mu}{\mu} = \dfrac{\sigma}{\sqrt{n\mu}}$. We also know that for our first example of homogeneous dispersal, $\mu = \sigma^2$, which is the same as saying that $\sqrt{\mu} = \sigma$. Substituting $\sqrt{\mu}$ for σ, in the CV formula we have $\dfrac{\sigma_\mu}{\mu} = \dfrac{\sigma}{\mu\sqrt{n}} = \dfrac{\sqrt{\mu}}{\mu\sqrt{n}}$. Then, dividing the numerator and the denominator by $\sqrt{\mu}$ and rearranging, we obtain $\dfrac{\sigma_\mu}{\mu} = \dfrac{1}{\sqrt{\mu n}}$.

The product μn (mean density times sample size) in the denominator represents the total number of individuals counted in the sampled quadrats. Thus, with a count of 160 individuals, the standard error must be $\dfrac{1}{\sqrt{160}} = 0.08$, regardless of whether we sampled with a size of $n = 16$ or a size of $n = 4$, or whether one set of samples was 25% of the other set (Table 5.2). This formula shows that although there is an appreciable gain in going from 80 to 160 individuals, very little is gained by counting more than 240 individuals. The value of the precision (Equation 4.10) would be a little higher (about 3%) with $n = 24$ instead of $n = 16$, because of the t values (Table 4.1). The researcher must decide whether the effort in sampling and counting a larger number of biological samples (24 instead of 16) is worth this slight gain in precision, and whether it is even feasible, given possible time and money constraints.

Where the mean equals the standard deviation. In this second case, instead of the mean's approaching the value of the variance (σ^2), it now approaches the standard deviation (σ). This situation is indicative of considerable spatial clumping, or aggregation. Once again we shall choose the first example so that $\mu = 10$ is the true mean density value. But in this example we shall let $\sigma^2 = 100$, so that $\sigma = 10$. For the second example, we shall let $\mu = 40$, $\sigma^2 = 1,600$, and $\sigma = 40$.

The situation has now changed rather dramatically from that in which we have spatial homogeneity. We can see in Table 5.3 that for any number of individuals counted, the coefficient of variation of the sampling distribution, $\dfrac{\sigma_\mu}{\mu}$, is much smaller for the smaller quadrat size (that is, a quadrat size for which $\mu = 10$) than for the larger

TABLE 5.2 When Mean Equals Variance, Precision of Confidence Interval Depends on Individuals Counted, Not Quadrat Size

SAMPLE SIZE (n)	STANDARD ERROR $\left(\dfrac{\sigma}{\sqrt{n}}\right)$	COEFFICIENT OF VARIATION OF SAMPLING DISTRIBUTION $\left(\dfrac{\sigma_\mu}{\mu}\right)$	INDIVIDUALS COUNTED
For $N = 100$			
8	$3.16/2.82 = 1.12$	$1.12/10 = 0.11$	80
16	$3.16/4.00 = 0.79$	$0.79/10 = 0.08$	160
24	$3.16/4.90 = 0.64$	$0.64/10 = 0.06$	240
32	$3.16/5.66 = 0.56$	$0.56/10 = 0.06$	320
For $N = 25$			
2	$6.32/1.41 = 4.48$	$4.48/40 = 0.11$	80
4	$6.32/2.00 = 3.16$	$3.16/40 = 0.08$	160
6	$6.32/2.45 = 2.58$	$2.58/40 = 0.06$	240
8	$6.32/2.83 = 2.23$	$2.23/40 = 0.06$	320

NOTE: The standard error $\left(\dfrac{\sigma}{\sqrt{n}}\right)$, coefficient of variation (Equation 4.9) of the sampling distribution $\left(\dfrac{\sigma_\mu}{\mu}\right)$, and individuals counted for $\mu = 10$, $\sigma = 3.16$ for sample size $n = 8$, 16, 24, 32 and for $n = 2$, 4, 6, 8, with $\mu = 40$ and $\sigma = 6.32$.

quadrat size (where $\mu = 40$). Notice that for $n = 8$ the coefficient of variation is the same in both examples, even though in one we counted 80 individuals and in the other 320. Again, let us examine why this is so.

Let us first write $\dfrac{\sigma_\mu}{\mu} = \dfrac{\sigma}{\mu\sqrt{n}}$. We know that for this example $\mu = \sigma$. So, substituting μ for σ in this formula and canceling, we obtain $\dfrac{\sigma_\mu}{\mu} = \dfrac{1}{\sqrt{n}}$. This is a rather amazing result because it is not intuitive. One would, rather, expect that the precision (CV) would be related to the variance, mean density, and sample size. Yet here, the math clearly tells us that the CV is a constant that depends only on the statistical sample size, and not the number of individuals counted.

When $n = 8$, the coefficient of variation is $\dfrac{1}{\sqrt{8}} = 0.35$ (Table 5.3) regardless of the number of individuals counted. We obtain the same coefficient of variation by counting 80 individuals as we do by counting 320. *Consequently, when $\mu = \sigma$, the best sampling strategy is to maximize n; that is, to take as many samples as possible, within reason.* However, because we will not know with certainty when designing our survey which

TABLE 5.3   When Mean Equals Standard Deviation, Precision of Confidence Interval Equals $\dfrac{1}{\sqrt{n}}$, Not Number of Individuals Counted

SAMPLE SIZE (n)	STANDARD ERROR $\left(\dfrac{\sigma}{\sqrt{n}}\right)$	COEFFICIENT OF VARIATION $\left(\dfrac{\sigma_\mu}{\mu}\right)$	INDIVIDUALS COUNTED
For $N = 100$			
8	10/2.83 = 3.53	3.53/10 = 0.35	80
16	10/4.00 = 2.50	2.50/10 = 0.25	160
24	10/4.90 = 2.04	2.04/10 = 0.20	240
32	10/5.66 = 1.77	1.77/10 = 0.18	320
For $N = 25$			
2	40/1.41 = 28.37	28.37/40 = 0.71	80
4	40/2.00 = 20.00	20.00/40 = 0.50	160
6	40/2.45 = 16.33	16.33/40 = 0.41	240
8	40/2.83 = 14.13	14.13/40 = 0.35	320

NOTE: The standard error $\left(\dfrac{\sigma}{\sqrt{n}}\right)$, coefficient of variation of the sampling distribution $\left(\dfrac{\sigma_\mu}{\mu}\right)$, and individuals counted for $\mu = 10$, $\sigma = 10$ for sample size $n = 8, 16, 24, 32$ and for $n = 2, 4, 6, 8$, with $\mu = 40$ and $\sigma = 40$.

of these cases we have, the optimal approach is to take more small samples rather than fewer large samples. It is responsible to err on the side of more and smaller quadrats (biological samples).

In the real world, the type of sampling gear is often of a fixed size, so that it may be difficult to adjust the quadrat or biological sample size. Then, too, sampling may be from a ship, by means of a long wire, so that taking many small samples is costly. These problems must, of course, be taken into account when deciding on the size of the quadrat or coring device and thus determining how many samples can reasonably be taken. In addition, field researchers may find themselves somewhere between the two cases, where, on the one hand, $\mu = \sigma^2$, and, on the other, $\mu = \sigma$. We more thoroughly examine $\sigma \le \mu \le \sigma^2$ in Chapter 6; nevertheless, the illustrations shown here indicate that while we may do just as well with a few larger quadrats, we will always do as well, or much better, with a larger number of smaller biological samples.

Finally, it goes almost without saying that when $n = 1$ we do not have a statistical sample, just an observation. Always replicate, and the more, the better.

Summary

1. When no assumption of Normality or any other distributional configuration can be made (perhaps because pilot information is unavailable), the Law of Large Numbers can be used to calculate the number of biological samples (n) required to be within k standard deviations of the true mean with any desired confidence (Equation 5.2). When $k = \sigma$, $n = 20$, with 95% confidence.

2. Assuming Normality of the sampling distribution of the mean density, yet still without any pilot or baseline data, the number of biological samples (n) required for any stipulated precision with any desired degree of confidence can be calculated in standard deviation units (Equation 5.5). When $k = \sigma$, $n = 4$, with 95% confidence.

3. Assuming Normality, but this time beginning with an estimate of the standard deviation calculated from a pilot study, the number of biological samples (n) required for any stipulated precision with any degree of confidence can be calculated (Equation 5.5).

4. Precision is defined as the standard Normal deviate (z or t, as appropriate) times the coefficient of variation (CV) of the sampling distribution of the mean. When the spatial distribution of individuals is homogeneous ($\mu = \sigma^2$), then $CV = \dfrac{1}{\sqrt{\mu n}}$. This indicates that the number of individuals counted is paramount, and thus quadrat (biological sample) size or area is up to the discretion of the investigator so long as n is large enough (at least 4) to avoid using very large values of t. When the spatial distribution is very heterogeneous ($\mu \approx \sigma$), then $CV = \dfrac{1}{\sqrt{n}}$. This indicates that the number of quadrats to be sampled is a very important decision.

5. In general, the best overall strategy is to maximize the number of biological samples (n). That is, sample many small quadrats instead of a few large ones.

6. The number of samples required so that an estimate of the mean density will lie within 5%, 10%, or 15% of the true mean is too large for most biodiversity or field ecology studies. Consequently, for sampling most natural populations, a precision of between 25% and 50% is all that is reasonably attainable.

Problems

5.1 For *Pouteria macrophylla* calculate the number of samples required, using $z = 2$, in order for the survey to produce results that will be within 75% (that is, 0.75μ) of the true mean. Do this by taking two sets of n samples at random, the first from Appendix 1 (where $N = 100$) and the second from Appendix 2 (where $N = 25$). Calculate the 90% confidence limits, confidence intervals, and precision for each case. Comment on the results.

5.2 Carry out the same exercise for *Astrocaryum macrocalyx*. Comment on the results.

5.3 The tertiary marine sediments at Calvert Cliffs, Maryland, were sampled for foraminifera. A quadrat of $1\,m^2$ was placed on the outcrop divided into $N = 100$ squares and 5 samples were then taken at random. To gain perspective on scale, we note that the foraminifera are shelled protozoans that average about 0.5 mm in diameter. This means that a 1-m^2 quadrat is equivalent to a 4-km^2 area for a tree with a diameter of 1 m. For the species *Buliminella elegantissima* the 5-quadrat sampling yielded 47, 119, 159, 234, and 352 individuals. For the species *Valvulinaria floridana* the sampling yielded 4, 23, 14, 24, and 36 individuals. For *Epistominella pontoni* the sampling yielded 0, 1, 1, 8, and 12 individuals. (All data are drawn from Buzas and Gibson 1990.) Calculate the number of samples (where $z = 2$) required in order for the density estimate to be within 50% of the true mean density for each species. If the n's differ, how would you resolve the conflict in optimizing the number of samples, since all the species are in a single biological sample and each sample will be processed?

6

SPATIAL DISTRIBUTION:
THE POWER CURVE

W E HAVE ALREADY DISCUSSED randomness from a sampling point of view in Chapter 2. Random sampling ensures that each of the possible samples has an equal probability of being chosen for enumeration. The procedure is strictly and quantitatively defined. Let us now examine the concept of random and nonrandom from another viewpoint; that of the individual organisms.

When we look at Figures 1.2 or 2.1, which visually portray the location of each of the 663 trees in a 1-hectare Bolivian forest plot, we notice that the individual trees represented by the dots are distributed throughout the hectare. From a qualitative assessment through visual inspection some people would conclude that the distribution in space (*the spatial distribution*) is random. Others would notice that some of the quadrats contain only one or two individuals and thus might conclude that the pattern is somewhat random or, perhaps, aggregated.

The common usage (nonstatistical) of the word *random* is imprecise. Usually, observers using this term mean that they see no readily apparent pattern. Ecologists refer to three kinds of spatial distributions: (1) random, meaning no pattern; (2) aggregated or clumped, meaning discrete groups such as cows in a meadow; and (3) even, meaning that the individuals are evenly spaced, like trees in an orchard. The spatial distributions referred to by biologists and geologists are visual descriptions and not probability distributions. In the biological and geological literature one often sees the phrase "random distribution" or "this species has a random distribution." In statistics there is no probability distribution designated as a "random distribution." Like the term *sample* discussed in Chapter 2, this point causes confusion between statisticians and biologists.

In this chapter we will not discuss in detail the applicability of various probability distributions for describing spatial distributions. Instead we show how the two statistical parameters discussed earlier, the mean density and the variance, when estimated at various densities, can be used to distinguish the three categories of spatial distribution of interest to ecologists: random, aggregated, and even. To do so, we will use the *power curve*, a plotted curve that simply raises the mean density to an exponent that then allows this quantity to be equated to the variance. The plot of the resulting equation will then be a straight line on a log–log scale; that is, a plot in which x values and y values are both logarithms. The lines for random, aggregated, and even spatial distributions are easily distinguished using this method.

SPATIAL RANDOMNESS

Imagine an individual's falling through currentless water or windless air toward some locality that has been divided into N cells or partitions. These cells or partitions could be like the quadrats shown in Figure 2.1 for the hectare of Bolivian forest, or they could be some other core or sample arrangement. If the medium itself exerts no influence on the individual, and if the individual imparts no direction to its fall, then a good deal of uncertainty is involved concerning any prediction as to which of the N cells the organism will fall into. The maximum amount of uncertainty will occur when the individual has an equal chance of landing in any of these N cells. In that case, the falling individual has N equivalent choices, and the probability of its ending up in any particular cell is $\frac{1}{N}$. Spatially, this means that within our designated area the individual exhibits no particular preference as to location.

This conceptual experiment can give us a simple, but very strict, definition of what constitutes random. Regardless of how the individual got there, for the individual, *random* is defined as having an equal probability of appearing in any of the N cells. Thus, the probability of an individual's being in a particular cell is $p = \frac{1}{N}$. Of course, variables are random, not individuals. Thus the variable of interest in this situation is the presence or absence of the individual in the cell. The variable is recorded as not one but a sequence or group of variables denoting success or failure (in this case, presence or absence), one for each cell. Each of these variables takes on a value of 1 if the organism falls into that particular cell and 0 otherwise.

Now suppose another individual falls into the zone of cells. Our definition of random requires that this second individual, likewise, has no preference for any particular cell. The occurrence of the previous individual that has already arrived cannot influence the second arrival. Each individual's fall and subsequent cell placement does not affect later arrivals; each placement is independent of all the others. Individuals can continue to arrive in an already occupied cell, provided that the cell is not filled so as to exclude new individuals or to interfere in any way with their arrival. In such circumstances individuals are said to be *distributed randomly* over the area in which the cells are located. The process of repeated, dichotomous occurrences of success and failure is called a *Poisson process*. The count of individuals in a particular cell after any given duration of a Poisson process is a random variable. The counts for all cells constitute a *Poisson distribution*. This distribution is discussed in more detail in Chapter 9.

We can continue this process until we have a total of n individuals distributed into N cells. We also have N Poisson distributions, one for each set of individuals in each cell. The number of cells that contain 0 individuals, 1 individual, 2 individuals, and so forth can easily be tabulated. One way of testing whether the individuals are spatially randomly distributed is to compare their observed tabulation with that expected from a Poisson distribution, which has the same number of individuals (n) and cells (N) as the sampled population.

This procedure of fitting (or comparing) a Poisson distribution to the observations is commonly followed in many studies of spatial distribution. The question of whether a group of individuals is distributed randomly over some space is important for both behavioral and environmental reasons. If the individuals really have no preference as to how they are arranged within the target population, then neither behavioral characteristics nor the environment have any effect on their distribution.

Thus, in studies of ecological spatial distribution the Poisson is usually considered before any other distribution. If the individuals are randomly distributed (cannot be distinguished from a Poisson distribution) within the area of interest, then we have a quantitative description of their distribution. The Poisson distribution has only one parameter; the mean. This single parameter adequately summarizes the entire distribution.

Scale is, of course, always important. Given a large enough area, no group of organisms is randomly distributed. For example, consider any ubiquitous terrestrial or marine species, on a global scale using degrees of latitude and longitude to define quadrats. Such species would not be found in a great number of quadrats. Consequently, not only the target population (species, genus, family, sex, or others) is important, but also the targeted environment.

Often, even in seemingly homogeneous environments, researchers find that observed spatial distributions are not random; they are not well fit by the Poisson distribution. In these cases, attempts are often made to fit other distributions, such as the Negative binomial distribution. These distributions are discussed in Chapter 9. For now, it is important to know simply that the fitting of a distribution to a set of observed data is not a well-defined procedure. There are many possible ways to group the observations, as well as a multitude of relevant aspects that introduce reservations for use with natural populations. In addition, although disagreement of the observed values with those expected from a Poisson distribution could appear to indicate a nonrandom arrangement, congruence of these values does not necessarily ensure that the observed distribution follows some random rule. The large number of possible distributions for fitting, as well as the fact that many of these distributions converge, makes their use unwieldy; that is, they are indistinguishable under certain conditions, such as when using large samples.

For these reasons, instead of calculating the probabilities of obtaining the number of cells (or samples) with 0, 1, 2, . . . individuals, we will choose a simpler route. We will consider only two parameters, both of which are now familiar: the mean, μ, and the variance, σ^2. The use of only these two parameters may not be totally encompassing for a comparative study because they do not completely describe the population distribution spread. However, by calculating the mean and the variance at various densities and then fitting the result with a power curve, discussed below, this method can indeed be used as a basis for ascertaining whether there is a *random distribution* of the individuals. Both alternatives (fitting a distribution or using only the parameters from a distribution) for ascertaining "randomness" of the distribution data are abstractions from the observations. However, not only is the latter procedure simpler than fitting a distribution to the observed spatial data but it leads more readily to some notable insight into the organization of natural populations.

For a Poisson distribution, the mean μ equals the variance σ^2. Equivalently, we can say that for a Poisson distribution,

$$\frac{\sigma^2}{\mu} = 1. \tag{6.1}$$

Equation 6.1 tells us that for a Poisson population the ratio of variance to mean is 1. If a natural population has a variance-to-mean ratio, also known as *relative variance*, of 1, then it is said to have *randomly distributed individuals*. However, it is possible to find a population with a relative variance equal to 1 whose distribution of individuals will not be best described by a Poisson probability distribution. For our purposes, a test of the distribution's randomness is not the main concern; we are most intent upon obtaining a standard against which we can measure the degree of aggregation in natural populations.

In general, scientific studies of natural populations have shown that the individuals rarely are randomly distributed. Many such studies seem to have ignored the influence of the size of the area sampled and the homogeneity of that area when ascertaining randomness. Nevertheless, the Poisson distribution is a useful standard against which departures from spatial homogeneity can be measured. The assumption is made that the amount by which the variance-to-mean ratio deviates from 1 expresses the extent of the departure from randomness. Two such departures are usually considered: spatial evenness and spatial aggregation, discussed in the following two sections.

SPATIAL NONRANDOMNESS

The first departure from a random situation involves a condition wherein $\sigma^2 < \mu$. In this circumstance the spatial distribution is often termed *even*, or *regular*. The individuals are evenly spaced, like trees in an orchard. Most often evenness is an attribute of artificial or regrown populations; but some species of marine invertebrates, such as the clam *Tellina* (Holme 1950) and a phoronid, a worm-like invertebrate (Johnson 1959), do exhibit behavioral characteristics generating spatial distributions that are well described as even. Some birds in Peru (M. Foster, pers. comm. 1997), as well as other species exhibiting territorial behavior, undoubtedly provide other examples.

The vast majority of organisms have the second type of departure from randomness, wherein $\sigma^2 > \mu$. This kind of distribution is usually referred to as *aggregated*, or *clumped*, and a host of statistical distributions have been used to describe this situation. Ecologists usually attribute aggregated distributions to environmental differences, which occur even on a micro scale, and/or to reproductive–social behavior. But scale or size of the sample area is also a prime determinant of aggregation. For example, if the target area is so large that a population exists in only one segment of it and not throughout, then the population will sample as aggregated. On the other hand, if we are interested only in the segment or cell in which the population occurs, and thus sample only within this segment, we may indeed find that the population is random. To further complicate matters, as we shall see, the reverse is also true: A population that is aggregated over the entire area of interest may appear random if the quadrat size chosen for conducting the sampling is small enough.

The synonymous terms *aggregated* and *clumped* are in some ways misnomers because they conjure up images of extreme aggregation, like cows in a meadow or fish in a school. For many organisms the aggregation is much more subtle. For these, the variance σ^2 is, indeed, greater than the mean μ; but the aggregation may not be distinguishable to the eye from a random pattern. In addition, as the mean increases, the relative variance also increases (that is, departs from 1). Alternatively, because rare species have decidedly low density values (equivalent to that of a much smaller cell, core, or quadrat used for species with higher densities), we could conclude that divergences from Poisson expectations are not significant. This would be a misconception. Indeed, most people find it difficult to distinguish visually between a pattern statistically determined to be random and one termed aggregated. Most observers expect random patterns to visually exhibit more evenness than they do.

Looking at Figure 1.2, the now familiar Bolivian plot 01, we recall that $\mu = 6.63$ and $\sigma^2 = 4.69$, for $N = 100$. This gives us a variance-to-mean ratio of 0.71. We still see a few patches with almost no trees, even though this plot is very homogeneous by nature's standards. Visual estimation of patterns as well as of numerical values is a high-risk business. This is why proper sampling, enumeration, and calculation of parameters are required in ecological investigations.

POWER CURVE

Let us return to the random situation for a particular group that exhibits a range of population densities over a series of biological samples. We begin by calculating a series of values (μ, σ^2) for each sample. Although the densities may differ from sample to sample, we find that $\sigma^2 \approx \mu$ (σ^2 *is approximately equal to* μ) for all of them. This means that all samples display a fairly random pattern. A plot of the variances, σ^2, against the means, μ, on an *arithmetic scale* (where the plotted numerical values are positive or negative, and units are equidistant) will result in an approximate straight line. This does not occur often in natural populations. Usually the data show some aggregation, and a plot of the sample values of mean density and variances on an arithmetic scale is at least slightly curved. Basically, a power curve occurs when the arithmetically scaled plot is not a very straight line but a plot of the same data on a logarithmic scale will give a straight line. We will examine how this can occur, and its implications for spatial distribution in the field.

Before we discuss the applications and uses of the power curve, we shall further explore the mathematical properties of straight lines. The familiar equation for a straight line (*linear equation*), in *slope-intercept form*, is

$$y = a + bx. \tag{6.2}$$

In this equation, b is called the *slope* of the line. The slope of a line is a measure of how much the y value increases for each unit increase in the x value. (Mathematically we say that the angle from the x axis to the line is the *inclination of the line*; the slope of the line is the *tangent* of that angle.) In addition, a is the y *intercept* (the place where the line cuts through the y axis). For our purposes we will be plotting the mean on the x axis and the variance on the y axis. Therefore, substituting the symbols for the mean and the variance into Equation 6.2, we get

$$\sigma^2 = a + b\mu. \tag{6.3}$$

When σ^2 is greater than μ, a plot of the variances versus mean densities will result in a curved line on an arithmetic scale. The points show a tendency to cluster tightly at small values and to fan out as larger values are examined. When this occurs, it is often useful to plot these same quantities, σ^2 and μ, on a log–log scale. That is, we transform to $\log(\sigma^2)$ for the y axis and plot against $\log(\mu)$ for the x axis. The result is often a straight line.

Data that yield a straight line on a log–log scale exhibit what is usually called a *power curve* or *power law*. The word *curve* gets confusing because the curve is evident only before the transformation to a log–log scale is made. A straight line on a log–log scale (that is, data in which σ^2 and μ were plotted on an arithmetic scale to yield a curving line) will exhibit the following relationship:

$$\sigma^2 = a\mu^b. \tag{6.4}$$

Here a and b are constants that must be determined by calculations with the observed data. By taking logs of both sides of the equation, we can write Equation 6.4 in the more familiar form of the straight line (*linear*) equation. Thus Equation 6.4 becomes

$$\log(\sigma^2) = \log(a) + b\log(\mu). \tag{6.5}$$

The major reason that Equation 6.5 is preferred over Equation 6.4 is that this equation, now in linear form, is easily solved for a and b on a calculator or computer by a variety of statistical and spreadsheet packages. It is very important to remember the relationship between Equations 6.4 and 6.5. The point of especial import is that the slope b is not transformed to a logarithm, but the intercept a is so transformed.

There are many ways that we could find values for a and b to be able to make the straight line go through the sets of values (μ, σ^2). Probably the simplest would be to plot the observed points, and then draw a line by sight that characterizes the relationship. From this line we could then read off values for the slope and the intercept. However, we prefer to be more scientific. The equation for the line should be obtained in an objective, replicable way. Even requiring objectivity, we still have a choice from among many possible methods by which to calculate these quantities. The usual method selected by statisticians and by most statistical computer packages uses the *least squares* criterion.

LEAST SQUARES CRITERION FOR THE POWER CURVE COMPUTATION

The least squares criterion requires that the a and b values for Equation 6.5 be computed so that the average of the squared errors between the observed data points and the values on the line is as small as possible (it is *a minimum*; that is, we *minimize the error*). To see this, recall that every point on the y axis (a value of σ^2, which we will designate as y_{x_i}) corresponds to a point on the x axis (a value of μ that we will designate as x_i). Now for any one point on the x axis, x_i, we will choose some value, y_{x_i}. This estimated or predicted value may not be the same as the corresponding observed value on the y axis; that is, y_{x_i} might not equal y_{x_i}. Thus there may be some error; our observed value may differ from our estimated value. This may happen regardless of the way we pick the value for y_{x_i}. We can write this error (which is the difference between an observed and a predicted variance) as $e = y_{x_i} - \hat{y}_{x_i}$. Remember also from Chapter 3 that we could have put the observed value last in this difference, but the order we show here is the usual or preferred notation.

The least squares criterion demands that we choose a and b in such a way that the average amount of this error squared over all the observations will be as small as possible, or, a minimum. This means that for the equation

$$\frac{\sum_{i=1}^{N}(y_{x_i} - \hat{y}_{x_i})^2}{N} , \tag{6.6}$$

the proper solution is to make this average a minimum over all N observations; that is, the smallest number possible. This directive is the least squares criterion. We shall use the least squares method again in Chapter 10 when we discuss regression.

Power Curve for Estimating Spatial Distribution

The power curve can be used for a variety of problems. We may be interested in the relationship of the variance to the mean for a single species. In this case we would need estimates of the parameters σ^2 and μ from varying densities. The Bolivian trees form a single frame captured at a single time and cannot provide such estimates.

However, we can examine the power curve for all 52 species in the Bolivian plot. We will use all these 52 species to develop a power curve. Then from the 52 species we will select 4 that occur at differing densities and use these to see how well the power curve we obtained can predict their variances.

Alternatively, the Bolivian study plot can also be considered from the standpoint of total trees, undifferentiated by species. We will see that the power curve calculated with the species data does not predict the variances for total trees very well. However, a power curve based upon data from a random sample of size $n = 8$ (from $N = 100$, Appendix 1) does a fair job of predicting the total tree variances.

Looking at the predictive value of these curves in another way, we show that a power curve based on the 100 quadrats (Appendix 1) does quite poorly in predicting the variances in this plot when the plot is divided instead into 25 quadrats (Appendix 2). But when we take a sample with $n = 8$ from the 25-quadrat arrangement, we obtain quite reasonable predicted variances. Finally, we look at a combined sample with data taken from both the 100-quadrat and the 25-quadrat examples.

All the examples included in the next sections are designed to illustrate the importance of carefully choosing the categories (individuals, species, etc.) that you wish to quantify and the scale on which the categories are to be measured.

Table 6.1 shows the quantities necessary to perform the least squares calculations for the equation with the means and the variances. To obtain the answers, it is only necessary to use the log(mean) as the x values and the log(variance) as the y values. For this computation, we require the square of the x values, and find the summations $\sum x^2$, $\sum x$, and $\sum y$ (bottom of the columns in Table 6.1). We also need to multiply the paired values in the x and the y columns to find $\sum xy$, called the *cross-products*. The formula for the intercept of the line, a, is

$$a = \frac{\sum x^2 \sum y - \sum x \sum xy}{N \sum x^2 - \left(\sum x\right)^2} \tag{6.7}$$

and the formula for the slope of the line, b, which gives us the average change in y for a unit change in x, is

$$b = \frac{N \sum xy - \sum x \sum y}{N \sum x^2 - \left(\sum x\right)^2}. \tag{6.8}$$

Using the sums in Table 6.1, we have from Equation 6.7 the following calculation:

$$\log(a) = \frac{[(135.6019) \cdot (-77.3392) - (78.3444) \cdot (134.6819)]}{[52 \cdot (135.6019) - (6137.8450)]} = \frac{64.2302}{913.4538} = 0.0703.$$

Notice that we have log(a) when we are doing the calculations to solve Equation 6.5, not a itself. Using Equation 6.8 we obtain

$$b = \frac{[52 \cdot (134.6819) - (-78.3444) \cdot (-77.3392)]}{[52 \cdot (135.6019) - (6137.8450)]} = \frac{944.3656}{913.4538} = 1.0338.$$

Therefore, we can write Equation 6.5 as $\log(\sigma^2) = 0.0703 + 1.0338 \cdot \log(\mu)$. However, the usual form that we find in the literature is Equation 6.4. That is, we use Equation 6.5, which is a linear equation, as the easier form for obtaining a solution; but we really want the values without the logarithms. Therefore, we must transform back again.

A look at Equation 6.4 shows that, conveniently, we need only transform the intercept a, because b has not been transformed. That is, there is a b in both Equations 6.4 and 6.5; whereas in Equation 6.4 we see the constant a, but in Equation 6.5 we see log(a). Therefore, because we know that log(a) = 0.0703, we solve for a (by using the *inverse* of the log, which is called the *antilogarithm*) to find $a = 1.1757$. Thus Equation 6.4 becomes

$$\hat{\sigma}^2 = 1.1757 \mu^{1.0338}. \tag{6.9}$$

TABLE 6.1 Least Squares Calculations of a and b for Equation $\log(\sigma^2) = \log(a) + b \log(\mu)$

SPECIES NUMBER	LOG(μ) x VALUES	LOG(σ^2) x^2	y VALUES	y^2	xy
1	0.4014	0.1611	0.4231	0.1791	0.1699
2	0.1173	0.0138	0.1379	0.0190	0.0162
3	−0.2076	0.0431	−0.1104	0.0122	0.0229
4	−0.6021	0.3625	−0.6064	0.3678	0.3651
5	−0.7213	0.5202	−0.6310	0.3981	0.4551
6	−0.7696	0.5922	−0.6966	0.4852	0.5361
7	−0.7696	0.5922	−0.7421	0.5507	0.5711
8	−0.7696	0.5922	−0.7421	0.5507	0.5711
9	−0.7696	0.5922	−0.6178	0.3817	0.4754
10	−1.0458	1.0936	−0.8480	0.7191	0.8868
11	−1.0458	1.0936	−0.9918	0.9837	1.0372
12	−1.0969	1.2032	−1.0287	1.0583	1.1284
13	−1.2218	1.4929	−1.2487	1.5593	1.5257
14	−1.2218	1.4929	−1.2487	1.5593	1.5257
15	−1.3979	1.9542	−1.4157	2.0041	1.9790
16	−1.3979	1.9542	−1.4157	2.0041	1.9790
17	−1.3979	1.9542	−1.4157	2.0041	1.9790
18	−1.5229	2.3192	−1.5361	2.3596	2.3393
19	−1.5229	2.3192	−1.3089	1.7132	1.9933
20	−1.5229	2.3192	−1.5361	2.3596	2.3393
21	−1.5229	2.3192	−1.5361	2.3596	2.3393
22	−1.5229	2.3192	−1.5361	2.3596	2.3393
23	−1.5229	2.3192	−1.5361	2.3596	2.3393
24	−1.6969	2.8865	−1.7077	2.9164	2.9014
25	−1.6969	2.8865	−1.7077	2.9164	2.9014
26	−1.6969	2.8865	−1.7077	2.9164	2.9014
27	−1.6969	2.8865	−1.7077	2.9164	2.9014
28	−1.6969	2.8865	−1.7077	2.9164	2.9014
29	−1.6969	2.8865	−1.7077	2.9164	2.9014
30	−1.6969	2.8865	−1.7077	2.9164	2.9014
31	−1.6969	2.8865	−1.4023	1.9664	2.3825
32	−1.6969	2.8865	−1.7077	2.9164	2.9014
33	−2.0000	4.0000	−2.0044	4.0175	4.0087
34	−2.0000	4.0000	−2.0044	4.0175	4.0087
35	−2.0000	4.0000	−2.0044	4.0175	4.0087
36	−2.0000	4.0000	−2.0044	4.0175	4.0087
37	−2.0000	4.0000	−2.0044	4.0175	4.0087
38	−2.0000	4.0000	−2.0044	4.0175	4.0087
39	−2.0000	4.0000	−2.0044	4.0175	4.0087

(continued)

TABLE 6.1   (*continued*)

SPECIES NUMBER	LOG(μ) x VALUES	LOG(σ^2) x^2	y VALUES	y^2	xy
40	−2.0000	4.0000	−2.0044	4.0175	4.0087
41	−2.0000	4.0000	−2.0044	4.0175	4.0087
42	−2.0000	4.0000	−2.0044	4.0175	4.0087
43	−2.0000	4.0000	−2.0044	4.0175	4.0087
44	−2.0000	4.0000	−2.0044	4.0175	4.0087
45	−2.0000	4.0000	−2.0044	4.0175	4.0087
46	−2.0000	4.0000	−2.0044	4.0175	4.0087
47	−2.0000	4.0000	−2.0044	4.0175	4.0087
48	−2.0000	4.0000	−2.0044	4.0175	4.0087
49	−2.0000	4.0000	−2.0044	4.0175	4.0087
50	−2.0000	4.0000	−2.0044	4.0175	4.0087
51	−2.0000	4.0000	−2.0044	4.0175	4.0087
52	−2.0000	4.0000	−2.0044	4.0175	4.0087
Sums	−78.3444	135.6019	−77.3392	133.9951	134.6819

NOTE: Least squares calculations were performed to obtain *a* from Equation 6.7, and *b* from Equation 6.8, for the equation $\log(\sigma^2) = \log(a) + b \log(\mu)$ for 52 species.

Having done this, we can estimate the error defined by Equation 6.6. We obtain a value of 0.0008 for the error for our example of 52 tree species in the Bolivian forest plot. No other straight line through these points will give a smaller value. This means that when we plot Equation 6.9 on the same graph as our observed data points, it will fit the points "best" in the sense that it will have the smallest amount of error. We say this is the "best" fit.

Minimizing the Error for All Species

Now for any given new mean density value, μ, we can estimate the variance, σ^2. We do this by way of Equation 6.9. Note that Equation 6.9 was estimated from the mean and the variance of each of the 52 species. Rather than listing all 52 species values, Table 6.2 shows the true mean and variance and the estimated variance from Equation 6.9 for just 4 species that have widely varying densities. The estimated values for the variance are close to the observed values for the rarer species (#3, #11, and #33), but are somewhat high for the most abundant species, *Scheelea princeps*. Alternatively, if we use Equation 6.9 to estimate the variance for total individual trees (where the true mean is known to be 6.63), we obtain $\hat{\sigma}^2 = 8.31$. The true variance, σ^2, for total trees, however, is $\sigma^2 = 4.69$ (Appendix 1). We should not be disappointed by this discrepancy between calculated and true variance of the trees, because in our calculation of Equation 6.9 we minimized the error for all 52 species at the same time,

TABLE 6.2   A Comparison of Estimated Variance from a Power Curve Against the True Variance for Four Arbitrarily Selected Species

SPECIES NUMBER	MEAN DENSITY (μ)	TRUE VARIANCE (σ_2)	ESTIMATED VARIANCE $(\hat{\sigma}^2)$
1	2.52	2.65	3.06
3	0.62	0.78	0.72
11	0.09	0.10	0.10
33	0.01	0.01	0.01
Total	6.63	4.69	8.31

NOTE: Species number refers to the identity numbers listed for Bolivian tree species in Appendix 1. Power curve is $\hat{\sigma}^2 = 1.1757\mu^{10338}$.

most of which have small means and variances. We did not minimize the error for all 663 individual trees. This shows that we must use the equation that was based upon the species data for predictions about species, not individuals. We must use the proper tool for the job. The power curve (Equation 6.9) will do nicely for most of the species in our $N = 100$ quadrat size. Had we wished to estimate the variance for total trees we should have selected an alternative approach, which we shall discuss in the next section.

Logarithmic Transformation

The power procedure is simple but involves one complication. When we transform to logarithms as a convenience only, and interest is in the untransformed (or raw) values, we obtain *inefficient estimates* of the values a and b. That is, the estimates will not have the smallest possible variance. This failure to obtain the smallest variance is not troublesome for our descriptive use of these values. Nor is it a problem in our Bolivian example, because the straight line on the log–log scale is such a good description (the error is so small). The straight line, or linear, equation describes the trend in the observed data quite well. However, when we discuss inferential uses of prediction, this inefficiency may become a concern. The alternative is to estimate the values for a and b by using *nonlinear estimation* methods. Although such nonlinear methods are now widely available in computer statistical packages, they are quite complicated to use correctly. For natural population purposes, the linear approach provides sufficiently (indeed, exceedingly) accurate estimates.

Spatial Distribution of Individuals

In the preceding section, we calculated the power curve and the average error for all 52 species of the Bolivian forest data that we have been using throughout this book. The resulting equation did poorly for total trees. But many times it is more useful to calculate a power curve for the total number of individuals, undifferentiated by species.

We randomly selected 8 statistical samples from the $N = 100$ quadrat data (Appendix 1). Thus we have obtained 8 means and 8 variances for total trees. From these data we estimated a power curve for total individuals over the entire plot (Figure 6.1). We fitted these 8 points with a power curve of the form

$$\hat{\sigma}^2 = 0.80\mu^{0.86}. \tag{6.10}$$

The values of 0.80 and 0.86 are the least squares estimates. We know that the true value of the mean density, μ, is 6.63. So we can obtain from Equation 6.10 a value for the estimated variance, $\hat{\sigma}^2$. The estimated variance of 4.08, which we obtain from the data collected in just 8 of 100 quadrats, and undifferentiated by species, is rather close to the true value of the variance, 4.69. Certainly, it is much closer than is the estimate of 8.31 (Table 6.2), which we estimated from the power curve generated from the data based on all 52 individual species in the study plot.

Comparison of Differently Sized Samples of Individuals: Scale Changes

Let us now consider what we can do if we have inventories or continuous monitoring at the same locality, but biological sample sizes were not standardized to size. We will again use the Beni plot 01 as our example. In this Bolivian plot we also use $N = 25$ quadrats instead of the earlier $N = 100$. For $N = 25$ the true mean density of individuals (trees) per quadrat, μ, of all tree species combined is calculated as 26.52 (Appendix 2). How well would Equation 6.10, which was based upon a random sample from $N = 100$ quadrats, do for an estimate of the variance of total individuals at the $N = 25$ quadrats scale?

Using Equation 6.10, which was based upon the 8 samples of trees from $N = 100$, we obtain $\hat{\sigma}^2 = 13.41$ as the estimated variance. Compare this with the true value, $\sigma^2 = 26.65$ (Appendix 2). The results are stark. When we sampled the $N = 100$ plot division and used our resultant power curve (Equation 6.10) to predict the variance for the entire 100-quadrat plot, our results were satisfactory ($\sigma^2 = 4.69$ compared to $\hat{\sigma}^2 = 4.08$). However, when we used Equation 6.10 to predict the variance over the same hectare plot that was now divided into 25 (larger) quadrats, our results were grim: Compare $\sigma^2 = 26.65$ with $\hat{\sigma}^2 = 13.41$. Clearly, we must use equivalent-sized samples for comparisons. Let us persevere.

Let us now take 8 random statistical samples of size $n = 4$ each from the $N = 25$ quadrat sample (plotted in Figure 6.1). In this way we will have 8 means and 8 variances for our calculations. Using the least squares solution, as before, we obtain a power curve of

$$\hat{\sigma}^2 = 3.73\mu^{0.61} \tag{6.11}$$

based on these 8 samples from $N = 25$. Now, substituting $\mu = 26.52$, which is the true mean density for the plot when divided into 25 quadrats, we obtain from Equation 6.11

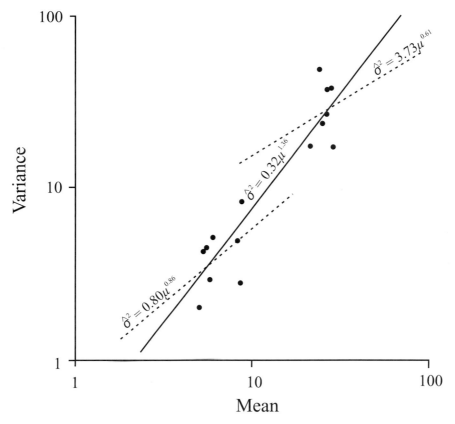

FIGURE 6.1 Power curves plotted on log axes for total Bolivian trees from $N=25$ quadrats, $N=100$ quadrats, and combined. The upper dotted line represents a curve based on $n=8$ from $N=25$ quadrats. The lower dotted line represents a curve based on $n=8$ from $N=100$ quadrats. The solid line is for all 16 points combined.

a value for the estimated variance: $\hat{\sigma}^2 = 27.55$. This value is quite close to the true value of $\sigma^2 = 26.65$. However, here again, if we try to compare and we estimate the variance from Equation 6.11 for $\mu = 6.63$ (from the $N = 100$ quadrat example), we obtain $\hat{\sigma}^2 = 11.82$, which is not close to the true value of $\sigma^2 = 4.69$.

What we have gained from these comparative examples is information on *scale*. Power curves generated from samples taken on one scale are not efficient for calculating estimates on a different scale. A power curve must be generated for each scale, or each quadrat size, considered.

Combining Samples of Different Sizes: Problems of Scale

An alternative and more general solution is to treat the 8 random samples from the $N = 100$ quadrat data and the 8 random samples from $N = 25$ quadrat data as a single statistical sample of $n = 16$. That is, we examine the possibility of having different subplots or biological sample sizes available from an area for which we require a density estimate. Using these data, we obtain a least squares solution of

$$\hat{\sigma}^2 = 0.32\mu^{1.36}. \tag{6.12}$$

Using Equation 6.12 and substituting $\mu = 6.63$, we obtain $\hat{\sigma}^2 = 4.19$ (the corresponding true value is $\sigma^2 = 4.69$). If we select $\mu = 26.52$, we get an estimate of $\hat{\sigma}^2 = 27.62$ (while the true value for $\sigma^2 = 26.65$). Equation 6.12 gives a reasonable estimate of the variance when the mean number of trees is in units (a scale of 0 through 10) or in tens (a scale of 10 through 100), whereas Equation 6.10 is applicable only for units, and Equation 6.11 is useful only for tens. That is, the general principle is that we must decide upon a reasonable scale and standardize these for the entire density study.

Figure 6.1 graphically illustrates why this is so. Only Equation 6.12 considers all of the points; that is, only Equation 6.12 considers the entire scale. The individuals in the Bolivian plot are fairly homogeneous. However, by considering the $N = 25$ quadrat size, as opposed to the $N = 100$, we have effectively examined the trees at a higher mean density. This has been achieved not by finding a stand with higher densities, but by changing the scale (size of quadrats or biological samples) of our observations. When calculating a power curve with variances and means, if we wish to predict the variance over a set of mean values that vary by more than one order of magnitude, the observed data set should include observations at those differing scales (over orders of magnitude, for example, units, tens, hundreds, etc.). Whereas Equation 6.12 does well for total individuals with means in units and tens, it does not do so well for individual species, most of which have means of less than 1. This example shows that consideration of scale is vital to a successful study of density. For prediction, field measurements must be on the same scale as that intended for the predictions. This congruency is especially necessary when using a power curve for which small differences make very noticeably large deviations in predictions.

EXAMPLES OF POWER CURVE CALCULATIONS FOR ORGANISMS

Thus far we have been using the Bolivian forest plot of 52 species and 663 trees for our survey of the utility of power curves in estimating variances for distinguishing spatial characteristics under circumstances of differing scales. In order to examine spatial distribution, Taylor (1961) and Taylor and colleagues (1980) calculated the slope and intercept (constants a and b, respectively, in Equation 6.4) for a wide variety of organisms. We selected some of their examples that well illustrate the variation obtained. We supplemented their power curves with a curve calculated by Vezina (1988) for a variety of marine benthic invertebrates. And, of course, we have the a and b values for the total Bolivian trees (Equation 6.12). All these examples are presented in Figure 6.2. For illustrative purposes, we shall examine these graphed lines, which represent the transformed power curves of each study organism, for the information they can provide on spatial distribution.

Most researchers who use power curves for spatial studies emphasize the characteristics of the slope b in their work. Let us consider an intercept approximately equal to 1 ($a \approx 1$). When $a = 1$, if the slope $b = 1$ also, then, substituting these values for a and

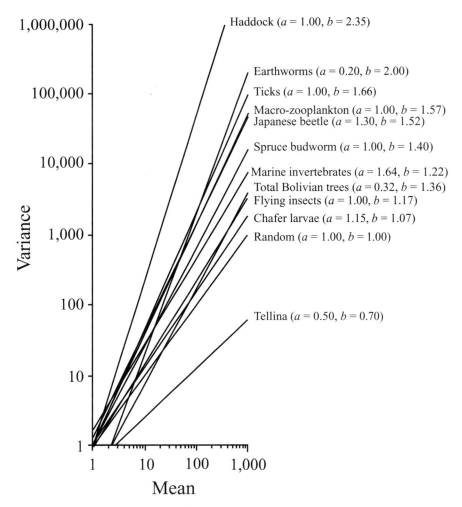

FIGURE 6.2 Power curves on log axes for a variety of organisms. (Data from Taylor 1961; Vezina 1988; Dallmeier and colleagues 1991.)

b into Equation 6.4, we get $\sigma^2 = \mu$. Because these two quantities, variance and mean density, are equal, or approximately equal, this means that $\dfrac{\sigma^2}{\mu} = 1$ (or almost 1). Achieving a value close to 1 for the relative variance, $\dfrac{\sigma^2}{\mu}$, suggests that the population could be Poisson distributed, and, hence, called random.

However, most often in natural populations, b is not equal to 1, but is greater than 1 (Figure 6.2). Thus we see from Equation 6.1 that the relative variance must be greater than 1; $\sigma^2 > \mu$. When $b = 2$, for example, a condition denoting a good deal of aggregation, then Equation 6.4 becomes $\sigma^2 = \mu^2$ (we still are examining $a \approx 1$), or $\sigma = \mu$ (see Chapter 5). Figure 6.2 shows that the majority of values for b for these natural populations are, in fact, between 1 and 2.

THE SLOPE AS AN INDEX OF AGGREGATION

The value of b has been used as an *index of aggregation*, or an *index of clumping* (McIntyre and colleagues 1984). Based on this index of aggregation, Figure 6.2 indicates that haddock is the most aggregated and that the evenly spaced *Tellina* is the least. At first sight, we might be encouraged to conclude that each group of organisms has an easily calculated value of b that remains invariant for that group. Mother Nature, however, seldom offers such simplicity. Wratten (1974) showed that for *Euceraphis punctipennis*, an aphid species, the value of b ranged from 1.17 to 2.26 in response to varying environmental conditions over a period of a few months. Thus a single species may vary its pattern of aggregation depending on the ecological conditions. Nevertheless, we might expect the overall spatial behavior of a species, measured over a considerable period of time under varying environmental conditions, to exhibit a characteristic or intrinsic value of b. Such an approach was adopted by Taylor and colleagues (1980), who calculated power curves from data on hundreds of species of aphids, birds, and moths gathered during observations spanning from 11 to 15 years at numerous sites over an area of 2,300 km^2 throughout Great Britain and in Europe. The amount of time and the spatial scale of their study were large enough to ensure no possibility of environmental uniformity. The power law (Equation 6.4) did indeed fit their data well for 95% of the species. The researchers concluded that each species has a characteristic exponent of b. Species within the same genus, however, have exponents that vary considerably. For example, in the genus *Aphis* (Figure 6.3) values of b range from 2.95 to 1.65. Numerous other examples are given in the analyses of Taylor and colleagues (1980).

The results obtained by Wratten (1974) for winged terrestrial animals, such as *Aphis*, are very similar to the results one of us obtained for *Ammonia beccarii*, a marine species of benthic foraminifera (Buzas 1970; Buzas and Severin 1993). In Rehoboth Bay, Delaware, the number of living individuals of *A. beccarii* was counted in 5 replicate samples from each of 16 approximately equally spaced (10 m apart) localities at a single sampling time. In the Indian River, Florida, the number of individuals was counted in 4 replicate samples at each of two localities (one grassy and one bare) about 1 m apart, at 17 sampling times, each a fortnight apart during 1978. Another grassy site in the Indian River was sampled with 4 replicates at each of 51 sampling times during 8 of the years spanning 1977 to 1988. The power curves relating the variance and the mean calculated from these observations are shown in Figure 6.4. The largest value of b (2.62) was obtained from observations made at a single time in Rehoboth Bay. Smaller, but remarkably similar, values were obtained at the grassy and bare sites sampled during 1978 in the Indian River. Finally, the samples obtained over the period of a decade in all seasons provided the smallest, and presumably characteristic, estimate of b.

THE IMPORTANCE OF THE INTERCEPT a

Therefore, while a species may occupy a landscape with a characteristic (or intrinsic) value for b, the estimation of that value evidently requires many years of sampling

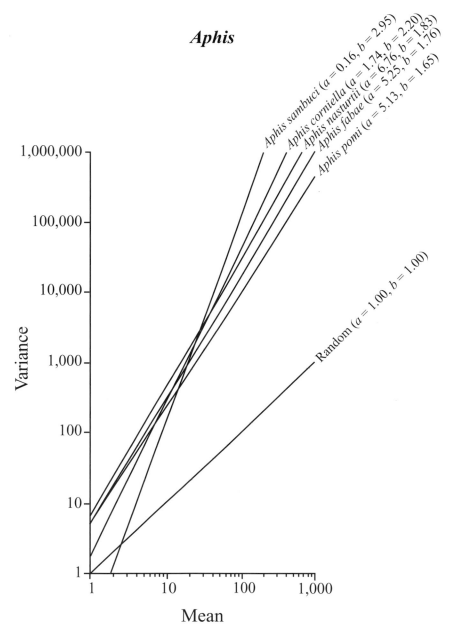

FIGURE 6.3   Power curves on log axes for *Aphis*. (Data from Taylor and colleagues 1980.)

over a wide variety of environmental conditions. A value of *b* so obtained is an ecologically important attribute of species behavior.

As indicated earlier, even if sampling is carried out over an extended range of space and time, the value of the constant *a* may vary considerably. Most researchers emphasize the constant *b* and ignore the intercept value *a*. Let us explore whether this is a wise choice (Figures 6.2, 6.3, and 6.4). For natural populations it is necessary to calculate

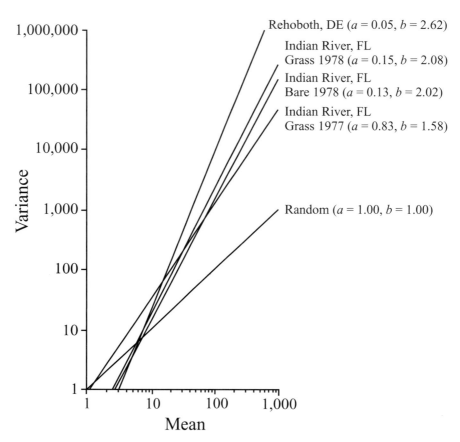

Ammonia beccarii

FIGURE 6.4   Power curves on log axes for *Ammonia beccarii*. (Data from Buzas 1970; Buzas and Severin 1993.)

both *a* and *b* values in order to distinguish a specific line. Mathematically, these quantities work in tandem, although this relationship is quite often ignored in the scientific literature.

When mean densities are high ($\mu > 100$), the lines (power curves) are arranged, mostly, as sequences according to the values of *b* (Figures 6.2, 6.3, and 6.4). At even higher densities, the relatively large mean, when raised to the *b* power, usually overshadows the importance of the value of the *a* constant in the equations. Consequently, the predicted value for the variance depends most heavily on the value of *b*. Even at these higher mean densities, however, with mean densities in the hundreds, the lines cross for earthworms and ticks, macro-zooplankton and Japanese beetles, total Bolivian trees, and flying insects (Figure 6.2). At a mean value of 1,000, the estimated variance, $\hat{\sigma}^2$, for total Bolivian trees (where $b = 1.36$) is still lower than that for marine invertebrates (where $b = 1.22$). This is because of the large difference in the 2 values of a. For this pair, the estimated value of the variance becomes larger for the trees only above a mean density of 100,000. What this means is that a is important.

In general, to achieve an equivalent estimated variance (the crossover of curves) for 2 different organisms, we require that as the value of the a constant increases by an order of magnitude, the value of b decreases by 0.5. For example, the estimated variance, $\hat{\sigma}^2$, at $\mu = 100$, $a = 0.05$, and $b = 2$, is equivalent to that obtained from the power curve with $a = 0.5$ and $b = 1.5$.

When mean densities are in tens or lower, even though the values of b for different organisms may be quite different, the resulting lines may cross because of the varying values of a, which now becomes more important. Under these circumstances, equations with large differences in the value of b may yield approximately the same values for the predicted variances.

So at lower densities the value of the a constant may contribute greatly to the predicted value for the variance. For example, for *Aphis sambuci*, Figure 6.3 shows $\hat{\sigma}^2 = 0.16\mu^{2.95}$, and for *A. pomi* $\hat{\sigma}^2 = 5.13\mu^{1.65}$. At $\mu = 14$, the variances, $\hat{\sigma}^2$, are 384.77 and 399.33, respectively. Moreover, at mean densities lower than 14, the predicted variance for *A. sambuci* will be smaller than that of *A. pomi* because of *A. sambuci*'s lower value of a, even though the value of b for this species is much larger, indicating spatial heterogeneity. Similar examples are apparent from the consideration of *Ammonia beccarii* shown in Figure 6.4. Therefore, it is most important that if the purpose is to examine the spatial configuration of a species, the determination of a power curve should be based on samples with densities that span more than one order of magnitude. This can be accomplished by sampling a variety of environmental conditions over a period of time, and then examining the entire curve. We can achieve this result also by varying quadrat size; that is, the biological sample size, as well as by sampling at various times.

A species that has an aggregated (heterogeneous) distribution may at lower densities appear to be randomly or homogeneously distributed. Consider *Ammonia beccarii* (Figure 6.4), where the equation to determine the variance is $\hat{\sigma}^2 = 0.15\mu^{2.98}$. This means that at $\mu = 6$, the estimated variance is given by $\hat{\sigma}^2 = 6.23$, with a relative variance (Equation 6.1) of 1.04. If we sampled at this density only, the species would appear to be randomly distributed, although because the value is $b = 2.08$, this is clearly not true. In this region of low densities, these distributions will appear as random. Because mean density varies with quadrat size as well as with environmental conditions over time, the scale (quadrat size) of the observations must always be taken into consideration when we are interested in spatial configuration.

USING THE POWER CURVE TO CHOOSE AN OPTIMAL SAMPLE SIZE

In Chapter 5 we stated that an investigator may have to rely on estimates of the variance, σ^2, calculated from a pilot study in order to judge the optimum sample size for conducting a survey of a natural population. However, if a value for the mean density is known or anticipated for a study area or target group, the power curve can be used for obtaining an estimate of the variance, σ^2. In turn, we can then use this value for σ^2 in Equation 5.5 to obtain a sample size estimate for a field project.

For example, let us presume that we wish to go back to Rehoboth Bay, Delaware, to conduct another investigation of species density of *Ammonia beccarii,* which was already sampled by Buzas (1970). We can use the results of the 1970 study at this same location (Figure 6.4) as a pilot survey and assume these data represent the best available for the relationship of the mean and the variance in this area. The estimate of the variance from the relevant equation is $\hat{\sigma}^2 = 0.054\hat{\mu}^{2.62}$. This, in turn, can be used to calculate the number of samples required at some predetermined confidence limit for any anticipated mean density μ. Let us say the anticipated μ is 10, then $\hat{\sigma}^2 = 0.054 \cdot 10^{2.62} = 20.84$ as we show in Table 6.3. Now assume that we wish to be 95% confident that our estimate of mean density will be within 50% (0.50) of the true value. From Chapter 5 on sample size determination, we know that we can use Equation 5.5 to obtain an approximate sample size. For this equation we have $z^2 = 1.96^2 \approx 4$, and the criterion of 50% of the true mean is written as $k = 0.50 \cdot \mu = 0.50 \cdot 10 = 5$ so that $k^2 = 25$. Now using Equation 5.5, $n = \dfrac{z^2\hat{\sigma}^2}{k^2} = \dfrac{4 \cdot 20.84}{25} = 3.33 \approx 3$ biological samples. Alternatively, if the anticipated $\mu = 100$, then $\hat{\sigma}^2 = 0.05 \cdot 100^{2.62} = 8{,}689.00$, and $n = \dfrac{4 \cdot 8689.00}{2500} = 13.90 \approx 14$ biological samples. Fourteen is thus the number of biological samples we would need if we thought that the mean density for *A. beccarii* was about 100.

If we were interested in examining the density of the same species in the Indian River, Florida, and if we used the results obtained from bare sand in 1978 for a pilot survey shown in Table 6.3, the equation we would use is $\hat{\sigma}^2 = 0.13 \cdot \mu^{2.02}$. Then, for $\mu = 10$, $\hat{\sigma}^2 = 0.13 \cdot \mu^{2.02} = 13.61$. We have then $n = \dfrac{4 \cdot 13.61}{25} = 2.18 \approx 2$ biological samples. For $\mu = 100$, $\hat{\sigma}^2 = 0.13 \cdot 100^{2.02} = 1{,}425.42$, and $n = \dfrac{4 \cdot 1425.42}{2500} = 2.28 \approx 2$ biological samples. This tells us that for a highly clumped spatial pattern, the higher the density per sample, the more samples we need to collect. When the spatial distribution is more homogeneous or even, fewer samples are needed for an equivalent amount of information.

TABLE 6.3   Using Pilot Studies on *Ammonia becarii* and Power Curves for Sample Size Estimation

	a, b IN POWER CURVE	n, VARIANCE $(\hat{\sigma}^2)$ FOR $N = 10$	n, VARIANCE $(\hat{\sigma}^2)$ FOR $N = 100$
Delaware 1972	0.054, 2.62	3, 20.84	14, 86.89
Florida 1978	0.130, 2.02	2, 13.61	2, 1,425.42
Florida 1977–1978	0.830, 1.58	5, 31.56	2, 1,199.72

If the results from Indian River, Florida, that were obtained during the sampling from 1977 to 1988 were used as the most characteristic values for the species, then we would use the equation $\hat{\sigma}^2 = 0.83 \cdot \mu^{1.58}$. Using this equation for $\mu = 10$,

$\hat{\sigma}^2 = 0.83 \cdot 10^{1.58} = 31.56$, and $n = \dfrac{4 \cdot 31.56}{25} = 5.04 \approx 5$ samples. If we anticipated $\mu = 100$,

then $\hat{\sigma}^2 = 0.83 \cdot 100^{1.58} = 1{,}199.72$, and $n = \dfrac{4 \cdot 1199.72}{2500} = 1.92 \approx 2$ samples, which we have summarized in Table 6.3.

In the first instance, for which $b > 2$, many more samples were required when $\mu = 100$ than when $\mu = 10$. In the second instance, for which $b \approx 2$, the same number of samples was required at both densities. And in the third, for which $b < 2$, fewer samples were required when $\mu = 100$ than when we anticipated that $\mu = 10$. If a good deal of effort is required in counting the individuals within the quadrats, or biological samples, then the higher densities will always require considerably more work. In the third instance, for example, when fewer samples are required at the higher density, we still have an average of about $n\mu = 50$ individuals to be counted for 5 biological samples when $\mu = 10$; when $\mu = 100$, we can expect about $n\mu = 200$ for the 2 biological samples.

CHOOSING QUADRAT SIZE WHEN POWER CURVE PARAMETERS ARE UNKNOWN

Suppose we have no information about the parameters of the power curve. Suppose, too, that we suspect aggregation and that we have the ability to adjust the quadrat size for our observations on density. What quadrat, or biological sample size (for mean density) should we choose to ensure consistency in the precision (see Chapter 4) of our density estimate?

To help with this decision, we constructed Figure 6.5. An estimate of σ was made for power curves with intercepts $a = 0.5$ and 1.5 and slopes $b = 1.5$, 2.0, and 2.5. Next we calculated, as a measure of precision, the coefficient of variation (CV), using the

equation $CV = \dfrac{\hat{\sigma}}{\mu}$, for values of μ up to 100. When $b = 2$, the CV's shown in Figure 6.5

are approximately equivalent (here $\sigma = a\mu$, see Chapter 5). When $b = 1.5$, the CV's decrease with increasing μ (better precision). At $b = 2.5$ the CV's increase markedly with an increase in μ (worse precision). Note that we showed the case of $b = 2.5$ only for $a = 0.5$; the precision is much worse for $a = 1.5$. However, around $\mu \approx 10$, the curve with $b = 2.5$ converges with the rest.

In summary, if we have no information on the spatial distribution, and especially sometimes if our organisms are suspected of exhibiting a good deal of spatial aggregation ($b \geq 2.5$), the best bet is to choose a quadrat size; that is, a biological sample size that will ensure that the mean density will be less than 10.

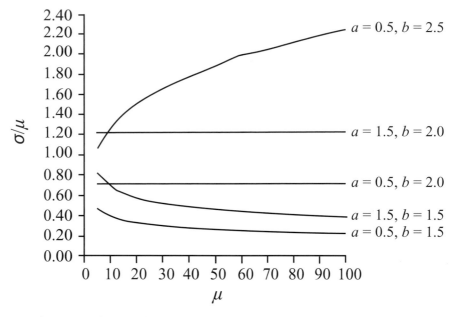

FIGURE 6.5   Plot of the coefficient of variation (σ/μ) against the mean density (μ) for different values of the constants (a and b) in the power curve equation. Note that the values converge for μ less than about 10.

SUMMARY

1. When $\sigma^2 = \mu$, the spatial distribution is deemed to be random, or homogeneous; when $\sigma^2 < \mu$, the spatial distribution is designated as even. Most commonly in nature, $\sigma^2 > \mu$, and the spatial distribution is called aggregated, or clumped.

2. The power curve, $\sigma^2 = a\mu^b$, is used to describe the relationship of the mean density to the variance. Its uses for natural populations are twofold: (1) to examine spatial distribution for individuals ungrouped, or grouped in any way, for example, by species, genera, and so on; and (2) to predict variance from an estimated mean density value calculated from data in a pilot study in order that we can choose an optimum sample size for a biodiversity or field study.

3. The least squares criterion allows for an estimate of each of the constants, a and b, in the equation for the power curve. This method gives values for a and b that will allow the predicted values to be closest to the observed values over the entire data set.

4. The exponent b can be used as an index of aggregation. When $b < 1$, the distribution is even; when $b = 1$, the distribution is random; and when $b > 1$, the distribution is aggregated.

5. While the exponent b can be regarded as an intrinsic characteristic of a species, to obtain a stable estimate of b for any species often requires sampling a variety of environmental conditions over a considerable period of time.

6. At high densities ($\mu > 100$), the predicted variances from power curves are often distinct.

7. At low densities ($\mu < 10$), the power curves converge, and the constant a may become more important in the determination of predicted variances than is the constant b.

8. At low densities aggregated populations may appear random. Therefore, when examining spatial configuration, determination of the power curve must be based on samples with wide-ranging densities. That is, either time or space or both must be varied.

9. Predicted variances from the power curve generated from data in a pilot study can be used to choose an optimum sample size for conducting a density study of a natural population.

10. If a good deal of spatial aggregation (high variance, $b > 2$) is suspected, and if the size of the biological sample can be adjusted, then selecting a size to ensure that $\mu < 10$ is a good choice to maximize precision.

11. If a spatial distribution is truly random, then varying quadrat size will not make any difference. In nature, however, truly random distribution is improbable, so it is important to carefully select the size of the quadrats into which the entire study area will be divided before the fieldwork is undertaken.

PROBLEMS

6.1 Take 8 random samples of 4 observations each of *Pouteria macrophylla* from Appendix 2 (where $N = 25$). Calculate a power curve for the variance and the mean. Using your equation, find the predicted variance for means of 0.62, 2.48, 10, and 100.

6.2 Take 8 random samples of 4 observations each of *Pouteria macrophylla* from Appendix 1 (where $N = 100$). Calculate a power curve for the variance and the mean. What is the predicted variance for means of 0.62, 2.48, 10, and 100?

6.3 Combine the two sets of samples from the $N = 25$ and $N = 100$ inventories taken in the previous two problems and calculate a power curve for the variance and the mean. What is the predicted variance for means of 0.62, 2.48, 10, and 100? Discuss the results of the first three problems.

6.4 At the 95% confidence level, calculate the number of samples required for confidence limits of $\pm 75\%$ for expected means of 0.62, 2.48, 10, and 100. Do this by first using the power curve formulas for $N = 25$ and then $N = 100$ quadrats from Problems 6.1 and 6.2. Based on what you have learned, what size quadrats would you use, and how many samples would you take?

6.5 In 1990, Ferrari and Hayek examined daytime samples in a seasonal survey of *Pleuromamma xiphias*, the looking-glass copepod. Although two different kinds of sampling gear (trawl and bongo nets) were used, corrections were made so that the data in their Tables 2 and 3, summarized and presented below, can serve as 4 replicates for the number of stage-6 female copepods (abbreviated as CVI) found in each of the months.

Month	Numbers of CVI
August	405, 404, 534, 506
October	61, 529, 994, 447
February	660, 792, 392, 1,083
April	728, 515, 167, 311
May	206, 519, 341, 281
July	444, 390, 830, 212

a. Calculate and fit a power curve to this data.

b. Discuss the spatial distribution of these looking-glass copepods.

7
FIELD SAMPLING SCHEMES

Expert judgment and experience are vital to the success of most fieldwork. However, before going into the field to make a determination that a sampling study is designed efficiently for the intended purposes, we must rely on some objective criteria of efficiency. Increasing the complexity of the design does not ensure its success or an increase in efficiency. The precision of the results depends heavily on how the sample is selected and how the estimated values are obtained. Choice of sample size is an important consideration, but we make the most effective use of available and usually limited resources when we use a set of objective criteria to select among alternative designs for placing the samples in the field study area.

When we can be sure that the sample will indeed be a *probability sample*; that is, a random sample, we can be confident that there will be an impartial way to measure the results. We therefore must ensure that the probability of any *sampling units* (the object of our random sampling effort; the units into which the target population is divided for sampling purposes) being included in the sample is known. If the basis for inclusion in a sample is judgment, regardless of how expert, we will not have a reproducible measure of our field study's usefulness.

In this chapter we define and discuss only probability sampling methods; that is, the methods by which sampling units are selected or obtained with known probabilities. We will also describe the field conditions for which one or another of the methods might be favored for natural population work and for which the resultant estimates should be the most precise. We begin with the most fundamental method of sampling, *simple random sampling*, sometimes also called *unrestricted random sampling*, and continue on to more elaborate schemes within the category of *restricted random sampling* methods. Here, "restricted" refers to conditions placed on the randomization of each of the sampling units. For example, our target area might consist of three distinct habitats or sediment types. Instead of selecting a random sample from the total target population, we could select a random sample from each of the three habitats or sediment types. Such a modification would place a restriction, or condition, on the simple random selection procedure.

SIMPLE (UNRESTRICTED) RANDOM SAMPLING

In statistical texts the emphasis on simple random sampling (srs) is considered unreasonable in practical application by many, if not most, field researchers for a variety of practical reasons. Figure 7.1, column A, gives an example of an srs of size $n = 4$ from our Bolivian plot. This figure lays out the 100 quadrats in a straight transect to provide a visual comparison of the sampling schemes that are discussed in this chapter. In field situations, an srs may be difficult or impossible to achieve.

For example, an oceanographic ship may proceed along a particular traverse or course, with stations located along it. The exact location of the stations may depend upon the ship's mission, and the biologist may have to obtain samples whenever it is possible, rather than when it is desirable. Similarly, a mammalogist may randomly select latitude and longitude values on a map grid for quadrat placement but then find that, once in the forest, the selected location is inaccessible. Terrestrial surveys locate samples along transects or traverses, but strict adherence to an srs of points or localities along the length might require backtracking over many kilometers. Overall, in many ecological surveys, for a wide variety of reasons, srs is an impossibility.

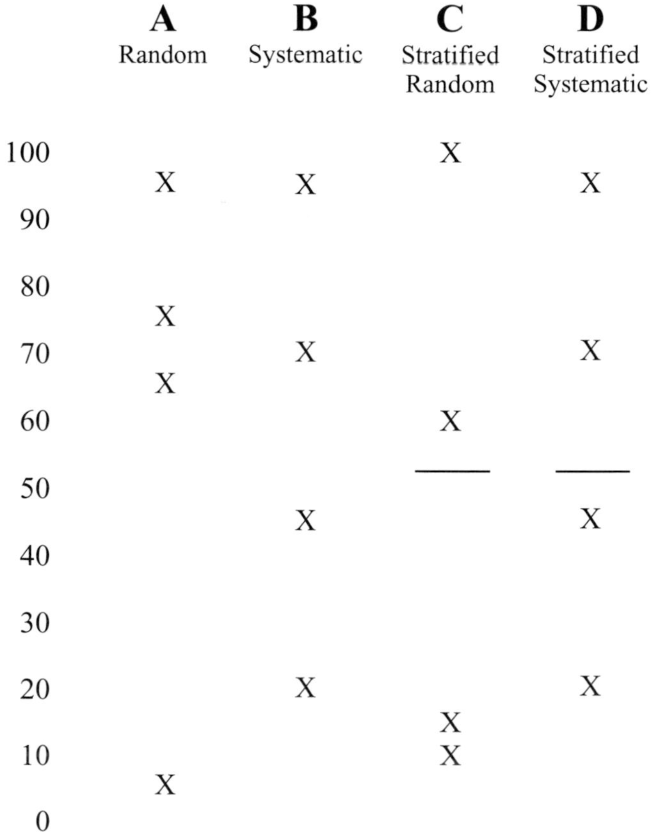

	A Random	**B** Systematic	**C** Stratified Random	**D** Stratified Systematic
100			X	
	X	X		X
90				
80				
	X			
70		X		X
	X			
60			X	
50			———	———
	X		X
40				
30				
20		X		X
			X	
10			X	
	X			
0				

FIGURE 7.1   (A) Simple random sampling (srs) along a traverse. (B) Systematic sampling along a traverse. (C) Srs along a traverse with two strata. (D) Systematic sampling of two strata along a traverse. The drawn line signifies the break between the two strata. The leftmost column can represent 100 quadrats laid out linearly.

The Quadrat as the Primary Sampling Unit

In discussions of density throughout the earlier chapters of this book, we obtained our data by counting the number of individuals within the confines of a sampling device called a quadrat, whose placement or selection was accomplished by simple random sampling procedures.

When we applied random sampling techniques, we selected the locations of the quadrats to be sampled, not the locations of the individuals. The selection of quadrats was made at random within the target area. That is, the *primary sampling unit* was the quadrat or space, not the individual organisms. The variable of interest is the count per unit space, not the count of the individuals.

In our Bolivian examples we were interested in obtaining an estimate of mean density, and our target population always consisted of the total number of quadrats ($N = 100$ or $N = 25$), not the 663 individuals. The total statistical sample size ($n = 8$ or $n = 4$) was the accumulation of all sampled quadrats, or biological samples. What we were interested in was the number of individuals per unit space, which we called density. However, this procedure often causes confusion in the communication between biologists and statisticians. When biologists tell a statistician that they are counting the number of individuals in some space, and justifiably show a great interest in the individuals, the statistician assumes that the individuals are the variable of interest. However, in studies of density, it is the quadrat, not the individual, that is the primary sampling unit. It is the quadrats, not the individuals that are randomly selected.

The Individual as the Primary Sampling Unit

Sometimes the variable of interest is the individual, and so the desired primary sampling unit is the individual, not the quadrat. Clearly, when our interest lies with individuals as the primary sampling unit, we usually cannot line up or list the total population of individual animals, plants, protists, or other organisms, and then (randomly or otherwise) select a sample. We are not sampling cards from a file. Randomly selecting individuals from a natural population is, however, possible for some species, such as elephants, in which individuals are large, or the Bolivian trees, once they are mapped.

Individuals as primary units are of interest only when we are estimating proportions (Chapter 8), size, biomass, or some related measure. Most often, however, the sampling of either a naturally occurring aggregation (for example, a breeding pond or tadpole swarm) or a contrived aggregation (for example, a core or quadrat) is the most efficient method of obtaining such proportional information on the individuals. That is, and for a variety of reasons, we cannot directly sample individuals. Instead, we sample a space and then simply work with the individuals that happen to occupy that space. In such a case, the quadrat, or other sampling device, is the object whose selection must comply with statistical sampling procedures, even though the individual is actually the focus of interest. Consider, for example, a researcher who wishes to sample the protozoa in a pond. It would be impossible to select individuals at random, because most are not even visible to the naked eye. In such instances, the investigator might remove small volumes of water from locations randomly distributed in the pond, and then work with the protozoa that happen to be in the samples.

The difference between the individual versus the quadrat as the primary sampling unit is not trivial, because the estimation of statistical parameters differs for the two methodologies. Biologists must always clearly designate the primary sampling unit, the variable of interest, and the purpose of their studies.

Cluster Sampling

When the primary sampling unit is the quadrat or other sampling device, yet interest is in distinctions among individuals (such as the proportional abundances), we are working with an indirect method of getting at the variable of interest. This indirect method is called *cluster sampling*. In cluster sampling of natural populations, the sampling device is the primary sampling unit and is called a *cluster* by statisticians.

When we design a cluster sampling strategy, there are some conditions that must be met for cluster sampling of natural populations in order for the results to be treated statistically. First, there can be no overlap in the clusters (quadrats). Because the fieldworker constructs and places the sampling device, this condition should not present a problem. Second, each organism must be in one and only one cluster. This is not a problem with most sessile organisms, but it may be a problem with rhizomes, ramets, or colonial invertebrates, and especially with highly mobile organisms. In such cases, some arbitrary rules are imposed to ensure that an individual (the term *individual* may need some arbitrary definition) is either in or out of a particular cluster.

There are two types of cluster sampling of importance for working with natural populations: single-stage (or simple) cluster sampling, and multistage cluster sampling. *Single-stage* (or *simple*) *cluster sampling* is done in one step. Clusters are selected, and every organism in each selected cluster is counted. Simple cluster sampling is most common, for example, when transects or quadrats are laid out for counting (or measuring proportions of) terrestrial vertebrates, large fossils, or botanical specimens. *Multistage cluster sampling*, in contrast, is done in two steps. The clusters are selected, and then a sample of organisms is taken from within each cluster (the cluster is *subsampled*). Thus every organism in the subsample is counted, but not every organism in the sample is counted. An example of a device for subsampling marine invertebrates is shown in Figure 7.2.

In nature it would be an anomaly if clusters each contained the same number of individuals. However, subsamples (see also Chapter 8) of equal numbers of individuals can be obtained intentionally from the clusters. In such cases, only the first "*x*" number of individuals in each cluster are sampled. The advantage of such a procedure is that it reduces work and time to practical and manageable amounts. The efficiency benefits of subsampling make it a common practice in sampling; for example, in studies of invertebrates, archeobotanical specimens, and sometimes insects. Multistage cluster sampling with subsampling is often the methodology of choice for very small

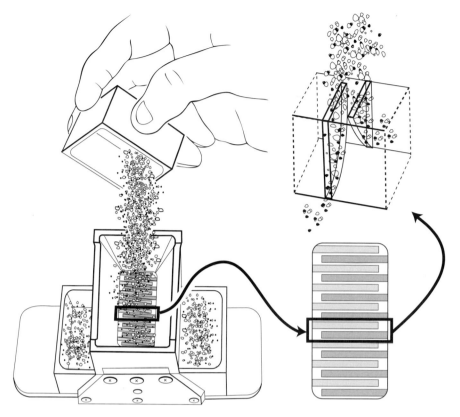

FIGURE 7.2   A subsampling device called a microsplitter. The device has multiple slides going in alternate directions to divide a poured sample into halves.

terrestrial and aquatic organisms and marine invertebrates, as well as archeobotanical studies of seeds or pollen.

While considering different types of sampling schemes in the rest of this chapter, we illustrate them with spatial units (quadrats or transects) as the primary sampling unit. The same methodology can be used with the individual as the primary sampling unit, or with the spatial unit as the primary sampling unit and the individuals as the object of interest.

SYSTEMATIC (OR ORDERED) SAMPLING

An alternative to simple random sampling that is used to avoid practical problems of applying srs schemes in the field is to gather samples at ordered intervals, predetermined in either space or time. Such a procedure is attractive because it has the advantage of spreading out the samples over the entire target area. For example, a transect may be cut through the marsh or rain forest. Then, say, 5 samples would be gathered along its length at approximately equal distances apart. More precisely, the total length of the transect may be divided arbitrarily into, say, 20 parts; a biological sample would be gathered at every fourth part to give a statistical sample of size 5.

Statisticians define this method of selecting samples at equal intervals or increments (with or without a random start) as *systematic sampling*. For biologists, the term *ordered sampling* is preferable to avoid confusion, because for them the word *systematics* has an entirely different connotation. However, the phrase *systematic sampling* is thoroughly ingrained in the statistical literature; so we use the two terms interchangeably.

Systematic sampling involves the implicit assumption that it is possible to arrange, or order, the target area in some reasonable way, and that the researcher can make use of this ordering to pick the sample systematically. In addition, most of the population variation can be assumed to be within the selected units (quadrats, transects, or traverses). There are two major questions posed by an investigator who chooses to select the sampling units in an ordered (systematic), rather than a random, way. First, what are the consequences for the estimators (for example, for density) when making ordered rather than random selections? Second, are there characteristics of the target population that might indicate that an ordered approach may actually be better (in some sense) than random selection?

Estimators for Systematic Sampling

To examine the first question (the consequences for the estimators of choosing a systematic rather than random method), recall that for simple random sampling the mean density and variance estimators are unbiased—no matter what the spatial distribution of the target population. Often, we have heard researchers justify their use of systematically selected, ordered samples by saying that their sample could have been selected as a random choice. This may be true, but it is not a justification; this fact, for the one observed sample obtained, is irrelevant to the discussion, as we shall see.

Let us examine an example of $N = 100$ and $n = 4$ from the Bolivian forest plot example we have been using throughout this book. We begin by selecting a starting quadrat randomly. Then we take 3 more (equidistant) biological samples along the remaining length of the 100-quadrat plot, as shown in Figure 7.1, column B. Because there are 100 quadrats, and we want 4 samples, we divide 100 by 4 to get 25. Then, we pick one random number between 1 and 25. From our table of random numbers we picked the number 20, so the first quadrat will be #20. The remaining 3 biological samples must then be quadrats #45, #70, and #95, each separated by 25 units from its neighbor. We now have 4 biological samples obtained at equal intervals from the possible $N = 100$ quadrats in the target population. This is thus an example of an ordered, or systematically selected, statistical sample of size $n = 4$ in $N = 100$.

Alternatively, we could have consulted Figure 2.1 as a good representation of the Bolivian plot, because it consists of 10 parallel transects of 10 units or quadrats each. A common field procedure would be to select one such transect (column) in this grid system; say, near the left of the plot, and to take 4 biological samples along its length. Then the field biologist might repeat this exercise on a second transect, chosen to be nearer the right side of the plot for a more even coverage. Figure 7.3(A) shows that, using both transects, we would have 8 samples, or a statistical sample of

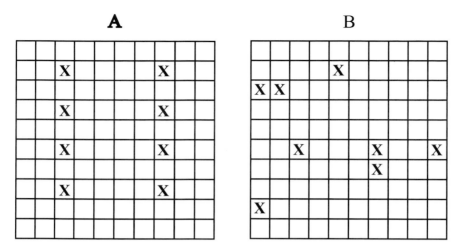

FIGURE 7.3 (A) Systematic sampling of Bolivian plot 01. (B) Simple random sampling (srs) of Bolivian plot 01 (same as Figure 2.2).

size $n = 8$ in a target population of $N = 100$. This is in contrast to the simple random sampling scheme we used to select $n = 8$ from $N = 100$ in Chapter 2, which is depicted in Figure 7.3(B).

Strictly speaking, the statistician would have more stringent requirements for a systematic sample. A true systematic sample would equally space all 8 samples over the entire plot. For example, a sample of size 4 would be an acceptable systematic sample of every 25th quadrat after a random start, but the sample of size 8 would not. We shall see the consequences of the alteration to the true systematic sampling scheme that is used by the fieldworker for natural population sampling.

The numbers of trees in each of our 8 ordered quadrats are 4, 9, 5, 7, 4, 6, 13, and 11. Their sum is $\Sigma x_i = 59$. If we use the formulas for random sampling presented in Chapters 2, 3, and 4, we would obtain the following. From Equation 2.2 we would calculate a mean density estimate of $\hat{\mu} = \dfrac{59}{8} = 7.38$. Equation 2.5 would yield an estimated standard deviation of $\hat{\sigma} = \dfrac{23.28}{7} = 3.34$ (dividing by $n - 1 = 7$). From Equation 3.6 we would calculate a standard error estimate of $\hat{\sigma}_{\mu} = \dfrac{3.34}{\sqrt{8}} \cdot 0.96 = 1.13$. From our two-tailed t-table (Table 4.1) with this small sample, we obtain $t_{0.95, 7} = 2.36$ and $t_{0.90, 7} = 1.89$. The confidence limits now can be calculated easily. Using Equation 4.7 at 0.95 confidence, we calculate $t\hat{\sigma}_{\mu} = 2.36 \cdot 1.13 = 2.67$, resulting in the limits (7.38 ± 2.67), or the interval $4.71 \le \mu \le 10.05$. At the 0.90 confidence level, $t\hat{\sigma}_{\mu} = 1.89 \cdot 1.13 = 2.14$, so that the confidence interval limits are 7.38 ± 2.14, and the interval is $5.24 \le \mu \le 9.52$. In both cases, we can see that the true mean, $\mu = 6.63$, lies well within the confidence limits.

Now, let us consider what is involved in analyzing this systematic sample as it was collected, and not as if it were randomly collected. That is, before we give equations for estimates for systematic sampling, let us examine the procedure in more depth.

Sampling Distribution for Systematic Sample Collection

It might appear that we could analyze our systematic sample just as if it had been an srs, merely by using the srs formulas to obtain a mean, a variance, and a standard error of the mean. But this is not so. We have seen (Chapter 3; sampling distributions) that the size of the standard error is an indication of how precise (close to the true value) the mean density estimate is. Accordingly, the precision of an estimate depends not on the one sample we obtain but on the set of all possible samples that could have been obtained. That is, for our case, it depends on the sampling distribution of the mean density.

The set of possible values of mean density for our 8 systematic samples is considerably smaller than the set of values for all possible random samples of size 8 from $N = 100$. For a random sample of $n = 8$, the appropriate sampling distribution contains 186,087,894,300 possibilities ($_{100}C_8$, where C refers to combination). We found this number by using methods of counting large numbers of combinations, or combinatorics (see Chapter 3).

Table 7.1 shows the sampling distribution for 8 samples taken systematically in the ordered fashion we selected using the 2-transect method depicted in Figure 7.3(A). Specifically, we make a random selection of one of the 5 columns (or transects) on the left of the plot. Then we take a corresponding transect on the right side, so that the two selected are 4 transects (columns) apart. Thus the possibilities for transect pairs are (1, 6), (2, 7), (3, 8), (4, 9), or (5, 10). After the 2 transects are selected, we must place the 8 quadrats in some ordered way within them. These can be placed in rows: (1, 3, 5, 7), (2, 4, 6, 8), (3, 5, 7, 9), or (4, 6, 8, 10). This gives a total of 20 possible ordered samples of size 8. Because there are only 20 possible samples of size 8 ordered in this way, this set of 20 samples constitutes our sampling distribution.

Remember from Chapter 3 that a sampling distribution contains the means calculated from each of the possible samples forming an observed distribution. Here we have described one possible ordered method of selecting samples from a target population arranged in a 10-by-10 grid system of columns, or transects (Figure 7.3(A)). For this system, we can describe all possible samples that can be formed by using the ordered method. Table 7.1 presents the 20 possible means and standard deviations drawn from these 20 possible sets. This tally represents a sampling distribution.

We can then calculate a mean for the sampling distribution as a whole. That mean of the entries in the sampling distribution is 6.51, and the standard deviation, which is also the standard error (by definition), is 0.85. We can compare the standard error of 0.85 to the value of 1.13 we got for the estimate that used our systematically collected samples, but was based on the formulas for simple random sampling (srs). The smaller the standard error, the narrower will be the confidence interval. For srs, we obtained a 90% interval of $5.24 \le \mu \le 9.52$. Using the systematic sampling error (standard error) from our sampling distribution, the 90% interval is $7.38 \pm 1.89 \cdot 0.85$, or 7.38 ± 1.61, which gives the narrower confidence interval of $5.77 \le \mu \le 8.99$. The narrower confidence interval tells us that this estimate is more precise for the systematic sampling than that for the srs. This is true even though the mean density estimate for the systematic sampling was actually a poor one. That is, we know the true value is

TABLE 7.1   The Sampling Distribution for 8 Systematic Samples from 2 Transects

Possibilities	Total Trees (y_i)	Mean (μ_i)	Standard Deviation (σ_i)
1	42	5.250	2.053
2	60	7.500	1.414
3	37	4.625	1.506
4	60	7.500	1.604
5	43	5.375	1.923
6	50	6.250	1.832
7	46	5.750	2.252
8	51	6.375	1.996
9	57	7.125	3.061
10	56	7.000	2.507
11	59	7.375	3.335
12	58	7.250	2.252
13	53	6.625	1.996
14	61	7.625	2.200
15	55	6.875	2.167
16	60	7.500	2.204
17	50	6.250	1.581
18	47	5.875	2.748
19	46	5.750	1.581
20	50	6.250	2.659

NOTE: The 20 possible sets of samples begin with the possibility of a random start from any of the first 5 transects on the left side of Bolivian plot 01, with $N = 100$. Total trees in all 20 samples combined is 1,041. $\frac{1.041}{20} = 52.05$ for the average number of trees per sample. $\sum_{i=1}^{20} \hat{\mu}_i = 6.51$ is the mean density of trees based on 20 samples of $n = 8$ each. $\frac{\sum_{i=1}^{20} (y_i - \mu_i)^2}{19} = 0.85$ is the standard deviation of 20 possible samples, or the standard deviation of the sampling distribution, which is called the standard error.

6.63, and we got an estimate of 7.38 in the previous section using Equation 2.2. Thus, if we use srs calculations for systematic samples, our interval may be less precise than if we use formulas specifically for systematic samples. Let us consider how we can obtain such formulas.

The Mean of a Systematic Sample

For the simple random sampling exercise we undertook (depicted in Figure 7.3(B)), the mean of the sampling distribution (that is, the mean of all possible random samples of size 8) came out as 6.63, which equals the true mean density, and tells us that the arithmetic mean is an unbiased estimator (Chapters 2 and 3). Note that the sampling distribution derived from our systematic samples yielded a mean of 6.51 (Table 7.1). This estimate is not exactly, but is nonetheless very close to, the true mean that we know to be 6.63. For systematic sampling, the rule is that the arithmetic mean will be an unbiased estimate only if the sample is spread over the entire population in equal increments. Also, the ratio of the number of possible quadrats to the sample size $\left(\dfrac{N}{n} \right)$ must be an integer in order for the resultant mean to be unbiased.

Generally, the closer we come to adhering to these two conditions, the closer we come to having an unbiased estimate. In the $n = 8$ case, the probability of selection over the plot is not equal everywhere. For example, we did not give the transects that were more or less than 4 columns apart a chance to be included in the sample (for example, columns 1 and 2 could not form a selected pair). Also, $\dfrac{100}{8}\left(\dfrac{N}{n} \right)$ is not an integer.

Regardless of the size of the sample, when we use systematic sampling the mean will usually be biased. However, this bias may not have serious consequences.

Although many sampling distributions will be as limited in size as that in our example, many (probably most) for natural populations are not small. Consider, for example, samples taken by a marine benthic specialist. It is common to find that the stations along a traverse may be as much as a kilometer distant from one another. Along this path, the ship stops, and a small sample of the bottom sediment is retrieved. To describe the sampling distribution for the set of possible systematic samples would require a countably infinite set of values, or at least a very, very large number of possibilities. Thus, when the relationship is one of a minuscule sampling area to a vast target area, the possible values for the sampling distributions of a set of systematic versus a set of random samples is many orders of magnitude beyond that in our example. The bias in the systematic mean is nonexistent in such cases, and the srs formulas can be used for the systematic calculations without problems. We would have no difficulty in assuming that the sampling distribution is Normal, unlike that for our Bolivian forest exercise, in which the number of possible sample means (the sampling distribution) is small, even when the sample size is large. Consequently, we can safely advise that there would be no notable reason in situations characterized by very large target populations to decline to use systematic sampling.

The Variance of a Systematic Sample

In the previous section we applied srs formulas to our systematic sample and compared the sampling distributions from the two types of sampling schemes: simple random and systematic. We did not calculate a variance estimate directly by a formula specifically for systematic samples. The reason is that this is not possible when we have only 1 systematic sample; that is, when we took only a single 2-transect (8-quadrat) sample of the $N = 100$ set.

For srs we know from Equation 3.4 that $\sigma_\mu^2 = \left[\dfrac{(N - n)}{(N - 1)} \right] \dfrac{\sigma^2}{n}$. We also have the advantageous result that we can estimate σ^2 (defined by Equation 2.3) with $\hat{\sigma}^2$ (defined in

Equation 2.5) from the observations in just 1 statistical sample. With this srs, we can estimate σ_μ^2 using Equation 3.6. Here, $\hat{\sigma}_\mu^2 = \left[\dfrac{(N-n)}{(N-1)}\right]\dfrac{\hat{\sigma}^2}{n}$. This exercise tells us that all we need to know is the variance calculated from just 1 random sample in order to be able to estimate the standard error (standard deviation of the sampling distribution) of the mean density when srs is used.

The problem with taking observations in an ordered (or systematic) way, and not in a random manner, is that the same result does not hold. That is, regardless of how large the statistical sample (n, number of quadrats, or biological samples) we have no systematic sampling formula to use to calculate an estimate of the variance that is generally useful for natural population sampling. The lack of a reasonable variance estimate is the most serious disadvantage of taking just a single systematic sample from a natural population. Without such an estimate, as we saw in Chapter 2, we have no way to judge how close our mean density's observed value is to the true value we are trying to estimate. We solve this problem by a simple conceptual change, discussed next.

REPLICATED SYSTEMATIC SAMPLING

Let us consider the 8 ordered samples in Figure 7.3(A) to represent not 1, but 2 systematic samples from the target population, each of size 4. That is, the ship does not have to go to sea again; the biologist does not have to finance a second trip. Merely thinking of the sample of 8 as 2 samples of 4 will suffice. This can be called *replicated systematic sampling*, or *replicated ordered sampling*.

This conceptual change allows us to calculate a *consistent* (as n increases, estimator and parameter converge in probability; they get closer with p = 1.0) *variance estimate* because we need more than one ordered set of observations to get such an estimate, and now we have it. A replicated systematic sampling has the added advantage of providing a more even coverage of the area than would most random samples of the same size. With this uncomplicated alteration we can now obtain an unbiased and consistent estimate of the sampling variance because we have replication. In order to be sure of this unbiased result, we really should have 2 random starts, one in each of the 2 transects, rather than the single random start imposed in both for our nonreplicated example. However, it is unlikely that this option will actually be under the control of the fieldworker in many cases, and any representative, selected starting point will suffice. Clearly then, an ordered sampling approach can be used with confidence, albeit still with caution. Many statisticians do not sanction a systematic sampling approach because of the uncertainty about possible trends or periodicities in the population data. We recommend it here, however, because as long as we are fully aware of its dangers, we can capitalize on the simplicity and thus work toward an increase, or at least not a decrease, in precision.

In practice, there are two situations, and two accompanying sets of formulas, of importance for systematic sample surveys of natural populations. First, if the target population is homogeneously (or randomly) distributed over the target area, or the time of interest, we can treat the result of a systematic sampling as if it were random.

The estimators to use are those for srs (Equation 2.2 for $\hat{\mu}$ and Equation 2.5 for $\hat{\sigma}^2$). These are also the estimators to use when the samples are minuscule in comparison with a very large target population, like cores along an ocean transect, or bacterial samples from a river. The difference is merely that the x_i observations are ordered and not random. Then we use Equation 3.6 or 3.7 to calculate $\hat{\sigma}_\mu^2$. Next we obtain confidence intervals by the methods presented in Chapter 4. However, this standard error estimate will be too big or too small (and we won't know which) unless the population is distributed homogeneously.

Second, for populations that are distributed sparsely and heterogeneously, unevenly, periodically, or along some gradient, more than one set of ordered observations must be taken. This is called *replicated ordered sampling*. If such a sample design is used, the estimate of mean density, based on the total sample of n observations, is obtained by adding together the separate observed means from each of the 2 or more ordered statistical samples and dividing by the numbers of such statistical samples, which takes the symbol m. Note that Figure 7.3(A), for example, depicts an m of 2; that is, 2 transects, if the second transect is treated as a replicate of the first. For m separate mean density estimates $\hat{\mu}_i$ obtained from the m replicates, we take

$$\hat{\mu}_{syst} = \sum_{i=1}^{m} \frac{\mu_i}{m} \tag{7.1}$$

and

$$\hat{\sigma}_\mu^2 = \frac{\sum_{i=1}^{m}(\hat{\mu}_i - \hat{\mu}_{syst})^2}{m(m-1)} \cdot \frac{(N-n)}{(N-1)}. \tag{7.2}$$

Equation 7.1 is an unweighted average of the separate, systematic sample means because we are replicating, or repeating, the same-size statistical sample. With Equation 7.2 the variation between the sample means is used to estimate the sampling variability of their means. Thus for replicated systematic sampling, the variance and the standard error can be estimated right from the observed data. We can then use this value (Equation 7.2) for calculating confidence intervals by inserting it into Equation 4.7.

Returning to our Bolivian example (Figure 7.3(A)), where we have 2 ordered samples of size $n = 4$, we can calculate the mean density estimate and its standard error by Equations 7.1 and 7.2.

For $i = 1, 2$, we calculate the estimated mean for each as follows:

$$\hat{\mu}_1 = \frac{(4+9+5+7)}{4} = 6.25$$

$$\hat{\mu}_2 = \frac{(4+6+13+11)}{4} = 8.50$$

so that

$$\hat{\mu}_{syst} = \frac{(6.25+8.50)}{2} = 7.375.$$

Then,

$$\hat{\sigma}_\mu^2 = (6.25 - 7.375)^2 + (8.50 - 7.375)^2 \cdot \frac{\dfrac{(100-8)}{99}}{(2 \cdot 1)}$$

$$= (1.266 + 1.266) \cdot \frac{92}{198} = 1.176.$$

Of course, as we have discussed, this example is rather unrealistic because of the plot's known homogeneity; the random sampling formulas would really be an appropriate choice for study plot 01. However, we only know this because we have the entire population. This omniscient knowledge is, of course, not usual for natural population surveys. Thus, unless we know with certainty that we have a homogeneous spatial distribution, we take a chance if we use the srs formulas when systematic sampling. There is only one exception in which we run no risk and that is if the target is large, and we are taking a very small number of samples.

TARGET POPULATION, SPATIAL DISTRIBUTION, AND SYSTEMATIC SAMPLING

The second question that we posed concerned possible characteristics of the target population suggestive of the need for, or advantage of, systematic over random samples. Examination of the properties of the estimators for the systematic method of sampling can lead us to a deduction related to this second question. That is, unless the target population is homogeneously distributed throughout the area of interest, the variability of a mean density estimate based on a systematic sample may differ considerably and inconsistently from the same estimate based on a random sample (of the same size). With patchy or sparse species distributions, the selection of systematic samples may not be useful, and the sampling distributions from ordered versus random samples may differ markedly.

Sampling over time, for example, can be a problem. Systematic sampling every hour over 24 hours would not be efficient or useful; for example, if the organisms swarm or become active, only at a certain time of day. Alternatively, a systematic sample every 24 hours would be very effective for this situation if the period of sampling coincided with activity. Only when the researcher has some prior information about a target population will it be useful to consider a systematic sampling scheme.

SAMPLE SELECTION STRATEGIES: RANDOM VERSUS SYSTEMATIC

Let us outline some strategies, based upon both mathematical statistical results and knowledge of sampling in natural populations, for the use of systematic samples instead of srs for natural population sampling plans.

1. If we select the first sample, quadrat, transect, or core randomly (a random start), the mean of a systematic sample is an unbiased, or nearly unbiased, estimate of the population density. A true random start is unusual except,

possibly, when sampling some terrestrial animals. If we take a very small number of biological samples, however, from a large target area, this bias is of no concern.

2. Whether the size of the sampling variance (the standard error) of a systematic sample is bigger or smaller than the size of the sampling variance from an srs depends upon the spatial distribution of the organisms in the target population. Because the spatial distribution may not be known prior to the survey, extreme care must be taken when using a systematic sampling scheme. If the spatial distribution is homogeneous, systematic estimators can be more precise than those from random sampling. If the spatial distribution is sparse or patchy, it may not be possible to tell whether the estimators from systematic sampling will be more, or less, precise. If the organisms are spread out along a gradient, or cline, systematic sampling along that gradient or cline is a wise choice.

3. For systematic and random samples of the same size,

   a. The use of a systematic strategy will not be disadvantageous in terms of variance reduction (precision of the density estimate) if the population being sampled is truly homogeneous (or random). As we saw from the Bolivian example, the two types (random and systematic) of sample selection methods are of essentially equivalent precision under conditions of a spatially homogeneous distribution. Statisticians usually say that "on average" this equivalency is true. Thus, although the systematic sample variance may not equal that for random sampling for any single, finite population and sample size, in general, the two are equivalent and can be used interchangeably under conditions of homogeneous or random distributions of the target population.

   b. The sampling variance of the mean density estimate will be smaller; that is, more precise or "better," for a systematic sample if the target population has a consistent trend along a gradient (for example, from shallow to deep water, or from lower to higher elevation) and if sampling is spread along the gradient. For example, when density increases with nearness to water, or abundance has some periodic relation with time, and samples are taken in equivalent periods, then the systematic sampling variance, $\sigma^2_{\mu-syst}$, will be smaller than the sampling variance from a random scheme $\sigma^2_{\mu-srs}$. In this case the mean density estimated with systematic samples can be expected to be more precise than if srs were to be used.

   c. If we take a single, systematic sample of any size, no matter how large, and use the variance estimator for srs $\left(\hat{\sigma}^2_\mu = \dfrac{\hat{\sigma}^2}{n} \right)$, we will usually seriously overestimate the true variance because most natural populations are not homogeneously distributed. This is true unless the target area is very large in relation to the size or number of samples.

   d. Finally, with one, single, systematic, statistical sample (our example had $n = 8$) with a random (or representative) start, we cannot get an estimate of the variance that is unbiased, or *consistent* (getting closer to a related parameter; that is, *converging in probability* as the sample size gets larger). The amount of bias in the estimate from the observed sample will be bigger or smaller than that of 1 sample collected with an srs scheme, and this difference will not be estimable from the survey results. Thus, we should never use an unreplicated systematic sampling method for natural population work.

The use of systematic, or ordered, sampling is ingrained in the protocol of field biologists who survey natural populations. This is a dependence based upon necessity and cost as well as convenience. Therefore, it would be of benefit to be able to sample in an ordered fashion without the problems just described. We can accomplish this by replicated systematic sampling methods.

Stratified Sampling

The Bolivian forest plot used as the routine example in this chapter and throughout this book is small in area as well as homogeneous. In a survey of a much larger area, or for an expanded period, we would probably not expect such homogeneity over the entire study. In natural population sampling the researcher may be sampling along some gradient with an accompanying large change in the distribution of the organisms under study.

For example, many marine organisms, especially invertebrates, exhibit a distinct zonation with depth. A traverse beginning near shore might typically encounter a very high density for a particular benthic species and a decreasing density of the same species with depth, while another species will exhibit the opposite pattern. The population density of many terrestrial and arboreal species may vary with elevation, time of day, month, or stage in the breeding cycle. Stations along the traverse may be quite close or a kilometer or more apart. Often the researcher will have only a vague idea of what distribution of species will be encountered or what distributional differences will exist along the transect. The sampling device (quadrat, core, net, transect, or other) used throughout the sampling will most often be fixed and of a standard size and construction. Usually, the researcher is interested in a complex of species rather than a single species. Consequently, this ordered sampling of natural populations along a gradient uses procedures that are quite different in precision from that of the equidistant samples laid out for the rather homogeneous target population of Bolivian trees.

When there is doubt about the constancy of target populations distributed within the time or space to be covered by the survey, it is usual to select samples within each zone, block, biofacies, grouping, or section. Alternatively, if we know that the spatial distributions do indeed differ in density over the target area or with time, then each such different grouping can be treated as a separate population for sampling purposes. When this occurs, we can obtain a much more precise estimate of the population mean (and total) density by explicitly recognizing the *stratification* (division of one target population into parts based on some known characteristic or variable) in our sample design.

Statisticians define this sampling scheme as *stratified sampling*. Occasionally, in the botanical literature, this type of sampling is called *representative sampling*. An advantage of this design is that the *strata* (or microhabitats, groups, or blocks) can be selected to cover important biofacies or habitats of interest (also called *domains of study* in the statistical sampling literature), for which separate sample information is needed. The advantages of stratified sampling are nearly always considerable, except when the population is known to be extremely homogeneous. The reduction of variance and the narrowing of the resultant confidence interval about the true mean density value are important attributes of this method.

Once again, we stumble over the problem of the same words being used with quite different meanings in different disciplines. For a geologist or stratigrapher, the words *stratum* and *stratification* have an altogether different and distinct connotation. This can be particularly confusing because the stratum of the geologist is often a good unit upon which to base a statistically stratified sample. Nevertheless, we shall continue using the term *stratified sampling* as defined here because it has become an accepted part of the literature for non-geologic field sampling.

Up to this point in this book, we have discussed methods that involve taking a sample from the entire population of interest. Stratified sampling takes samples instead from *subpopulations* of the total target population. It requires prior knowledge of a limited (usually 2 to 5) number of zones, habitats, sections, subareas, subgroups, or biofacies (or even sexes, races, age classes, and groups defined by behavioral, morphometric, or descriptive variables) within the target population. Successful use of stratified sampling also depends on accurate identification of these strata, each of which is treated as a separate, independent population. The strata can be based on prior determination of distributional differences, or can be determined by means of supplementary information on the organisms. Even stratification for convenience, based on, for example, a state or local boundary or a height above ground or sea level, can be useful.

In many cases there is not merely one purpose for the field trip, nor only 1 species to be collected. Thus there may be no optimal decision among possible variables by which to stratify. However, a stratified design still can be more advantageous than srs.

An intuitive notion that the organisms are distributed preferentially along a gradient is enough of a basis upon which to select a stratified sampling scheme. However, a decision is needed about where and how to separate the different areas along that gradient. The boundaries of each area are called *cutpoints*. Cutpoints must define nonoverlapping, distinct, and exhaustive groups or areas for study. This does not mean that the organisms of interest must be distributed spatially into distinct habitats or areas, only that the cutpoints can be clearly defined for statistical purposes. For example, using 2 microhabitats distinguished by "on the ground" and "in the leaf litter" may allow for confusion. In contrast, using "1,000 to 2,000 meters above sea level" versus "2,001 to 3,000 meters above sea level" is clear, especially for the inexperienced.

A separate sampling is performed independently within each *stratum* by using any of a number of different sample designs. In this way, most of the variation in the population should be between (or among) the strata, rather than within any single stratum. Essentially, this *among-strata variation* is eliminated by the sampling because samples are taken from each of the strata separately. The variance of any estimate of mean density based on stratification thus will not incorporate the among-strata variability. This says that the variance of a sample of size n, spread over strata, will most usually be considerably reduced from the variance of a density estimate obtained from n randomly selected observations.

COMPARISON OF SYSTEMATIC AND STRATIFIED SAMPLING

The description of stratified sampling invites comparison with the systematic sampling scheme discussed earlier in this chapter. We shall examine the similarities to help lead us in a decision process for selecting the most appropriate sampling method for a particular study. For this comparison, Figure 7.1 will be a central focus.

Systematic Samples as Implicit Strata

Figure 7.1, column B, illustrates that an ordered sampling method for selecting a sample of size 4 ($n = 4$) actually divides the target population of 100 quadrats into 4 implicit strata, or groups, of size 25 $\left(= \dfrac{100}{4} \right)$ each. The number of strata is given the symbol k. Beginning with a random start point (selecting a random number between $k = 1$ and $k = 25$) between quadrats #1 and #25, each group consists of the same number of possible samples; that is, 25, from which we select only 1. In our example, quadrat #20 is between quadrats 1 and 25; quadrat #45 is in the interval of quadrats (26, 50); #70 is in (51, 75), and #95 is in (76, 100). Each of the 4 statistical observations, or biological samples, is equidistant; in this case, 25 quadrats apart. Thus the mechanics of systematic sampling can be viewed in Figure 7.1, column B, as equivalent to stratified sampling with a sample size of 1 ($n = 1$) selected per stratum.

Systematic Sampling for Stratum Detection

In the example just given, each stratum has the same 1/4 probability of being selected, and we choose a single biological sample spread evenly over the target area from each stratum. From this, we see that another advantage of systematic sampling is that this design can sometimes uncover hidden strata or biological population changes in the sample area. In systematic sampling, however, as we have seen, if we increase or take larger and larger samples, the variance does not necessarily decrease, unless we have information on the distribution of the organisms; for example, that they are distributed on a gradient. If systematic samples uncover implicit strata, then there will be a variance decrease if subsequent samples are taken within each stratum. All these points illustrate, once again, that caution must be used when the sample design calls for ordered, or systematic, sample selection.

Use of Supplemental Species Data from the Target Population

Supplementary information (environmental, behavioral, sexual, familial, and so on) on the target population can be used in two ways. The first is to form a basis upon which a systematic sample can be selected. In this method there is the inflexibility of sampling from the entire, often large, target area. Often, independent information is still needed on each of the target groups. For example, we can systematically take cores along a transect that runs from shallow to deep water, and our interest will then be in the density with depth differences. Such information is unattainable with a single, statistical, even large-sized sample, as well as with systematic sampling, because we cannot traverse the depth and measure variation at each depth unless we stratify.

The second use for supplemental data on individuals in the target population is to allow us to decide how best to apportion the target population into distinct strata. For example, we would decide to take samples from shallow water and then additional samples from deep water if we knew that a species was found in both, but we suspected varying density. Stratification would thus allow for independent information to be obtained from each depth. With systematic sampling we would be limited to equal sampling across all implicit or explicit strata, but with stratification we can obtain samples of larger or of unequal sizes, and thereby increase the precision. Stratified sampling is of particular value, of course, when the primary purpose of the survey is to compare the differing strata (biofacies, habitats, etc.).

The usual application of stratified sampling involves taking at least 2 samples, either in a random or an ordered way, from within each stratum. In this way we obtain at least 1 replicate sample within each stratum. Therefore, we would expect stratification to provide a more precise mean density estimate for the total population than would systematic sampling, which has only 1 sample per *stratum*. However, this expectation is justified only for certain distributions or spreads of the target population. Let us examine some of the situations in which stratified sampling is clearly superior to systematic sampling.

Sampling Within Each Stratum

If we use srs to select the samples from within each stratum, the design is called *stratified random sampling*. This type of stratified sample, pictured in column C of Figure 7.1 for the example of 2 strata ($k = 2$), is collected the same way as an srs, except that the choice of sites for the fieldwork is made separately and independently in each stratum; not from the entire target area or population. Alternatively, if ordered sampling is used within a stratified framework, this design is called *stratified systematic sampling*. It is pictured in column D of Figure 7.1, again for 2 strata.

Note that the placement of samples in columns B and D of Figure 7.1 is exactly the same. However, in column D we have identified 2 strata, and so they are each treated separately in the calculations for obtaining the mean density, the variance, and the precision before results are obtained for overall estimates. For example, if a biologist identified 2 habitats; say, under rocks and in leaf litter, for the target amphibian species *Plethodon cinereus*, there would be 2 strata in the forest target area. Or, if a marine researcher identified 2 habitats by use of contrasting sediments, there would likewise be 2 strata within the marine target environment. Each habitat or microhabitat, then, could be randomly sampled, or counts could be obtained by taking ordered samples within each habitat to produce the desired observations.

Yields from differing sections of the same stratum tend to be more alike than yields from different strata, regardless of the target taxon. That is, after all, the whole point of choosing a stratified approach and for deciding how to apportion the boundaries between strata. For example, if we stratify into breeding pond and leaf litter, we would expect there to be a very similar high density to be found in repeated samples from the breeding pond on a summer evening, as well as a similarity among the leaf litter samples at the same time. We would not expect a similarity in the counts obtained between these 2 microhabitats. This within-stratum resemblance will occur if time is taken to ascertain that the strata are well-defined, mutually exclusive, and appropriate with respect to behavior, unified habitats, environment, or other characteristics. The precision of the density estimator increases as the variance within the stratum decreases, which, in turn, will ensure that the variance between strata will increase.

The variance of the density estimate, based upon a stratified sampling design, depends on only the *within-strata variance* (that is, each separate variance in each stratum). The overall variance does not depend on, or involve, any heterogeneity that is between or among the strata. Therefore, we want to know whether simple random or systematic sampling, performed within each of the strata, will give the smallest within-stratum variance. The answer is not straightforward. Whether stratified random or stratified ordered sampling will prove to be more precise depends on which of the two methods is the more precise within each stratum. Because each stratum's sampling is performed independently, the conditions for precision (that is, homogeneity or heterogeneity and size in the target population) are the same as those discussed under the section on systematic sampling of a single, ungrouped target population.

STRATIFICATION AND THE SIZE OF THE TARGET POPULATION

Many biologists recognize the value of stratification for sampling work. However, a full understanding of the construction and use of the estimates derived from a stratified sampling project is not as widespread. When, for example, samples are taken within each habitat or biofacies, a density or other estimate for each separate stratum may (incorrectly) be treated as if it were unbiased. Then, by (incorrectly) making use of a theorem in mathematical statistics that (correctly) states that the sum of unbiased estimates is itself also unbiased, the biologist obtains a population estimate by adding the individual density estimates and averaging over all strata, habitats, or biofacies. The problem with this approach is that the stratum estimates are not always unbiased for natural population sampling.

In order to obtain unbiased estimates from a stratified sampling scheme for finite populations, we must know both N, the total population size of either the individuals or quadrats (cores, transects, or biological samples), and N_i, the total population size from each stratum. Three possible scenarios cover situations for most natural population sampling when stratification is a possibly useful sampling scheme. In each, we shall examine the use of stratification and the properties of the statistical estimates.

The most common situation faced by field researchers is that the sampling device can sample only a relatively minuscule area compared to the target area. Does stratified sampling make sense under such conditions? Here scale becomes very important. If the target population is the Bolivian forest plot 01 we have been using, and if we stratify the $N = 25$ quadrats, say, then we are not sampling a small area from a large target but the entire target population. If, however, the 25 quadrats are sampled from plot 01, but the target population actually is a much greater amount of the surrounding forest, perhaps many hundreds of km², then the quadrats become a very small proportion of the vast target area. This situation usually applies with small organisms, such as the invertebrates and protists found in marine sediments or soil, but it could just as well pertain to trees if the area of interest is an expansive forest. In such cases, regardless of the numbers of organisms obtained, the values of N (total sample size) and N_i (total sample size per stratum) need not be known or estimated, and all estimates will be both consistent and unbiased. When values of N become irrelevant to the work, this is a direct consequence of the scale factor on the formulas. In such cases the target area; for example, the ocean transect or forest from which the samples would be selected is so large that it contains a nearly infinite number of possible biological samples. Overall, if we structure sampling work in a stratified way when the scale of the target population is huge compared to the area sampled; that is, when N and N_i cannot be known we can obtain consistent and unbiased estimates.

A second situation often encountered in the field is that of a relatively small target population, for which we plan to use quadrat sampling. Such conditions quite commonly pertain to studies of larger organisms that live in a relatively small area. Consider, for example, fish in a pond. The total number of possible quadrats (N) can indeed be estimated from a map or an areal photo, as they cover an area of finite size. This is the same situation we would have if we stratified the $N = 25$ quadrats from the Bolivian plot 01, and regarded this 1-hectare plot as our entire target population. In such situations, we can readily figure out or estimate both N and the number (N_i) of possible quadrats to cover each stratum. These values of N and N_i must be known for the target area before beginning the sampling of each stratum if we wish to form a total population estimate. Without this information we cannot obtain unbiased estimates from the data collected during stratified sampling, and we cannot make decisions on sample allocation. Thus, when sampling for density, it is possible to get unbiased strata, as well as total estimates, provided that the target area is finite and that the number of quadrats within each stratum bears some real relationship to the total number of quadrats into which the target population as a whole is divided.

Finally, the third situation that may confront a researcher conducting field sampling of natural population is when the primary sampling unit is the individual organism, not the quadrat. (This situation will be discussed more fully in Chapter 8.) In such cases, the quantities we would need for a stratified study are N, the total number of organisms in the target population, and N_i the total count of such organisms in each of the strata. Usually, these counts will be unknown and not estimable. Thus, the statistical estimates would be biased. Overall, a situation in which the primary sampling unit is the individual organism would not seem to be amenable to stratified sampling.

But there are ways to finesse things. *Double sampling* is one way to deal with the problem of coming up with an estimate of N and N_i to use when the sample is not a quadrat or transect or core but the individual organism in a finite population. Double sampling is a method by which the target populations of size N, and the strata sizes,

N_i are first estimated, and then sampled. Double sampling is used when these population values are unknown but stratification is suspected to be an efficient method for reducing the variance. Basically, one first samples to produce a large pilot study, from which N and N_i can be estimated. Then a second, smaller sampling from the initial larger sample is undertaken and used as the data for calculating density and variance. Double sampling is usually not a practical approach for fieldwork with most of the world's taxa because the first step is to gather a large sample from each stratum by which to estimate the unknown population sizes. After these estimates are obtained, a smaller random sample is selected from the first sample in order to estimate the parameters of interest. Double sampling has advantages, but there are two reasons that make this methodology impractical. The most obvious reason is that the likelihood is quite low of obtaining and identifying a large sample of organisms in a cost-effective manner for a majority of taxa. The second reason is that the primary sampling unit in surveys of most natural populations is usually not the organism but the sampling device. Luckily, sampling with a device or by quadrat is the other alternative to being able to get estimates of N and N_i.

Thus, we have three possible field situations. The first occurs when the target area or population is effectively infinite, or large enough to overwhelm efforts to sample any reasonable portion of it. In this case, there is no need to have values for N or N_i sizes, and stratification can prove useful. The second and third situations occur when the target area or population is finite in size. Then, the primary sampling unit can be either a quadrat (biological sample) or the individual organism. To summarize these two different field conditions, stratified sampling makes sense for most taxa only in the second situation. It does not work effectively when the primary sampling unit is the individual organism. Double sampling can correct for some of the problems of the latter case, but it is often impractical for microscopic or very small organisms of even moderate complexity.

EXAMPLE OF STRATIFICATION

Let us pretend that we have reason to believe there are 2 soil-type categories, or strata, in the Bolivian example (Appendix 2), and that the target population is divided into just 2 subareas that each contain 3 rather large quadrats. In this conceptual organization, let us select these 2 sets, or 2 strata: quadrat numbers 1, 6, and 11 will form stratum #1, and quadrat numbers 2, 3, and 4 will form stratum #2. These two groups, shown in Table 7.2, will become our target population, with $N = 6$. Notice the use of a capital N

TABLE 7.2   Mean and Variance Calculations for a Total Population of $N = 6$, and Its Division into 2 Strata

	$N = 6$	$N_1 = 3$	$N_2 = 3$
Quadrats	(30, 30, 31, 24, 24, 20)	(30, 30, 31)	(24, 24, 20)
Mean density	$\mu = 26.5$	$\mu_{str1} = 30.33$	$\mu_{str2} = 22.67$
Variance	$\sigma^2 = 16.6$	$\sigma^2_{str1} = 0.2176$	$\sigma^2_{str2} = 7.1289$

means that $N = 6$ is the entire target population. This example, of course, is unrealistically small, but the arithmetic will be clear and it will provide for a manageable teaching tool.

For this example, stratum #1 contains $N_{str1} = 3$ (with counts of 30, 30, 31), and stratum #2 contains $N_{str2} = 3$ (with counts of 20, 24, 24), where the subscripts indicate the stratum number. These 6 quadrats, considered as an entire population, yield a mean density of $\mu = \dfrac{159}{6} = 26.50$ (from Equation 2.1) and a variance $\sigma^2 = 16.60$ (Equation 2.3). For this stratified population, using the formulas for population mean (Equation 2.1) and variance (Equation 2.3), we obtain $\mu_{str1} = 30.33$ (based on a total count of 91 trees) and $\sigma^2_{str1} = 0.22$. For stratum #2 we calculate $\mu_{str2} = 22.67$ (based on a total count of 68 trees) and $\sigma^2_{str2} = 7.13$.

Let us now take a sample of size 4 from the total target population of 6 quadrats. Table 7.3 lists the samples selected. First, we will pick 4 quadrats at random from the 6 considered as a single target population. For srs, we randomly select 4 quadrats from the 6, without regard for the subdivisions into strata. The 4 we selected randomly were quadrats #2, #1, #6, and #11. Taking these 4, we obtain counts of 20, 30, 30, and 31 individuals, respectively. The sample mean from our srs is thus $\hat{\mu}_{srs} = \dfrac{111}{4} = 27.75$ (from Equation 2.2), which we know is unbiased (but not equal to μ for this particular sample). This random sample has a variance estimate of $\hat{\sigma}_{srs} = 26.92$ from Equation 2.5, and this gives a standard error estimate of $\hat{\sigma}^2_{\mu} = \dfrac{26.92}{4} = 6.73$.

Now, for comparison, we want an srs whose total size is 4, so we select $n = 2$ from each of the 2 strata or subpopulations. For this srs with equal sample sizes of $n_1 = n_2 = 2$ (to keep a total size of 4), we randomly select quadrats 1 and 6 for the first strata and quadrats 2 and 4 for the second. Then we obtain counts of 30 and 30 from stratum #1 to give $\hat{\mu}_{str1} = \dfrac{(30 + 30)}{2} = 30.00$, $\hat{\sigma}^2_{str1} = 0$. We also obtain counts of 20 and 24 from

TABLE 7.3   Mean and Variance of Simple Random Sample (srs)

	SRS	STRATA		TOTAL STRATIFIED SAMPLE
	$n = 4$	$n_1 = 2$	$n_2 = 2$	$n = 4$
Quadrats	(20, 31, 30, 30)	(30, 30)	(20, 24)	
Mean density	$\hat{\mu}_{srs} = 27.75$	$\hat{\mu}_{str1} = 30$	$\hat{\mu}_{str2} = 22$	$\hat{\mu}_{str} = 26.00$
Variance	$\hat{\sigma}^2_{srs} = 26.92$	$\hat{\sigma}^2_{str1} = 0$	$\hat{\sigma}^2_{str2} = 8$	$\hat{\sigma}^2_{str} = \dfrac{1}{2} = 0.50$

NOTE: srs is $n = 4$ from $N = 6$ in Table 7.2 and for strata 1 and 2.

stratum #2 to give $\hat{\mu}_{str2} = \dfrac{(20 + 24)}{2} = 22.00$, $\hat{\sigma}^2_{str2} = 8.00$. We used srs separately in each stratum, so the estimators are merely those from Chapters 2 and 3. Then, the formula for the mean of the sample of size 4 is obtained by combining the individual means from each stratum. This formula is

$$\hat{\mu}_{str} = \frac{1}{N} \Sigma^N_{i=1} N_i \hat{\mu}_{str_i}. \tag{7.3}$$

Remember, we don't need limits on the summation sign when it is clear what we are adding. For example, here we must be adding over all the terms from $i = 1$ to $i = N$ because it is the only subscript used in the formula. Notice in this formula also that multiplying by $\left(\dfrac{1}{N}\right)$ is equivalent to dividing by N, so that Equation 7.3 is an average. In this formula, the symbol on the left side that we are solving for represents the mean per unit for the complete stratified sample. The right side shows that this total estimate is obtained by multiplying the number of units available in each stratum by the observed mean from that stratum, and dividing by the number of all possible quadrats; that is, the total statistical sample size, N.

If the biological samples are taken independently within each stratum (with no requirement that sampling be srs within each stratum), the overall variance of $\hat{\mu}_{str}$ (as an estimate of the total population mean), is given by

$$\hat{\sigma}^2_{str} = \frac{1}{N^2} \Sigma^N_{i=1} N_i^2 \hat{\sigma}^2_{str_i}. \tag{7.4}$$

Equation 7.4 shows that this variance estimate depends only on the individual stratum variances. This means that only the variance of each stratum changes; N and N_i are constants for the sample. Thus, if we perform srs in each stratum, then each stratum mean and variance estimate is unbiased and consistent, and the overall estimates will also have these properties. Therefore, for stratified random sampling, Equation 7.4 becomes

$$\hat{\sigma}^2_{str} = \frac{1}{N^2} \Sigma^N_{i=1} \left[\frac{(N_i^2 \hat{\sigma}^2_{str_i})}{n_i} \frac{(N_i - n_i)}{N_i} \right]. \tag{7.5}$$

However, if our sample size is very small compared to the known or suspected size of the target population (*sampling fraction*), this formula simplifies even further to

$$\hat{\sigma}^2_{str} = \frac{1}{N^2} \Sigma^N_{i=1} \left[\frac{(N_i^2 \hat{\sigma}^2_{str_i})}{n_i} \right]. \tag{7.6}$$

For our example, in which we have 2 strata, the unbiased estimate of mean density is the properly weighted combination of the stratum means,

$$\hat{\mu}_{str} = \frac{1}{6} \cdot [3(22) + 3(30)] = \frac{1}{6} \cdot 156 = 26. \text{ Then, } \hat{\sigma}^2_{str} = \frac{1}{6^2} \cdot \left[3(3-2) \cdot \frac{\hat{\sigma}^2_{str1}}{2} + 3(3-2) \cdot \frac{\hat{\sigma}^2_{str2}}{2} \right] =$$

$\frac{1}{36} \cdot 18 = \frac{1}{2}$ is also a weighted combination of the stratum estimates. Table 7.2 summarizes the results of the mean density and the variance estimates for each of the 2 strata and for the target population as a whole.

For this very small population (just an $N = 6$ portion of the Bolivian hectare), stratification has been quite effective in reducing the variance from 26.92 to $1/2 = 0.50$. One reason stratification has helped is that the quadrat counts within each stratum were quite similar, but quite different from those in the other stratum.

Intuitively, we can understand another reason for this useful decrease in the variance. Taking an srs of size 4 from this population, we have $_6C_4 = 15$ entries in the sampling distribution of the mean, whereas for the stratified sample of size 4 with 2 from each stratum, we have $_3C_2 \cdot _3C_2 = 3 \cdot 3 = 9$ entries. For a more realistic population size and sample size combination, the sampling distribution for the stratified sample will usually contain considerably fewer values than that for srs and will result in a lower standard error than srs. However, there must be reason to believe that subdividing the population has biological importance before we can use stratification effectively. Thus, we increase the precision of our estimate of mean density by decreasing its variance when we stratify the population.

CAUTIONS FOR STRATIFICATION

Two cautionary points are important to consider when deciding whether to use stratified sampling. The stratum means may turn out to be biased estimates of the subpopulation means from each stratum. That is, $\hat{\mu}_{str1}$ is not always an unbiased estimator for μ_{str1}. The bias disappears, however, if the sample size selected in each stratum is the same proportion of the stratum's target population. The bias also disappears if we are dealing with an infinite or extremely large target population. The disappearance of bias when dealing with large target populations has to do with the weights that were used to construct the estimates. That is, if $\frac{n_i}{N_i}$ is the same for all i strata, then the estimates are unbiased. The ecological importance of this rule is that we cannot always expect to be able to estimate mean density of each habitat in the most efficient way if we also want to estimate density for the entire population, unless we are dealing with an infinite or countably infinite situation (for example, sediment cores or oceanographic samples).

The second cautionary point can be gleaned by taking another look at our $N = 6$, $k = 2$ stratification exercise. Let us reflect on the true value of the population mean (26.5) when compared with the estimated mean derived from srs (26) and with stratification (27.75). Of course, the true mean will be unknown to us in actual field practice, but the discrepancy that shows up in our example may be a bit startling. Many times in the literature, the researcher will incorrectly show that the formula used to obtain the overall mean is merely a sum; that is, $\hat{\mu}_{str} = \frac{\hat{\mu}_{str1} + \hat{\mu}_{str2}}{2}$. In our example, using this incorrect summing method, we would have obtained $\frac{22+30}{2} = \frac{52}{2} = 26$. Not only is this the same as for srs but it is closer to our true population value of 26.5. What is the problem?

Recall that it is the formula or estimator that is unbiased, not each sample value. In addition, a review of the definition of unbiased implies "on the average," or when we replicate a large number of times. We have, however, picked just one very small ($n = 4$) sample, which has the added complication of having identical counts as some of the observations in both strata. (Refer to Figure 7.2 to see that 1 stratum contains 2 quadrats with counts of 30, and the other stratum contains 2 quadrats with counts of 24.) What we just showed was that with random sampling, either on the entire or on the subdivided population, we always run the risk of getting a poor (unrepresentative) sample. In particular, however, notice that despite the small sample size, the random sample picked up only 1 value from stratum #1 and 3 from stratum #2, whereas the stratified sample was forced to divide the total sample size equally between the strata. For a larger population then, and as long as stratification is applicable, stratified sampling will always give a much more representative picture of the total area, and therefore, of its counts, than will srs.

DESIGN INTERRELATIONSHIPS

Statisticians see many of the sampling schemes discussed in this chapter as interrelated methods. Because many sampling texts do not employ unified notational systems throughout the discussions of sampling topics, field researchers may be led to believe that each sampling scheme is unique and unrelated. Even with clear and concise sampling texts, decisions are difficult when picking an appropriate scheme to use for a specialized field investigation. Thus, there are major problems involved when biologists try to understand and adapt statistical theory to the sampling of natural populations.

Let us consider the earlier example in which the biologist lays out a transect on the forest floor, or the benthic specialist or oceanographer travels to deeper depths along a straight path. Let us say there are to be 4 locations (stations, quadrats, zones, localities, etc.) to be visited along a gradient or linear trail, either fortuitously or with the specific purpose of examining the microhabitats. What type of design is optimal? What criteria should be used to come to a decision on the "best" sampling design, given the constraints in such a study? Remember, we must think of the fieldwork problems as well as the statistical design, or we may not be able to analyze the resultant survey data.

In order to find an optimal design, we will consider and list the methods that the biologist might use in the field, taking into consideration the constraints, such as forest impenetrability, ship's time and expense, and imprecision of latitude and longitude markings in the field, among others. We shall outline treatments for coping with these problems by choice of sampling method and design, focusing on those that are common in the scientific literature.

A very common procedure, as the ship goes from shallow to deep, or the cline increases in altitude, is that only 1 biological sample is taken at just some of the stops along the way. Either a core is dropped over the side, observational methods for larger organisms are made in one point count or dive (possibly observing in an equidistant sphere, centered at a selected point), or a quadrat is laid out or dropped and all organisms within it (or within, below, and above it) are counted. How does this protocol fit with our statistical sample designs? Let us assume there were 4 such biological samples obtained along the traverse or transect. We could correctly describe the actions of the biologists and the statistical methodology by any one of the following scenarios.

1. The 4 samples constitute one set of 4 independent, single, biological samples. That is, the researcher took 1 biological sample ($n = 1$) at each separate locality and will use it to describe that locality. The implicit assumptions, or the questions of interest, are the existence and characteristics of 4 separate target populations. The choice of the 4 points for sampling may have been the equivalent of random selection. Or the choice may have been systematic, with the 4 samples taken at approximately equal distances along the trail or path.
2. The samples could be considered as 4 biological clusters, samples, or statistical observations composing 1 statistical sample of size 4 ($n = 4$), collected with a systematic sampling design. Under this scenario, an implicit assumption is that there is 1 target population of interest.
3. The sampling is 1 single, systematic, but stratified sample taken over 4 strata, with a sample of only $n = 1$ in each stratum. Although similar to column D of Figure 7.1, this design assumes that there are 4 target populations of interest.

Each of these descriptions can be correct, and could be examined with a quantitative approach. However, each has its problems. The first design is the one that is quite common within a wide segment of the biological research community. The individuals in the 4 single samples are identified to species, with each sample counted and plotted separately. Interpretations of locality differences in density or community structure are made. By using the data in this manner, there is no possibility of obtaining an overall (combined) density estimate, nor are there any variance estimates. There is no replication.

The second possibility, which supposes a systematic sampling design, is preferable to the first in one obvious way. The statistical sample size is now greater than 1; it is $n = 4$. Therefore, if we selected the first position at random, we can calculate an estimated value for mean density of the target population of interest, such that the value will be nearly, if not completely, unbiased. This scenario, however, still leaves us without a consistent variance estimate, because all the data collected are still regarded as a single, unreplicated, systematic statistical sample. We have obtained no quantitative information on the spatial distribution of the organisms. There is still only 1 observation at each locality, so that independent information for each place cannot be considered, even if there is an indication of an underlying gradient or of differing biofacies. However, if this were an example in which the sampler had been placed in only one of an extremely large number of possible places (for example, along the Atlantic seacoast), then this scheme could be examined as a random sample of size $n = 4$, which would yield unbiased estimates for both mean and variance.

For the third interpretation, which is as a stratified sampling, the researcher must have been able to identify the localities as 4 distinct habitats or target populations. Some particular species may be of interest and its density is known or suspected to be different at each locality, perhaps because of a pilot study. Sampling within each stratum could be either random or systematic (ordered). The latter would apply if each subsequent sampling were taken at a fixed distance from the previous. Thus, because there is only 1 biological sample taken in each stratum, this design is not better than, but equivalent to, systematic sampling. However, it does have an advantage over scenario #2 in that it is representative of the total population. This is because it does provide a sample from each of the different groups. For the biologist, there is only a conceptual distinction between this scheme and the first one proposed. In many cases the decision as to whether to call it systematic or stratified may be made post hoc. It might be called systematic if the expected differences between the habitats or localities failed to materialize when quantitative sampling took place. On the other hand, if expectations were met and the sampling is regarded as stratified and, in addition, when the primary sampling unit is the count, not the quadrat or core, no unbiased estimates of the entire population can be obtained. If the core or quadrat is regarded as the sampling unit, this caveat will not apply.

We now have to make a decision from among these three alternatives, each of which has practical difficulties. Our sampling is complete, but how do we regard the data for statistical treatment? Should we treat the sampling as 4 single biological samples (a stratified sample of $n = 1$ at 4 localities), as systematic ($n = 4$), or as stratified ($n = 4$, with $n = 1$ per stratum)? The optimal solution is to treat the data as if it were 1 single sample of size 4 collected with a systematic design. This is scenario #2, as discussed earlier. By proceeding in this way, we can obtain an estimate of mean density that can be replicated, although we still cannot calculate a consistent variance estimate. Indeed, after some further information is obtained (for example, from museum collections or the research literature), we may be able to analyze these data as a random sample of size 4, as we have discussed. Also, we can discuss each separate position, or locality, in a qualitative or descriptive way. So, we have satisfied two purposes, depending on the biological question being asked. However, the amount of fieldwork that was involved appears to have been wasted to some extent because we cannot analyze the data in a way that could give us the most useful information. We chalk up the exercise to experience, and vow to do a better job on the next trip by deciding on a sampling methodology before we go into the field.

How would we do a better job? What changes would we make in the sampling method? Let us assume we have the same possibilities, but we have taken the advice of this book and are now determined to sample in a way that includes replicates. This means that instead of taking only a single sample at each site, we obtain at least 2 or maybe even 4 biological samples at each of the 4 localities.

One more decision needs to be made: Do we undertake the replication using a single-stage or multistage approach? That is, do we count all individuals we obtain in the core or quadrat, or do we subsample? There is no general decision rule for deciding whether sampling schemes with replication are best analyzed as a single- or multistage design. Multistage designs have the added complication that variability can be separated into two parts when individuals (secondary sampling units) are distributed among a group of sampled primary units (that is, the quadrat, core, cluster, and so on). We must recognize the variation among the primary biological samples themselves, as well as the variation of the individuals within these primary units, and accept the fact that the formulas for multistage design will be more complicated than for single-stage designs.

In a single-stage design we count all the individuals within each selected primary unit. Then, as we described, the sampling variance of the mean density depends on only the first source of variation; that of the primary units. For a multistage design, however, we subsample from the cluster (core, quadrat, or biological sample), and we use estimates of the primary unit's mean density in the formulas in place of the true population means. We thereby increase the variance. This forced increase that is the outcome of multistage design depends on the second source of variation; that is, the variation of the organisms within the quadrats, cores, or biological samples. In both the single- and multistage cases, if we had replicated (that is, taken at least 2 or even 4 cores or quadrats in each place) we could have made the best use of our resources.

CHOOSING THE BEST SAMPLING DESIGN

Any decision concerning which sampling scheme is best for our goal of sampling natural populations is made even more difficult because minor variations in the organization and execution of the field study will surely introduce even further complications and enlarge the list of possible statistical designs. How can we evaluate the design possibilities, adapt their advantages, and minimize their disadvantages for our unique biological requirements?

We have seen that unrestricted srs is not usually the easiest or most efficient method to use in field or museum study. The precise methods of restricted random sampling, whether applied in a systematic, stratified, or cluster sampling design, depend for their increased precision on two basic points. First, to eliminate the possibility of extremely unrepresentative samples, and the attached important source of variation, each restricted method spreads out the samples across the population. Second, the rules require that we sample more where it is most useful (which, in turn, is more cost-effective). For example, it is easier to get a large sample size if we cut the population into strata, and sample some strata thoroughly, than it is if we just wander into a large target area and find any target organisms or places we can. What this means is that if we have enough pilot information to lead us to decide to stratify, we should sample from each of the strata, not just a few. This decision to sample from all strata accomplishes two important things. First, sampling from each stratum eliminates an important source of variation because the largest part of the variability is between each of these groups, not within each stratum. Second, *sampling intensity* (sample size relative to target population size) can be increased. This increase can occur even though there are now multiple target areas because, relative to one total target area or population, sampling in the subareas or strata combined can increase, owing to ease of accessibility of each stratum. An increase in sampling intensity offsets the problem of restricting samples to a few localities, which could result in loss of precision.

An opinion survey of sampling statisticians would never provide consensus approval of any one "best" method for surveying natural populations. Nor will we make a recommendation that, overall, one and only one method is generally better than the rest. However, for researchers who aim to minimize effort for maximal reward, we do have a suggestion. Without a doubt, there must be replication of biological sampling. You may or may not wish to stratify your target population; you may or may not wish to sample systematically; you may or may not wish to randomly place quadrats. *But you should always replicate.*

With this in mind, let us revisit our $n = 4$ transect. Figure 7.4 shows what we regard as the optimal method for undertaking the sampling of this transect. We plan 2 transects, with 4 sites to be sampled on each. At each of the 4 sites, at least 4 replicates are taken. (Except in the case when the biological samples were collected for an entirely different purpose, and the researcher received them as an afterthought, it is of prime importance to collect more than 1 biological sample at each location.)

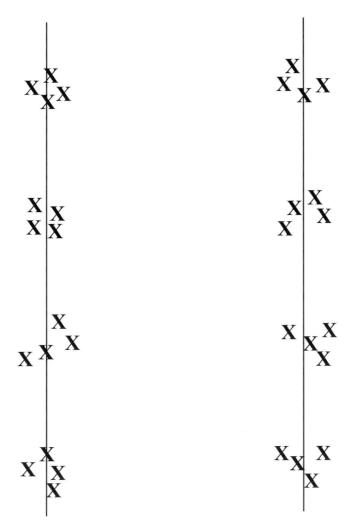

FIGURE 7.4   An example of replicated systematic sampling, with 4 replicates at each location/ stop.

As we hope to have shown you, systematic sampling can effectively determine implicit strata as well as serve as stratified sampling if subpopulations can be identified. That is why our chosen "optimal" method portrayed in Figure 7.4 includes equidistant stops for the replicates. Remember, too, that clusters or groupings of individuals allow for the unbiased estimation of means and variances, while stratification for collection of individual organisms alone does not. This is why our suggested method portrayed in Figure 7.4 shows random placement of quadrats (cores, biological samples) at each stop along the transect.

Overall, the most effective choice for natural population surveys, in the absence of other specific needs or particularized hypotheses, is the delineation of strata at specified intervals throughout the target area, with replicates taken at each point of interest within each stratum. This is the essence of Figure 7.4. Our general recommendation, thus, is to use

1. Stratification with replicates taken in each stratum, or
2. Replicated systematic sampling. That is, at each stop in the ordered selection process, replicates (at the very least 2, and preferably at least 4) should be taken (see Chapters 4 and 5).

By using either of these two recommended sampling schemes, one can expect to obtain information on the population characteristics within distinctive habitats, as well as to calculate total estimates with effective statistical properties over the entire range. In the field, it is safe to use systematic sampling only when the fieldworker is sufficiently acquainted with the organisms to feel confident that no periodicities (like swarming or increased densities with gradient) in their numbers (over space or time) exist. This caveat is a major concern because neither the statistical sample nor its variance will yield any indications of this biological condition, should it exist. The periodicity will not be apparent in the calculations (compare with Figure 1.1). The results will be very poor, although the researcher won't know it. The greatest increase in precision or decrease in variance will occur with a systematic sample that has been taken over a target population in which there is a high correlation in individual counts between adjacent quadrats (*serial correlation*).

SUMMARY

1. When sampling for density, the quadrat (biological sample) is the preferred primary sampling unit.

2. When sampling for relative abundance (species proportions), biomass, and such, the individual is the preferred primary sampling unit.

3. Cluster sampling uses a device (for example, a quadrat) as the primary sampling unit even if the individuals are of primary interest.

4. The most usual alternative to simple random sampling (srs) is systematic sampling. In systematic sampling, the samples are taken at ordered (often equal) intervals.

5. There is no overall disadvantage to the use of systematic sampling if the target population is homogeneous or extremely large. The sampling variance will be smaller (more

precise) for a systematic sample if the target population exhibits a consistent trend along a gradient.

6. We cannot calculate an estimate of the variance that is useful from a single systematic sample, no matter how large. Replicates are essential.

7. Replicated systematic sampling enables the calculation of a useful variance because it yields a consistent and unbiased variance estimate.

8. Distinct target populations (zones, sites, biofacies, etc.) within a larger sampling area are called strata. Stratified sampling can usually reduce the variance in the sample and narrow the resultant confidence limits about the mean. Either srs or stratified systematic sampling can be used to sample the strata. Caution must be exercised because the stratum means can be biased estimates of the subpopulation means in each stratum.

9. For sampling natural populations we recommend stratification, with replicates taken in each stratum, or replicated systematic sampling. If a traverse or transect is sampled systematically at each stop, at least 2 (preferably 4) replicates should be taken. By using either of these sampling schemes one can obtain information on the separate habitats as well as being able to calculate total estimates.

8

SPECIES PROPORTIONS:
RELATIVE ABUNDANCES

FAUNAS AND FLORAS are often summarized, or characterized, by the percentage of their components. The components, or categories, may range from the species level to family level or even higher in the taxonomic hierarchy. For example, a subtropical forest might be characterized by the percentage of palm trees contained therein. The marine fauna of the Arctic can often be identified by the percentage of a relatively few key species. Alternatively, interest may lie in sexual dimorphism, in proportions of adults versus larvae, or in some other community-level subdivision. Such succinct summaries of faunas and floras are important in both ecological and paleoecological studies. This chapter provides information on the individuals, not as they occupy space or occur within plots but as they relate to one another and to the total target population.

PROBABILITY OF A RANDOMLY SELECTED INDIVIDUAL BELONGING TO A SPECIES

Usually, n individuals are selected from the study area, and these individuals are grouped, or identified, into categories of interest. In this chapter we use "species" to represent any category into which we could place the organisms; for example, genus, family, sex, or age. Here we have an important departure from our previous sampling scheme, which focused on population density. For species proportions, we are interested in the individual organism, regardless of whether the organisms were collected as part of a biological sample or as individuals. In the biological approach to proportions, the *sampling unit* is taken to be the individual. The sampling unit is no longer the quadrat, core, or subplot, which were the sampling units deployed in the estimates of population density in the previous chapters.

Let us begin with a hypothetical area that contains S species, each of which is represented by N_i individuals. The *total number of individuals* is then

$$N = \sum_{i=1}^{S} N_i. \tag{8.1}$$

This formula shows that the symbol N represents the sum of all the N_i individuals, included in all S ($i = 1, 2, \ldots, S$) species. The subscript i is used because there can be a different number of individuals in each species, and there are S different species over which we must add or sum. The true proportion, or *population proportion*, of the ith species is given by

$$P = \frac{N_i}{N}.$$ (8.2)

Notice that we have focused on a single species; that is, species i. Because our interest is in only this one species, we do not need to subscript the proportion; we could, but it would add confusion. For species i, if we take a sample of size n, the *observed proportion* of this species most often is called its *relative abundance*. Here, the observed proportion of the ith species is

$$p = \frac{n_i}{n}$$ (8.3)

where n_i is the number of individuals of species i that were observed in the sample. This proportion is often multiplied by 100 and expressed as a percent.

Notice that in this chapter we are not using the notation $\hat{p}$ for the estimate, but merely p, while P denotes the population value. Our failure to include the caret in the notation for an estimate of proportion is intentional: We wish our notation to correspond with the most common notation in both statistical and biological references. It is standard practice to signify estimates of population density with a caret, but not to do so for estimates of proportions.

The value of p, as calculated in Equation (8.3), is referred to by various authors as *percent species*, *relative abundance*, *percentage abundance*, *fractional abundance*, and a variety of other terms. These proportions are popular because, unlike density figures, they are circumscribed. By circumscribed we mean that the sum of the proportions must add to 1.00, or 100%, so that the range is constricted. Consequently, when an investigator is examining a list of species from different areas, or from the same area over time, the numbers are easily understood (they always range between 0 and 1 or 0 and 100). The large, wide-ranging variation inherent in density measures is more difficult to visualize.

If the n individuals were selected at random from N, then the sample proportion, or relative frequency (as given by $p = \frac{n_i}{n}$), can be thought of as the *estimated probability* that any particular individual selected at random belongs to the ith species. In statistics, the equating of relative frequency and probability can be done because of the statistical result called the *Law of Large Numbers*. For example, if species i makes up 50% of a fauna, then an individual chosen at random from this fauna would have an approximate probability of $p = 0.50$ of belonging to species i. In other words, on the average (or, in the long run) we could expect that every second individual selected ($0.50 = 1/2$, or 1 of 2) would belong to this species i. Thinking about species proportions in this probabilistic way (that is, as a probability associated with the individual's selection), rather than as a proportion of the fauna, helps clarify our perception of what is meant when a biota is characterized in terms of percentages. Although this notion is intuitively clear, researchers seldom think of species proportions or percentages in this helpful way.

BERNOULLI TRIALS

When our interest is focused on a particular species, say, species i, any individual selected at random from the total number of individuals will either belong to species i or will not. These are the only two possibilities. By convention, we will use the symbol p again, this time as the chance that an individual belongs to species i (called a *success*); the symbol q is the probability that the individual does not belong to that species (called a *failure*). This reuse of the letter p should not be confusing because, as we shall show, there will be a basic relationship to unite these seemingly distinct uses.

Because we are investigating an "either/or" possibility (that is, a *dichotomous situation*), there are no other possibilities besides the two (finding or not finding the selected species). Thus p and q, which are always numbers between 0 and 1, must add to 1, or 100%. In turn, it must be true that $q = 1 - p$. That is, the probability that we do not find the species of interest equals 1 minus the probability that we do find it.

Natural populations often contain a very large (or assumed *countably infinite*) number of individuals, so we can sample any size target population by removing the individuals taken with each sample without fear of affecting the results in subsequent samples. We do not need to return the individuals to the target population before we take the next sample. The statistical term for this is *sampling without replacement*. For a smaller natural population, however, sampling without replacement might introduce a bias in later samples. We can avoid this problem sometimes by using tagging procedures. After the first sample is measured and tagged, the individuals are returned to the population. Already-tagged individuals may turn up in subsequent samples. This method of tagging and return is equivalent to *sampling with replacement*.

In both situations the population of choices is quite large, and we can easily assume that the probability of observing or obtaining an individual of a selected species is constant (*independent*) during our sampling period. This type of "yes" (success) or "no" (failure) sampling condition, when carried out repeatedly one time after another, such that the probabilities remain the same (*stationary*) for each try, may be referred to as a *Bernoulli trial*. The entire procedure of selecting a group of n of these dichotomous (yes/no) observations is termed a *stationary Bernoulli process*. The method by which a stationary Bernoulli process is achieved is *Bernoulli sampling*, also known as *Binomial sampling*. In small, limited populations (for example, a swarm of copepods, bees, amphibian larvae, or a school of fish), each sample taken under a sampling without replacement approach can decrease the total population size to a noticeable degree. In such cases, the values for p and q can certainly change, and so it is called a *nonstationary Bernoulli process*.

THE BINOMIAL DISTRIBUTION

Under a set of assumptions that includes having a dichotomous situation and a constant probability of success, we can set up this sampling problem as an application of the Binomial distribution. For any selected values of p (probability or chance of a success) and n (sample size), we can calculate the probability of finding exactly k individuals belonging to a particular species of interest; say, k individuals of our favorite species i. Earlier, we wrote n_i as the number of individuals of species i, and now we are using k for the same quantity. The difference is that we are now talking about only one single species; our species i. Prior to this we had wanted to talk about any one of the total number of species. This switch from n_i to k is meant to reduce confusion by simplifying terms. It allows us to drop the subscript and use just the letter k (but we could not reuse just n, of course).

Let us suppose we want to estimate the proportion of the frog *Rana boreas*. In the field we select a sample of size $n = 5$, and with each selection assumed to be independent of each other, we assume that the probability of finding a member of this species rather than a member of another species remains constant. That is, the chance of finding an individual of this species in our designated habitat, on average, is always the same, or it doesn't vary noticeably, during the time of our fieldwork. We can expect to obtain anywhere from 0 to 5 individuals of *R. boreas* in our sample of size 5 (5 observations). We are not interested in the order in which these specimens are obtained, but we would like to know how many possible results we can get, and the chance of our obtaining a certain number of specimens of *R. boreas*. A simple mathematical rule tells us that in a sample of size n we will have 2^n different sequences of success or failure in finding *R. boreas*. Table 8.1 lists those possibilities. Because this table shows all possible results, it constitutes a sampling distribution (Chapter 3). Just as with our other sampling distributions, we can assign a probability to each possibility. This probability will equal 1 divided by the total number of possibilities. Accordingly, each sequence, of the 32 possible, has the same probability of $\dfrac{1}{2^5} = \dfrac{1}{32}$.

Since the concern is with how many individuals of *R. boreas* we found in our 5 observations, we are not interested in the arrangement of these possibilities. If we failed to find *R. boreas* on the first 4 tries, and then found a specimen on the last try, we would have a total of 1 specimen. Yet if we found a specimen the first time we looked and then found no more, we would still have only 1 individual specimen (FFFFS = SFFFF). Thus, we can summarize Table 8.1 to eliminate redundancies. This is done in Table 8.2.

The Binomial Distribution of Counts

Conveniently, for the type of Binomial sampling introduced in the previous section, there is a general rule that allows us to calculate the probability of obtaining any number of successes (individuals) without having to count them as we did in Table 8.1. This rule also lets us calculate the probability (pr) of obtaining k successes in n Bernoulli tries, when we have a constant probability p of success at each try. Note that we chose to use pr for probability here to avoid the confusion that would result if we used P for the probability as we have done in the past and then P also for the population value. Note that throughout this chapter we use pr to indicate the probability of an event, also because the use of yet another "p" would be confusing. The rule can be written

$$pr(k \text{ individuals}|n, p) = {}_nC_k p^k q^{n-k} \tag{8.4}$$

TABLE 8.1 All 32 (= 2^5) Possible Results for a Sampling of 5

NUMBER OF SUCCESSES	POSSIBLE SEQUENCES OF SUCCESSES (S) AND FAILURES (F)	CHANCE OF OCCURRENCE OF THE SEQUENCE
0	FFFFF	1/32
1	SFFFF	1/32
1	FSFFF	1/32
1	FFSFF	1/32
1	FFFSF	1/32
1	FFFFS	1/32
2	SSFFF	1/32
2	SFFSF	1/32
2	FSSFF	1/32
2	FFSSF	1/32
2	FSFSF	1/32
2	FFFSS	1/32
2	FSFFS	1/32
2	SFSFF	1/32
2	FFSFS	1/32
2	SFFFS	1/32
3	SSSFF	1/32
3	FSSSF	1/32
3	SFSSF	1/32
3	FFSSS	1/32
3	FSFSS	1/32
3	SFFSS	1/32
3	SSFSF	1/32
3	FSSFS	1/32
3	SSFFS	1/32
3	SFSFS	1/32
4	SSSSF	1/32
4	SSFSS	1/32
4	SSSFS	1/32
4	SFSSS	1/32
4	FSSSS	1/32
5	SSSSS	1/32

NOTE: Success (S) is, for example, finding *Rana boreas*, failure (F) is not finding. This table contains all possibilities for a sample of $n = 5$. Therefore, it represents a sampling distribution for the random variable "number of successes."

TABLE 8.2   Possible Results for a Sampling of Five, Without Regard to Order

NUMBER OF SUCCESSES (INDIVIDUALS, k)	NUMBER OF WAYS TO OBTAIN k INDIVIDUALS	CHANCE OF OBTAINING EXACTLY k INDIVIDUALS IN SAMPLE OF SIZE 5
0	1	1/32
1	5	5/32
2	10	10/32
3	10	10/32
4	5	5/32
5	1	1/32
Totals	32	1.00

where $q = 1 - p$ (or $p + q = 1$). This equation, read from left to right, tells us that the probability (pr) of finding k individuals in a target area from which we are taking a sample of size n, and for which there is a constant probability (p) of observing, capturing, or selecting an individual from the target group, is given by the formula on the right side. We have already discussed the combinatorial notation, $_nC_k$ (see Chapter 3). In Equation 8.4, $_nC_k$ is called a *Binomial coefficient*. The term *coefficient* is used in the arithmetic sense that it is a multiplier of the other terms. An alternative way this coefficient is written (see Chapter 3) is with

$$_nC_k = \binom{n}{k}.$$

Both sides of the equation are, according to Chapter 3, expressions that are equal to $\frac{n!}{k!(n-k)!}$. This latter formula gives us the number of ways to obtain k individuals from a sample of n individuals, because we don't care about the order in the sample in which we find our individuals. Although Equation 8.4 may look complicated, it is quite simple to work through. The left side is merely shorthand notation for stating the particular probability (when we know n and p) that we want to calculate; the right side gives us the way to calculate the number that is the probability. For example, suppose we want to know the chance that we find exactly two specimens of R. *boreas* in our sample of 5 when we don't know the probability. The Binomial rule gives us

$$pr(2|5, p) = {_5C_2}p^2q^3 = \frac{5!}{2!\,3!}p^2q^3 = 10p^2q^3.$$

A look at line 3 in Table 8.2 shows that we have used the rule successfully to obtain the count of 10 for the number of possible samples containing exactly 2 specimens. Table 8.3 applies this Binomial rule to our entire sample of size 5.

TABLE 8.3   Probability of Obtaining Exactly k Individuals, Using the Binomial Rule

NUMBER OF SUCCESSES (INDIVIDUALS, k)	NUMBER OF WAYS TO OBTAIN k INDIVIDUALS	CHANCE
0	1	$_5C_0 p^0 q^5 = 1 p^0 q^5$
1	5	$_5C_1 p^1 q^4 = 5 p^1 q^4$
2	10	$_5C_2 p^2 q^3 = 10 p^2 q^3$
3	10	$_5C_3 p^3 q^2 = 10 p^3 q^2$
4	5	$_5C_4 p^4 q^1 = 5 p^4 q^1$
5	1	$_5C_5 p^5 q^0 = 1 p^5 q^0$
Totals	32	1.00

NOTE: Number of successes is the number of individuals obtained in a sample of size 5. Chance is the probability of obtaining exactly k individuals. Probability of obtaining exactly k individuals from a sample of size 5 by the Binomial rule is $pr(k$ individuals $n, p) = {_nC_k} p^k q^{n-k}$.

How does Table 8.3 compare with Tables 8.1 and 8.2? We can see that the Binomial coefficient gave us exactly the same numbers (1, 5, 10, 10, 5, 1) as when we counted each of the possibilities. However, in Table 8.1 we said that each arrangement had an equal chance. That meant that in our Bernoulli trials, each time we made an observation or looked for a specimen of R. boreas, we actually were assuming that the probability of seeing such a specimen was 1/2, or that R. boreas comprised about 50% of the target fauna (so that for $n = 5$ we had $(1/2)^5$, or 1/32). This fact shows convincingly that it is not realistic to use a mathematical tool without considering the applicability of the assumptions, both explicit and implicit. The Binomial rule (Table 8.3) is more general because it allows us to specify a realistic probability for the species. This probability could have been obtained, or at least estimated, from a literature search or from past fieldwork in the same area and under the same conditions. (In previous chapters we introduced this notion of the importance of a pilot study before the fieldwork.) For example, let us assume that from past fieldwork and some museum searches we thought that the declining population of blue heron (Florida caerulea) now represented only about 20% of the total avian fauna in our target area. We then would set $p = 0.2$ and $q = 0.8$ in our Binomial rule. Then, to figure the chance of obtaining exactly 3 specimens of F. caerulea in a sample of size 5, we would have

$$pr(3|5, 0.2) = {_5C_3} 0.2^3 0.8^2 = 10(0.008)(0.64) = 0.0512.$$

This result of 0.0512 means that we have about a 5% chance, or about 5 chances in 100, of getting exactly 3 specimens of F. caerulea in our 5-specimen sample, if the population of this species is really about 20% of the avian population.

One of the points that we have just reaffirmed by this series of examples is that the count of individuals can be looked at as a random variable. That is, in a sample of size n, we can find anywhere from 0 to n specimens of species i (successes). This random variable has a distribution in samples of size n that is called the *Binomial distribution* (or, less frequently, *Bernoulli distribution*).

To summarize, we refer to each observation of a particular genus, species, sex, age, and so on in our sample as the result of a *Binomial trial* (or *Binomial experiment*) because of the following reasons. First, there are only two possible outcomes at each time; either we find one of what we are looking for or we don't. Second, the probability of finding an individual of the particular group (and also of not finding it) remains the same over the entire sampling period. We have discussed field situations in which this would be a reasonable assumption. Third, the observations are considered independent of one another. We have noted that the Binomial is appropriate in field situations in which the natural population is large enough that each individual observation can safely be said to be independent of any other. We have also said that the condition of independence can be simulated by sampling the exact same population by replacing the sampled individual before the next subsequent sampling; one example being the mark-and-release method. We are not referring here to capture–recapture models, which are not based on Binomial models.

We can repeat this Bernoulli trial situation n times, which is the equivalent of taking a sample of size n, and the probabilities of each outcome can be calculated and tabulated. The result is called a *Binomial distribution*. The Binomial is a *discrete distribution* since the random variable k (which is the number of finds, individuals, or successes) is a *discrete random variable* that takes on the values $k = 0, 1, 2, \ldots, n$, for a sample of size n. The use of the word *discrete* refers, as it did in Chapter 3, to the fact that the possible values for the random variable, and for its tabulation into a distribution, are discrete. The probability of getting any particular number of individuals in the sample is easily obtained by using the Binomial rule.

We need to examine a few more characteristics of this Binomial distribution before we turn to our main problem of obtaining information on species abundance or proportion by use of this distribution. Recall that when we discussed the t distribution, in Chapter 4, we pointed out that this was really a *family of distributions*, each member of which is distinct for each different sample size (n). Similarly, the Binomial distribution is a family of distributions. But here we must know p and n to determine any member of this family. Second, since each member distribution is distinct for each combination of p and n, we say that p and n are the *parameters* of the Binomial distribution. This relationship is expressed by the symbol $bin(n, p)$. For example, Figure 8.1 shows the graphs of two examples: the first with $n = 5$ and $p = 0.5$, the second with $n = 5$ and $p = 0.2$, along with some other possible values for p. Visually, the graphs appear very different. However, there is at least one similarity among all 12. In each case the probabilities increase as k (the number of successes) increases, until a maximum is reached; then the probabilities decrease. For comparison, consider both Figures 8.1 and 8.2. Regardless of sample size, these pictures show only one *mode*, or highest point, somewhere between $k = 0$ and $k = n$, with probabilities gradually decreasing on either side of this highest point.

The exact placement of this high point, and the shape of the graph or histogram itself, depends upon the values of n and p. In fact, the point of highest probability (highest point on the y axis, or the mode) is always located at the value of np, if this is an integer. In our example with $p = 0.5$ and $n = 5$, this highest point on the y axis

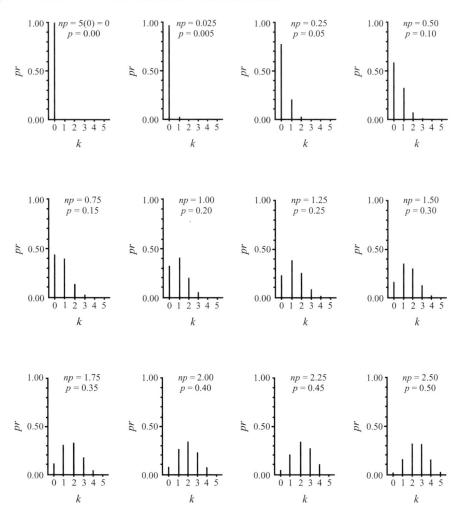

FIGURE 8.1   Plots for the Binomial distribution with $n=5$ and $p=0.00$ to 0.50. The x axis is the number of individuals (k); the y axis is the Binomial probability (pr); np is the mean.

is $5(0.5) = 2.5$, which is not an integer. In this case, the highest relative frequency is close to 2.5, even though this value is not a possible number of individual specimens. A result from mathematical statistics states that it is always true that $-q < (c - np) < p$, where c represents the value with highest probability. A most useful interpretation of this high point is that the mode is also the estimated mean value, or

$$np = \hat{\mu} \tag{8.5}$$

for the Binomial distribution, when the random variable is the count of individuals (number of successes in the sample). For the population with N total individuals and constant probability P, we have

$$NP = \mu. \tag{8.6}$$

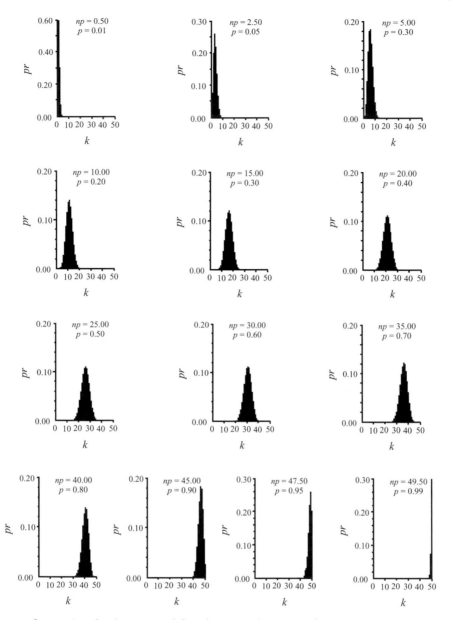

FIGURE 8.2   Plots for the Binomial distribution with $n=50$ and $p=0.01$ to 0.99. The x axis is the number of individuals (k); the y axis is the Binomial probability (pr); np is the mean.

In addition, the variance for the Binomial when the random variable is the count of individuals is calculated as

$$\sigma^2 = NPQ \qquad (8.7)$$

where the population value $Q = 1 - P$. The estimate of the variance in the sample; that is, for sample size n and constant probability of success p, is

$$\hat{\sigma}^2 = npq \qquad (8.8)$$

with $q = 1 - p$. We can use these facts to form confidence limits in a later section of this chapter.

The Binomial Distribution of Proportions

Thus far, we have been looking only at the number of successes, or individuals found. However, most often the biologist is interested in the *proportion of successes*, not the simple count of the number of successes in the sample. By this we mean interest lies in a particular species or group as a part of an ensemble.

How can we use the information on counts, finds, or successes, and their distribution to help with the problem of determining these proportions? Again we turn to mathematical statistics and use a remarkably helpful result. It has been proven that for sampling from stationary Bernoulli processes (which we have in sampling situations for natural populations), the random variable of the number of successes has exactly the same distribution as the random variable of the proportion of successes in a total sample of fixed size n. In the notation we have been using we can write

$$pr(k \text{ successes } | n, p) = pr\left(\frac{k}{n}\middle| n, p\right). \tag{8.9}$$

That is, k successes in a sample of size n gives a proportion, or relative abundance, of $\frac{k}{n}$. For our blue heron example, we found that we had $pr(3 \text{ successes} | 5, 0.2) = 0.05$. In other words, the chance of finding 3 blue herons in a sample of 5 birds is only 0.05.

Just as we did with estimates of mean density, we are interested in placing confidence intervals about our estimates of relative abundance. When we examine museum records or organize a preliminary pilot study and calculate, for example, that species i makes up 50% of the total taxa, how confident are we that this estimated value is within some stated percentage of the true value? To answer such questions we require estimates of the mean, variance, and standard error of the proportion under study. Recall that the variable x_i in the density chapters was the count of the number of individuals in a quadrat, or sample area. In the present case, the sampling units are the individuals rather than the quadrats. As we select individuals at random, each one either belongs to species i or it does not. Hence, we can assign a 1 to an individual belonging to species i (a success) and a 0 to an individual that does not belong (a failure). The variable x_i then becomes a sequence of 1's and 0's over the entire sample. For example, for species i we might obtain 00111 (instead of but equivalent to the notation FFSSS in Table 8.1) for our sample of 5, indicating that we found 3 individuals. We can calculate the required parameters for the Binomial distribution for the proportion of successes either by the formulas given in the density chapter (Equations 2.2, 2.4, and 2.5) or by the application of a result in mathematical statistics. This latter approach is a simpler method, as we shall demonstrate.

First, let us use the Chapter 2 formulas to calculate a mean and variance. We begin with Equation 2.2 for calculating the mean density estimate from samples taken. Here the mean density estimate is given as

$$\hat{\mu} = \frac{1}{n} \sum_{i=1}^{n} x_i$$

where n is the total number of sampling units (quadrats or individuals) selected at random, and x_i is the abundance of each in the chosen space. But now, x_i is either 0 or 1 for each observation, rather than the number of individuals, or count (as it was for the density estimates); as such, it is now called an *indicator variable*. An indicator variable is simply a variable that indicates whether an event does or does not occur by taking on the values 1 and 0, respectively. For example, for species i in our sample of 5, $x_1 = 0$, $x_2 = 0$, $x_3 = 1$, $lch_4 = 1$, and $x_5 = 1$. The mean of these 5 observations is $\frac{x_1 + x_2 + x_3 + x_4 + x_5}{5} = \frac{0+0+1+1+1}{5} = \frac{3}{5}$, which is also our relative abundance. If we let k represent the number of individuals of the ith species found in the sample (number of successes), then $k = x_1 + x_2 + \cdots + x_n = \sum x_i$ and we can write

$$\hat{\mu} = \frac{1}{n} \sum_{i=1}^{n} x_i = \frac{1}{n}(k) = \frac{k}{n}.$$

Recall, however, that in Equation 8.3 this value of $\frac{k}{n}$ (which is $\frac{n_i}{n}$ for species i) is the sample species proportion, p. Thus we have the convenient result for the Binomial that the arithmetic mean is equal to the probability of a success, or of finding an individual. Thus,

$$\hat{\mu} = p \tag{8.10}$$

when the random variable of interest is the proportion of successes; that is, the species relative abundance. The sample mean is indeed the sample proportion, and p is an unbiased, consistent estimate of the proportion P of the entire target population.

The variance of the distribution of the proportions in the sample is then

$$\hat{\sigma}^2 = \sum_{i=1}^{n} \frac{(x_i - \hat{\mu})^2}{n} = pq \tag{8.11}$$

where q (remember) is the probability that the individual does not belong to the species (and p is the probability that it does). Just as we multiplied Equation 2.4 by $\frac{n}{n-1}$ to counteract a known bias, so the equation for calculating an unbiased estimate of the variance of the distribution of the proportions in the sample is

$$\hat{\sigma}^2 \left(\frac{n}{n-1} \right) = pq \left(\frac{n}{n-1} \right). \tag{8.12}$$

For the entire target population, the respective parameters are $\mu = P$ and $\sigma^2 = PQ$.

The estimated sampling variance of the species proportion, $\hat{\sigma}_p^2$, with the fpc (finite population correction factor from Chapter 3), is

$$\hat{\sigma}_p^2 = \frac{\hat{\sigma}^2}{n}\left(\frac{N-n}{n}\right) = pq\left(\frac{N-n}{N(n-1)}\right) \tag{8.13}$$

where N is the total number of individuals in the target population. Ignoring the fpc (we can do so because N is usually very large and $\frac{n}{N}$ approximates 0.05), we obtain the more useful formula for natural population sampling,

$$\hat{\sigma}_p^2 = \frac{pq}{n-1}. \tag{8.14}$$

Equation 8.14 is true because we know from Equation 3.5 that the estimated sampling variance $\hat{\sigma}_p^2$ must be the variance divided by the sample size, $\frac{\sigma^2}{n}$. By substitution from Equation 8.12 we get

$$\hat{\sigma}_p^2 = \frac{\hat{\sigma}^2}{n} = \frac{1}{n}\left(\frac{n}{n-1}\right)pq = \frac{pq}{n-1}.$$

The standard error of the species proportion is, then, the square root of the sampling variance, or,

$$\hat{\sigma}_p = \sqrt{\frac{pq}{n-1}}. \tag{8.15}$$

Usually $\hat{\sigma}_p^2$ is written simply as $\frac{pq}{n}$, not $\frac{pq}{n-1}$, and this introduces very little error, provided n is large. The estimate $\frac{pq}{n}$, however, unlike the sampling variance for density, is *biased* even with an infinite population size. Nevertheless, for the sake of simplicity and in keeping with convention, the shorter form, $\frac{pq}{n}$, is usually used. Thus, we can obtain all the sample statistics important for our purposes just from knowing the sample proportion and the sample size alone if the Binomial is applicable.

In summary, for the Binomial distribution when n individuals are selected at random, and the random variable is the species proportion, we have the density estimate, $\hat{\mu} = p$. The variance is given by $\hat{\sigma}^2 = \left[\frac{n}{n-1}\right] \cdot pq$. The standard error is given by $\hat{\sigma}_p = \sqrt{\frac{pq}{n}}$.

Examining the Parameters of the Binomial Distribution

We now know that n and p are the parameters of the Binomial distribution, and that the shape of the distribution varies as n is increased or decreased while p is kept constant. Additionally, the shape changes when p is changed but n is not. Figure 8.2 displays the Binomial distribution when p is varied with a constant sample size of $n = 50$. Figure 8.1 portrays the Binomial distribution when p is varied with a constant sample size of 5. A comparison of these figures shows that when p is very small (near 0.00) or very large (near 1.00), regardless of what the sample size n is, the shape is distinctly different from situations in which p is closer to 0.5. For those cases in which shapes are different depending on the value of p, mathematical statistics shows us that some continuous distributions can fit the observed data just as well as the discrete Binomial provides a fit. That is, for certain values of n and p we can use continuous distributions to calculate confidence intervals. Although the confidence intervals will not describe the exact probabilities; that is, they won't be *exact intervals* all the time (exact intervals occur only when we use the Binomial probabilities themselves), the continuous distributions whose probability values are used to approximate Binomial probability values (*approximation distributions*) will be much easier to calculate. For example, routines for the Poisson and Normal approximations are available in all the common statistical computer packages. These approximation distributions offer the big advantage that for large sample sizes their results yield confidence intervals that are (practically) indistinguishable from the intervals we would get by completing the tedious Binomial calculations. There are computerized packaged programs that will compute the Binomial for large sample sizes, but beyond a certain size, they too will use an approximate computing algorithm (for example, the *Incomplete Beta*). Therefore, although we could compute confidence limits about the species proportion using the discrete Binomial itself all of the time, in practice, the approximate (*asymptotic*) formulas are usually used. Note that the exact Binomial confidence intervals will rarely be symmetric, whereas the confidence intervals will always be of equal width on either side of the estimate when the Normal approximation method is used. The figures in this section show that the shape changes in the Binomial are apparent for sample sizes as small as 5. Thus, with a bit of caution in the presentation of the results, we can use the mathematical statistical results for continuous distributions in our preparation of the confidence intervals for relative species abundances.

Let us examine some of the salient features of these graphs in order to determine the most reasonable continuous distribution to fit to the Binomial observations. Fundamentally, there are two situations.

Fixed n and varying levels of p. We can see in Figures 8.1 and 8.2 that the shapes of the distribution graphs change as p changes. For p near either of its limits (0 or 1), each graph shows merely a single peak at $k = 0$ or $k = n$. As p increases slightly, more peaks are added to the *tail*, but they are never as large as the first largest peak, or mode. As p continues to increase, the picture gains a second tail, and the shape becomes more balanced. Indeed, when $p = 0.5$, the distribution is symmetric about $\frac{n}{2}$. (This is true if

n is even; if n is odd, the mode is the average of the two central points.) In Binomial distributions, the mean and mode are always within 1 unit (of k) distant from each other. If np is an integer, the mean equals the mode. The farther p is from 0.5, in either direction, the more asymmetrical the shape, or the more skewed, which means that one tail gets elongated.

Fixed p and increasing sample size n. When we examine the same p value but for different sample sizes, we see that the whole graph moves toward the right on the bottom axis; it also becomes more balanced, or flatter, and it widens; that is, the spread increases. The move along the bottom axis can be seen by looking at the mean np. Each time n increases by 1, the mean (the central tendency) moves a distance of p. It makes sense that if we want to fit the plot with a distribution, we would like it to stay still. Therefore, using the random variable, say, K (the count of the number of successes), we could substitute the value $k - np$ for k. This is the same procedure we used to obtain a z score in Chapter 3. In this case, the result is that μ is replaced with $\mu - np$, or 0. We could also use the random variable X as the proportion (rather than K, the count) and substitute $(x - p)$. In turn, this would keep the distributions centered at 0, the *origin*, and make it easier for us to fit a continuous distribution.

The degree of flattening increases; that is, the mode (and the mean) decreases, as n gets bigger. In fact, if we had plotted an abscissa (bottom axis) with $\dfrac{1}{\sqrt{n}}$ and the other axis (side axis or ordinate) as the height of our mode, we would get a straight line. The points on the line for $p = \dfrac{1}{2}$ would have coordinates $\dfrac{1}{\sqrt{n}}$, bin(n, $n/2$). Knowledge that the mode, and each value on either side of the mode, will decrease *inversely with* $\sqrt{n}$ (that is, *directly with* $\dfrac{1}{\sqrt{n}}$) will help us to obtain a reasonable fit to the Binomial.

Finally, we have seen that for any probability distribution, the sum of the ordinates (the probabilities) is always 1. Therefore, as the distribution flattens and probabilities decrease when n increases, the picture must look more spread out along the axis. The standard deviation is a measure of this spread (when p is constant, we can say this spread is *proportional to n*) and the range is always 0 to n. We learned from Tchebychev's theorem (Chapter 2) that there is very little probability in the tails of any probability distribution, and the Binomial is no exception. We know, then, that as n increases the standard deviations in the family of Binomials will increase in proportion to $\sqrt{n}$. It would be easier to fit the Binomial if the spread did not increase. Therefore, to keep the variance constant when we fit a continuous distribution we could take our substitute value of $k - np$ (or $x - p$) and divide by the appropriate variance (as we did in Chapter 3). This would result in a *standardized Binomial variable* for the number of successes, $\dfrac{k - np}{npq}$, or for the proportion of successes, $\dfrac{x - p}{\frac{pq}{n}}$, each of which has mean of 0 and variance of 1.

THE POISSON APPROXIMATION

Gathering all this information together, we can see why there are two different approximation distributions that will fit the Binomial. First, when p is near 0 and n is large, the Binomial is fit with a *Poisson distribution*. The Poisson is a distribution that is skewed to the right, just as are the plots of the Binomial when p is small. The Poisson is useful when p is near 0, n is large, and np is less than about 5. But, as evident in Figures 8.1 and 8.2, the fit would actually not be bad for $n = 5$ if p were small. The Poisson distribution, as we discussed in Chapter 6, is also used alone, without reference to the Binomial, to describe some sampling situations. However, we shall illustrate here the comparison (or fit) of the Poisson distribution to a Binomial data set.

For the Poisson, the mean and the variance are equal (see Chapter 6), which in fact greatly simplifies the work. The probabilities for a Binomial random variable that are of interest for natural population work are then the probabilities of k successes for a Poisson given by

$$pr(k \text{ successes}) = \frac{e^{-\mu}\mu^k}{k!}.$$

We can then set

$$pr(k \text{ successes}) = pr\left(X = \frac{k}{n}\right) = pr(K = k \text{ successes})$$

from the Binomial. In addition, we know from Equation 8.5 that $\mu = np$ for the Binomial on counts, so we can write

$$pr(k \text{ successes}) = pr(K = k \text{ successes}) = pr\left(X = \frac{k}{n}\right) = \frac{e^{-np}(np)^k}{k!}. \quad (8.16)$$

In Equation 8.16 the symbol e is a constant equal to 2.7183 . . . , which is the base of the natural logarithms. The formula on the right side of Equation 8.16 is the formula we use for determining a Poisson distribution. In contrast, the left side looks quite similar to the notation (Equation 8.9) for the Binomial distribution, except that we omitted the parameters n and p. This resemblance is because, for the Binomial, $k = 0$, 1, 2, . . . , n; whereas for the Poisson, there is a continuous distribution, $k = 0, 1, 2, ...$ There is no stopping point. As to the right-side formula, we merely substitute the value $np = \mu$ into the formula for the Poisson distribution to obtain our Equation 8.16 result. Any hand or computer calculator will compute this easily once np is known. Table 8.4 illustrates the computations. The *recursive formula of* Equation 8.17, which is just Equation 8.16 rewritten in a set of repetitive steps, is easier to use quickly than Equation 8.16:

$$pr(K = k \text{ successes}) = \frac{\mu}{k} pr(K = k - 1 \text{ successes}). \quad (8.17)$$

TABLE 8.4 Fit of a Poisson Distribution to a Binomial

$X = K$	BINOMIAL	POISSON	RECURSIVE FORMULA FOR FITTING POISSON
0	$1 \cdot 0.10 \cdot 0.9^{10} = 0.349$	0.369	$p_0 = e^{-1} = 1/e$
1	$10 \cdot 0.1^1 \cdot 0.9^9 = 0.387$	0.368	$p_1 = (np/1) \cdot p_0 = 1e^{-1}$
2	$45 \cdot 0.1^2 \cdot 0.9^8 = 0.194$	0.190	$p_2 = (np/2) \cdot p_1 = (1/2)1e^{-1}$
3	$120 \cdot 0.1^3 \cdot 0.9^7 = 0.057$	0.063	$p_3 = (np/3) \cdot p_2 = (1/3) \cdot p_2$
4	$210 \cdot 0.1^4 \cdot 0.9^6 = 0.011$	0.016	$p_4 = (1/4) \cdot p_3$
5	$252 \cdot 0.1^5 \cdot 0.9^5 = 0.001$	0.003	$p_5 = (1/5) \cdot p_4$
6	$210 \cdot 0.1^6 \cdot 0.9^4 = 0.0001$	0.0005	$p_6 = (1/6) \cdot p_5$
7	$120 \cdot 0.1^7 \cdot 0.9^3 = 0.0^59$	0.0^48	$p_7 = (1/7) \cdot p_6$
8	$45 \cdot 0.1^8 \cdot 0.9^2 = 0.0^64$	0.0^59	$p_8 = (1/8) \cdot p_7$
9	$10 \cdot 0.1^9 \cdot 0.9^1 = 0.0^80$	0.0^61	$p_9 = (1/9) \cdot p_8$
10	$1 \cdot 0.1^{10} \cdot 0.9^0 = 0.0^91$	0.0^71	$p_{10} = (1/10) \cdot p_9$

NOTE: For the Poisson, $np = 10 \cdot 0.1 = 1$. For the Binomial, $n = 10$, $p = 0.1$, and $q = 0.9$, with the random variable, $X = K$, being the count of the individuals. The notation 0.0^51 means 0.000001.

This formula is used after we find $p_0 = e^{-np} = \dfrac{1}{e^{np}}$ for our observed or desired value of np.

THE NORMAL APPROXIMATION

The second method of fitting the Binomial is with the *Normal distribution*. This choice should be evident from an inspection of Figures 8.1 and 8.2 (notice the symmetric bell shapes), for the cases where p is in the middle of its range, even when n is rather small. The general rule for use of the Normal distribution is that n should be large and p not small. As we have mentioned, use of the words *small* and *large* is sometimes disconcerting to users of statistical methodology. To the mathematical statisticians themselves, however, these qualitative terms pose no problem because their interest is not with real data. We shall see that there are methods for determining sample sizes to use for the approximate fit, but the basic rule is to understand that the approximation is only good if the data are well fit. How can this be known? Either we must have a predetermined rule that sets actual limits or we must plot. A picture of the data, when possible, is always a requisite for good statistical practice.

By examining Figures 8.1 and 8.2, we have already seen the advantage of using the standardized Binomial random variable. That is, instead of using the number of successes k, we used its *standardized form*:

$$z_k = \frac{k - np}{\sqrt{npq}}.$$ \hfill (8.18)

However, for our problem of estimation of relative abundance, interest is in the random variable that is the sample proportion $p = \dfrac{k}{n}$, rather than in the number of successes. Thus, the standard Normal random variable of concern is

$$z_p = \frac{\left(\dfrac{k}{n} - p\right)}{\sqrt{\dfrac{pq}{n}}}. \tag{8.19}$$

This involves a calculation that is quite simple by hand or with a computer. A result from mathematical statistics (*DeMoivre's theorem*) tells us that as n gets larger, the variables z_k or z_p may be used as a *standard Normal random variable*. That is, all we need to do is perform the subtraction and division, and then the answer can be looked up in a standard Normal table to obtain its probability. We are then assured that this result is quite close to the true, or exact, value we would get had we made the Binomial calculations. To show the effect of using a standard Normal approach for the calculation, we present Table 8.5. This table compares the exact Binomial probabilities with those derived from the Normal approximation without correction, and also with two

TABLE 8.5 Fit of Normal Distribution (With and Without Correction) to the Binomial

		NORMAL APPROXIMATION		
COUNT (k)	BINOMIAL bin(10, 0.5)	UNCORRECTED	CONTINUITY CORRECTION	NORMALIZED CORRECTION
0	$1 \cdot 0.5^0 \cdot 0.5^{10} = 0.0010$	0.0008	0.0020	0.0019
1	$10 \cdot 0.5^1 \cdot 0.5^9 = 0.0098$	0.0057	0.0112	0.0112
2	$45 \cdot 0.5^2 \cdot 0.5^8 = 0.0434$	0.0289	0.0435	0.0434
3	$120 \cdot 0.5^3 \cdot 0.5^7 = 0.1172$	0.1030	0.1145	0.1147
4	$210 \cdot 0.5^4 \cdot 0.5^6 = 0.2051$	0.2635	0.2045	0.2049
5	$252 \cdot 0.5^5 \cdot 0.5^5 = 0.2461$	0.5000	0.2482	0.2487
6	$210 \cdot 0.5^6 \cdot 0.5^4 = 0.2051$	0.2365	0.2045	0.2049
7	$120 \cdot 0.5^7 \cdot 0.5^3 = 0.1172$	0.1030	0.1145	0.1147
8	$45 \cdot 0.5^8 \cdot 0.5^2 = 0.0439$	0.0289	0.0435	0.0434
9	$10 \cdot 0.5^9 \cdot 0.5^1 = 0.0098$	0.0057	0.0112	0.0019

NOTE: For $n = 10$, $p = 0.5$. The Binomial is calculated for each $k = 0, 10$. The Normal distribution is calculated for each $z_k = \dfrac{(k - np)}{\sqrt{npq}}$. The continuity correction is the difference between the cumulative Normal calculated for $k - 1/2$ and $k + 1/2$. The probability of this effectively truncated Normal, from $-1/2$ to $+ 10\frac{1}{2}$ is 0.998, whereas the probability of the Binomial is 1. Therefore, in column 5 (normalized correction) we can take each fitted probability and normalize it for a further correction by multiplying by 1/0.998.

different correction methods. The correction method called a *continuity correction* can be used to compensate for the fact that the Binomial is a discrete distribution and the Normal, with which the Binomial is approximated, is a continuous distribution. We shall investigate the effect of a continuity correction when we discuss confidence intervals for the proportions.

The important point for biologists is the determination of when to use the Normal approximation instead of the Binomial. The most practical rule is that when $np > 5$ and $n > 30$, the Normal approximation will serve nicely. The translation of the rule means that when the tails contain very little of the probability, the Normal is the same as the Binomial to a useful number of decimal places. Otherwise, use the Binomial calculations. The Poisson may yield a good approximation of Binomial confidence intervals when p is small. However, theory shows that the Normal can also approximate or fit the Poisson. Overall, it is much more efficient for work with natural populations to always use the exact Binomial calculations, or at least to do so until the Normal approximation takes over and becomes useful. This recommendation holds regardless of the size of the p value. Many of the older texts, and even some recent ones, offer different recommendations, which were developed for ease of calculations before computers and packaged computer programs. In cases in which we recommend the Normal, it may be almost as simple to use the Binomial; the difference will be unnoticeable for the accuracy required for work with natural populations.

CONFIDENCE LIMITS FOR SPECIES PROPORTIONS

When our Binomial scenario is applicable, confidence limits can be calculated for proportions using the same methodology already presented for density calculations in Chapter 4. From Equation 4.2 we can construct a confidence interval based upon our sample estimate $p = \dfrac{k}{n}$ for the true proportion P in the population. The confidence interval is written

$$p - z_p \sigma_p \leq P \leq p + z_p \sigma_p \tag{8.20}$$

where z_p from Equation 8.19 is the Normal deviate for the proportion, corresponding to the desired level of confidence. For example, when $z_p = 1.96$, Equation 8.20 gives a 95% confidence interval. However, in trying to use this equation to form a confidence interval with our sample data we encounter added complications. These are

1. The variable $p = \dfrac{k}{n}$ (or k) is discrete, and the Normal is continuous.
2. When P is close to 0 or 1, the distribution of p (or k) is skewed.
3. The formula for $\sigma_\mu = \sigma_p$ involves the unknown parameter P, unlike the formula for the standard error when we considered the density.

To overcome these problems in the simplest way, we perform a series of steps. First, in Equation 8.20, we merely use our estimate for σ_p, or $\hat{\sigma}_p = \sqrt{\dfrac{pq}{n-1}}$, although the biased form $\sqrt{\dfrac{pq}{n}}$ will also be found in the literature. The confidence interval for P, then, is

$$p - z_p \sqrt{\frac{pq}{n-1}} \le P \le p + z_p \sqrt{\frac{pq}{n-1}}. \qquad (8.21)$$

When the fpc is used, the confidence interval becomes

$$p - z_p \sqrt{\frac{pq}{n}\left(\frac{N-n}{N-1}\right)} \le P \le p + z_p \sqrt{\frac{pq}{n}\left(\frac{N-n}{N-1}\right)}. \qquad (8.22)$$

Next, we can use a continuity correction in Equation 8.22 to adjust for the fact that the variable $\dfrac{k}{n}$ is discrete and that the Normal is continuous. The confidence interval then becomes

$$\left(p - \frac{1}{2n}\right) - z_p \sqrt{\frac{pq}{n}\left(\frac{N-n}{N-1}\right)} \le P \le \left(p + \frac{1}{2n}\right) + z_p \sqrt{\frac{pq}{n}\left(\frac{N-n}{N-1}\right)}. \qquad (8.23)$$

It should be clear that the continuity correction actually decreases the size of the confidence interval. However, as the example to come will show, the correction is probably not worth the effort.

MINDING P'S AND Q'S

Before returning to our Bolivian forest plot for an example, we ought to mind our p's and q's. We will examine the effect these two quantities have on the standard error of p to gain some insight into application of the statistical nuances of "small" and "large" sample sizes.

Many ecologists have an intuitive notion that a species comprising, say, 90% of the total number of individuals can be more precisely estimated (less error) than a species comprising, say, 10%. In other words, it is thought that a very abundant species (90%) has a narrower confidence interval than a rarer species (10%). On the other hand, we have heard some researchers express exactly the opposite opinion. Based upon what we have described about the Binomial distribution, let us examine the size of the standard error of p as the maximum is approached.

For the calculation of the confidence limits, the values of n and z_p are chosen at the outset and are, therefore, constants for these calculations. That is, after the fieldwork, we know our sample size (n), and when analysis begins we select a particular value of z_p. Consequently, in each situation, the values of p and q are the only values that will be seen to vary across the species sampled. Thus p and q are the basis of the variation in the confidence limits.

Figure 8.3 is a plot of p against $\sqrt{pq}$ (not the standard error $\sqrt{\dfrac{pq}{n}}$, because n is constant for our problem). The largest value of $\sqrt{pq}$ occurs at 0.5, with standard deviation

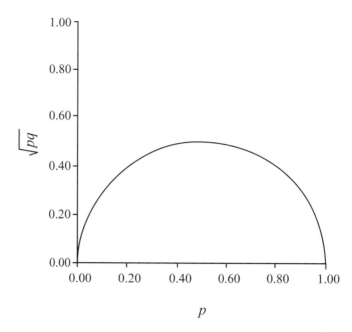

FIGURE 8.3   Plot of species proportion estimated probability (p) versus $\sqrt{pq}$, where $q = 1 - p$.

(from Equation 8.11) of $\sqrt{0.5 \cdot 0.5} = \sqrt{0.25} = 0.5$. As this figure shows, the values of $\sqrt{pq}$ do not vary appreciably from p approximately in the range of about 30% to 70%.

At $p = 0.3$, the standard deviation is $\sqrt{0.3 \cdot 0.7} = \sqrt{0.21} = 0.46$. The largest deviation, or error, then is for those species or categories in the middle (species of medium abundance), not at the ends (rare or abundant species). Notice that the curve is symmetrical and drops precipitously at the two ends. If a species is calculated to comprise 99.9% of the population, researchers are seldom concerned with placing confidence limits on such a high value. Whether the true value lies between, say, 99.85% and 99.95% is of little importance. Rare species are, however, often of great concern, and some investigators fret over confidence limits for a species calculated to comprise, for example, only 0.1% of the total. As Figure 8.3 shows, we have the same problem at both ends. In statistics texts, this problem is called the *degree of fit* of the Normal approximation, which improves as p comes closer to 0.5.

For example, if $n = 1,000$ and $p = 0.001$, and if we choose a 95% level of confidence so that $z_p = 1.96$, then using Equation 8.20 we obtain

$$z_p \sqrt{\frac{pq}{n-1}} = 1.96 \sqrt{\frac{0.001 \cdot 0.999}{1000}} = 0.00006.$$

Hence, from Equation 8.20, the confidence interval is $0.0009 \le P \le 0.001$. This simply tells us that P, the true proportion of this species, is very small. Thus the whole exercise becomes meaningless. The meaninglessness of this trivial exercise was intuitively clear at the other end of the distribution, when the most abundant species was considered; but it is not so obvious to those seeking information on rare species. As the sample size, n, increases, the situation becomes worse because the value of p approaches the value of $z_p \hat{\sigma}_p$. Consequently, if n is large and only a single individual of a species is encountered, as is often the case, confidence limits contribute nothing to the characterization of the fauna or flora.

AN EXAMPLE OF SPECIES PROPORTION: ESTIMATION FROM THE BOLIVIAN TREES

In the Bolivian forest plot, Dallmeier and colleagues (1991) identified 52 tree species and plotted on x–y coordinates the position of each of the 663 individuals. The true species proportions are given in Appendixes 1 and 2. As an illustration of the closeness with which we can estimate the actual species proportion when we apply the Binomial methods, we will take random samples of individuals from this population and calculate confidence limits.

We begin by using the x–y coordinates given in Dallmeier and colleagues (1991) to obtain some differently sized samples of n randomly chosen individuals. The sample sizes selected are $n = 25$, 100, 200, and 300 individuals. In general, if the ratio of $\dfrac{n}{N}$ is less than about 5%, which is the case in most natural population surveys, the fpc need not be used. In the example given here the smallest sample size is $n = 25$ individuals and the total number of individuals is $N = 663$. Hence, $\dfrac{25}{663} = 0.037$, or about 4%. However, for $n = 100$ individuals, we get an fpc of $\dfrac{100}{663} = 0.15$, or 15%. For $n = 300$ we are sampling nearly half the total, so we will use the fpc in the example given here as an illustration.

Table 8.6 gives the true proportion for each species, as well as the estimate of the proportion and the confidence limits obtained from samples of 25, 100, 200, and 300 individuals. The calculation of confidence limits for *Scheelea princeps* for the sample of size $n = 100$ is presented here in the text to illustrate the procedure. The estimate of p, calculated from the sample of 100, is 0.50, so that $q = 1 - 0.50 = 0.50$, and pq is estimated to be 0.25. From Equation 8.13, we have

$$\hat{\sigma}_p^2 = \left[\frac{N - n}{(n - 1)} \right] \cdot N \cdot pq = \left[\frac{663 - 100}{100 - 1} \right] \cdot 663 \cdot 0.25 = \frac{563}{66637} \cdot 0.25 = 0.0086 \cdot 0.25 = 0.0021 \,.$$

The standard error of p is $\hat{\sigma}_p = \sqrt{0.0021} = 0.0463$. If we require 0.95 confidence, then $z_p = 1.96$ and $z_p \hat{\sigma}_p = 1.96 \cdot 0.0463 = 0.0907$, or 0.09. The confidence interval is then 0.50 − 0.0907 ≤ P ≤ 0.50 + 0.0907, or 0.4093 ≤ P ≤ 0.5907. The use of the continuity correction (Equation 8.23) would change these values only in the third decimal place; that is, to the interval 0.4043 ≤ P ≤ 0.5957.

As we indicated earlier, the correction often amounts to very little. Had we ignored the fpc and the continuity correction, then we would have $\hat{\sigma}_p = \sqrt{\dfrac{0.25}{100}} = 0.05$, and

TABLE 8.6 True Bolivian Species Proportion, Estimated Proportions, and 95% Binomial Confidence Limits

Species	True p	$n = 25$ p	$z_p\hat{\sigma}_p$	$n = 100$ p	$z_p\hat{\sigma}_p$	$n = 200$ p	$z_p\hat{\sigma}_p$	$n = 300$ p	$z_p\hat{\sigma}_p$
1. *Scheelea princeps*	.38	.48	±.20	.50	±.09	.45	±.06	.39	±.04
2. *Brosimum lactescens*	.20	.20	±.16	.14	±.06	.18	±.04	.21	±.03
3. *Pouteria macrophylla*	.09			.06	±.04	.08	±.03	.10	±.02
4. *Calycophyllum spruceanum*	.04	.04	±.04	.04	±.02	.05	±.02		
5. *Heisteria nitida*	.03	.04	±.08	.02	±.02	.02	±.02	.03	±.01
6. *Ampelocera ruizii*	.03			.01	±.02	.005	±.008	.03	±.01
7. *Astrocaryum macrocalyx*	.03	.04	±.08	.03	±.03	.02	±.02	.02	±.01
8. LIANA	.03	.08	±.11	.02	±.02	.02	±.02	.02	±.01
9. *Salacia*	.03			.02	±.02	.02	±.02	.02	±.01
10. *Sorocea saxicola*	.01							.01	±.01
11. *Trichilia pleena*	.01	.04	±.08	.02	±.02	.02	±.02	.01	±.01
12. *Aspidosperma rigidum*	.01			.01	±.02	.02	±.02	.01	±.01
13. *Pithecellobium latifolium*	.01	.04	±.08	.01	±.02	.005	±.008	.01	±.01
14. *Swartzia jorori*	.01					.005	±.008	.01	±.01
15. *Guapira*	.006							.003	±.004
16. *Triplaris americana*	.006					.01	±.01		
17. LEGUM	.006			.02	±.02	.01	±.01	.01	±.01
18. *Acacia loretensis*	.005					.005	±.008		
19. *Cupania cinerea*	.005					.005	±.008	.003	±.004
20. *Gallesia integrifolia*	.005			.01	+.02	.005	±.008	.003	+.004
21. *Pera benensis*	.005					.01	±.01	.003	±.004
22. *Spondius mombin*	.005			.01	±.02	.005	+.008	.003	±.004
23. *Swartzia*	.005			.01	±.02	.01	+.01	.003	±.004
24. *Erythroxyl*	.003							.01	±.01
25. *Eugenia*	.003			.01	±.02	.005	±.008		
26. *Garcinia brasliensis*	.003							.003	±.004
27. *Inga cinnamomea*	.003							.003	±.004
28. *Inga edulis*	.003								
29. *Pithecellobium angustifolium*	.003			.01	±.02	.005	±.008	.003	±.004
30. *Simaba graicile*	.003			.01	±.02	.01	±.01		
31. *Virola sebifera*	.003			.01	±.02	.005	±.008	.003	±.004
32. *Xylopia ligustrifolia*	.003								
33. *Byrsonima spicata*	.002					.005	±.008	.003	±.004
34. *Copaifera reticulata*	.002							.003	±.004
35. *Cordia*	.002					.005	±.008		
36. *Drypetes*	.002							.003	±.004

(Continued)

TABLE 8.6   (continued)

SPECIES	TRUE p	$n = 25$		$n = 100$		$n = 200$		$n = 300$	
		p	$z_p \hat{\sigma}_p$	p	$z_p \hat{\sigma}_p$	p	$z_p \hat{\sigma}_p$	p	$z_p \hat{\sigma}_p$
37. *Faramea*	.002								
38. *Ficus trigona*	.002								
39. *Genipa americana*	.002								
40. *Guazuma ulmifolia*	.002			.01	±.02	.005	±.008		
41. *Inga*	.002								
42. *Luehea cymulosa*	.002					.005	±.008		
43. *Maytenus ebenifolia*	.002					.005	±.008		
44. *Miconia*	.002	.04	±.08	.01	+.02	.005	±.008		
45. *Platymiscium*	.002	.04	±.08	.01	±.02	.005	±.008		
46. *Randia armata*	.002							.003	±.004
47. *Salacia*	.002								
48. *Toulicia reticulata*	.002							.003	±.004
49. *Unonopsis mathewsii*	.002							.003	±.004
50. *Vochysia mapirensis*	.002								
51. −4	.002								
52. −5	.002								

NOTE: The sample unit is individual trees. CL $= p \pm z_p \hat{\sigma}_p$.

$z_p \hat{\sigma}_p = 1.96 \cdot 0.05 = 0.098$, so that $0.4020 \leq P \leq 0.5980$. In all cases we missed the mark because the true proportion is $P = 0.38$, and the confidence interval is (0.40, 0.60). The most influential factors in this "miss" of the confidence interval seem to be our poor initial estimate of P (that is, we got $p = 0.5$) and the possibility of non-independent contents of our samples. This value of $p = 0.50$ is an unbiased estimate, but it is 0.12 away from the true mean, which we know to be 0.38. We were simply unlucky.

All in all, this exercise provides an excellent example of an often overlooked fact in fieldwork in which the cost of obtaining samples may be quite high. If we have only one estimate of a parameter, it is widely understood that the larger the sample size, the narrower the confidence interval. However, as we can see from the example, a large sample size (and thereby a narrow confidence interval) is of no use if our one initial estimate is a poor one. A look at any of the sampling distributions that we have tabulated in this book shows that there will always be the possibility of large variability in our estimates. This is one more reason why more than one estimate is useful; replication is vital for good scientific work.

A BIOLOGICALLY REPRESENTATIVE SAMPLE SIZE

In preparing to sample for species proportions, field researchers often struggle with the question of how many individuals must be sampled in order to produce results that are "representative." This is an ambiguous question, and consideration of Table 8.6 provides us with an equally ambiguous answer: "It depends."

The first consideration, which is always of paramount importance in field study, is the amount of work required to secure n individuals of the particular group in question. The more we count, the more "representative" the sample becomes. However, regardless of the work involved, perhaps 25 were all that could be obtained. We can see that in our $n = 25$ individuals for our (and $N = 663$) Bolivian tree example, coming up with a database of just 25 proved to be intolerably unrepresentative. Species comprising as much as 9% of the population were not encountered at all. Another sample of 25 might encounter these common species but miss others. The confidence limits for a sample of $n = 25$ individuals indicate that not only are we likely to miss some relatively abundant taxa, but also that the species proportions for the ones we do encounter have very wide confidence intervals; that is, we shouldn't trust our point estimates to be close to the true values. At $n = 100$, however, the situation is much better. Only 2 species having proportions of 0.01 (that is, 1% of the total species) were missed; the confidence limits narrow considerably. At sizes of $n = 200$ and $n = 300$, the confidence limits are within a range most researchers would consider reasonable. Note that, in going from 200 to 300 individuals, the decrease in confidence limits, and thus the change in the width of the confidence interval, is small. Consequently, the amount of additional work required to achieve a not very large improvement in results should be considered before one opts for the more costly field survey. In our particular sample of $n = 300$, the estimates of P were remarkably close to the true value.

For species having a true proportion of less than 0.01, a good segment of them will not be sampled at any of the chosen sample sizes. We recall that for any species i, the probability (p) that a randomly chosen individual will belong to species i is given by $p = 1 - q$. For the Binomial distribution the probability that n individuals will each belong to species i is $p^n = (1 - q)^n$. Thus the probability that none will belong to this particular species must be $1 - p^n$, since all the probabilities must add to 1. Consequently, the probability that at least 1 individual from n randomly chosen individuals will belong to species i is $1 - q^n$. In the Bolivian plot there are 20 species represented by only 1 individual each. Since this hectare plot is our population, the true species proportion, P, for each of these is $\dfrac{1}{663} = 0.0015$ (0.002 in Table 8.6). For a size of $n = 25$, $1 - q^n = 1 - 0.9985^{25} = 0.037$. That is, if we choose 25 individuals at random from the 663 in the population, and if species i has a true proportion of $P = 0.0015$, then the probability that an individual of species i will be among the 25 individuals in our sample is 0.037. At a sample size of 100 individuals, we have $1 - 0.9985^{100} = 0.14$; for the sample of size 200, we get $1 - 0.9985^{200} = 0.26$. Finally, at a size of 300 individuals, the probability is $1 - 0.9985^{300} = 0.36$.

Table 8.6 shows what we might expect from any n value chosen for our survey; namely, that we increase the number of rare species encountered as we increase sample size. But the same rare species are not always encountered in different samples. Although we have sampled nearly half the population at $n = 300$, only 32 of the 52 species occurred in that sample. Interestingly, in this case, 32 also occurred in the $n = 200$ sample. In fact, if we were to apply statistical theory, we could say that we would never be able to sample anywhere near all the rare species. The *Borell–Cantelli Lemma* (Cramér 1946) states that for any countably infinite series of events (in our case, the selection of individuals of rare species), with associated probabilities, under a mild condition (assumption) that is reasonable, that after some point of greater and greater sampling only a finite number of our species can ever be found or seen. Here the serendipity involved with the sampling of rare species manifests itself.

STATISTICAL SAMPLE SIZE FOR RELATIVE ABUNDANCES OF SPECIES

The equation for the confidence limits of a proportion can easily be manipulated to allow for the calculation of the sample size needed to attain the desired interval size and the degree of confidence we wish to place on it. Similar to what we did in Chapter 5, and ignoring the fpc, we can designate the error we are willing to make when using p instead of the true value P as $d = z_p \sqrt{\dfrac{pq}{n}}$. Squaring both sides we obtain $d^2 = z_p^2 \left(\dfrac{pq}{n} \right)$. Transposing provides the desired solution; namely

$$n = \frac{z_p^2 pq}{d^2}. \tag{8.24}$$

The observant reader might have noticed that we used the biased form of the standard error, $\sigma_p = \sqrt{\dfrac{pq}{n}}$, instead of the unbiased form, $\sigma_p = \sqrt{\dfrac{pq}{n-1}}$ from Equation 8.15. Use of Equation 8.15 would have increased our estimated sample size by only 1, and since Equation 8.24 gives an approximate size, we chose the more common form to solve for n.

As an example, let us say we wish to be 95% confident that the error will be no more than ± 0.05 for a species with an estimated proportion of $p = 0.20$; for example, *Brosimum lactescens* in the Bolivian plot. How many total individuals would we have to sample to achieve this level of precision and confidence? We have $z_p = 1.96$, so, $z_p^2 = 3.8416$, $p = 0.20$, $q = 0.80$, and $pq = 0.1600$. Then for ± 0.05 as the desired interval width, $d = 0.05$ and $d^2 = 0.0025$. From Equation 8.24 we find $n = \dfrac{3.8416 \cdot 0.1600}{0.0025} = 245.86$, or 246 individuals are required. Note that in Table 8.6, for a sample size of $n = 200$ and $p = 0.18$, the confidence limits were $p \pm 0.04$. The slight discrepancy arises from the use of the fpc in the calculation of Table 8.6. To illustrate further the relationship between some desired confidence limits, of width d and the number of individuals needed to obtain them, we constructed Table 8.7.

Table 8.7 reveals that for a small gain in our interval estimate; that is, a narrowing of the interval, a very large premium must be paid in the numbers of individuals that must be sampled. As pointed out earlier, and as Figure 8.3 shows, the largest value of the standard error occurs at $p = 0.50$, although only a little difference in the standard error exists between 0.30 and 0.50. When p is in the range of 0.30 to 0.50, in order to decrease d from 0.05 to 0.03 we are required to obtain three times as many individuals.

TABLE 8.7   The Number of Individuals (n) Required to Estimate Species Proportions at a 95% Confidence Level

p	pq	$z_p^2 pq$	d	d^2	n
0.05	0.0475	0.1825	0.05	0.0025	73
			0.03	0.0009	203
0.10	0.0900	0.3457	0.07	0.0049	70
			0.05	0.0025	138
			0.03	0.0009	384
0.20	0.1600	0.6147	0.07	0.0049	125
			0.05	0.0025	246
			0.03	0.0009	683
0.30	0.2100	0.8067	0.07	0.0049	165
			0.05	0.0025	323
			0.03	0.0009	896
0.40	0.2400	0.9220	0.07	0.0049	188
			0.05	0.0025	369
			0.03	0.0009	1,024
0.50	0.2500	0.9604	0.07	0.0049	196
			0.05	0.0025	384
			0.03	0.0009	1,067

NOTE: A confidence level of 95% is assumed throughout. p = species proportion; pq = mean; $z_p^2 pq$ is divided by different values of n to show how the width of the confidence interval will change slightly, but the corresponding estimated sample size will increase greatly; $z_p = 3.8416$; $q = 1 - p$; $n = \dfrac{z_p^2 pq}{d^2}$; $d = z_p \sqrt{\dfrac{pq}{n}}$; $d^2 = \dfrac{z_p^2}{\frac{pq}{n}}$; width of the confidence interval is $p \pm d$; d = the error we are willing to make when using p in place of P with confidence level z_p.

If we are interested in a particular proportion, we can easily choose from Table 8.7, or we can calculate the number of individuals required for the desired interval estimate size. Usually, however, researchers are interested in establishing species proportions for the entire group under consideration. If we are willing to accept a confidence limit of ± 0.07 for the worst case, namely, $p = 0.50$, then about 200 individuals will do. If we insist on the tighter confidence limits of ±0.05, we will require nearly 400. Depending on the work required, 200 to 400 individuals should satisfy most investigators' requirements, regardless of the value of N, which is usually very large.

This n of 200 to 400 is an enormous number of individuals for some taxonomic groups, such as many of the vertebrates we might wish to sample. For groups of organisms where only a few individuals can be sampled, we must simply accept very wide confidence intervals. The columns showing confidence limits for $n = 25$ in Table 8.6

illustrate the difficulty inherent in making estimates of species percentages based on only a small n.

CLUSTER SAMPLING FOR SPECIES PROPORTIONS

As we have indicated, researchers often prefer species proportions over densities. Species proportions are preferred because they are easily calculated from a single biological sample; they are circumscribed (that is, they range only between 0 and 1); and they allow for a quick comparison of one area with another. We have shown that 200 randomly chosen individuals will provide us with a margin of error (that is, the width about the estimate, not the true value P) of ± 0.07 in the worst case (when $p = 0.50$) at the 95% confidence level. Most researchers are quite content with this amount of precision and go happily about their work. Unfortunately for these investigators, depending on their intention, they may be operating under a false sense of security.

The selection of n for the number of individuals to be randomly sampled was easily accomplished in the Bolivian example because each individual in the entire plot was mapped in advance. This kind of knowledge is obviously not available for fieldwork. The target population in this example is the entire plot, and the primary units sampled are the individuals, and they are easily sampled. Likewise, in some situations, individual organisms can be sampled directly in the field from a relatively large area.

Often, however, the quarry are smaller organisms such as small mammals, birds, fish, amphibians, or insects, or the host of invertebrates or protists in soil, water, or sediment. Usually these individuals cannot be sampled directly; that is, one by one. Instead, a sampling device is commonly used. A single biological sample, such as a quadrat, transect, net, or core, may often contain hundreds to thousands of individuals, depending upon the target taxon. Field or laboratory extraction of these individuals, performed through a variety of techniques, often allows us to pick individuals at random from the biological sample. Here the situation is analogous to the Bolivian example; the entire plot is the target population and the individuals are the primary units. This is the kind of situation we have been investigating thus far in this chapter. Calculation of Binomial confidence limits are appropriate and are applicable, by inference, to the biological sample that, in this case, is the target population. Thus such confidence limits apply only to the biological sample. Essentially, we are placing confidence limits on our laboratory or field procedure.

Alternatively, researchers may often assume that the biological sample is representative of a much larger area, which is the real subject of their interest (the actual desired target population). There are now really two levels of inference: the first from the sample of individuals to the available target population; the second from the available target population to the larger, desired target population. Researchers may (wrongly) treat the counted individuals as though these individuals were selected at random from the larger target population, and they often use the confidence limits obtained accordingly. Herein lies the rub. Individuals have actually not been chosen at random from this larger population, because individuals were not the primary sampling units (Chapter 7). Instead, all the individuals were counted in a specific area or space. The primary sampling unit is thus the quadrat, not the individual. Unlike some of the smaller corrections we have introduced for the fpc or continuity, this misconception of what the sampling unit really is can be a potential source for serious error. When we choose a quadrat, we must accept the individuals contained therein.

In statistical terminology, the required methodology is called *cluster sampling for proportions*. That is, we designate or choose the *n* quadrats at random, not *n* individuals. As we have discussed (Chapter 7), quadrat sampling can be a form of cluster sampling, but it does not have to be. In fact, the form of quadrat sampling we did to obtain density estimates was not an example of cluster sampling. In our quadrat sampling, using the Bolivian forest data, we were not selecting individuals at random from the entire population of 663; we were selecting *n* quadrats at random from an entire population of 100, or in some cases, 25 quadrats. Therefore, usually each of the *n* clusters contained a different number of individuals. We may, for example, be interested in the proportions of sexes, species or genera, or age groups in the target population.

Let us begin our investigation of cluster sampling for calculating species proportions by examining just one of the *n* quadrats, in order to acquire some notation to describe the problem. We will consider our favorite quadrat, or biological sample: *i*. In this quadrat, we will find a total of m_i individuals of all species, and this number will include a_i individuals of the species, group, or type in which we have an interest. (We introduce entirely new symbols here, *m* and *a*, in order to distinguish them from *n* and *N*.) From the entire set of all *n* quadrats sampled (of the *N* possibilities in the entire target population), we will have a total of $m = \sum m_i$ individuals. It should be clear that, for any particular species or grouping in this i^{th} quadrat, there is a basic proportion,

$$p_i = \frac{a_i}{m_i}.$$
(8.25)

That is, the number of distinct individuals that we found of the species of interest is divided by the total number of individuals from all species that we found in this single quadrat, or cluster. This is the proportion of the species of interest in our single quadrat.

Continuing on for the entire set of all *n* quadrats, or clusters, from which we obtain our entire sample, there is an overall proportion of interest. We can write this proportion as

$$p = \frac{\sum_{i=1}^{n} a_i}{\sum_{i=1}^{n} m_i}.$$
(8.26)

This formula tells us first to add all the individuals of species *i* that we found in each of the *n* quadrats; this is the numerator. The denominator is the total of all individuals of all species found in the *n* quadrats. Equation 8.26 is an *estimate (p) of the true proportion (P)* for our particular species in the target population. However, this estimate, unlike that for simple Binomial sampling, is biased; that is, we can be sure that in repeated samples *p* will not equal *P*. The bias will be more serious for highly clumped, nonhomogeneous target populations, when some quadrats provide very large or very small, extreme means. The reason there is bias is that the calculation of the variance of the proportions does not take into account the variation between the sampling units (the randomly selected quadrats). These quadrats are, in effect, *replications* and must be considered. Remember, however, that if the quadrats were selected in an ordered way (Chapter 7, systematic sampling), this variance would not be included in our final variance estimate. If we let

$$\hat{\mu}_m = \frac{\sum_{i=1}^n m_i}{n} = \frac{m}{n},$$ (8.27)

then this formula for estimating the total mean will represent the total of all individuals (*m*) divided by the number of quadrats (*n*). Therefore, $\hat{\mu}_m$ will also represent *the average number of individuals per quadrat* in the entire sample. With these quantities, we can now write the formula for the variance of the value we derived for (species) proportion in Equation 8.25. The sample variance is calculated as

$$\hat{\sigma}^2_{pclus} = \frac{\sum_{i=1}^n m_i^2 (p_i - p)^2}{n(n-1)\hat{\mu}_m^2}.$$ (8.28)

The subscript *pclus* (which stands for cluster sampling methods for ascertaining *p*) is used to avoid confusion with $\hat{\sigma}^2_p$ given by Equation 8.13.

The square root of the variance, $\hat{\sigma}_{pclus}$, is the standard error. In this case it is the *standard error of the species proportion* for cluster sampling. In this chapter we shall also use the notation $\hat{\sigma}_{pbin}$ for clarity, to distinguish the standard error from Binomial sampling of individuals from the quantity determined for clusters.

To illustrate, we return to the Bolivian plot. When the plot is divided into 25 quadrats (Appendix 2), the mean number of trees per quadrat is about 26 $\left(\frac{663}{25} = 26.52 \right)$.

In order to make the results obtained from cluster sampling simpler to compare with the results obtained using Binomial sampling, we shall pick 3 different sample sizes at random from Appendix 2. First we select a sample of just 1 single quadrat; our second sample will have 4 quadrats; and the last sample will contain 8 quadrats. We would expect these selections to yield about 26, 100, and 200 individuals, respectively. How were these estimates obtained?

We know the overall mean density estimate is about 26. Because this is an average, we make the assumption, which is better for homogeneously distributed populations than for aggregated, that, on average, each quadrat will contain this same number of individuals. Thus, we have an estimate of 26 individuals in just the 1 quadrat. For our samples with multiple quadrats, we have $4 \cdot 26 = 104$ individuals, and $8 \cdot 26 = 208$ individuals, which we rounded to 100 and 200, respectively, for convenience. Now, for 1 quadrat, Equation 8.28 cannot be used because with only 1 quadrat or cluster there is no variance, and so only the Binomial estimate of the standard error can be calculated. This is another illustration of why replication is so important.

To alleviate any anxiety Equation 8.28 might produce, we will illustrate the calculation of the confidence limits for the species *Scheelea princeps* from the data obtained from 8 random quadrats selected from Appendix 2. Table 8.8 displays the elements necessary for the calculation. We have from Equation 8.28 that the variance of the overall proportion based upon the 8 quadrats is

$$\hat{\sigma}^2_{pclus} = \frac{172.3860}{(8)(7)(631.26)} = \frac{172.3860}{35350.56} = 0.0049$$

and its square root, the standard error of the proportion, is $\hat{\sigma}_{pclus} = 0.0698$. Then, to obtain the 95% confidence limits we use Equation 8.21. That is, we multiply the standard error by 1.96 to obtain $p \pm 0.1369 \approx p \pm 0.14$. Had we used the Binomial calculation, then, from Equation 8.14, $\hat{\sigma}^2_{pbin} = \frac{pq}{n-1} = \frac{0.2239}{200} = 0.0011$, where $q = 1 - 0.3383$; and from Equation 8.15, $\hat{\sigma}_{pbin} = 0.0335$. It is important to point out that in Equations 8.14 and 8.15 the letter n is the total number of individuals composing the sample, whereas in Equation 8.28, n refers to the total number of quadrats. Now, multiplying $\hat{\sigma}_{pbin} = 0.0335$ by 1.96, we obtain $p \pm 0.0656 \approx p \pm 0.07$, which is half the value of ± 0.14 that we get when using $\hat{\sigma}_{pclus}$. Consequently, when confidence limits are attached to our estimates, we have $0.20 \leq P \leq 0.48$ for the *cluster confidence interval*, whereas for the Binomial confidence interval we obtain $0.27 \leq P \leq 0.40$. We know, of course, that the true proportion for this plot is 0.38 (Appendix 2), and so both intervals contain the true value.

TABLE 8.8 Calculation of the Variance for Cluster Sampling of the Species Proportion for *Scheelea princeps*, Bolivian Plot 01

QUADRAT	I	a_i	m_i	m_i^2	p_i	$(p_i-p)^2$	$m_i^2(p_i-p)^2$
1	1	0	30	900	0[1,2]	0.1144	102.96
14	2	6	20	400	0.3000	0.0015	0.60
3	3	8	24	576	0.3333	0.000025	0.0144
18	4	13	23	529	0.5652[1,2]	0.0515	27.2435
4	5	6	24	576	0.2500[2]	0.0078	4.4928
19	6	14	37	1,369	0.3784	0.0016	2.1904
10	7	14	24	576	0.5833[1,2]	0.0600	34.5600
23	8	7	19	361	0.3684	0.0009	0.3249
		$\Sigma = 68$	$\Sigma = 201$				$\Sigma = 172.3860$

NOTE: $N = 25$ quadrats; $n = 8$. The number of the quadrat is i, $i = 1 \ldots 8$. Values for calculating Equations 8.26, 8.27, and 8.28 are given by $p = \Sigma a_i / \Sigma m_i = 68/201 = 0.3383$; $\hat{\mu} = \Sigma \frac{m_i}{n} = \frac{201}{8} = 25.12$; and $\mu^2 = 631.26$, respectively.

[1] These values of p_i lie outside the interval estimates for p based upon the cluster methods.
[2] These values for p_i lie outside the interval estimates for p based upon the Binomial.

For this example of estimating species proportions based on a sample of 8 quadrats from a target population of 25 quadrats, we showed that the cluster formula gave a confidence interval about the *total proportion* (that is, the true species relative abundance, *P*) based upon an estimate calculated from the sample of size 8. The confidence interval for *P* was based upon the value of *p* calculated from Equation 8.26, not any single species proportion in a single quadrat or observation, p_i (Equation 8.25). This means that the confidence interval was based on all quadrats sampled. However, it is quite common to find that a researcher selects only one area, habitat, or quadrat upon which to base the investigation of a much larger population. This would be equivalent to using one of the p_i's as the total estimate of *P*. That means that we would be basing our research and our estimate on a sample of size 1. Cast in this light, researchers should be careful indeed to ensure that they do not slip into confusing the individual organism as the sampling unit when their data set was really collected with the quadrat as the sampling unit.

Let us examine the set of individual quadrat proportions for the Bolivian forest plot that has been divided into 25 quadrats, out of which we have randomly selected 8 for sampling (Table 8.8). Three of the p_i's obtained (those for quadrats 1, 18, and 10) lie outside the cluster confidence limits for the total species proportions, and four of them (1, 18, 10, and 4) are outside the Binomial limits. By looking at Table 8.8, we can see why. The random selection of quadrats produced some unlucky choices. In quadrat 1 the most abundant species, *Scheelea princeps*, is not even present at all, and in quadrats 18 and 10 this species is about 20% more abundant than the true proportion of $P = 0.38$.

Table 8.9 shows the confidence limits (CL) that were then calculated separately for each species, using the quadrat data. Confidence limits were calculated both with the Binomial formula (Equation 8.20) and by the cluster formula (Equation 8.28) for estimating standard errors. For the more abundant species, the cluster confidence limits are usually larger. The two sets of limits become approximately equal, as might be expected, for the rarer species. For those species making up 1% or less of the fauna there is virtually no difference between the two calculation methods. The number of species encountered in the quadrat sampling was similar to the numbers encountered when individuals were sampled at random. Of the 52 species in the plot, 7 were encountered in the single quadrat sample with 19 individuals; 24 species in the 4 quadrats contained 104 individuals; and 31 species were in 8 quadrats, with 201 individuals. Although the confidence limits calculated using the cluster formulas are the appropriate limits for the quadrat sampling of the Bolivian plot, Table 8.9 indicates that the Binomial limits almost always give confidence limits that bracket the true proportion (except for *Brosimum lactescens*, with 104 individuals). We should keep in mind, however, that the Bolivian tree plot is quite homogeneous. Other, less homogeneous, data sets will not behave as well. An illustration is in order.

A FOSSIL EXAMPLE OF CONFIDENCE INTERVAL ESTIMATION FOR RELATIVE ABUNDANCE

Here we will undertake an exercise that uses data other than our (usual) Bolivian forest trees. In the Bolivian data set the total content of the quadrat is counted. Here we introduce an example to highlight the difficulties encountered when species are not

TABLE 8.9 Binomial and Cluster 95% Confidence Interval Calculations for Three Sample Sizes from Bolivia

		$n = 1$		$n = 4$			$n = 8$		
	True	p	bin	p	bin	clus	p	bin	clus
Species	p	$m = 19$	$z_p\hat{\sigma}_p$	$m = 104$	$z_p\hat{\sigma}_p$	$z_p\hat{\sigma}_p$	$m = 201$	$z_p\hat{\sigma}_p$	$z_p\hat{\sigma}_p$
1. *Scheelea princeps*	.38	.37	±.22	.42	±.10	±.10	.34	±.07	±.14
2. *Brosimum lactescens*	.20	.26	±.20	.12	±.06	±.09	.25	±.06	±.09
3. *Pouteria macrophylla*	.09	.10	±.14	.06	±.04	±.07	.08	±.04	±.06
4. *Calycophyllum spruceanum*	.04	.10	±.14				.04	± .03	± .03
5. *Heisteria nitida*	.03	.05	±.10	.06	±.04	±.06	.03	±.02	±.04
6. *Ampelocera ruizii*	.03			.02	±.03	±.02	.01	±.02	±.02
7. *Astrocaryum macrocalyx*	.03			.04	±.04	± .05	.02	±.02	±.02
8. LIANA	.03			.04	±.04	± .05	.03	±.03	±.04
9. *Salacia*	.03			.05	±.04	+.02	.02	±.02	±.02
10. *Sorocea saxicola*	.01						.01	+.01	±.01
11. *Trichilia pleena*	.01			.01	±.02	±.02	.01	±.01	±.01
12. *Aspidosperma rigidum*	.01			.01	±.02	±.02	.01	+.01	±.01
13. *Pithecellobium latifolium*	.01			.02	±.03	±.02	.005	±.01	±.01
14. *Swartzia jorori*	.01						.01	±.01	±.01
15. *Guapira*	.006			.02	±.03	±.04	.01	±.01	±.02
16. *Triplaris americana*	.006								
17. LEGUM	.006			.01	±.02	±.02	.01	±.01	±.01
18. *Acacia loretensis*	.005	.05	±.10				.01	±.01	±.01
19. *Cupania cinerea*	.005						.01	±.01	±.02
20. *Gallesia integrifolia*	.005			.01	±.02	±.02	.01	±.01	±.01
21. *Pera benensis*	.005						.005	±.01	±.01
22. *Spondius mombin*	.005			.02	±.03	±.02	.01	±.01	±.01
23. *Swartzia*	.005								
24. *Erythroxyl*	.003						.005	±.01	±.01
25. *Eugenia*	.003								
26. *Garcinia brasiliensis*	.003								
27. *Inga cinnamomea*	.003			.01	±.02	±.02	.005	±.01	±.01
28. *Inga edulis*	.003			.01	±.02	±.02			
29. *Pithecellobium angustifolium*	.003								
30. *Simaba graicile*	.003			.01	±.02	±.02			
31. *Virola sebifera*	.003								

(continued)

TABLE 8.9   *(continued)*

	TRUE	n = 1		n = 4			n = 8		
		p	bin	p	bin	clus	p	bin	clus
SPECIES	p	$m = 19$	$z_p\hat{\sigma}_p$	$m = 104$	$z_p\hat{\sigma}_p$	$z_p\hat{\sigma}_p$	$m = 201$	$z_p\hat{\sigma}_p$	$z_p\hat{\sigma}_p$
33. *Byrsonima spicata*	.002			.01	±.02	±.02	.005	±.01	±.01
34. *Copaifera reticulata*	.002	.05	±.10				.005	±.01	±.01
35. *Cordia*	.002						.005	±.01	±.01
36. *Drypetes*	.002								
37. *Faramea*	.002								
38. *Ficus trigona*	.002						.005	±.01	±.01
39. *Genipa americana*	.002								
40. *Guazuma ulmifolia*	.002								
41. *Inga*	.002						.005	±.01	±.01
42. *Luehea cymulosa*	.002								
43. *Maytenus ebenifolia*	.002						.005	±.01	±.01
44. *Miconia*	.002								
45. *Platymiscium*	.002								
46. *Randia armata*	.002			.01	±.02	±.02			
47. *Salacia*	.002								
48. *Toulicia reticulata*	.002			.01	±.02	±.02			
49. *Unonopsis mathewsii*	.002			.01	±.02	±.02			
50. *Vochysia mapirensis*	.002								
51. −4	.002			.01	±.02	±.02	.005	±.01	±.01
52. −5	.002			.01	±.02	±.02			

NOTE: The sampling unit is quadrats, where $N = 25$ quadrats in the Bolivian forest plot; 95% confidence limits $(CL = p \pm z_p\hat{\sigma}_p)$ are calculated by using $\hat{\sigma}_p$ from the Binomial and cluster methods.

homogeneously distributed and to highlight the importance of selecting the appropriate formula for confidence interval construction. When subsampling is used (for example, when organisms are microscopic) and the quadrat's contents are sampled, the cluster method must be used.

In the Chesapeake Bay region, rich fossiliferous beds of Miocene sediments are exposed at Calvert Cliffs, Maryland. On the horizontal surface of Bed 16 a square template 1 m on a side (1 m² quadrat) was placed on the exposure (Buzas and Gibson 1990). This quadrat was divided into 100 equal-sized subquadrats, and 5 random samples of sediment (each of 10 ml) were retrieved. The number of individuals of each species of foraminifera was enumerated.

To gain perspective, we note that given the diameter of a foraminiferal shell (about 0.5 mm), the area delineated (1 m²) is equivalent to a 250,000 m² area for a tree with a diameter of 0.25 m, which is the average breast-height diameter of the trees in Bolivian plot 01. Recall that the area for the Bolivian plot 01 is only 10,000 m². The foraminifera, like a host of other smaller organisms, require only a very small areal sample in relationship to their own size. A biological sample for smaller organisms, such as foraminifera, is often required to represent a much larger area relative to the size of the organism than is a sample used to sample larger organisms. For example, for smaller organisms a few replicate samples, each only 10 cm² in size, might be used to develop density and proportion estimates for an area that is several km²; whereas for larger organisms a few replicate samples, each 10 m², might be used to develop population estimates for the same amount of territory. Provided that the organisms are distributed throughout the area investigated, this reduction in n/N for small organisms presents no problem.

When a standard-sized sample is taken (in this case, 10 ml) of very abundant organisms, the number of individuals in the sample is often too numerous to count. Consequently, the researcher must subsample the sample. Ideally, individuals can be chosen at random from the biological sample to form a subsample. In our Miocene foraminiferal example, this was accomplished by pouring the sample through a splitter, which allows each individual to fall through in one of two possible ways. The first split, then, randomly divides the individuals in the sample so that their total is now $\frac{1}{2}$ the original. The $\frac{1}{2}$ sample can then be divided again so that $\frac{1}{4}$ is obtained, and so on. The 5 samples in our example were split into the fractions $\frac{1}{4}, \frac{1}{8}, \frac{1}{4}, \frac{1}{4}$, and $\frac{1}{4}$. (We introduced the $\frac{1}{8}$ fractionation for one sample simply because it contained so many more individuals than the other samples.) The total number of individuals counted in each subsample were 398, 433, 421, 316, and 563, respectively.

Notice that if we wished to estimate the average density per 10 ml, we would need to multiply these totals by the denominator of the fraction to obtain the total number of individuals per 10 ml. Our estimate for the number of individuals in each 10 ml sample would then be 1,592, 3,464, 1,684, 1,264, and 2,252. The estimate for the mean density is $\hat{\mu} = 2,051.20$, which is determined by $\frac{1592 + 3464 + 1684 + 1264 + 2252}{5} = \frac{10256}{5}$.

The variance is $\hat{\sigma}^2 = 750,427.20$, and the standard deviation is $\hat{\sigma} = 866.27$. Unlike the homogeneous Bolivian example, the very large variance relative to the mean indicates considerable spatial heterogeneity in our foraminifera fossils (see Chapter 6). Thus we would expect that the Binomial estimate of standard error based on the sampling of individuals would be highly inaccurate.

Because we are interested in species proportions, either the individuals selected at random from the sample (those in the subsamples) or the projected total may be used in the calculation of variance and confidence limits. We choose the smaller number counted in the subsamples because it is more manageable and was actually observed.

Table 8.10 shows the pertinent data and the proportions and confidence limits calculated for the Miocene foraminifera. The column labeled $2(\sigma_{pbin})$ is the rounded value of $z = 1.96$ times the standard error, and shows the 95% confidence limits for each of the single biological samples. These confidence limits represent the same sample methodology as those calculated for the Bolivian plot 01 when individuals were sam-

TABLE 8.10   Calculations of Binomial and Cluster Confidence Intervals for Species
Proportions of C. *lobatulus*

a_i	m_i	p_i	$2(\sigma_{pbin})$	$2(\sigma_{pclus})$
172	398	0.68	±0.05	NA
137	433	0.32	±0.04	NA
211	421	0.50	±0.05	NA
119	316	0.38	±0.05	NA
193	563	0.34	±0.04	NA
$\Sigma = 932$	$\Sigma = 2{,}131$	$p = 0.44$	±0.02	±0.13

NOTE: The number of individuals of *Cibicides lobatulus* (a_i), total number of individuals (m_i), and species proportion of each biological sample (p_i) from $n = 5$ biological samples in Bed 16 at Calvert Cliffs, Maryland. Confidence intervals are calculated from the formulas $CL_{bin} = p_i \pm 2(\sigma_{pbin})$ and $CL_{clus} = p_i \pm 2(\sigma_{pclus})$.

SOURCE: Data from Buzas and Gibson 1990.

pled at random from the entire plot. They are germane for the 10 ml sample, but not for the 1 m² template. This is because they are calculated for individuals chosen at random from 10 ml, not from the 1 m² sample, and inference is weakened.

If we use the Binomial estimate for the confidence limits for all 5 samples together (2,131 individuals and estimate of $p = 0.44$), we obtain an interval for *Cibicides lobatulus* of $0.42 \leq P \leq 0.46$. This yields a confidence limit of $CL = p \pm 0.02$ for our estimate of $p = 0.44$. At this point, the intuitive feeling that an experienced researcher has about Mother Nature should take over and exclaim that this is too good to be true. $CL = p \pm 0.02$ is just too good to be a real-world possibility. Indeed, that intuition is correct. So let us examine Table 8.10.

Our confidence interval using the cluster standard error estimate is $0.31 \leq P \leq 0.57$, yielding a confidence limit of $p \pm 0.13$. The difference between a confidence limit of ± 0.02 and ± 0.13 in this example is considerable, and the preferred choice is apparent when we look at the observed p_i's. One of the p_i's (0.68) lies outside the cluster confidence limit, and all the p_i's lie outside the Binomial limit. The total number of individuals in each biological sample is over 300, which is an accepted standard for micropaleontological counts (Buzas 1990). Because so many individuals were counted, this can lull a researcher into a false sense of security. Often only 1 biological sample with several hundred individuals is used to estimate the species proportions for the population, ignoring the need for replication. In Table 8.10 we include Binomial confidence limits for each biological sample. These all have widths of about ± 0.05, and none of them produce confidence intervals that would include our best estimate of $p = 0.44$. Clearly, confidence intervals using the Binomial on a single core (or quadrat) are highly misleading. Once again, we stress the need for replication. Always take replicates.

We conclude that using Binomial confidence limits on data for which the individuals were not sampled at random; that is, in which subsamples were taken from a single biological sample, can lead to serious errors in calculating relative abundances, or proportions, for taxa in the middle range of p. When p is small ($p < 1\%$ for this example), the widths of the confidence limits for these rare species are about the same as the value of the proportion, and so either the Binomial or cluster-based estimates will provide useful information (see Figure 8.3).

A NOTE OF CAUTION

Another facet to keep in mind when using species proportions is that they represent the relationship of a part (the number of individuals in the particular group or species) to a whole (the total number of individuals in the entire sample). Thus there are two key values that are subject to the inherent difficulties encountered with any derived relationship. Poor results may ensue if the estimates for either value change greatly.

As we have seen, the species proportion for a particular species is given by $p_i = \dfrac{a_i}{m_i}$, where a_i is the number of individuals of the species i counted in a quadrat (the ith quadrat), and m_i is the total number of individuals in the ith quadrat counted in the sample. If a_i is relatively constant from quadrat to quadrat, yet m_i changes greatly, the corresponding species proportion will also change greatly. This inconsistency bodes poorly when, for example, we have a situation in which the density of our species of interest is the same in, say, 2 biological samples, and yet its proportion varies greatly between the 2.

Data from the fossil beds at Calvert Cliffs, Maryland, provide us with a striking example of this problem. Table 8.11 shows that the total mean density for all foraminiferal species in Bed 16 (2,940 total individuals in $n = 45$ biological samples) is an *order of magnitude* (differing by a multiple of 10) larger than in Bed 18 (766 in $n = 35$ biological samples). For *Cibicides lobatulus* the species proportion given in percentages, as well as the mean density, decreases substantially from Bed 16 to Bed 18. For the other two species under consideration, the situation is quite different. The mean

TABLE 8.11   Mean Densities and Species Proportions for Three Foraminiferal Species

SPECIES	BED 16		BED 18	
	MEAN DENSITY	%	MEAN DENSITY	%
Cibicides lobatulus	1,295	44	31	4
Bolivina paula	577	20	263	34
Buliminella elegantissima	272	9	337	44
Total all species	2,940	766		

NOTE: Data taken from Miocene strata, Beds 16 and 18 at Calvert Cliffs, Maryland.

SOURCE: Data from Buzas and Gibson 1990.

density for *Bolivina paula* decreases from Bed 16 to Bed 18, while the species proportion increases by 14%. For *Bulimenella elegantissima*, the mean density increases somewhat, but the species proportion increases by 35% from the first to the second bed. Obviously, if we are concerned with increases or decreases in species abundances, we must first consider the focus of our study. Are we primarily interested in getting good estimates for densities (absolute abundances) or for proportions (relative abundances)? The term *abundance* is often used extensively in ecological as well as paleoecological reconstructions, and the researcher should always keep in mind what is being considered. Had we only considered, as is often the case, the few hundred individuals picked from the samples and their standard error, σ_{pclus}, the fact that the very large change in proportions was due to a large change in density would have gone unnoticed.

Depending on the objectives of the research, either density and/or species proportions may be the appropriate measures. But the investigator should always realize that the proportions are once removed from the actual observations made in the field, namely the species counts per unit of space, or species density.

Summary

1. If n individuals are selected at random from a target population of size N, and n_i is the number of selected individuals that belong to the ith species (group), then the proportion $p = \dfrac{n_i}{n}$ is the estimated probability that an individual selected at random belongs to species i.

2. When sampling individuals at random, the individual, not the quadrat, is the primary sampling unit.

3. In samples of natural populations, either the individuals or a sampling device may be the primary sampling unit; different rules may apply for each sampling design method.

4. The probability of finding exactly k number of individuals belonging to species i in a sample of size n can be obtained by using the Binomial distribution (Equation 8.4).

5. Given the species proportion p and the number of individuals n, the variance of p can be calculated (Equation 8.14); the largest error of estimation occurs at $p = 0.5$; that is, for those species that are neither very rare nor very dominant.

6. When p is small and n is large, the Poisson distribution (Equation 8.16) approximates the Binomial. A researcher can opt for the Poisson distribution under these circumstances without risking improper calculations. However, for practical purposes with natural populations the Binomial itself is quite simple to calculate and the answer is exact.

7. The exact Binomial calculations are quite readily done on personal computers, avoiding the need for approximations. However, if the sample size is large, the statistical programs for Binomial calculations may themselves drop into an approximation mode.

8. When p is not small and n is large ($np > 5$; $n > 30$) the Normal distribution approximates the Binomial, and so the Normal distribution (Equation 8.18) can be employed instead.

9. Confidence intervals can be calculated for species proportions using Equations 8.20, 8.21, 8.22, or 8.23.

10. Even when an estimator of a species proportion is known to be unbiased; that is, when it is known to be equal, on average, to the true proportion in the target population, any one estimate can still be quite deviant from the true relative abundance value, and the resultant confidence interval will be affected. Therefore, replication is essential in natural population fieldwork.

11. The number of individuals required in a sample (regardless of N) to achieve a specified degree of confidence in the relative abundance estimate can be calculated from Equation 8.24. For $n = 200$ at $p = 0.50$, the confidence limits are $CL = p \pm 0.07$. For $n = 400$ at $p = 0.50$, the confidence limits are $CL = p \pm 0.05$. Notice that a doubling of the sample size achieves only a slight gain in precision (decrease in width).

12. Sampling quadrats and counting the individuals contained therein is called cluster sampling when we are interested in a species attribute, such as relative abundance.

13. When quadrats are sampled for relative abundance estimation, they are the primary sampling units. Confidence limits using the Binomial formulation are not valid in this form of sample. The Binomial procedure is relevant only to surveys in which the individual, not the quadrat, is the sampling unit.

14. The variance for proportions using cluster sampling is given by Equation 8.28. Confidence intervals can then be calculated by using this variance estimate.

15. Confidence intervals determined from either the Binomial or the cluster methods of calculation will essentially be equivalent at low species proportions; that is, when $p < 0.01$, or about 1%.

16. The use of proportions presents inherent difficulties. For example, if the number of individuals of species i remains constant when counts are made of each of the n quadrats or cores, but the total number of individuals of all species changes greatly from one sample to the next, the corresponding species proportion will also change greatly.

17. Comparisons of relative abundances across surveyed areas or fauna must also state sample sizes and counts in addition to proportions or percentages.

PROBLEMS

8.1 Thirty individuals were chosen at random from Bolivian plot 01. Five belonged to *Pouteria macrophylla*. Calculate the species proportion of *P. macrophylla* and the Binomial confidence interval using that estimate. Does the true proportion of this species fall within the interval's limits?

8.2 From Appendix 2 take 4 quadrats at random. Calculate both the Binomial and the cluster confidence limits for the species proportions of *Pouteria macrophylla*. Comment on your results.

8.3 For the 4 randomly selected quadrats of Problem 8.2,

a. Calculate the proportion of each species per quadrat $\left(p_i = \dfrac{a_i}{n_i}; i = 1, 2, 3, 4 \right)$.

b. What is the total proportion of *Pouteria macrophylla* found over all quadrats? Show that $p = \dfrac{\sum a_i}{\sum m_i}$ does not yield the same result as $p = \sum n p_i$ in Equation 8.26.

8.4 Researchers working with small organisms often encounter a problem of having too many individuals to count, even when small samples are taken. Consequently, they resort to counting some fraction (a subsample) of the original sample. You should now be statistically sophisticated enough to realize that selecting the first few hundred individuals encountered from a tray containing the entire sample is a very poor procedure. Ideally, individuals should be selected at random from all the individuals encountered in the sample. This too may prove difficult because most organisms are not like lottery balls that can be rotated and popped out one at a time (this procedure might not be random either!). Therefore, some sort of splitting device that allows each individual to go into one container or another is desirable (as in Figure 7.1). In this fashion, the original sample can be split into $\frac{1}{2}$. The $\frac{1}{2}$ can then be split again so that one obtains $\frac{1}{4}$ and so on.

Buzas and Gibson (1990) followed this procedure in their study of fossil foraminifera at Calvert Cliffs, Maryland. Here are the results from replicates at a single locality for the species *Bulimenella elegantissima*:

SAMPLE	SAMPLE SIZE n	(SPLIT)	$n_{B.\ ELEGANTISSMIA}$
G-1	450	$\frac{3}{4}$	179
G-2	318	all	120
G-3	237	all	81
G-4	412	$\frac{1}{2}$	182
G-5	448	$\frac{1}{2}$	206

We are interested in calculating confidence limits for the species proportion (relative abundance) of *B. elegantissima*.

a. Calculate the Binomial confidence limits for each of the 5 samples, using first the observed and then the estimated number of individuals. (For example, the estimate for G-4 is twice the number observed; that is, 182·2 = 364.)

b. Calculate the Binomial confidence limits for the entire set taken together for observed and estimated individuals.

c. Calculate cluster confidence limits for the observed and estimated number of individuals.

d. Comment on the results.

9

SPECIES DISTRIBUTIONS

O NE OF THE MARVELOUS ASPECTS of descriptive statistics is that, providing we know (or can assume) the statistical distribution, a large array of natural population data can be completely summarized by a few parameters. Throughout this book we have used two parameters, the arithmetic mean (for example, mean density and mean relative abundance) and the variance, as summary descriptors of a natural population. For example, the data in Appendix 1, which lists all the species and individuals within a 10-quadrat plot from the Bolivian Beni Reserve, can be condensed into the single (univariate) values of $\mu = 6.63$ individuals per quadrat (for mean density) and $\sigma^2 = 4.69$ (for the variance, which is the square of the standard deviation, σ). These two parameters describe the distribution of 663 individual trees within the entire plot. Likewise, in Chapter 2 when we took the randomly selected sample of just 8 out of 100 quadrats, ignoring the separate species and focusing on the total individuals, we estimated population values of trees at Beni Reserve that were unbiased and, in practical terms, not far off the mark. Our estimate of mean density and variance were $\hat{\mu} = 6.00$ and $\hat{\sigma}^2 = 5.14$.

In this chapter we turn our attention to a new level of detail. We are not just interested in how 663 trees are distributed in a forest plot; we are interested in how the 52 species of 663 trees are distributed. For example, let us look at only 3 species from our Bolivian plot, numbered in Appendix 1 in order of abundance: *Scheelea princeps*, species 1 (the most abundant); *Brosimum lactescens*, species 2; and *Calycophyllum spruceanum*, species 4. We could use the notation in Table 9.1, whereby the numbers 1, 2, and 3 used as *subscripts* simply distinguish the 3 species in our example. This numbering is not meant to be indicative of the abundance ordering of these 3 species in Appendix 1.

COVARIANCE

If our concern is with each species separately, a tabular arrangement would give us the desired information. However, if we are interested in the interaction or interrelationships of these 3 species within the subplot of 8 quadrats, or within the entire 100-quadrat plot, as representative of some community structure, there is one more quantity required for our study. This quantity is called a *covariance*. That is, we are interested not only in how the count for each species varies, but in how each covaries

TABLE 9.1 Notation for Mean and Variance for 3 Species of Trees in Plot 01 of the Bolivian Forest Reserve

SPECIES	MEAN DENSITY	VARIANCE
Scheelea princeps	$\mu_1 = 2.52$	$\sigma_1^2 = 2.650 = \sigma_{11}$
Brosimum lactescens	$\mu_2 = 1.31$	$\sigma_2^2 = 1.374 = \sigma_{22}$
Calycophyllum spruceanum	$\mu_3 = 0.25$	$\sigma_3^2 = 0.248 = \sigma_{33}$

NOTE: The notation for variance when a single species (the ith species) is covaried with itself can be written as either the square of the standard deviation (σ_i) or as a matrix element, with notation σ_{ii}.

SOURCE: Derived from data in Appendix 1.

with one another. For any pair of species (say, i and j) in the target population, the statistical covariance is a parameter that can be defined as

$$\sigma_{ij} = \frac{\sum_{i=1}^{N} \sum_{j=1}^{N} (X_i - \mu_i)(X_j - \mu_j)}{N}. \tag{9.1}$$

Equation 9.1 is the first time that we have used a double summation ($\Sigma\Sigma$). This indicates merely that we are forming the parts dealing with i (the first species of interest) and j (the other species of interest) separately, but simultaneously. This formula says that we compute the means of the 2 groups or species (i and j) separately, then subtract each species mean (μ_i or μ_j) from the respective observation (X_i or X_j) on that species (that is, the mean from N quadrat counts of species i, and the mean from the N observations of species j). Each pair of terms is multiplied together. Next, all pairs or products are added to obtain a final sum. The final sum is divided by N, which is the number of total quadrats for which we have counts. Notice the similarity of Equation 9.1 with Equation 2.3.

Table 9.1 shows that there are two ways to notate the covariance of a single species with itself. The variance of *Scheelea princeps*, for example, is notated as σ_{11} or σ_1^2. Introducing the second notation is important because we can incorporate it into the matrix notation.

To explain this usage, let us consider only 1 species; that is, species 1. We introduce it here only to make it easier to understand the mathematics of covariance applied to 2 separate species. Let us substitute the single species notation from Table 9.1 into our Equation 9.1. This substitution yields

$$\sigma_{11} = \frac{\sum_{i=1}^{N} (X_i - \mu_1)(X_i - \mu_1)}{N}.$$

We can work with this equation as a simple algebraic relation. We no longer need the double summation symbols because both quantities use only the subscript i (they are the same), where i represents the quadrat for counts of *S. princeps*. Since the quantity within each set of parentheses is the same, multiplied together it is just squared, and we could instead write

$$\sigma_{11} = \frac{\sum_{i=1}^{N}(X_i - \mu_1)^2}{N}.$$

A comparison of this equation with Equation 2.3 shows that the covariance of a species with itself is just the species variance. That is, in statistical notation, $\sigma_{11} = \sigma_1^2$, and we have gained nothing of practical value except an insight into the fundamental mathematics of covariance, which will be used in simplifying the following section.

Notice that both forms of notation are used in Table 9.1. We can thus rewrite the variances for the set of 3 species as $\sigma_{11} = 2.650$, $\sigma_{22} = 1.374$, and $\sigma_{33} = 0.248$. This leads us to the insight that what is missing for our 3 species' interaction study is σ_{12}, σ_{13}, and σ_{23}, the covariances of each species of interest with each of the others. That is, we no longer have a simple, univariate, statistical situation when we are interested in the study of more than 1 grouping, or species. We have a multiple species; that is, a *multivariable*, or *multivariate*, situation. These terms may or may not be used interchangeably in all statistical applications, but in this situation the distinction is unimportant.

MATRIX NOTATION

In statistical work we put all this information about covariance together, not in a table, but in a *matrix*. Statisticians use *matrices* to summarize all the information for multivariate or multivariable situations. For example, we could write the column of the 3 species' means from Table 9.1 in matrix notation as follows:

$$\mu = \begin{pmatrix} \mu_1 \\ \mu_2 \\ \mu_3 \end{pmatrix}. \tag{9.2}$$

The boldface notation indicates that μ is a matrix, while the 3 individual means within the parentheses are its *elements*. We can also form a matrix with elements that are the variances and covariances. The notation for the *variance–covariance matrix* is

$$\Sigma = \begin{pmatrix} \sigma_{11} & \sigma_{12} & \sigma_{13} \\ \sigma_{21} & \sigma_{22} & \sigma_{23} \\ \sigma_{31} & \sigma_{32} & \sigma_{33} \end{pmatrix} = \begin{pmatrix} \sigma_1^2 & \sigma_{12} & \sigma_{13} \\ \sigma_{21} & \sigma_2^2 & \sigma_{23} \\ \sigma_{31} & \sigma_{32} & \sigma_3^2 \end{pmatrix}. \tag{9.3}$$

That is, the variances are written down the *diagonal*, and the covariances are the *off-diagonal elements*. Notice that for this example, the covariances on either side of the diagonal are equal, for example, $\sigma_{12} = \sigma_{21}$, or in general, $\sigma_{ij} = \sigma_{ji}$. We simply filled in the redundant values to make a *square matrix*.

Clearly, μ and Σ each have a different number of elements. To show how many elements are in a matrix we talk of the *order* of the matrix. The 3 arithmetic means that are elements of μ are written in one column because we copied them from Table 9.1. We could also notice that this 1 column has 3 rows, or 3 entries, in it. Therefore, an easy way to write the *order of a matrix*, or the number of elements, is not as a total but as the number in each of its rows and columns. For example, we write $\mu_{3 \times 1}$. The arrangement in which we write the numbers is not arbitrary. We must read the order as number of rows by (or times) number of columns. In this way we can distinguish it from $\mu_{1 \times 3} = (\mu_1\, \mu_2\, \mu_3)$, which also has 3 elements, but which has been rewritten to have 3 columns and 1 row (or 1 entry) in each. Similarly, we would write $\Sigma_{3 \times 3}$, which shows that the variance–covariance matrix has an order of 3×3 (and therefore has $3 \times 3 = 9$ elements). Finally, an alternative term for a matrix with only 1 row or only 1 column is a *vector*. Therefore, we call μ the *mean vector*.

Just as for the univariate case when we used two parameters to summarize the population information, we now use 2 matrices (or a matrix and a vector) to describe the multivariate case. Although there are still 2 quantities, we must recognize that there are now more than two parameters as elements. Fortunately, all we need do is substitute into Equations 9.2 and 9.3 the sample estimators μ_i, $\hat{\sigma}_1^2$, and $\hat{\sigma}_{ij}$ to obtain our desired parameter estimates $\hat{\mu}$ and $\hat{\Sigma}$.

With the inclusion of the covariances, we obtain an amount of information drawn from our observed data that is not as simple to distill as when we were dealing with only univariate situations. For our 3 species, Table 9.1 is a reasonable and widely used presentation method for the separate information on each species. However, we would choose to work from Table 9.2 if we wanted to explore the distribution patterns or investigate possible interrelationships among the species or groups. In most cases when there are 3 or more groups, it is impossible to evaluate species or group interactions and spatial patterns or distributions from mere visual inspection of the numbers. Fortunately, there are a large number of statistical techniques that pertain to the variance–covariance matrix, which allow us to distill this information even further. Application of these sophisticated mathematical statistical techniques, many of which are still in the developmental stages, is only possible with computer assistance. Thus, only in the past decade has it been possible to solve ecological or paleoecological problems involving multiple group or species interactions, spatial pattern recognition, or distributional modeling. However, many ecological problem areas still rely solely on univariate solutions.

An aspect of natural populations ripe for advancement with multivariate approaches is the study of characteristics of abundance. For example, in our forest plot at the Beni Reserve, we counted $N = 663$ individuals represented by $S = 52$ species. Even the most casual naturalist notices that the individuals are not distributed evenly among the species. In order to describe, and to make inferential statements about, the meaning of this multiple species data set, we need advanced statistical methodology that incorporates both the species interaction and the spatial pattern information available in the data shown in Appendix 1.

Let us again consider our 3-species example. From Appendix 1 we formed the (52 X 3) matrix of data shown in Table 9.2. In order to describe the abundance of species observed

TABLE 9.2 Counts of Three Species of Trees in Bolivian Plot 01

	SPECIES		
QUADRAT	S. princeps	B. lactescens	C. spruceanum
1	0	1	0
2	0	5	0
3	0	4	0
4	0	5	0
5	1	1	0
6	2	2	0
7	0	4	0
8	2	3	0
9	0	1	0
10	3	1	0
11	2	1	0
12	3	1	0
13	2	0	0
14	1	0	0
15	3	2	0
16	0	1	1
17	3	2	0
18	0	0	1
19	2	2	0
20	4	3	0
21	1	3	0
22	2	1	1
23	4	3	0
24	3	0	0
25	3	0	0
26	4	0	0
27	3	0	0
28	7	2	0
29	6	1	0
30	5	0	2
31	4	3	0
32	2	1	1
33	4	0	0
34	4	2	0
35	4	2	0
36	1	2	0
37	4	1	0
38	2	2	0

(continued)

TABLE 9.2 *(continued)*

	SPECIES		
QUADRAT	*S. princeps*	*B. lactescens*	*C. spruceanum*
39	6	0	0
40	2	3	0
41	1	0	0
42	2	2	1
43	3	0	0
44	6	1	0
45	4	0	1
46	3	0	0
47	4	0	0
48	2	1	0
49	3	2	0
50	2	0	0
51	1	2	0
52	2	1	0
53	2	0	0
54	0	1	1
55	1	0	1
56	3	2	0
57	2	1	1
58	1	2	0
59	4	0	0
60	5	1	0
61	4	0	0
62	2	2	0
63	2	1	0
64	3	0	0
65	2	0	0
66	0	3	0
67	1	1	1
68	0	2	0
69	2	1	0
70	4	2	0
71	5	1	1
72	2	2	0
73	2	1	0
74	3	4	1
75	3	2	1

(continued)

TABLE 9.2   (*continued*)

| QUADRAT | SPECIES | | |
	S. princeps	B. lactescens	C. spruceanum
76	6	1	0
77	3	0	1
78	3	1	1
79	4	1	1
80	5	1	2
81	2	2	0
82	1	2	0
83	2	1	0
84	0	2	0
85	3	1	0
86	2	0	0
87	3	0	0
88	3	0	0
89	1	0	0
90	1	1	1
91	0	3	2
92	5	1	0
93	5	2	0
94	4	1	0
95	4	1	0
96	1	2	1
97	3	3	0
98	2	1	1
99	2	0	1
100	2	1	1
Mean	2.52	1.31	0.25
Variance	2.650	1.374	0.498

SOURCE: Appendix 1 for $N = 100$.

in the field, the qualitative notions of *abundant*, *common*, and *rare* are often used. To summarize these ideas quantitatively, the results of Table 9.2 could be presented as

$$\mu = \begin{pmatrix} 2.52 \\ 1.31 \\ 0.25 \end{pmatrix} \quad \Sigma = \begin{pmatrix} 2.650 & -0.496 & -0.030 \\ -0.496 & 1.374 & -0.058 \\ -0.030 & -0.058 & 0.248 \end{pmatrix}.$$

It is probably too much to expect that a single class of matrices for μ and Σ could describe all populations. However, there may be different identifiable classes (that is, various unique patterns of elements in Σ or patterns of Σs across different environments) for different taxa and/or environments. This area is still a rich source for exploratory research.

HISTORICAL PERSPECTIVE

The problem of investigating the abundance data for biological meaning first appeared in the scientific literature about 50 years ago, when neither advanced statistical techniques nor computers were available for biological adoption. Before these analytical aids, the data of Table 9.2 had to be treated in a univariate fashion; in effect, ignoring information available on the extent to which the groups or species covaried, and ignoring the distributions and spatial interrelationships. Instead, the count data were ordered by the number of species represented by 1 individual, 2 individuals, and so on, up to the most abundant group. This ordering can be done regardless of the taxonomic group considered (species, genus, and so on), provided that the group size is large. We can of course illustrate this style of ordering for our Bolivian plot. For all taxonomic groups, only a few species are abundant; most are judged rare. However, as with natural populations of large mammals, if the number of species in the group under study is quite limited, this will not be possible.

Historically, the sample abundance data were not presented as in Table 9.2 but were organized pictorially into a *species abundance distribution*, in which $x = 1, 2, \ldots, n$ individuals are contained in $y = 1, 2, \ldots, s$ species. Fifty years ago it made sense to present this same data in a univariate summary form, such as in Table 9.3, which presents the species abundance data for our Bolivian plot. Notice that when we use this tabular form we no longer include the information on each quadrat, or the counts themselves. We have sacrificed that information for data on the number of individuals in summary form.

Our observations were obtained in the field as the number of individuals of each species within each quadrat. Table 9.3 alters the presentation of that data to show the summarization over all quadrats of the number of individuals occurring within each species. Because *frequency* is the number of occurrences of a certain type of event, Table 9.3 is an unusual presentation in that it is actually a *frequency of frequencies*. This layering of frequencies has been, and continues to be, a source of confusion when summarizing the results of surveys of natural populations.

For this data organization, an attempt was made to find a fit with some univariate model. Figure 9.1 portrays the collapsing and loss of information inherent in this older methodology. As can be seen, the mean vector, or simply the total number of individuals observed for each species, is the sole input for univariate modeling. There is no variance–covariance matrix, so interrelationships and distributional pattern is ignored and lost. Thus a multitude of biologically important scenarios may be described by a single model.

We shall examine the univariate approach because it persists in the literature today, and because it is the basis for obtaining some of the important and commonly used diversity measures (Chapters 12–17). When we discuss each model in

TABLE 9.3 Observed Number of Species Having $i = 1, 2, \ldots, n$ Individuals in Bolivian Plot 01

ABUNDANCE, n (# INDIVIDUALS IN STUDY PLOT)	NUMBER OF SPECIES, s (AT THAT ABUNDANCE LEVEL)
1	20
2	9
3	6
4	3
5	0
6	2
7	0
8	1
9	2
17	4
18	0
19	1
25	1
62	1
131	1
252	1

turn, we will define each letter used. However, be aware that, just as in statistical and mathematical texts, a single letter may be reused with a different meaning in a different model. For example, for both the Geometric and the Harmonic series the letter k is standard usage, but it is defined differently for each series. This overlap in usage is a necessity for mathematics (because of the interrelationships of the series and distributions), but it may be quite confusing unless it is kept in mind throughout.

SERIES AND SEQUENCES

Most often, if we have taken a large enough look at, or sample of, a natural population, we will detect an orderliness. The picture or graph of the *sample* or *observed species abundance distribution* (e.g., Table 9.3) will be in the form of a so-called *hollow curve* (Figure 9.2). The sample is usually restricted to a particular taxonomic group (like trees, isopods, fish, or birds) in a particular area. In Figure 9.2 we can see that for our Bolivian tree data most of the species are represented by only 1 individual, some fewer have 2 individuals, and so on in a decreasing manner. Because there are a finite, or countably infinite, number of individuals taking on a limited number of species values, we are dealing with a discrete random variable; that is, one represented by integers. (The topic of discrete random variables is discussed in Chapter 3.) The distribution of

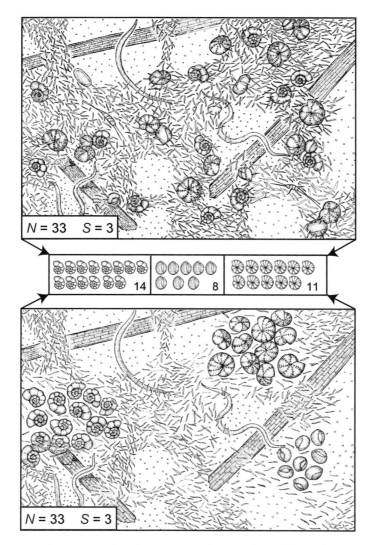

FIGURE 9.1   Both representations have $N = 33$ and $S = 3$. When only N and S are used, the information on the different spatial arrangements is lost.

this random variable is, in turn, a discrete distribution; that is, the actual data are mapped as discrete points although, as we shall see, a curve might be fitted for the convenience of analysis.

The abundance literature contains unnecessary admonitions to consider fitting only discrete distributions because of the data's discreteness. However, we believe that the convenience (and statistical acceptability) of considering a discrete variable to be continuous argues for the practice of fitting the discrete data with a continuous distribution. For species abundances, our justification for considering the fit of a continuous distribution is that in most natural populations the range of possible abundance values is quite extensive and the probability that any particular species will have any particular number of individuals is exceedingly low.

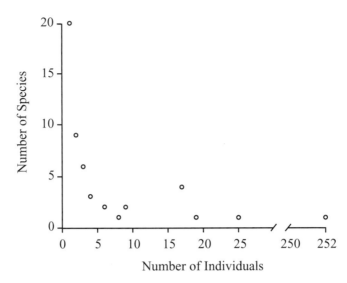

FIGURE 9.2   Hollow curve of Bolivian abundance data. Number of individuals by number of species, plotted with $x = 1, 2, \ldots$ individuals and $y = 1, 2, \ldots$ species.

Let us examine the observed species abundance distribution of Table 9.3. Based upon the values in this table, Figure 9.2 depicts a typical abundance curve. Its shape expresses certain important characteristics of the observed abundance data. It has a very long tail in one direction (*positive skew*). The length of the tail is, in part, determined by the size of the sample, but it is also related to the density, or number of species observed. The curve is cut off (*truncated*) near the y axis (there are no x values of 0 or less). This truncation expresses the obvious fact that it is not possible to observe any species represented by fewer than 1 individual. We now proceed to attempt to find a curve to fit to the data; that is, we need to find a theoretical curve with the key characteristics that most closely matches the natural ordering of the data points.

Series of numbers as well as a variety of both discrete and continuous probability distributions have been proposed in the literature to fit the observed species abundance distributions. This topic became fodder for biologists as well as statisticians. Biologists saw that when the field data on species and individuals were accumulated and presented, a curve with a familiar shape usually resulted. Some biologists thus tried to deduce a plausible, general, ecological theory that would explain the prevalence of this common pattern of abundance distribution among species. Statisticians, in contrast, are enamored of situations giving rise to families of curves, and so they attempted to describe the generality of the pattern by way of probability theory, not ecological theory. The result was that massive effort was spent in generating descriptive curves and providing biological scenarios (assumptions and interpretation) by which these curves could be construed as fundamental to the species abundance representation in nature. As a reasonable consequence, further effort was expended in argument to proscribe the "best" curve, or that distribution that could best describe the biological reality and could be differentiated from all other possible distributions. This presumably fundamental distribution pattern could then be examined for its implications in nature. In other words, both biologists and statisticians reviewed a lot of theory and data sets in an attempt to find the best curve that could serve as the generic rule from which causal explanation could then be posited.

While this flurry of curve fitting was under way, some authors looked at the numbers of possible individuals listed along the x axis and focused on the (undisputed) fact that only discrete values were possible. They saw the numbers of species on the y axis as another set of discrete terms corresponding to this first set. Consequently, they sought discrete mathematical series of numbers (instead of probability distributions) for fitting the observed data. The mathematical series approach may appear to be simpler than an attempt to curve fit. One merely calculates, term by term, and uses these values as the expected numbers of species for each incremental increase in number of individuals. The ease of execution is hampered, however, by some attendant drawbacks, as we shall show. Not least among the drawbacks is that this procedure is not what statisticians call "curve fitting," or fitting a distribution. However, there is a nice relationship between some series of such numbers and a related probability distribution.

In mathematics a group of specific, related, ordered terms (say, $x_1, x_2, x_3, \ldots, x_n$) can be called a *sequence*. One way to model observed abundance data is to find a formula that generates a sequence of numbers that will fit (or come close to) the actual values observed. These generated numbers are then used as the *expected species values*. When the terms in the sequence are summed, the result is called a *series*. For example, the notation $x_1 + x_2 + x_3 + \cdots + x_n$ represents the process of summing the n individual terms. In many instances the sum of the terms of a sequence has a nice, neat formula, which well expresses the series. When such a formula can be found, the series is called a *convergent series*. One example of a convergent series (that you can use to amaze friends) is instructive here: Simply find the sum of the first few whole numbers. For example, we shall find the sum of the first 10 whole numbers (integers). We can write the left side as $1 + 2 + 3 + \cdots + 10$; this sums to 55. Now 55 is also the answer to this formula: $\dfrac{10 \cdot 11}{2}$. The formula turns out to be generic. For any sequence of whole numbers beginning with 1, the formula $(1 + 2 + 3 + \cdots + n) = \dfrac{n(n+1)}{2}$. That is, the series expression on the left side equals the right-hand-side formula. Another way to say this is that the sum of all the terms on the left can be found by using, or *expanding*, the formula on the right; thus both sides are making the same statement. *Series expansions* can be useful for fitting observed data values with expected values, as we shall demonstrate.

The results of trying to provide either a series explanation or a distributional explanation of the observed abundance curve have been mixed. Many distributions and series have been shown to fit reasonably well or to model certain sets of observations, but no one model can describe all abundance data. Even small, homogeneous groupings of a single taxon are not well described by a single model. Different field sampling methods can give rise to differently shaped curves, even for the same taxa in many cases. In addition, a mathematical distribution that can be fitted to an observed sample may not be the most appropriate distribution to fit the target population.

In this chapter we present some of the most well-known mathematical abundance models that are used to fit the distribution curves of field data. We shall discuss their intrinsic meaning and their mathematical and biological implications. Advantages and disadvantages will be detailed by providing numerical examples. However, three caveats must be made at the outset. First, there can never be any absolute and inarguable answer to the question of which distribution provides the best fit to any particular data set. We can only hope to identify a distribution that, with high probability, is likely to have produced the observed data. We shall make this point again in Chapter 17 when we discuss alternative methods of fitting. The fit might then help us, with high probability, to characterize an underlying relationship or a process that seems to be at work in determining population abundances for our target population. But because the multivariate facets of the data are being suppressed by our chosen univariate approach to modeling, any such characterization will not be unique. A second caveat is that no one can use just 1 unreplicated sample data set (or even a few sets) and hope to ascertain whether a theoretical distribution will actually describe a natural target population. This is especially true for small samples. A final caveat is that the process by which the distribution is supposedly "fit" to the data by the biologist does not always adhere to what a statistician would find acceptable. In this chapter we describe the historical methods and present alternatives as well.

The examples we present in this chapter are designed to lead the reader through a group of possible descriptors of the abundance curve and to provide insight into the consequences for biological interpretation when we choose one or another of these probability distributions. The most obvious place to begin is to explain how one goes about finding any model, appropriate or inappropriate, to launch the search for the best fit. How does one start with a series of numbers or points on a graph and end up with a formula that, when expanded, provides a good match?

The selection of a particular set of numbers or a probability distribution to fit to the observations usually is based upon two factors. First, there is a group of relatively well-known series and probability distributions devised by mathematical statisticians, and their shapes are easily found in elementary probability texts. These are the common ones that are compared with any observed shape obtained from a wide range of applications. The second factor directing the quest pertains to what is known about the ways that combinations of (random) events take place. Mathematical statisticians have proved that certain mechanisms (in nature or human-made) are related to particular series, or probability distributions, under defined conditions. Biologists can then use this knowledge to interpret such conditions and mechanisms in biological terms, and to consider the implications. Once a potentially good model is selected, what remains to be done are the technical computations of fitting that model to actual observed data and examining how well the 2 shapes match. If the match is poor, then select and test another model.

Let us go through the selection and/or fitting of some of the most commonly suggested theoretical distributions. We will use the data in Table 9.3 as an example, while pointing out general principles.

HARMONIC SERIES

One series of numbers that is known to be applicable to a wide variety of problems in which the individuals tend to be aggregated into a few of the discrete x-value possibilities (number of individuals) is the *Harmonic series*. Early on, the Harmonic series was explored for its potential to fit many actual data sets of biological abundance (e.g., Corbet 1941; Preston 1948). A Harmonic sequence of values can be written

$$x, \frac{x}{2}, \frac{x}{3}, \ldots, \frac{x}{n}, \frac{x}{n+1} \ldots,$$

where, for our purposes, x is the number of species represented by 1 individual. The ellipsis (three dots) at the end of the formula indicates that the Harmonic series is actually an *infinite series* of numbers. Because we need to use only a certain number, say n, of these terms to fit our abundance data, this formula becomes

$$x, \frac{x}{2}, \frac{x}{3}, \ldots, \frac{x}{n}.$$

The series formed from this finite sequence of numbers is called a *Truncated series* because it includes terms only up to the nth. Truncation is important here because only in this way is the series *convergent*; only in the Truncated series is there a finite sum. If we saw that this Harmonic series produced a good fit with the observed data, we would say that the abundances were *Harmonically distributed*. This terminology applies even though we don't write a probability distribution in specific terms.

GEOMETRIC SERIES

Although the Harmonic series seemed to provide a reasonable fit for some biological distributions, it did not work very well for many of them. So, the search was on for a series that would provide a better fit for more of the data sets. Alterations of, and adjustments to, the Harmonic series were tried because the Harmonic values declined too sharply when compared to many observed data sets. The *Geometric series* stretches out the hollow curve of a Harmonic series. A Geometric series is mathematically defined quite generally as

$$a + ax + ax^2 + ax^3 + \cdots + ax^{n-1}$$

where a is any constant number greater than 0, and $n = 1, 2, \ldots$. The formula for the sum of this Geometric series is

$$\frac{a(1 - x^n)}{1 - x}$$

where x cannot be equal to 1, because the denominator would then equal 0 (because $1 - x = 1 - 1 = 0$) and we would have the problem of division by 0. Note that the Geometric series corresponds to a probability distribution, which we shall discuss.

Why would biologists get the idea that a Geometric series might be applicable to their work? Why did they think that actual data sets of natural populations might be best expressed by the formula for the Geometric series? Basically, biologists enthusiastic about the curve-fitting prospects of the Geometric series were interpreting mathematical assumptions into biological terms and hoping for the best. Let us explain.

Let us say we are performing a set of Bernoulli trials (catching or observing individuals of some species; see Chapter 8). We ask the question: What is the probability that we first see an individual of this species after we have observed a certain number of individuals of other related species? That is, we are interested not in the distribution of p (the probability; see Chapter 8) but in the distribution of k, which is the observation number (statisticians call this the *trial number*) on which we first obtain a "success" (observe or catch the organism). The distribution of k is called a *Geometric distribution*, and the results obtained for the expected number of species are the terms of the Geometric series, if the frequencies (y values) decrease in a geometric progression as the x values increase. The distinction between a Harmonic and a Geometric progression is that the frequency values of the former decline arithmetically, whereas the frequency values of the latter decline geometrically. A Geometric distribution graphs as a hollow curve with a more extended hollow than does the Harmonic distribution.

The formula for a Geometric distribution is rather simple to obtain. If we define p as the probability of a "success," then the probability that we first observe an individual of the desired species, or that we get our first success on the first try (that is, when $k = 1$) with some probability p, is

$$pr(1 - p) = p.$$

This expression, like all mathematical statements, can be read from left to right like a sentence. Let us use some numbers for ease of understanding. If, for example, we think we know that the chance that we will see an amphibian specimen of, say, *Rana pipens* is 30%, we could write

$$pr(1, 0.30) = 0.30.$$

This expression reads: The probability (pr) that we observe our first success, or a specimen of the species of interest, on our first try (when $k = 1$ and $p = 0.30$), is merely the probability of seeing that species, or 0.30. As we have discussed in Chapter 8, this value (30% in this case) is the relative abundance. For example, for the tree *Scheelea princeps* the relative abundance would be 0.38 for our Bolivian study plot.

The probability that we do not get our desired specimen until the second observation would be

$$pr(2, p) = (1 - p)p.$$

For *Scheelea princeps*, the probability that we get a specimen on the second try, or that in a sample of 2 individuals, the second would be a *Scheelea princeps* specimen is

$$pr(2, 0.38) = 0.62 \cdot 0.38 = 0.24.$$

Can you see that in order to have our first success (p) on the second try, we absolutely must have a failure ($1 - p$) on the first try? When this is clear, the formula becomes quite obviously correct. If we continue in this way, we obtain a very simple formula or rule for the Geometric distribution,

$$pr(k, p) = (1 - p)^{k-1} \cdot p,$$

which can be read that we have obtained $k - 1$ failures before our first success, which occurred on the kth try. It is as easy as that.

May (1975) described a possible scenario for which the *idealized community abundances* could adhere to, or be described by, a Geometric series. He stated that for this to occur the community ecology must be dominated by one unique factor. In addition, there must be straight, hierarchical divisions of the niche, with each species, in turn, commanding a constant fraction of the available space. Cohen (1968) also presented an ecological scenario for geometrically distributed population sizes to hold. He posited that population growth can be viewed as a sequence of Bernoulli trials. (Recall that this means p is the probability an individual is added to the population; q is the probability it is not added; these are the only 2 choices. Although this is not our usual usage of these 2 terms, within the limits of Bernoulli sampling, Cohen can define p and q any way he wants.) These trials continue until the first failure, when the total population size will be well described by this sequence of geometrically decreasing numbers. The other assumption is that the probability that an individual is added is independent of the existing size of the population. This assumption may be unrealistic for most natural populations, but it is not necessarily of major consequence to the fit of the model.

NEGATIVE BINOMIAL DISTRIBUTION

Another approach to ascertaining distributional patterns in a data set is to create a series that represents the number of failures before our first, second, or k^{th} success. That is, how many observations would we need to make before we could expect to encounter 1 individual of the target species? In mathematical terms, we are interested in the random variable ($n - k$) that is derived by subtracting the number of successes from the number of sampling units in our data set. This random variable has a *Negative binomial distribution*. The name is apt because there can be a clear relationship of the Negative binomial with the more common Binomial discussed in Chapter 8.

Description of Over-Dispersion

The curve of a Negative binomial distribution is relatively easy to calculate. Not surprisingly, the Negative binomial distribution has been recognized as a probability distribution in the statistical literature since the 1700s. However, there are many other underlying model situations besides the number of failures that this distribution can describe or be based upon. The Negative binomial distribution has been used for a wide variety of applications, including biological problems. For example, it was devised to deal with the *over-dispersion* problem (*clumping*) common in abundance observations. That is, when the variance is considerably larger than the mean density (Chapter 6), or when the tail of the observed data is very elongated or heavily skewed, a Negative binomial solution for curve fitting can be helpful.

A Negative binomial distribution is well defined by two parameters: the mean μ and an exponent (in the mathematical equation for this distribution) $k > 0$. Once we know these two parameters, we can easily find the variance, which is defined as

$$\sigma^2 = \mu + \frac{\mu^2}{k}.$$

There can be more than one way to write (that is, to describe) the parameters of a distribution. These alternative descriptions of the same distribution are called *parameterizations*. For the Negative binomial, sometimes instead of using k and μ, we write the distribution with parameters μ and σ^2. In the other distributions that we already discussed, such as the Normal, Binomial, and Poisson, the variance σ^2 was always the *shape parameter*, but here the parameter k serves that function. From the variance formula just presented we see that the smaller k gets, the larger the variance becomes. The larger k is, the smaller $\frac{\mu^2}{k}$ becomes until it approaches 0, and, in turn, the variance gets closer to the mean ($\sigma^2 = \mu + 0$, or $\sigma^2 = \mu$).

Relationship to the Binomial Distribution

The Negative binomial is, as we might expect, related to the (positive) Binomial discussed in Chapters 3 and 8. The relative frequencies, or probabilities, can be described by a Binomial with a negative ($k < 0$) index. This is because we are dealing with failure to observe, rather than success. In any Binomial, "positive" or "negative," we expand the formula (that is, multiply and write out all the terms). We begin the usual Binomial with $(q + p)^n$, which is called the *kernel*. Following through, the "positive" Binomial probabilities may be obtained from the terms of the Binomial series expansion:

$$(q + p)^n = q^n + nq^{n-1}p + \frac{n(n-1)}{2!}q^{n-2}p^2 + \cdots + \frac{n(n-1)(n-2)}{k!}q^{n-k}p^k + \cdots + p^n.$$

Here, p ($0 < p < 1$) is the constant probability; that is, the same probability throughout the entire study of a "success" in each trial; $q = 1 - p$ is the probability of a "failure" on a trial; and n is the number of independent trials or observations. We usually know n and then estimate p. The mean for the positive Binomial is np and the variance is npq, as is presented in Chapter 8.

In the Negative binomial, we expand a kernel of the form $(q - p)^{-k}$. That is, the probabilities (that $x =$ some value) can be obtained from the terms in the series expansion:

$$(q - p)^{-k} = q^{-k}\left[1 + k\left(\frac{p}{q}\right) + \frac{k(k + 1)}{2!}\left(\frac{p}{q}\right)^2 + \cdots + \frac{k(k + 1)\cdot(k + x - 1)}{x!}\left(\frac{p}{q}\right)^x + \cdots\right].$$

In this case $p > 0$, $k > 0$, and $q = 1 + p$. Notice that in this parameterization, since the only requirement on p is that it be greater than 0, p does not correspond to a probability. This can lead to confusion. The mean for this specification of the Negative binomial is kp, and the variance (pqk) is $kp + kp^2$. However, unlike the positive Binomial, the Negative binomial presents a problem of estimating both p and k at the same time.

The Negative binomial distribution is often suggested as a possible curve-fitting solution for dealing with the problem of clumping (over-dispersion), which is usually evidenced by a very elongated tail on the histogram of observations, as in Figure 9.2. The larger the variance in relation to the mean, the more over-dispersed the population. When we write the parameters of the Negative binomial as kp for the mean and $kp + kp^2$ for the variance, it is easy to see that the variance must exceed the mean. Thus it is indeed reasonable to try to compare or fit the Negative binomial to an over-dispersed data set.

Parameter Estimates

For a Negative binomial in which the parameters are μ and k, these may be estimated from the sample frequency distribution by the statistics $\hat{\mu}$ and $\hat{k}$. As might be imagined, the estimation of the mean μ is quite simple. Even with a hand calculator, we could get a good estimate from the formula

$$\hat{\mu} = \frac{\sum_{x=0}^{n} xf_x}{n} \tag{9.4}$$

where f is the frequency of individuals at each point x $(x = 1, 2, 3, \ldots, N)$, and N is the total of all sample points or individuals. Notice in Equation 9.4 that we are now going to use N as our sample, not n. This is because N is commonly used in the biological literature on species abundance. This should present no problem as long as the definition is clear. For our Bolivian example the average number of individuals (per species) from Table 9.3 can be calculated

$$\hat{\mu} = 1(20) + 2(9) + 3(6) + 4(3) + 6(2) + 8(1) + 9(2) + 17(4)$$

$$+ 19(1) + 25(1) + 62(1) + 131(1) + 252(1)$$

$$= \frac{663}{252} = 2.06.$$

Notice we didn't need to include all the 0's in the formula because they were multiplied and therefore always result in terms equal to 0, which do not contribute to the sum. However, we do count them among the number of terms. The 0's are not "missing data" for statistical work, but are observations: we looked and found none, or 0. That is why the denominator is 252 individuals and not 13 terms.

Estimation of k is more difficult than estimation of the mean. Quite a few formulas have been proposed for directly calculating an estimate of k (see, for example, Bliss and Fisher 1953). We shall describe the two most popular formulas, which are known as moment estimators and maximum likelihood estimators.

Before we obtain an estimator, we must calculate the variance estimate, $\hat{\sigma}^2$, for the data in Table 9.3. This is the same estimate as given in Chapter 2, as Equation 2.5, but it is rewritten for grouped data:

$$\hat{\sigma}^2 = \frac{\sum_{x=0}^{n} x^2 f_x - \frac{\sum_{x=0}^{n}(x f_x)^2}{N}}{N-1}. \tag{9.5}$$

A point on the calculations is important here. Equation 9.5 is called a *computational formula*. It is a formula, devised before computers and good hand calculators were available, that makes the computations easier. However, when working on a computer this formula should not be used because, although it is algebraically equivalent to the Chapter 2 formula, the answers will not be exactly the same each time. Therefore, in the computer era, Equation 2.5 is the correct formula to use. The reason for this difference in accuracy is attributed to the numerical accuracy of the computer computations. Thus we introduce the formula in Equation 9.5 here for illustrative purposes only.

The first estimate of k we shall call k_1. This is the simplest estimate. It is called the *moment solution*, or *moment–method solution*, and it was suggested for large samples by Fisher (1941). The formula to obtain the *moment estimate* of k is

$$k_1 = \frac{\hat{\mu}^2}{\hat{\sigma}^2 - \hat{\mu}}. \tag{9.6}$$

Using the Bolivian data set we have

$$\hat{\sigma}^2 = \frac{\begin{array}{c} 20(1)^2 + 9(2)^2 + 6(3)^2 + 3(4)^2 + 2(6)^2 + 1(8)^2 + 2(9)^2 \\ + 4(17)^2 + 1(19)^2 + 1(25)^2 + 1(62)^2 + 1(131)^2 + 1(252)^2 - \dfrac{663^2}{252} \end{array}}{251} = 340.00.$$

Using these two estimates, for the first two moments (parameters) in the formula for k, we have

$$k_1 = \frac{2.06^2}{(340.00 - 2.06)} = 0.01256.$$

Notice that we provided separate estimates of k and μ; there was no simultaneous solution method. Fisher (1941) defined conditions under which this moment solution for k is not efficient (statistical *efficiency* indicates an estimate with a relatively small variance). Equation 9.6 is reasonable to use for calculating the moment estimate of k in any observed data set when

1. μ is small, and $k/\mu > 6$.
2. μ is large, and $k > 13$.
3. μ is of intermediate size, and $(k + \mu) \cdot (k + 2)/\mu > 15$.

There are no definitions for "small," "intermediate," or "large" when we are interested in the moment formula's application. Nevertheless, the choice is usually quite clear after the value of k is compared to each of the calculations in the inequalities. For the Bolivian example we would not expect k_1 to be a good estimate because $\dfrac{(0.01256 + 2.06) \cdot (0.01256 + 2)}{2.06} = 0.00610 < 15$, using rule 3, and the other inequalities $(\dfrac{k}{\mu} > 6k$ and $k > 13)$ do not hold.

An alternative to the moment method of estimating the exponent k for a Negative binomial is the *maximum likelihood* (ML) *estimation* method devised by Fisher and given in Bliss and Fisher (1953). ML differs from the method of moment estimation in that it is an iterative method by which to compute a parameter estimate. That is, a first value is used to get successively better values, each closer to the final desired answer. ML is a rather involved and time-consuming method if worked by hand or with a calculator. However, it is an extremely practical estimation method on computer. Major statistical packages in computer software usually include ML because it is such an important tool for mathematical statisticians.

The ML method can obtain estimates of the Negative binomial exponent by finding the value that maximizes the *likelihood* of a sample (that is, the probability of the sample, given the parameters). The estimates obtained are guaranteed to have all the statistically desirable properties that we need for natural population study. The calculations can be performed by hand, following Bliss and Fisher (1953), but many statistical packages, like SYSTAT and SAS, will arrive at an exact ML estimate very quickly.

For our Bolivian example the ML estimate for k is 1.700835. Table 9.4 shows some of the calculations to obtain this estimate. We performed all the calculations in a computerized spreadsheet, which provides for instantaneous recalculations. The intermediate trial value in the example is just one of the many possibilities. The same final answer can be obtained with any number of different intermediate values that might be tested along the way. By using such a computer package, we can arrive at an estimate very quickly, regardless of how many trial values are needed.

These two methods (moment and ML), when used for estimating the exponent k of a Negative binomial, result in disparate estimates for our Bolivian data set. For the moment estimate we get $k = 0.01256$; from ML estimation we obtain $k = 1.700835$. Such differences can affect the fit of the theoretical distribution to any observed

TABLE 9.4 Example of Calculations for Maximum Likelihood Estimate of the Negative Binomial Exponent k, Using the Bolivian Data Set

INDIVIDUALS	SPECIES X	ACCUMULATION $A(X)$	CALCULATIONS (FOR EACH OF THE 3 ESTIMATES OF k)		
			$A(X)/0.01259 + X$	$A(X)/0.95 + X$	$A(X)/1.700835 + X$
1	20	32	31.6030655	16.41026	11.84819
2	9	23	11.42823071	7.79661	6.214814
3	6	17	5.64304134	4.303797	3.616379
4	3	14	3.489044401	2.828283	2.455781
5	0	14	2.792984024	2.352941	2.089292
6	2	12	1.995822079	1.726619	1.558273
7	0	12	1.711215305	1.509434	1.379178
8	1	11	1.372844634	1.22905	1.133923
9	2	9	0.998606389	0.904523	0.841056
10	0	9	0.898871018	0.821918	0.769176
11	0	9	0.817248669	0.753138	0.708615
12	0	9	0.749215821	0.694981	0.656894
13	0	9	0.691639462	0.645161	0.61221

14	9	0	0.642280925	0.602007	0.573218
15	5	4	0.333054456	0.31348	0.299386
16	5	0	0.31225488	0.294985	0.282473
17	4	1	0.235120405	0.222841	0.213894
18	4	0	0.222067269	0.211082	0.203037
19	4	0	0.210387239	0.200501	0.193229
20	4	0	0.199874479	0.190931	0.184325
Sums	52		71.97979619	49.53028	41.26317

NOTE: Calculations were performed in Microsoft Excel, using the formulas by Fisher in Bliss and Fisher 1953. The iterative procedure uses the moment estimate (0.01256) as the initial estimate. The formula for the maximum likelihood estimator (MLE) is $z_i = \sum[A(X)/k_i] - N \cdot ln[1 + (\hat{\mu}/k_i)]$, where $A(X)$ is the accumulated frequency in all species containing more than X individuals; k_i is the trial value for k on the ith iteration. Our $[\hat{\mu}] = 2.06$ for study plot 01 with $N = 52$ species. We obtained an intermediate estimate of 0.95 and a final estimate of $k = 1.700835$. Using our initial estimate of 0.01256, we obtained $\sum A(X)/(k_i + X) = 71.98$. The formula for z is then $52 \cdot ln(1 + 2.06/0.01256) = 265.5132$; $z_1 = 265.51 - 71.98 = 193.53$. This is positive, so the next value of k (that is, k_2) must be bigger than k_1. We can pick any value; our selection was $0.95 = k_2$. Using $k_2 = 0.95$ and repeating the steps above, we obtain 49.53. Then the entire formula gives $52 \cdot ln(1 + 2.06/0.95) = 59.96814$, and $z_2 = 59.97 - 49.53 = 10.44$, which is positive, so we pick another k bigger than the last. We selected $k_3 = 1.700835$ to obtain $52 \cdot ln(1 + 2.06/1.700835) = 41.26313$, and $z_3 = 41.26 - 41.26 = 0$ (actual value was 0.000041). Our goal in this MLE is to obtain a $z = 0$. Since this is as close to 0 as 6 decimal places can get us, we choose to stop and use k_3 as our final MLE value.

abundance curve. For most cases of heterogeneous data, when the variance is larger than the mean, the two distinct estimates for k will be highly disparate. In fact, the larger the variance in relation to the mean, the more likely the 2 values will be considerably different. In such circumstances, the ML estimate should always be preferred. However, these estimates can also be disparate when data are homogeneously distributed, as with our Bolivian trees. In general, ML estimation is the preferred method of estimating parameters.

Problems of Fit: A Bolivian Example

Let us use these two estimates of k to fit our Bolivian data set. The usual procedure found in the ecological literature for fitting follows that of Bliss and Fisher (1953), because it is here that Fisher first devised the ML estimate for the Negative binomial parameter k. Recall that we have an estimate of the mean of 2.06 and an ML estimate of $k = 1.700835$. The probability, which we will call P_x, that a species will contain

1, 2, . . . individuals, where $p = \dfrac{\mu}{k}$, is given by

$$P_x = \frac{(k + x - 1)!}{x!(k - 1)!} q^{-k}\left(\frac{p}{q}\right)^x.$$ (9.7)

Alternatively, we could let $\dfrac{p}{q} = \dfrac{\mu}{k + \mu}$ so that after this substitution Equation 9.7 gives us the alternative expression

$$P_x = \frac{(k + x - 1)!}{x!(k - 1)!}\left(1 - \frac{\mu}{k}\right)^{-k}\left(\frac{\mu}{\mu - k}\right)^x.$$ (9.8)

This probability, P_x, for any selected species, can be multiplied by S, the total number of species observed, to find the expected frequency of species with x individuals. Table 9.5 gives the expected frequencies for our Bolivian data set. For example, for the case of $x = 1$ we obtain

$$P_1 = \frac{(1.700835 + 1 - 1)!}{1!(1.700835 - 1)!}\left[1 + \frac{2.06}{1.700835}\right]^{-1.700835}\left[\frac{2.06}{2.06 + 1.700835}\right]^1.$$

Continuing with these calculations and plotting the results gives the curve fit in Figure 9.3. The calculations needed for the fitting of a Negative binomial involve some noticeable problems that are overlooked in the literature. Clearly, the exponent k (which represents the shape parameter) does not have to be an integer, just a positive number. This non-integer value is a component in the factorials (discussed in Chapter 3), which we defined there only for integers. This means that for this curve-fitting procedure, many calculators and computer formulas for $n!$ cannot be used to obtain the values for the computation because they work only when n is an integer. However, Bliss and

TABLE 9.5 Negative Binomial Calculations for Expected Frequencies of Bolivian Tree Data with 52 Species

INDIVIDUALS x	OBSERVED	PROBABILITY, P_x (THAT A SPECIES WILL CONTAIN x INDIVIDUALS)	EXPECTED FREQUENCY $(52 \cdot P_x)$	
			MLE = 1.700835	MLE = 2.0
1	20	0.241602	12.56	12.81
2	9	0.178711	9.29	9.75
3	6	0.120757	6.27	6.59
4	3	0.077734	4.04	4.18
5	0	0.048547	2.52	2.56
6	2	0.029698	1.54	1.51
7	0	0.017896	0.93	0.87
8	1	0.010661	0.55	0.50
9	2	0.006294	0.32	0.28
10	0	0.003689	0.19	0.16
11	0	0.002150	0.00	0.09
12	0	0.001246	0.00	0.05
13	0	0.000719	0.00	0.03
14	0	0.000414	0.00	0.01
15	0	0.000237	0.00	0.01
16	0	0.000136	0.00	0.00
17	4	0.000077	0.00	0.00
18	0	0.000044	0.00	0.00
19	1	0.000025	0.00	0.00
20	0	0.000014	0.00	0.00

NOTE: Calculations performed in MATHCAD. Negative binomial P_x, for $x = 1, 2, \ldots, 20$ (Equation 9.7) with Maximum Likelihood Estimate (MLE) $k = 1.700835$. This gives $p = 2.06/k = 1.21117$; $q = 2.21117$. Also included is a rounded MLE of 2.0.

Fisher (1953) obtained an answer despite this obstacle. The reason they could get an answer is that the definition we gave for a factorial is correct, but limited.

A more general expression can be written, but it involves complicated ideas from calculus (the evaluation of a *gamma function* as a *definite integral*). For their work, Bliss and Fisher (1953) sometimes were able to evaluate this integral. If they couldn't evaluate it for technical reasons, then (because their calculations were performed before the availability of hand calculators and computers), they used another approach. They might have used an *approximation formula*, which is a formula used to give a close but not exact answer. The well-known approximation formula to compute a factorial is called *Stirling's approximation*. This is given by

$$n! \approx (2\pi)^{1/2} n^{n+0.5} e^{-n}.$$

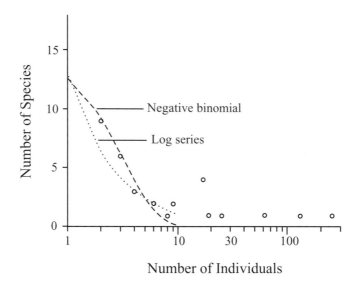

FIGURE 9.3   Negative binomial and Log series fit to Bolivian tree data.

with no restriction that n must be an integer, and where the symbol $\approx$ means "is approximately equal to." In this formula $\pi = 3.14159\ldots$ (an infinite or non-repeating decimal) and $e = 2.71.\ldots$ This e is also the base of the natural logarithm. Both quantities are found on most calculators and computer packages. Actually, Stirling's approximation formula for a factorial is surprisingly good. For example, we know $4! = 4 \cdot 3 \cdot 2 \cdot 1 = 24$, whereas by Stirling's formula we get $(2\pi)^{1/2} 4^{4+0.5} e^{-4} = 23.506$, or 24. Also, $20! = 2.433 \cdot 10^{18}$, while the approximation gives $2.423 \cdot 10^{18}$. Turning to the Bolivian data set, for $x = 1$ we need to evaluate $(1.700835)!$ and $(0.700835)!$. Stirling's formula gives

$$(2\pi)^{1/2} 1.700835^{2.200835} e^{-1.700835} = 1.47257$$

and

$$(2\pi)^{1/2} 1.700835^{1.200835} e^{-0.700835} = 0.8115797.$$

Using these 2 values, we obtain an expected frequency for $x = 1$ of
$$P_1 = \left(\frac{1.47257}{1} \cdot 0.8115797 \right) \cdot (0.259331 \cdot 0.547751) = 0.257740. \text{ Then } SP_1 = 52 \cdot 0.257740 =$$
13.4, or 13. This value is the expected frequency of species with 1 individual. A value of 13.4 is close to the value of 12.56 (both round to 13) that we obtained (when evaluating the gamma function) with MATHCAD in Table 9.5 or Figure 9.3. For our Bolivian data set, the actual observed frequency value is 20. That is, 20 of the 52 species are represented by just a single individual, which means that while both formulas produced similar results, they both resulted in underestimates for this particular case.

What if we were unable to use Stirling's approximation? If we were to use only the available formulas in which n must be an integer, we would be required to round the value of k to the nearest whole number, so that 1.700835 would become 2. Thus the first term in Equations 9.7 or 9.8 would be $\dfrac{2!}{1!\,1!} = 2$, and the value for P_1 would then equal 0.246252. Multiplying the value for P_1 by 52 would give an expected frequency of 12.80509, or 13 when $x = 1$. This rounding procedure did not make a noticeable difference in the fit of the Negative binomial for P_1. Furthermore, when we perform all the calculations we find that rounding has very little influence on the expected fit to the observed frequencies. However, it is risky to rely on rounding without first testing its efficacy. Each example should be checked.

The curve-fitting methods we have demonstrated that use the Negative binomial are reasonably straightforward in application. We first compute estimates for the mean and then use Equation 9.6 to obtain an estimate of k. For nonhomogeneous data, the moment estimate method for calculating k is not usually efficient. But we can use the moment estimate as a first trial value in our ML procedure, and then iterate toward a better fit. If we consider, however, the analogy with the positive Binomial, we uncover one additional prominent problem. For the Binomial we know that $p + q = 1$. For the Negative binomial, in the form usually used by biologists and specified by Bliss and Fisher (1953), Fisher showed that $p = \dfrac{\mu}{k}$ and $q = 1 + p$. In our example this gives $p = \dfrac{2.06}{1.700835} = 1.21$, obviously not a probability (which must be between 0 and 1). How do we deal with this situation?

The Negative binomial can be rewritten (this would be a different parameterization) with the requirement that we have a Bernoulli success probability parameter, which would take on values only between 0 and 1. However, the formula most often found in ecological treatments does not use this restriction. Rather, the most commonly used parameterization in the ecological literature is to assume independent Bernoulli trials and let $\dfrac{1}{q}$ be the probability of success. Then, $\dfrac{p}{q}$ would be the probability of failure. The probability that the Negative binomial variable equals x ($x = 0, 1, 2, \ldots$) would be written as

$$P_x = \left(\frac{1}{q}\right)^k \frac{k(k+1)\cdot\ldots\cdot(k+x-1)}{x!}\left(\frac{p}{q}\right)^x. \tag{9.9}$$

Equation 9.9 is merely a rewrite of Equation 9.7. This is the probability that $x + k$ trials are needed to obtain k successes or specimens of a particular species of interest. Here k would be a positive integer, not merely a positive number. When k is a positive integer, the probability distribution is called a *Pascal distribution* (although Pascal and Negative binomial are sometimes imprecisely used interchangeably).

Overall, we must be aware of the problem of parameterization changes and the possible discrepancy in the formula when using a Negative binomial to fit the observed data. For example, if we run a program to fit the Negative binomial to our Bolivian data, we get $k = 1.00$ and $p = 0.0859$, where the parameter p is a probability.

CONVERGENCE OF NEGATIVE BINOMIAL
AND OTHER DISTRIBUTIONS

As we change the value of the exponent k, the Negative binomial becomes indistinguishable from other important distributions that have been used for fitting abundance data. The range of values that k can take on is from 0 to infinity, although this can be extended (for example, see Engen 1974) from −1 to infinity. In this context, the word *infinity* implies that it is possible for k to be any very large number. There does not appear to be any biologically meaningful upper limit that we can specify as universal. If $k = 1$ in Equation 9.7, for example, we have pq^{n-1}, which is another way to write the Geometric distribution discussed earlier in this chapter. As the value for k increases, the entire distribution will take the shape of (or will approach) the Poisson distribution. The Poisson is, in this way, a *limiting form* of the Negative binomial.

For most taxa and for most data sets, the Poisson distribution does not provide a reasonable fit to the observations. For our Bolivian example of the abundance curve we have estimated a mean of 2.06 and a variance of 340.00. However, for a Poisson distribution (discussed in Chapter 6) we know that these two parameters (the mean and variance) must be equal. Thus we can see, without the necessity of further calculation, that it is highly unlikely that the Poisson will provide a satisfactory fit. The poorness of fit of the Poisson is usually attributed to the clumped dispersion pattern of individuals within species. Mathematically, however, the two distributions (Negative binomial and Poisson) are related, and this relationship can be seen in a variety of ways.

The terms *compound distribution* and *generalized distribution* are common in the literature. The Negative binomial distribution, among others we shall discuss in this chapter, is an example of a generalized distribution. A generalized distribution is one that is similar to, but a more complicated form of, some other distribution. The Negative binomial can be shown not only to approach a Poisson in the limit but to be an extension; that is, a *generalization* or a more complicated form, of the Poisson. The difference is that in a Negative binomial the mean density μ (which is the parameter of the Poisson) is not constant but varies continuously. That is, we can get a Poisson-shaped distribution when equal-sized samples are taken from homogeneous material or populations. We can think of this as counting individuals from equally abundant species, or of taking equal samples from each of a number of species. If unequal samples are taken (our data set of size n included samples of unequal size from each observed species), or if the population is heterogeneous, the value of the parameter in the Poisson would vary from sample to sample, or from species to species, and this would result in description by a new distribution.

One problematical assumption first used in methods of curve fitting was that the data set had been generated by successive, independent, and equal samples taken from a homogeneous population. The Poisson distribution came closer to field reality because it took into account how the number of individuals varied in the different samples. In practice, almost any field sample will contain individuals from species of unequal abundance. We saw this with the 52-tree species in Appendix 1. Thus the usual type of field sample is equivalent to taking different-sized, or unequal, samples of each species. Under these circumstances the data set is not best characterized as a single Poisson distribution, but as many such distributions, one for each species. Another way to state this is that the counts from the different species are *independent Poisson variates*; this occurs when all the individuals have small and independent chances of being captured. The probability of catching k individuals of any designated species is given by the Poisson probability $\dfrac{e^{-\mu}\mu^k}{k!}$, where μ is the expected number for any 1 species. If we make an assumption that the mean of the Poisson itself (the values for each of these species) follows some distribution, it is possible to derive new distributions.

This description of multiple, related Poissons is an example of a *compound distribution*. A compound distribution (often improperly termed a *generalized distribution*, even though these two can be equivalent sometimes) has a formula, or distribution, that involves a parameter that is itself a random variable with its own distribution. Depending upon which other distribution we compound with the Poisson, we get a new and different distribution. A Negative binomial can be shown to be a compound Poisson. Such a compounding is a special case of a *distribution mixture*. In such a distribution we are mixing or putting together two or more distributions. This mixing of distributions can be useful, for example, when each species has different capture or sighting probabilities over the sample area.

As a method for summarizing the observations with two parameters, such as the mean density μ and the exponent k, the Negative binomial will suffice. However, agreement of the observed and expected frequencies from this distribution cannot justify its basis as a characteristic of the target population. Many different mathematical models or interpretations can lead to the same Negative binomial. Biological interpretation is, therefore, not unique. A single data set or even a few data sets can be insufficient to distinguish among the different members of this family of distributions.

BROKEN STICK DISTRIBUTION

In 1957 MacArthur published an influential paper introducing an alternative model for fitting abundance data. This model came to be known as the *Broken Stick distribution* because of its proposed description of niche partitioning (or breaking). MacArthur's motivation was grounded, at least in part, in dissatisfaction with mathematical model fitting. His approach was not to find and fit some well-known statistical model; it was based, rather, upon his theories concerning what was occurring in the biological community.

MacArthur put together ideas on the arrangement and relative sizes of niches. Even though it was a new model for biologists, it was still a probabilistic model. However, the paper clearly shows that MacArthur's primary aim was to describe a hypothesis rather than to fit a probability curve. Despite his focus on niche partitioning, he still had to be concerned with the congruence of the observed and the expected frequencies as well as the interpretation of the mathematical assumptions into realistic biological terms.

The formula for the Broken Stick distribution is

$$N_i = \frac{N}{S}\sum_{n=i}^{S}\left(\frac{1}{n}\right) \tag{9.10}$$

where N is the number of individuals, S represents the number of species, and n is just used as an index ranging from i to S. This formula is not a probability specification. The term N_i represents the number of individuals in the ith (first, second, etc.) most abundant species. For very large numbers of species, an easier way to obtain the result is a modified version of the Broken Stick distribution, which yields an approximate formula:

$$N_i \approx \frac{N}{S} ln\left(\frac{S}{i}\right).$$

Note that we shall use the Broken Stick distribution for illustrative purposes in later chapters on biodiversity assessment because it is the simplest distribution for generating the actual numbers needed for the problem at hand. May (1975) states that a Broken Stick distribution can be expected when an ecologically homogeneous group of species randomly divides some resource (the "stick") among its members. In statistical terms, the Broken Stick model assumes that a given amount of mass (the "stick") is broken simultaneously.

Basing a model upon a biological hypothesis did not solve the problem of portraying the abundances of natural populations, even for small, homogeneous groups. This is because not many examples of reasonable fit of the Broken Stick model exist in the abundance literature. Usually the abundant species are too abundant, and the rare ones, too rare. One noticeable problem seems to be the assumption of simultaneous breakage (species intrusion of a resource); it is more biologically reasonable that breakage would be sequential, and so lead to a different model. In addition, the Broken Stick model can be shown to be merely a special case of the Negative binomial distribution. Therefore, nothing appears to be gained by choosing this specialized model over the more general one, or by beginning with a biological specification instead of a probabilistic model.

Logarithmic Series

Let us try another scheme for ascertaining a mathematical formula that can closely fit the abundance curves of natural populations. What if we examine the distributions to ascertain the discrepancies between the mathematical logic and the biological need? For example, an observant reader might ask, "What happens when the Negative binomial exponent k is exactly 0? Would this yield an altogether different distribution?" This very question was raised by a famous mathematical statistician: R. A. Fisher.

Fisher and colleagues (1943) first fit a Negative binomial to 2 large sets of butterfly and moth data. They assumed that this Negative binomial was the result of a compounding or combining of two other distributions (a Poisson and a Gamma). Fisher saw that the parameter k in the Negative binomial is an *inverse measure* of the variability (that is, as one goes up, the other decreases) of the different Poisson means from each of the species. If k is very large, these means will be approximately the same, and the overall distribution will approach a Poisson, as we described earlier. As k gets closer and closer to 0, the heterogeneity increases. It is in this sense that k has been called an *intrinsic property* of the sampled population. That is, it can be thought of as a measure of aggregation.

In this same tri-part paper, Fisher hit upon one of the most successful attempts at fitting observed ecological abundance data. The result of his work is now called a *Logarithmic series*, in which the formula for the *Logarithmic series distribution* came later. In the classic part of this paper, Fisher found that the Negative binomial fit the Malayan butterfly data published by Corbet ($N = 3{,}306$ individuals belonging to $S = 501$ species) and the English Macro–Lepidoptera observations of Williams ($N = 15{,}609$ and $S = 240$). When fitting this distribution, Fisher found that the exponent k was statistically indistinguishable from 0. Consequently, he investigated this situation when k could get closer to 0 (take on smaller and smaller fractions or decimal values, or *approach a limit of 0*). This new distribution is called a *special case*, or a limiting form, of the Negative binomial. It is similar to fitting a Negative binomial truncated at $x = 1$. However, a *Truncated Negative binomial* distribution (suggested in Brian 1953) is considerably more complicated mathematically and is quite difficult and tedious to fit. This is the special case in which the exponent k is assumed to be equal to 0. In this case, if we fit the Negative binomial, the value of k will be so minute as to be indistinguishable from 0 for all practical purposes, so we eliminate it from under the curve. (We must use the parameterization with p and k to see this, however.) In biological terms this truncation means that we are assuming that the number of unobserved species is unknown and that the series of values (on the x axis) cannot contain fewer than 1 individual per species.

As we reason through these different distributions, we see that, in practical application, there is little upon which to base a selection decision. And, as larger samples are taken, what little difference there may be among these fitted distributions becomes less and less evident. Whether we fit a compound distribution or just one of its component distributions appears to be of little import when samples are reasonably big and biologically representative of the target population.

The Logarithmic series name derived from the fact that the terms are obtained from a *series expansion* of the quantity (or formula for the series sum) of $-ln(1 - x)$. The terms are $-ln(1 - x) = x + \dfrac{x^2}{2} + \dfrac{x^3}{3} + \cdots + \dfrac{x^n}{n}$. Recalling the terms in the Harmonic series, we can see that if $x = 1$ then the 2 series are identical. Therefore we say that the limiting form of the Logarithmic series, when $x = 1$, is the Harmonic series. Actually, this expression for the Logarithmic series is not strictly what was used for expansion. The value for x is determined from a distribution once we have obtained values for N (number of individuals) and S (number of species), and a new constant α (*alpha*, a *proportionality constant*) becomes an important part of the distribution.

Fisher (1943) used the series expansion

$$-\alpha \, ln(1 - x) = \alpha x + \frac{\alpha x^2}{2} + \frac{\alpha x^3}{3} + \cdots + \frac{\alpha x^n}{n}$$

to obtain the necessary terms. The right side of this expression shows that the expected number of species containing 1, 2, ... , n individuals will equal the terms in the sequence αx, $\dfrac{\alpha x^2}{2}$, and so on. Here x is a constant that is always a number less than or equal to 1, and greater than 0. Alternatively, we could write that the probability of a count; that is, the logarithmically distributed random variable, taking on the value k is

$$P_k = \frac{\alpha x^k}{k} \tag{9.11}$$

for $k = 1, 2, \ldots$. In addition, $P_0 = 0$, but this term is omitted. Equation 9.11 reveals that the expected number of species with k individuals is proportional to x^k/k (the proportionality constant being α), where x is a parameter taking on values between 0 and 1 (and depending upon N and S). Consequently, for the Logarithmic series (*Log series*) the total number of species, S^*, is given by the sum

$$S^* = \frac{\sum \alpha x^k}{k} = -\alpha ln(1-x) \tag{9.12}$$

and the total number of individuals is given by

$$N^* = \frac{\alpha x}{1-x} \tag{9.13}$$

so that for any collection in which values of N and S are known, the terms of the series can be calculated, once the constants α and x are determined with the help of the equations for S^* and N^*.

Why might we need Equations 9.12 and 9.13 if we already have our observed totals for N and S? When we perform the calculations to fit the distribution, we shall show that S^* and N^* are useful checks on the values we obtain for our estimates α and x. The method of estimation for α is not just a simple substitution formula like it was for some of the other distributions. Unfortunately, we instead have to cope with a *simultaneous solution*. That is, we use both equations together to obtain a value for α, and it is rather troublesome to get the answers. Tables have been devised by both Fisher and colleagues (1943) and Williams (1943). However, the availability of computers today makes the troublesome solution no trouble at all. Fisher gave the relationships

$$x = \frac{N}{N + \alpha} \tag{9.14}$$

and

$$\alpha = \frac{N(1-x)}{x} \tag{9.15}$$

and he provided an equation for $\dfrac{N}{S}$ that eliminates x. The equation

$$\frac{N}{S} = \frac{e^{\frac{S}{\alpha}} - 1}{\frac{S}{\alpha}} \qquad (9.16)$$

was used in calculating the table given in Fisher's paper. This equation looks simple enough, but note (or try it) that it cannot be solved by writing this formula in a spreadsheet or database program, for example.

The solution to Equation 9.16 is called *iterative*. That is, one comes to a solution not in a single calculation but by way of successive, more accurate approximations, each of which makes use of the preceding approximate solution. To use this equation to solve for α, we need the capability of using trial values in a program such as MATHCAD. After we solve for α, we use Equation 9.14 to solve for x by simple substitution. Then we make the substitutions for the values of S^* and N^* in Equations 9.12 and 9.13, respectively, to see how close we came to the observed totals. If we are close to the observed values for S and N, we know that the estimates are good. How close is a matter of taste, but any reliable statistical package uses a very small difference, or *tolerance*, say, between 2 and 5 decimal places.

Since all we really need is to find 1 value, α, by iteration and the other, x, by substitution, the Logarithmic series distribution is actually a single parameter distribution. Another way to see that it involves only one parameter is to recall that the Log series is a Negative binomial (with parameters μ and k), but with k eliminated in practice, or set at 0. The median of this distribution is $\dfrac{0.56146}{\sqrt{(1-x)}}$, where the constant in the numerator is e^{-c}. The symbol c is called *Euler's constant* (which will surface again in our chapters on biodiversity).

An Illustration of Calculating α for the Bolivian Data

Once again, we illustrate using the data set for the Bolivian forest plot. Here $N = 663$ and $S = 52$. Using Equation 9.16, we solve for α with the aid of the MATHCAD program. We obtain $\alpha = 13.2139$. Then, using simple substitution and Equation 9.14, we calculate that $x = \dfrac{663}{(663 + 12.2139)} = 0.980459$. Using these values for α and x in Equation 9.12 for S^*, we obtain $S^* = -13.2139 - ln(1 - 0.980459) = -13.2139 \cdot -3.935240 = 52.000$. Using Equation 9.13 for N^*, we obtain $N^* = \dfrac{13.2139 \cdot 0.980459}{1 - 0.980459} = \dfrac{12.955687}{0.019541} = 663.0002$. These results indicate that we have very good estimates for α and x, because the calculated values of S^* and N^* are so close to the observed totals. Note that we used 6 decimal places for the constant x. Had we rounded off to $x = 0.9805$ and used this value, then $N^* = \dfrac{12.9562}{0.0195} = 664.4220$. Six or 7 places for x should be used routinely. Calculators and computers retain many places even when not displayed, so the calculations should present no problem.

Having obtained estimates of α and x, the series can now easily be calculated using the right side of Equation 9.10. The first term, the number of species expected with $n = 1$ individual, is $\dfrac{\alpha x^1}{1} = \alpha x = 13.2139 \cdot 0.980459 = 12.96$. Here we see that $\alpha\,(=13.21)$ is a value close to the number of species expected to be represented by 1 individual, because x is just slightly less than 1. The next term for $n = 2$ is $\dfrac{\alpha x^2}{n} = \dfrac{13.2129 \cdot 0.980459^2}{2} = 6.35$, and so on. The most abundant species, *Scheelea princeps*, has 252 individuals and, needless to say, the calculation of the entire series of 252 terms using only a calculator would be tiresome. While the process is easily carried out by computer, usually only α and the first few terms are of interest anyway because that is where we find most of the species.

Class Interval Construction

Table 9.6 shows the observed number of species with 1, 2, . . . , 252 individuals (from Appendix 1) and the expected number calculated from the Log series. This table shows the observed data both in the form presented in Table 9.3 and with the data grouped into class intervals. The grouping was done after the curve fitting was completed on the ungrouped observations, merely as an aid to the representation. The results indicate that the observed number of species represented by 1, 2, or 3 individuals, when combined, is more than we would expect from a Log series (35 vs 23). However, notice the similarity of fit between the Log series and the Negative binomial in Table 9.5. After $x = 9$, species occur sporadically so that the data are classed into intervals of unequal widths. The intervals are unequal in size on an arithmetic scale, and based upon a power of 2. They were selected in this way because the number of species represented by only a few individuals is not sporadic; had we chosen classes with a power of 10, say, we would not concentrate on natural population data where it is best or most informative. The base of the natural logs, e, would probably suffice, but the base 2 is most often used in the ecological literature because of Preston's work on the Log normal (a later section in this chapter).

If we had taken the number of individuals found at the boundaries of our chosen intervals and divided them equally into the groups on either side, as is often customary, we would have split data. For example, for the case of the boundary at 2 individuals, half of those species with 2 individuals would go to the class interval (1, 2) and half to the interval (2, 4). To avoid this problem of split data we used *open-ended limits*, or *half-open intervals*. Hence the class interval (1, 2) contains all observations greater than or equal to 1 but less than 2; that is, just 1. The usual notation is to use a bracket ([ ) to mean the value is included and a closed parenthesis ( )) to show the open-end value is not in the interval. Then, [2, 4) contains all observations greater than or equal to 2 but less than 4, and so on, as shown in Table 9.6. The classes are, of course, arbitrary and others could have been chosen. Just as in the lowest or more abundant classes, so the rarest class, [128, 256), has more species observed than expected (2 vs 0.33). However, here we are confronted with very small numbers. Curiously, the class [64, 228) has no observed species. The class [16, 32) has 6 species, which is more than in the previous

TABLE 9.6 Comparison of Raw and Grouped Observed Data with Log Series Fit

| n | RAW DATA | | GROUPED DATA | | | |
	OBSERVED	EXPECTED	INTERVAL NOTATION	GROUPED	OBSERVED	EXPECTED
1	20	12.96	[1, 2)	1	20	12.96
2	9	6.35				
3	6	4.15	[2, 4)	2, 3	15	10.50
4	3	3.05				
5	0	2.39				
6	2	1.96				
7	0	1.64	[4, 8)	4–7	5	9.05
8	1	1.41				
9	2	1.23				
.						
.						
.						
15	0	0.65	[8, 16)	8–15	3	7.72
16	0	0.60				
17	4	0.56				
18	0	0.51				
19	1	0.48				
.						
.						
.						
25	1	0.32				
.			[16, 32)	16–31	6	6.03
.						
.						
62	1	0.06				
.						
.			[32, 64)	32–63	1	3.82
.			[64, 128)	64–127	0	1.60
131	1	0.01				
.						
.						
.						
252	1	0.00				
.			[128, 256)	128–255	2	0.33

NOTE: The data set is drawn from Appendix 1, where $N = 663$, $S = 52$, $\alpha = 13.2139$, and $x = 0.980459$.

2 classes. Although this value fits the number predicted for the Log series (6.03), it does not follow an orderly and always *decreasing progression*. That is, it does not follow the kind of *monotonically decreasing progression* of values required by the Log series. For those species represented by more than 4 individuals, except for the [16, 32] and [128, 256] classes, the observed number of species are fewer than we would expect. In general, the rarest species are too rare and the most abundant are too abundant. The latter category, however, depending on how we define "most abundant," contains only 2 or 3 species. The data set is small compared to the Malayan butterflies ($N = 3,306$; $S = 501$) and the English lepidoptera ($N = 15,609$; $S = 240$). The Log series has an advantage over some other distributions for which the population form or distribution when sampled may be different. With a Log series distribution of the population, the samples can be recognized as conforming to the Log series, whereas the Negative binomial, for example, when sampled, can yield another (a Chi-square) distribution for the sample. In general, it is almost impossible to recognize a unique population distribution from a single sample. This is true especially when the sample is small.

In the foregoing exercise we calculated the Log series using the entire data set from the Bolivian plot 01 (Appendix 1). The results obtained from a sample of the entire set ought to prove instructive. Now, what if we sample just a portion of the target population? In the section of Chapter 8 on cluster sampling for proportions, we sampled at random 8 quadrats from a target population divided into 25 quadrats to ascertain species proportions. The quadrats chosen are shown in Table 9.7 and the data are available in Appendix 2. In all, there are $N = 201$ individuals represented by $S = 31$ species in these selected quadrats. Using Equation 9.15, we obtain a value of $\alpha = 10.2432$, and from Equation 9.13 a value of $x = 0.951510$. From Equation 9.12, $S^* = -10.2432 \ln(1 - 0.951510) = 30.79$.

From Equation 9.13, $N^* = \dfrac{10.2432 \cdot 0.951510}{1 - 0.951510} = 201.00$. Satisfied with the accuracy of our estimation of α and x, the series can be calculated using Equation 9.10.

Table 9.7 shows the results for our sample of 8 quadrats. The outcome is similar to that for the entire data set. Compared to the expected values for the Log series, there are too many rare species and too many abundant species. Because there are so many abundant species, this leaves too few species in the middle. The value of α for our sample of 8 quadrats is $\alpha = 10.24$. According to Fisher and colleagues (1943) the variance of α is

$$\sigma_\alpha^2 = \frac{\alpha^3 \left[(N + \alpha)^2 \ln\left(\dfrac{2N + \alpha}{N + \alpha} \right) - \alpha N \right]}{(SN + S\alpha - N\alpha)^2}. \tag{9.17}$$

Equation 9.17 was derived by Fisher for large samples only. In 1950 Anscombe provided an equivalent formula, using different notation, or symbols, for the quantities. Anscombe's formula, using our own notation and definitions of α and x, is

$$\sigma_\alpha^2 = \frac{\alpha \ln(2)}{\left[\ln\left(\dfrac{x}{1 - x} \right) - 1 \right]^2}. \tag{9.18}$$

TABLE 9.7   Comparison of Observed and Expected Number of Species from a Log Series

	RAW DATA		GROUPED DATA			
n	OBSERVED	EXPECTED	INTERVAL NOTATION	GROUPED	OBSERVED	EXPECTED
1	11	9.75	[1, 2)	1	11	9.75
2	12	4.64				
3	1	2.94	[2, 4)	2, 3	13	7.58
4	1	2.10				
5	2	1.60				
6	0	1.27				
7	1	1.03	[4, 8)	4–7	4	6.00
8	1	0.86				
.			[8, 16)	8–15	1	4.32
.						
17	1	0.26				
.			[16, 32)	16–31	1	2.43
.						
49	1	0.02				
.			[32, 64)	32–63	1	0.82
.						
68	1	0.01				
	.		[64, 128)	64–127	1	0.10

NOTE: The data set is drawn from Appendix 2, with 8 randomly selected quadrats sampled out of a total of 25. The quadrats sampled are 1, 3, 4, 10, 14, 19, and 23. For this sample, $N = 201$, $S = 32$, $\alpha = 10.2432$, and $x = 0.95150$.

Although Equations 9.17 and 9.18 were found to be algebraically equivalent by Anscombe, with the use of computers and their numerical accuracy, the answers will not be exactly the same. For example, for our previous example with $N = 1,500$, $S = 150$, $\alpha = 41.494$, and $x = 0.973$, Fisher's Equation 9.17 has a value of $\sigma_\alpha^2 = 3.885$. However, for this same example, Equation 9.18 gives $\sigma_\alpha^2 = 4.295$. This similarity is close enough for natural population work, especially for $x \geq 0.90$. If we were to calculate both Equations 9.17 and 9.18 for increasingly smaller N and S values, however, Fisher's would cease to be a reasonable estimate if x were quite low at the same time. Fisher's will give an unrealistically large variance for some small N and S, while Equation 9.18 will still appear to give a respectable value. A complete examination of the conditions under which we get discrepancies between these estimates is unnecessary, but we do know that as x becomes small, Anscombe's (1950) formula can yield a higher value than will Fisher's. However, this knowledge is of little help because as we demonstrate in later chapters, the Log series may not be an appropriate model under circumstances in which x is small. Indeed, despite a small standard error, or variance, σ_α^2 should not be used when $\dfrac{N}{S} \leq 1.44$, or $x \leq 0.50$.

Additionally, Anscombe (1950) also provided a formula for obtaining an estimate of a *Maximum Likelihood Estimate* (MLE). As we discussed in the Negative binomial section, an MLE has very desirable statistical properties, if it exists. However, this formula, which is

$$\sigma_\alpha^2 = \frac{\alpha}{\ln\left(\dfrac{x}{1-x}\right)}, \tag{9.19}$$

may appear simple to use, but it has difficulties that make it not entirely worthwhile. First, it is not a true MLE with all the attendant statistical properties, but a biased estimate of an MLE. For natural population work this means we would be using a number with an unknown or indeterminate amount of bias. Taylor and colleagues (1976) give an alternative form of Equation 9.19 that provides approximately, but not exactly, the same numerical values as Anscombe's (1950) MLE would yield. Again, computer numerical accuracy is the reason for the discrepancy. Both their estimate and Equation 9.19 are usually larger than Fisher's, but there are exceptions. For example, for $N = 210$ and $S = 82$ (which is reasonable for some insect species), Fisher's estimate is 30.204, while Anscombe's (1950) MLE estimate is 111.109, and Taylor and colleagues (1976) show a value of 29.866. In fairness, Anscombe (1950) gave an alternative formula (his Equation number 6.6) for certain cases like this one, and that formula gives a value of 34.237. Overall, throughout the recommended range of x, we recommend the large-sample formula proposed in Fisher and colleagues (1943), numbered here as Equation 9.17.

For the value of the variance of α in our sample of 8 quadrats, we obtained 1.48. Its square root, the standard error (remember, this is not called a standard deviation because it is the square root of the sampling variance of a population value) is 1.22. For 95% large-sample confidence limits (Chapter 4) we multiply by 1.96 (because the distribution of α has been shown to approach a Normal distribution). We thus obtain $\alpha = 10.24 \pm 2.39$, using Equation 9.17, giving a confidence interval of $7.85 \leq \alpha \leq 12.63$. We recall the true value of α for the entire plot 01 is actually 13.21, which is slightly above the upper confidence limit. Although a sample of 8 out of the total of 25 was entirely satisfactory for obtaining mean density and variance estimates of the target population, it is not large enough to safely determine a distribution of species that are not found in a homogeneous pattern, or that are over-dispersed.

Log Normal Distribution

The Need for a Logarithmic Transformation

Figure 9.2 shows a highly skewed pattern of points. When the sample data suggests that the population may be highly skewed, statisticians often suggest transforming the scale to logarithms. A logarithm is the power to which a base (that may be any positive number) must be raised to equal the number. If the base is 10, we have $10^n = S$, or $n = \log_{10} S$. Thus if $S = 100$, then, because we know that $10^2 = 100$, we have $2 = \log_{10}(100)$. This transformation may eliminate or reduce the skew, and a Normal distribution may then be fit to the data successfully. Let us transform the x values in Figure 9.2 to logarithms. We change the numbers of individuals 1, 2, . . . , 252 into logarithms: log(1), log(2), . . . , log(252). The result of this transformation is presented pictorially in Figure 9.3. In this figure we chose natural logarithms (base e, where $e = 2.71 \ldots$), which is abbreviated ln as usual. An equivalent picture would result if we had chosen (common) logs base 10.

A Problem of Terminology

A look at Figure 9.3 reveals that when we talk of the Log normal in the study of abundances, we mean that the analysis is done on the logarithms of the data, with the subsequent fitting of a Normal distribution. This figure also shows that we have a picture that ends at $x = 1$. Of course, we saw the similar problem with the Negative binomial. The problem is that we cannot fit a Normal distribution, which is a continuous distribution extending from very large positive to very large negative numbers. We would, in effect, need to fit only about half of a Normal distribution to the abundance curve. When we fit a distribution that has some cut-off point, we say we have a *truncated distribution*; in this case a *Truncated Normal distribution* would be the appropriate terminology. Although it was simple to truncate a series in the first section of this chapter, truncation of a probability distribution is a very complicated procedure to undertake, either by hand or by computer.

Mathematical statisticians have derived another very complicated distribution, called the *Log normal distribution*, that can be fit to the raw data to accomplish the same purpose as fitting the log-transformed data with a Normal distribution. We would need a truncated version of this also. Statistically, there is a direct relationship between the two choices, but the problem that the biologist then faces is that these two, seemingly equivalent, choices, both referred to as the *Log normal*, result in different answers (estimates). Luckily, the preferred choice has been to transform the observed data and fit a Normal distribution.

Preston's Fit of a Log Normal Distribution

Preston (1948) was the first to use the idea of transforming the observations to a logarithmic scale and then fitting a Normal distribution to the transformed data set. He used the theoretical Normal frequencies to graduate the observed frequencies, which had been transformed to logarithms and grouped. This means that he tried to get the expected values, based upon a Normal distribution, for each interval of species abundance, one at a time, by using a method of approximation. Preston's method was quite insightful, but empirical and computationally intensive, not mathematical. After changing to logs, he used an approximation method to the desired Normal distribution probabilities. The method has been called a *discrete approximation method*. The meaning of this is that one undertakes an approximation for each required value separately, not that one uses a discrete distribution. Grundy (1951) showed that Preston's method of discrete approximation for the Log normal did not yield a satisfactory fit when fitted probabilities were near 0. On the other hand, Bliss (1965) showed that this approximation method, although cumbersome and rather unrealistic from a biological perspective, was mathematically adequate, at that time, when used to fit observed data.

Remember that the observed data set was truncated; it had no observations for $x = 0$ individuals. In order to deal with the truncation that he observed, Preston (1948) introduced the idea of a *veil line* to separate the rarer species from those that were more abundant, or the sampled area from the unsampled. This led to the suggestion that the frequency distribution formed from a sample, and arranged in classes based on the logarithm, resulted in what has been called a *Truncated Log normal*.

Preston's approach can attain a reasonable fit when samples are large. Just how large a sample is necessary to achieve a Log normal fit is a problem both mathematically and biologically. Statistically, no matter the sample size, we can say only that this particular distribution could fit. But with a single observed data set we cannot rule out the possibility that some other distribution also fits, or that a more complicated or generalized form of the Truncated Log normal distribution may provide a better fit. May (1975) points out that both theory and observation show the Log normal should fit when a number of unrelated factors govern the way species accumulate. The fit of the Log normal has been shown reasonable for light-trap samples and large samples in which the target population is not well defined, and this may point to the truth of May's statement. If we were to take increasingly larger samples from a population that appeared to be distributed in a Log normal fashion, based on a small sample, the curve would become S-shaped (*sigmoid*) and flatten out as it approached the limit of the total number of species in the population, just like many of the distributions that we have examined.

Preston's Rationale and Interval Grouping

Why did Preston (1948) use an approximation for fitting a curve to the abundance distribution of the data set? There are two basic reasons. First, he was experimenting with the ideas and assumptions involved with the theoretical distribution and the biological data. Second, because the field data are discrete, a fit with individual, separate probabilities would seem a sensible choice. Unfortunately, there is no explicit formula (*closed form*) to obtain individual probabilities or expected numbers from a Log normal distribution. There is thus no simple formula Preston could substitute to get the probabilities needed.

Preston saw species abundance or commonness as a relative concept, which we have also pointed out. This means that he could group his data into classes in any manner he chose in order to form a frequency distribution. He chose, for his convenience, the grouping illustrated in Table 9.6, in which the midpoint of each species class is twice that of the preceding group. This method is equivalent to a transformation to a logarithm with a base of 2. Preston's base-2 approach to the Log normal is no longer convenient, because statistical and graphics packages generally use logarithms with a base of 10 (common logarithms) or e (natural logs, $e = 2.71\ldots$). After the transformation into log classes (called *octaves* by Preston, for the musical reference where each octave, or set of 8 notes, involves a doubling of the vibration frequency), Preston looked at each interval or class in the frequency distribution and fit, by eye, a small portion of the Normal probability curve. This inconvenient and inaccurate step is no longer necessary either, with the availability of computers. In fact, even in Preston's time, it might not have been necessary because he supplied the fact, although not the reference, that Singleton in 1944 had derived the appropriate MLE's.

Log Normal Fit to Bolivian Data

As an example, let us use Preston's (1948) method to fit our Bolivian abundances. Note that the intervals (0, 1), (1, 2), (2, 4), . . . are actually equal-sized intervals of width 0.3010 on the log 2 scale. Notice also that Preston's method of placing half the observed individuals into adjacent intervals actually forces the peak or modal class to be along the axis farther than merely at the first interval for species with 1 individual (Table 9.8). If we, however, use the open-ended intervals discussed earlier, we alter the shape of the curve (Table 9.9). Table 9.8 shows the intervals chosen by Preston. The endpoints of adjacent intervals coincide. For this reason his calculations are constrained. For example, to find the observed abundance for the first octave (0, 1), he took the 20 species with 1 individual and placed 10 of these into the next interval (1, 2). The mode, or highest observed octave total abundance, is 14.5 in octave 2, the *modal octave.*

When the octaves are determined and the modal octave found, the modal octave is given a coded value of 0. The value of 0 is then given to a new coded variable denoted by R, which is needed for the approximation. If we were to plot this new variable on an x axis, those observed abundances, or octaves, that are to the left of this modal amount would have negative numbers. Those octaves to the right would bear positive integers. This system is illustrated in the fourth column of Table 9.8. The next column is labeled R^2, which is the square of each of the R-coded values. The sixth column is the base-2 logarithm of each observed abundance value. In addition, we need the sum of the R^2 values as well as the average, or mean value, of the column of logarithms.

To obtain a logarithm to the base 2, one usually cannot push a button on a calculator or computer. However, almost any calculator or computer today easily computes logarithms base e or base 10. Therefore, we use the simple expedient of a constant multiplier. The formula to obtain the logarithm base 2 of any number, say x, is

$$\log_2(x) = \frac{ln(x)}{ln(2)}$$

where $ln = \log_e$, or the natural logarithm.

With the completion of these calculations, we are now able to estimate the Log normal parameters. The proportionality constant (a) is estimated by Preston (1948) with the formula

$$\hat{a} = \sqrt{\frac{\log_2\left(\frac{S_{MO}}{S_{R_{max}}}\right)}{R_{max}}} \qquad (9.20)$$

TABLE 9.8   Preston's Log Normal Approximation, Using His Method of Forming Octaves

OCTAVES	INDIVIDUALS PER SPECIES	%	R	R²	$\log_2[n_R]$	EXPECTED S_R (WITH $S_{MO} = 21.13$)	EXPECTED S_R (WITH $S_{MO} = 14.5$)
1	0–1	10	−1	1	3.3219	14.48	13.81
2	1–2	14.5	0	0	3.8580	15.20	14.5
3	2–4	12	1	1	3.5850	14.48	13.82
4	4–8	4	2	4	2.0000	12.52	11.95
5	8–16	2.5	3	9	1.3219	9.83	9.38
6	16–32	6	4	16	2.5850	7.01	6.69
7	32–64	1	5	25	0	4.54	4.33
8	64–128	0	6	36	0	2.67	2.54
9	128–256	2	7	49	1.9635	1.42	1.36
	Count: 52				Means: 3	15.67	1.9635

NOTE: The column S_R is the number of species in the Rth octave. The S_{MO} is the number of species in the modal octave, which can be either the observed value or a calculated estimate. $R = \log_2(S_i/S_{MO})$, for S_i the number of species in the ith octave. Note that the $\log_2$ of any number, say n, equals $ln(n)/ln(2)$. This frequency table contains 0 in the $R = 6$ interval, so we had to set the value of $ln(0)$ to 0. If we had used $ln(R + 1)$, the expected frequencies would have been much more discrepant from the observed. Expected frequencies are $S_R = S_{MO} + \exp(-a^2 \cdot R^2)$, where the parameters are the dispersion constant $a = 0.2199$ and $S_{MO} = 15.20$. The farthest octave from the mode is $\sqrt{[\log_2(SnMO)/SR = 7]}$. The count is 52 species. The total number of species theoretically available for observation is 122.49 (using expected $S_{MO} = 15.20$) and 116.9 (using observed $S_{MO} = 14.5$).

where S_{MO} is the observed number in the modal octave (14.5 for our example), and S_{Rmax} is the observed number of species in the octave with the largest R value (2 in our example). That is, R_{max} (and R_{min}) is the expected position of the rarest (most abundant) species. This number is found in the octave farthest from the mode. The R^2_{max} is the largest R^2 value in the column, which again is the one most distant from the mode. There is more than one way to calculate this estimate from the worksheet in Table 9.8. However, one of these ways is *not* to take $\dfrac{3.954196}{1.584963}$ and divide this ratio by 49 and then take the square root. The problem is that the logarithm of a division problem (a numerator/denominator) is not equal to the logarithm of the numerator divided by the logarithm of the denominator. The algebraic formula $\log\left(\dfrac{c}{d}\right) = \log(c) - \log(d)$ shows that the quantity we want is the difference of the logarithms. Thus, $\hat{a} = \sqrt{\dfrac{(3.954196 - 1.584963)}{49}} = 0.21989$. The answer we would get by the incorrect method is close (0.20087) but not correct. The value for $\hat{a}$ as shown by May (1975) is related to S_{MO} as well as to R_{max} and R_{min} by the formula

$$\hat{a}R_{max} = \sqrt{ln(S_{MO})}. \tag{9.21}$$

TABLE 9.9 Preston's Method, Using Half-Open Intervals

OCTAVES	INDIVIDUALS PER SPECIES	S_R	R	$-R^2$	$\log_2(S_R+1)$	EXPECTED S_R (WITH $S_{MO} = 18.06$)	EXPECTED S_R (WITH $S_{MO} = 20.0$)
1	[1, 2)	20	−1	1	4.3919	16.66	18.45
2	[2, 4)	15	0	0	3.9069	18.06	20.00
3	[4, 8)	5	1	1	2.3219	16.66	18.45
4	[8, 16)	3	2	4	1.5850	13.07	14.48
5	[16, 32)	6	3	9	2.5850	8.73	9.67
6	[32, 64)	1	4	16	0	4.96	5.49
7	[64, 128)	0	5	25	0	2.40	2.66
8	[128, 256)	2	6	36	1.0000	0.99	1.09
Count = 52	Means:	2.5	11.50	1.9651			

NOTE: The parameters of the fitted distribution by Preston's formulas are $a = 0.2842$, and then we get $S_{MO} = 18.0595$. As in Table 9.8, because the seventh octave contains 0, we substituted $0 = ln(0)$ in the calculations. Expected frequencies and total number of species are computed using both the observed and the expected values for S_{MO}, according to Preston's method. Total number of species theoretically available for observation is 112.66 (using expected $S_{Mo} = 18.06$) and 124.75 (using observed $S_{MO} = 20.00$).

The value obtained for $\hat{a}$ is almost always a number near 0.20. Many prominent researchers thought that this represented a hidden, but important, biological phenomenon. However, any sample in the size range of about 20 to 20,000, the most common range for natural population sampling, will yield such a value.

After obtaining $\hat{a}$, we next estimate the modal number of species n_{MO}. For this we need the averages of the $R \cdot R = R^2$ column and of the $\log_2(S_R)$ columns. We will use the notation μ_{R_2} and $\mu_{\log(n_R)}$. The formula is

$$S_{MO} = \exp(\mu_{\log(S_R)} + \hat{a}\mu_{R^2}). \tag{9.22}$$

Note that exp is a commonly used notation for raising e, the base of the natural logarithms, to a power. That is, $e^x = \exp(x)$. For our example, this quantity is $\exp[1.9635 + (0.220)^2 \cdot (15.667)^2] = 15.1961$.

Finally, we have the necessary ingredients for the expected frequency calculations. The formula for these quantities is

$$S_R = S_{MO} + \exp(-\hat{a}^2 R^2) \tag{9.23}$$

where S_R is the expected number of species in octave R. Because there is an R on both the left and the right of this equation, this tells us to change the value for each octave. That is, for example, $S_1 = 15.1961 + \exp(0.220^2 \cdot 1^2) = 14.4788$, and then $S_2 = 15.1961 + \exp(0.220^2 \cdot 2^2) = 12.5238$, and so on. It is also possible to use S_{MO} as the observed abundance in the modal octave. For our example this count is 14.5, and the expected frequencies using this value are given in the second column of the table. Actually, the fit is a little better with this value than with our estimate, as might be expected.

Table 9.9 shows that we repeated these calculations for the same forest data, but this time summarized into open-ended intervals rather than using Preston's (1948) method that pertains to Table 9.8. Although we obtained an estimate of $\hat{a} = 0.2842$, which appears much larger than that for the data presentation in Table 9.8, the expected frequencies when using S_{MO} are quite similar in both cases. It is only when we use $S_{MO} = 20$, our observed mode, that we see a difference in the expected frequencies. Notice that in this case the Log normal still has a mode whose position does not correspond to the observed mode in octave 1.

Density Function Fit

Preston's (1948) method for fitting abundance data is thus, at best, cumbersome. Alternatively, let us fit the distribution to the logarithms in a more accurate manner. If we can assume that the logarithm of our counts is a random variable that is distributed Normally, with mean μ and variance σ^2, then for any count greater than 0 (that is, $x > 0$, which is called the *domain* of the variable) we can use the formula for the *density function*. First we transform all the counts into logs (we use *ln*) and then calculate the mean and variance of the transformed counts. We then use the density function

$$\frac{1}{x(2\pi\sigma^2)^{\frac{1}{2}}} e^{-\frac{1}{2\sigma^2}(\ln(x)-\mu)^2}.$$

We can calculate the expected value for each individual (x) by this formula. Often, to avoid the use of exponents in a denominator, negative exponents are used. That is, we define $x^{-1} = \frac{1}{x^1}$, for example. Then, it is common to find the density function written as

$$(2\pi\sigma^2)^{-\frac{1}{2}} x^{-1} exp\left(-\frac{1}{2}(\ln(x)-\mu)^2\sigma^{-2}\right).$$

For example, in the Bolivia plot 01 the values of x are 1, 2, . . . , 252. Each of the values so obtained is then multiplied by the number of species ($S = 52$) to obtain the number of species expected to be represented by 1 individual, 1 individuals, and so on.

The procedure just explained may seem formidable, but it does show that for this Log normal distribution the mean is

$$\hat{\mu} = e^{\mu+\frac{\sigma^2}{2}} = exp\left(\mu + \frac{\sigma^2}{2}\right) \tag{9.24}$$

and the variance is

$$\hat{\sigma}^2 = e^{\sigma^2+2\mu}\left(e^{\sigma^2} - 1\right) = exp(\sigma^2 + 2\mu)\cdot(exp(\sigma^2)-1). \tag{9.25}$$

In addition, the mode is $e^{\mu - \sigma 2}$, while the median is e^μ. Notice that we are using ln, or the natural logarithm (we could have used log base 10 as well, or we could use a multiplier: $log_{10}x = \frac{\ln(x)}{\ln(10)}$, for example).

Using this density function method, we can fit the Bolivian forest data with a Log normal distribution (Table 9.10). For the rare species, the fit is quite good; however, for the abundant species the fit is not so good. Comparison of Tables 9.5, 9.6, 9.9, and 9.10 demonstrates just how difficult it is to fit each of the observations within an entire data set using any particular distribution.

LOG TRANSFORMATION PROBLEMS

For the Bolivian forest data set, the mean of the log-transformed counts (change to logs, multiply by frequencies and average) that we used was 12.74. This is not the value we would get if we took the natural logarithm of the raw mean of 2.06, which would be $\ln(2.06) = 0.7227$. Similarly, when we transform from the mean of the logs back to the sample, the transformation is not straightforward. We do not obtain the arithmetic mean of the counts in the sample if we change back from the mean of the logs. Rather,

TABLE 9.10 Observations from Bolivian Forest Data Set Fit to a Log Normal Distribution Curve, Using the Density Function

	RAW DATA		GROUPED DATA		
n	OBSERVED	EXPECTED	INTERVAL	OBSERVED	EXPECTED
1	20	18.93	[1, 2)	20	18.93
2	9	8.83			
3	6	4.65	[2, 4)	15	13.48
4	3	2.71			
5	0	1.69			
6	2	1.12			
7	0	0.77	[4, 8)	5	6.29
8	1	0.55			
9	2	0.40			
.					
.					
.					
15	0	0.09	[8, 16)	3	1.99
16	0	0.07			
17	4	0.06			
18	0	0.05			
19	1	0.04			
.					
.					
.					
25	1	0.02			
.			[16, 32)	6	0.43
.					
.					
62	1	0.00			
.			[32, 64)	1	0.06
.			[64, 128)	0	0.01
.					
131	1	0.00			
.					
.					
.					
252	1	0.00			
.			[128, 256)	2	0.00

we obtain the *geometric mean* of the original sample. (Instead of adding the counts and dividing by N, we would multiply all the counts and raise the result to the power $\frac{1}{N}$.) This geometric mean will always be less than, or *underestimate*, the arithmetic mean of the population, which is the usual quantity of interest. If we calculated the sample mean of the original data but had done the calculations on the logarithms, the estimate would be consistent (that is, it would always underestimate), but it would not be *efficient* (it would not have the same, but a larger, variance than when using the estimate based on logs). Similarly, the variance of the raw counts will not efficiently estimate the population variance when the Normal distribution, after log transformation, is used only as a convenience and interest is in the raw counts.

What this means is that when we fit the Normal distribution to the logarithms of the counts, we have two choices. We can transform back again to obtain an interpretation of the original counts, x, or we can discuss and interpret the results directly on the logarithms of the counts. More simply, the Normal may be a *convenience curve* (for example, when you transform to protect against an assumption violation for a statistical test) or it may be a *reference curve* (*interpolation curve*) for the original data. On what basis are these decisions to be made?

Most importantly, we must ascertain if the log transform can be assumed to depict the result of a *stochastic process* (a process that is not deterministic because it incorporates a random variable or random element) whose interpretation fits a generalized theory about the variable (the count) of the natural population under study. That is, does the logarithmic transform have an intrinsic meaning in the population? If so, the logarithmic quantity (log x) is used in place of the original variable, or count, as the variable of interest. For example, May (1975) stated that because populations tend to increase geometrically, not arithmetically, the "natural" variable is the log of the population density. If this is a palatable assumption, then the researcher would want to transform the observed values to logs and fit a Normal distribution with ultimate interpretation of the logs directly.

THE CANONICAL LOG NORMAL

Preston is credited with introducing more than his particular fit of a Log normal distribution, which we have already described. With his 1962 paper, Preston introduced the *Canonical Log normal*, which is a distribution based upon his *canonical hypothesis*. He investigated the relationship between the number of species per octave and the number of individuals in the species per octave. He found that each distribution could be fit by a Log normal and that the two curves had modes a fixed distance apart. That distance is

$$R_{\text{MO}} - R_{\text{N}} = \frac{ln(2)}{2\hat{a}^2}. \tag{9.26}$$

In this equation, R_{MO} is the modal octave of the species curve, and R_{N} is the modal octave of the individual's curve, while $\hat{a}^2$ is given by Equation 9.20. A simple way to write this relationship between the two curves is first to see that for each of the R octaves, it must be true that

$$I = N \cdot S \qquad (9.27)$$

where I represents the number of individuals, N the number of individuals per species, and S the number of species.

Preston compared these two Log normals in terms of, or at the point of, R_{max} (which is the octave for the last, most abundant species) and R_N (which is the modal octave for the individual's curve). This comparison affects a parameter change (re-parameterization) from the two parameters a and S_{MO} to the two parameters a and γ. The parameter change is accomplished by

$$\gamma = \frac{R_S}{R_{max}}. \qquad (9.28)$$

The canonical hypothesis of Preston is that for Equation 9.28, if $\gamma = 1$ (that is, if R_N and R_{max} coincide), then the Log normal species curve is called *canonical*. The term *canonical* applies because once N is specified, everything else is determined.

ESTIMATING TOTAL SPECIES FROM LOG NORMAL

A peculiar problem arises in natural population studies when the biologist is interested in estimating all the species in the target area. Knowing the number of species in the sample and obtaining a statistical estimate for the target population is important, but not sufficient.

Let us denote this total by the symbol S_T. Preston (1948) used the symbol S for this term, but we wish to minimize confusion here with our previous usage of S. The sample drawn from the target population will contain s species. From this sample, we will know for a fact that these s species must be in the population, but we will know nothing about how many species there are that we did not see or capture. Many have written that a major advantage of the use of the Log normal model is that (unlike some distributions, such as the Logarithmic series) the Log normal allows the investigator to estimate this total S_T in the catch area or population. For example, when using the Truncated Log normal distribution, the area under the curve is set equal to the number of species in the sample, s. To get an estimate of S_T, one calculates the area under the associated non-truncated, or complete, Log normal curve. To achieve this area calculation, we use the formula

$$S_T \approx \frac{S_{MO}\pi^{\frac{1}{2}}}{\hat{a}}. \qquad (9.29)$$

For our Bolivian example, $S_T = 170.26$ when using our estimates of $\hat{a} = 0.2199$ and $S_0 = 21.13$, or 116.88 for the same $\hat{a}$ but with an observed value of 14.5 as an estimate

(Table 9.8). Actually, this ability to come up with an estimate is equivalent to an artifact. Indeed, we can make an estimate of this total from our sample, but the estimate is bound to be misleading. For most taxa and sampling methods, the small proportion of species usually sampled rarely allows for accurate estimation of this total. This is true even if we could assume the Log normal is the correct model; that is, it is the best model for describing the data set. Even though S_T can be estimated mathematically and biologically, the actual result or estimated number is not satisfactory for any but the crudest purposes.

CRITIQUE OF THE VARIOUS ABUNDANCE MODELS

The Normal distribution fit to the logarithms of the observed data is a model with a very long tail. Many investigators have found that the Log normal distribution provides a poor fit with natural data sets because it has much more of an extreme skew than the observed data. Apparently, extremely large data sets (such as samples of moths taken by light traps and catches of other kinds of invertebrates) can be reasonably well fit by the Log normal. These same observation sets, however, can usually be well fitted by other distributions, too. Unlike the Log series, the Log normal model does not require that there must be more species represented by only 1 individual than any other category. The length of the tail; that is, the number of abundant species, may be related to the site and the combination of individuals encountered as well as to the sampling method. The Log normal appears to be useful when the standard deviation of the observations is approximately proportional to the size of the observations.

The Log normal is one of the most difficult curves to fit among any of those proposed for abundance data. It is complicated enough that even statisticians have had difficulty. For example, a 1965 publication by Bliss reported an analysis of insect trap data and advocated the use of a special compound form known as the Poisson-Log normal. However, it appears that he fit the usual Log normal. In 1950 Anscombe called this Poisson-Log normal form a *Discrete Log normal*. (He did not mean, however, that it was a discrete distribution.) In contrast, Cassie (1962) contended that the Poisson-Log normal was a new distribution. In large samples these variants are equivalent. Indeed, Bulmer (1974) showed that the compound Poisson-Log normal was formally equivalent to the Broken Stick distribution (which is the Negative binomial when $k = 1$) if breakage occurs sequentially.

Preston's methods were insightful in 1948 but now are unnecessarily computationally intensive and inaccurate by today's standards. Applying a Truncated Normal distribution to logarithmic data is time-consuming by hand, at the very least, and quite intractable for most researchers. With a computer program that does most of the work, it is almost straightforward, but the mathematics for any truncation curves are very advanced. With computerized statistical packages it may or may not be possible to program it in. The computations involved in fitting a truly Truncated Log normal (not Preston's approximation of such) are extremely complex. There are, however, other considerations that must be balanced against computational complexity. Usually, only very large samples (e.g., light-trap sampling of coleoptera, core sampling of microinvertebrates) have shown reasonable fit, as we already mentioned. If some other generalized distribution is also tried, the fit of the two is usually indistinguishable in repeated samples. Thus, if the purpose in fitting the abundance data is to be able to deduce some generalized rules at work in the population, a major problem confronts us if we cannot even uniquely distinguish with which curve we are working. If, however, the purpose is to provide a simple description of the abundance data by means of a small number of parameters, we can proceed to fit a simpler curve.

By this point, it may seem to the reader that this search for the best mathematical statistical model to fit the curve of the abundance distribution is a search for a will-o'-the-wisp, an exercise in futility. We have tried fitting theoretical distributions as exercises in mathematics with subsequent biological interpretation, and we have reversed the process and tried to devise biological reasoning first. Neither approach has worked; that is, neither approach has given us a unique model. We have also tried using simple number series as the expected numbers of individuals, but achieved little success. Aren't we really showing that no universal conclusion can be reached on how best to fit a theoretical distribution to species abundance data? The answer is "It depends."

For the statistician, any distribution that gives a reasonable fit, within some defined level of tolerance, would be fine. Maybe with enough data on butterflies, for example, the Log normal or the Log series might give the better fit most of the time. For example, at two stations where foraminifera were sampled (station 1 had $N = 7,745$, $S = 115$; station 2 had $N = 10,899$, $S = 117$), Buzas and colleagues (1977) showed that the Log series fit better than the Log normal. Yet if we give the statistician data sets drawn only from core samples for certain types of worms or only bat observations, the Negative binomial or perhaps the Harmonic series may provide the best fit for the observed, but relatively sparse, abundance data. The size of the collection and not the taxon appears to be important. The scientist needs a firmer basis.

The important question for natural populations is not whether the fit is "good," but whether it describes a characteristic inherent in the population of interest. Thus far we have shown that no single probability distribution, only one of a group of distributions, may be possible, perhaps because we have simplified, collapsed, or lost some of the essential biological information. Is one distribution better than another for all field samples of a particular taxon? The answer appears to be "no." Does "better" imply unique, or under all circumstances? The answer to this, too, is "no." Is one distribution preferable for a particular size of sample or type of sampling method that results in a larger or smaller catch? Here the answer appears to be "probably." The importance of a large sample is swamped by the consequences of an inability to sample the specific target population in a representative manner. For example, it may be true that the sheer numbers obtained by light-trap samples are so huge that one particular distribution will fit most readily. However, the ability of a light trap to gather specimens from a multitude of habitats and areas with undefined boundaries suggests that a specific target population is not being sampled. This method leads to observed data sets that may actually be best considered mixed populations rather than samples of some larger biological unit or area.

Now let us examine where all this analysis has taken us. We have seen that it is a mistake to use just one model and hope, from that, to obtain a unique and trustworthy interpretation of the biological or ecological assumptions. Nor can we hope that there is a uniquely best approach to the abundance distribution of any particular data set or taxon. Regardless of whether we begin with mathematical, statistical, or biological hypotheses and assumptions, the chosen approach will not guarantee a unique answer because we are suppressing some (often different) vital biological information in each. Therefore, we conclude that the search for one curve, and the selection of a most appropriate curve for all field situations with different (or even the same) taxa is hopeless. This is not to say that we can't do something quite useful in our search for a model. We can fit the observed abundance data with a particular curve, and we can select one distribution that will help to describe the particular field situation, just as we use the Normal for density and the Binomial for proportions. We could recommend, as did Koch (1969), that if inspection suggests no biological interpretation or mechanism for a particular distribution, try another one. Which one to pick? One could, of course, test all possibilities and see which is best for each particular situation encountered. However, as we hope this chapter has shown, the fits of many, if not most, distribution models will, in practice, turn out to be essentially equivalent. This convenience owes to the fact that the unavoidable error in field sampling is large enough to overcome any small deviance in the fit of a particular model. Many choices may be useful for practical purposes. For example, if you fit a Negative binomial when k is very near 0, you will have the equivalent of a Log series. (However, you would have done a tremendous amount of unnecessary work by choosing the Negative binomial.)

For large data sets the Log normal and Log series appear most reasonable. The Log series is closely related to the Negative binomial; it is discrete, and it offers a more elegant approach than the does the Log normal. Moreover, the constant α (the parameter of the Log series) is widely used as a measure of species diversity, as we discuss in Chapter 13. In practice, many researchers do not bother using the Log series for fitting the observed data; they use only the α parameter. Overall, we prefer the Log series over the Log normal when fitting large data sets. For smaller data sets one of the other distributions may fit better. Keep in mind, however, that the true shape of a distribution is often only apparent with a large amount of data. When plotted, even small samples from a known Normal distribution often do not appear Normal to the eye. Consequently, large data sets with a well-defined target population are most amenable for the study of species distributions.

SUMMARY

1. One way to fit a theoretical distribution to a data set is to calculate the moments, or parameter estimates, from the observed data and then to equate them to the moments, or parameters, of the theoretical distribution. This is called the method of moments.

2. A mathematical statistical procedure for calculating estimates of parameters, called the maximum likelihood (ML) method, gives estimates (MLE's) that are more efficient (have smaller variance) than those given by other methods, including the method of moments.

3. It is always preferable to use the ML formula for the parameter estimate, even if, as in the Negative binomial, a rounded number must be used for calculating the factorials.

4. This chapter describes all the most common theoretical distributions used as models for fitting abundance data. Among these, the Negative binomial is generalizable enough to cover most situations. If its exponent k equals 1, the Negative binomial is equivalent to the Broken Stick distribution with sequential breakages; if k is very large, the Negative binomial becomes a Poisson distribution; if k is near 0, the Negative binomial approaches a Logarithmic series distribution. In addition, if k equals -1, the tail of the Negative binomial is extremely skewed and thus can replace the Log normal.

5. The Negative binomial distribution is a widely applicable distribution for curve fitting of abundance data sets, but for most cases the Negative binomial requires computationally intensive procedures to obtain the parameter estimates and to make the fit.

6. No fit to observed abundance data is ever exact, especially for field samples of small or intermediate size. It is highly unlikely that one can gain singularly applicable information on population distribution by use of any of these models, at least without replication.

7. When a large number of species are so rare that they have a small or infinitesimal chance of being in the sample, this situation is well represented by the limiting form of the Negative binomial, when k approaches 0. The easiest expression of this is to use the Log series distribution to fit the observed abundance distribution.

8. The Log series distribution may well produce a fit as good as any of the other (more complicated) distributions, and Chapters 12, 13, and 14 show its added advantage when diversity studies are of interest.

9. The actual population distribution must be known if one wants to be sure that the resultant probability statements are exact. Unfortunately, only very rarely can one recognize this distribution simply by contemplating a sample. In most cases, however, it is safe to use an approximation.

10. This chapter shows that for most, if not all, field sampling of natural populations, a researcher can never know the true population abundance distribution by utilizing any of these methods (although see Chapter 14 for a discussion of SHE analysis). Therefore, trying to match observed with expected frequencies from a particular probability distribution will be an exercise in futility if its purpose is to gain population distribution information for any biological reason.

11. If a distribution fits the data reasonably well, and if that probability distribution is well behaved, easy to manipulate, and already tabled (or simple to calculate), a researcher will be able to gain knowledge by using simple summary estimates, and possibly confidence intervals.

12. The number of parameters it takes to well-define a theoretical distribution is an important concept. The more parameters a distribution needs to be well-defined, the more leeway there is in the fit to describe differing situations and anomalies. However, when we use a distribution with a higher number of parameters (say, 3 for the Log normal versus 2 for the Negative binomial), the mathematics gets considerably more complicated and curve fitting becomes much more difficult. Thus we should pick the compromise that does a reasonably thorough job of describing the data.

13. When we use the term *Log normal* in ecological studies, we mean a Normal distribution fit to the logarithms of the raw data. However, we do not simply equate the mean and standard deviation of the logarithms to the Normal parameters of mean and standard deviation.

14. The Log normal distribution fits many observed data sets reasonably well. Preston's (1948) approximate method for fitting the complicated Log normal to a data set is not necessary, nor is it an accurate enough fit in today's world of computer computations, which allow for the use of the density function method.

15. The Log series fits many observed data sets quite well. The Log series distribution differs from the Log normal in that it is simple to fit, and it needs no truncation methodology to fit natural population data. In addition, it is elegant and confidence intervals can be constructed using Normal z values.

PROBLEMS

9.1 Take the first 25 integers and perform on them the following operations:

a. Write the integers as a sequence.

b. Write the integers as a series.

c. Write the series expansion for these integers.

d. Write a formula for the sum of these 25 integers and find the value of that sum.

9.2 Appendix 5 presents a subset of bat counts published by Handley in 1976 as "Mammals of the Smithsonian Venezuelan Project." This data set was then used by Ralls and colleagues (1982) for finding correlations between three possible measures of size in neotropical bats. The subset of counts presented in Appendix 5 represents those voucher specimens preserved in alcohol in the Smithsonian collections.

a. Fit a log series distribution to this data set.

b. Discuss the relationship and importance of Equation 9.14 to the fit you obtain.

c. Compute a confidence interval for the Log series parameter for this data set.

9.3 The following counts for *Hyla crucifer* and *Rana sylvatica* were obtained over the 16 weeks of the Heyer 1976 study of anuran larvae.

Hyla crucifer: 0, 0, 0, 0, 0, 22, 81, 34, 101, 77, 141, 91, 56, 31, 0, 0
Rana sylvatica: 0, 0, 0, 0, 2, 23, 30, 45, 29, 26, 16, 7, 3, 0, 0, 0

a. Find the mean density and variance for each species.

b. Find the covariance of the 2 species.

c. Place your findings (values) in a V–CV matrix, as presented in Equation 9.3.

10

REGRESSION:
OCCURRENCES AND DENSITY

F IELD RESEARCHERS WHO SAMPLE natural populations may be constrained by time or cost limitations, so that the most that can be achieved is the compilation of a species list. In addition, when observational techniques are subject to excessive variability, it may be more advantageous merely to collect presence/absence data than to attempt collecting abundance values. This is the case, for example, for many secretive or highly mobile animals, or when interobserver error is high because of unequally trained or fatigued observers. Even for initial surveys with well-behaved organisms, such as trees or sessile organisms, a compilation of a species list from many localities may be just as useful as, and much less costly than, a labor-intensive enumeration.

Provided there are a large number of localities, such compilations will invariably reflect a pattern similar to that observed with the individual counts (when there are a large number of taxa and/or localities). That is, some species occur at many localities, others at some or very few, and most at one locality only. This information can provide valuable insight about the target area that is quite often overlooked or undervalued by researchers. In this chapter we show that there is a close relationship between density and occurrences of species. In addition, this chapter provides information on the kind of statistical regression analysis that allows one to capitalize on this relationship and that is worthwhile for understanding topics in subsequent chapters.

For many studies, the compilation of a species list is just as useful as, but less costly than, gathering data on species abundance. Foremost among the advantages of data on occurrences in an area is the use of the species list as an alternative to a *pilot study*. Instead of performing a field pilot study or preliminary field inventory, the museum records or species lists of the target locality compiled by other workers can be examined and summarized. The cost savings are achieved by (1) eliminating field trip expenses, (2) reducing time, (3) making secondary use of prior work already paid for, and (4) maximizing use of established museum collections and prior scientific work.

RELATIONSHIP OF DENSITY AND OCCURRENCES

Occurrence data can tell the researcher quite a lot about the target area, without the work, cost, and trouble of observing or counting individuals. Let us first examine the relationship between the number of individuals per species in a community and the number of localities or quadrats in which those individuals occur.

For example, even a casual look at abundance data for the entire Beni Reserve plot 01, divided into 100 quadrats (as in Appendix 1), reveals a relationship between density or abundance and the number of quadrats in which a taxon occurs. If we consider all species in our example plot, which is very homogeneous, there is at least 1 individual of 1 species or another in each and every quadrat. The most abundant species, *Scheelea princeps*, occurs in 87 of the 100 quadrats—more than any of the others. At the other end of the distribution are the very rare species, such as *Unonopsis mathewsii*, which is represented by only 1 individual; it can occur, of course, in only 1 quadrat. Thus abundance and density have some relationship to occurrence of species.

The first question to be asked is "Can we predict mean density just from knowing the number of times the species occurs in our biological samples?" It would be extremely useful to substitute knowledge of easily obtainable species occurrences from the literature, museum collections, or the field, for knowledge of species density, which is considerably more costly to obtain.

In order to predict species density using only the data of occurrences, three criteria must be satisfied:

1. We must be able to observe or find some statistical relationship that could lead to a reasonable amount of predictability between species density and species occurrences.
2. We must be able to assess the strength of this predictive relationship with some quantitative measure.
3. We must be able to form some relatively simple, but accurate, rule for the actual prediction of species density from just the species occurrences.

Descriptive Statistics of Regression

Thus far, this book has dealt with problems of inferring population characteristics from samples. In statistical application there is another aspect of inference. We could also be interested in the problem of predicting an individual case, based upon some observed information. The method used for making statistical predictions is called *regression*. In this chapter we are interested in the *descriptive statistics of regression* that can provide predicted information on 1 variable by using the known data of a second.

This term, *regression*, first appeared (without explanation) in Chapter 6, when we needed to find the constants *a* and *b* to fit the power curve to the density data. In that chapter, we used the power curve to *predict* the variance. We simply used the data we had collected to infer a relationship outside the actual collection. In Chapter 6 we could have used knowledge of the mean density to increase our knowledge about the variance; the extent to which we knew something about the mean density could reduce our uncertainty concerning the variance, which is the quantity being predicted. We even noted in that specialized, but unrealistic, case that *perfect prediction* was possible. That is, the *x* variable (mean density) could tell us everything about the *y* variable (variance). This opportunity for perfect prediction occurs when the distribution is strictly Poisson ($\sigma^2 = \mu$). Otherwise, we relate σ^2 to μ by a *functional relationship* called the power rule.

To make a prediction of density based on occurrences data, one first needs to understand the requirements for prediction. When interest is in prediction, the x variable is called the *independent variable*. The y variable is termed the *dependent variable*, because its value is related to, or depends upon, the value we choose along the x axis. We need a rule for predicting density from information on occurrences, and we intend to treat this rule as if it expressed the true relationship. We must also have a way to assess how well this rule works, or its *goodness of fit*. We can measure the fit by assessing how well the rule serves us when we try to obtain the density by using the occurrences. Thus we can use data sets in which we know both density and occurrences, and compare how well the observations match the predictions obtained from the rule. Usually, a rule will not give perfection, but the fit may be good enough that it will have practical ecological advantages. Of course, after applying this rule to many data sets in which we achieve reasonable fit, we can also examine the *errors of prediction* to gather information for revising the rule to make predictions more accurate.

As we discussed in Chapter 6, the very simplest predictive rule is of the form

$$y = a + bx \tag{10.1}$$

where b represents the slope of the line and a denotes the y intercept, or point on the y axis at which the straight line crosses. This equation represents a *linear function* (a *function* here means that we predict only one y from one x, even if there is more than 1 observed y value for 1 value of x). A *linear equation*, or linear function, can be used as a *linear regression* rule. This is not only the simplest but the most widely applicable equation for prediction in natural populations, and it often gives quite good approximations to other more complicated rules.

Properties of Occurrences as a Predictor Variable

The term *observed proportion of occurrences* signifies the relationship of the number of occurrences to the number of biological samples. The proportion of occurrences is thus p_o = number of occurrences/number of samples.

It can be useful to plot the mean density (the y variable) against the proportion of occurrences (the x variable). This arithmetic plot is a *scattergram*. A scattergram (for example, Figure 10.1) shows that the species have a tendency to group rather closely at small values of both x (occurrences) and y (mean density) and then spread out for larger values of these variables. When such a pattern is observed on an arithmetic scale it is an indication that a change to a logarithmic scale may be useful. Such a change will allow us to fit a linear rule or equation to the variables p_o and mean density, just as we did in Chapter 6 for the variance and mean density. A plot on a log–log scale (either base 10 or base e can be used) of all the data in Bolivian plot 01 is shown in Figure 10.2.

Clearly, there is some linear relationship between these two variables, mean density and proportion of occurrences (also called *frequency*, especially in the botanical literature) when plotted on this log–log scale. So, we could write the linear rule,

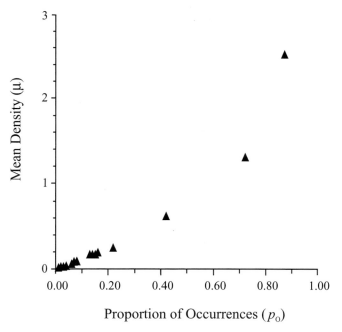

FIGURE 10.1 Arithmetic plot of mean density and proportion of occurrences for tree species in Bolivian plot 01. (Data from Appendix 1.)

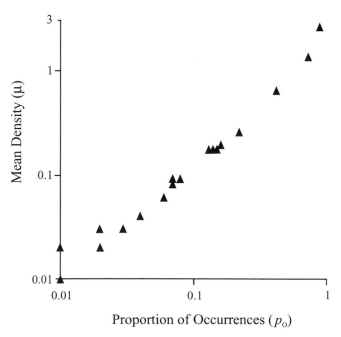

FIGURE 10.2 Log–log plot of mean density and proportion of occurrences for tree species in Bolivian plot 01.

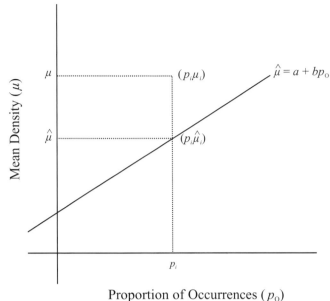

FIGURE 10.3   Plot of Equation 10.2 illustrating the distance between observed and predicted values.

$$\log(\hat{\mu}) = \log(a) + b\log(p_o). \tag{10.2}$$

In this equation a and b are constants that must be determined to make sure that the prediction is, in some sense, "best."

Figure 10.3 shows a plot of Equation 10.2. It provides a visual basis for arriving at an intuitive sense of which substitutions for the constants a and b would provide the "best" predictive equation. If we are then going to proceed as if this predictive rule were true, then it is important to ensure that the distance between the observed and predicted values (the error) of the mean density be the smallest possible, or a minimum. Just as in Chapter 6 then, we want the errors over the entire set of values to be minimized. And, just as in Chapter 6, the least squares criterion is the method of choice.

OBSERVED TREND

It may seem confusing that, as an example, we would be interested in predicting a value for the mean density (obtaining a value for μ) when we already know the observed mean density for our Bolivian plot. In fact, we are discussing the development of the regression approach to gain intuition concerning the relationship between our two variables: the mean density and the species occurrences. Strictly speaking, what we have drawn in Figure 10.3 is a *trend line* for the particular observations we have collected. The method of least squares regression will be used when we have some information on occurrences but are unable, or unwilling, to collect data on density.

The *observed trend* will help us obtain estimates of species density when we cannot glean this information from fieldwork or other sources. Regression methods apply when we want to use the data collected to predict a mean density value for the same or a related locality or observational area for which all that is known is which species occurred.

REGRESSION AND CORRELATION

Confusion exists in the research literature regarding the proper use of regression techniques. The most consequential problem concerns the need for assumptions about the collected data when using regression for descriptive purposes. In truth, we do not need to make any assumptions about the form of the distribution curves for either the species occurrences or the density. In addition, we do not need to assume any knowledge of the variability of the density measures within the localities on which we have collected occurrence data. The confusion exists because we do need some assumptions when we are attempting to infer the nature of an actual relationship in the population. If instead we use the regression equation to describe and examine the unobserved data in the same area as that in which we obtained our species list, then we are on perfectly safe statistical grounds.

Table 10.1 shows the calculations necessary to compute the values of slope and intercept for the Bolivian forest example, using the *normal equations* presented in Chapter 6: Equations 6.7 and 6.8. The equation obtained is

$$\log(\hat{\mu}) = 0.4641 + 1.105 \cdot \log(p_o).$$

Note that for regression, the slope b is called the *regression coefficient*, or sometimes the *beta* (β) *coefficient*. The symbol b is its estimate.

When using the equation to discuss the influence the occurrences have on the mean density, or when using knowledge of occurrences to gain knowledge of the mean density, we are in the realm of *regression* methods. However, if we are interested in observing the occurrences and the density for an area and, in turn, describing how these two quantities vary together over a set of samples or target areas, we are dealing with *correlation* methods. These two statistical methods are so highly interrelated that confusion is, for the most part, unavoidable. One source of confusion is the very close theoretical, statistical relationship between these two concepts. In addition, correlation can be used alone as well as used as an aid to interpreting regression. For the prediction of density in this chapter we are interested only in the latter use of correlation, as an evaluator of prediction. Let us now look at the relationship.

Evaluation of Prediction Error

As portrayed in Figure 10.3, we cannot expect that any regression line will explain all the variation in the observations; the line shows just an average relationship between x and y. But the observations hardly, if ever, lie along a perfectly straight line. Thus estimates of density based upon occurrences cannot be expected to be precisely cor-

TABLE 10.1 Calculations for Slope (b) and Intercept (a) of the Regression Equation for Predicting Density from a Data Set of Species Occurrences

Species	$y = ln(\mu)$	$x = ln(p_o)$	x^2	y^2	xy	Mean Density Observed	Mean Density Predicted ($e^{a+bln(x)}$)
1	0.09243	−0.1393	0.0194	0.8543	−0.1287	2.52	1.42
2	0.2700	−0.3285	0.1079	0.0729	−0.0887	1.31	1.26
3	−0.4780	−0.8675	0.7526	0.2285	0.4147	0.62	0.61
4	−1.3863	−1.5141	2.2926	1.9218	2.0990	0.25	0.30
5	−1.6607	−1.8326	3.3584	2.7580	3.0434	0.19	0.21
6	−1.7720	−1.9661	3.8656	3.1398	3.4839	0.17	0.18
7	−1.7720	−1.8971	3.5991	3.1398	3.3616	0.17	0.20
8	−1.7720	−1.8971	3.5991	3.1398	3.3616	0.17	0.20
9	−1.7720	−1.8971	4.1625	3.1398	3.6152	0.17	0.17
10	−2.4080	−2.6593	7.0717	5.7982	6.4034	0.09	0.08
11	−2.4080	−2.6593	6.3793	5.7982	6.0818	0.09	0.10
12	−2.5257	−2.6593	7.0717	6.3793	6.7166	0.08	0.08
13	−2.8134	−2.8134	7.9153	7.9153	7.9153	0.06	0.07
14	−2.8134	−2.8134	7.9153	7.9153	7.9153	0.06	0.07
15	−3.2189	−3.2189	10.3612	10.3612	10.3612	0.04	0.05
16	−3.2189	−3.2189	10.3612	10.3612	10.3612	0.04	0.05
17	−3.2189	−3.2189	10.3612	10.3612	10.3612	0.04	0.05
18	−3.5066	−3.5066	12.2960	12.2960	12.2960	0.03	0.03
19	−3.5066	−3.9120	15.3039	12.2960	13.7177	0.03	0.02
20	−3.5066	−3.5066	12.2960	12.2960	12.2960	0.03	0.03
21	−3.5066	−3.5066	12.2960	12.2960	12.2960	0.03	0.03
22	−3.5066	−3.5066	12.2960	12.2960	12.2960	0.03	0.03
23	−3.5066	−3.5066	12.2960	12.2960	12.2960	0.03	0.03
24	−3.9120	−3.9120	15.3039	15.3039	15.3039	0.02	0.02
25	−3.9120	−3.9120	15.3039	15.3039	15.3039	0.02	0.02
26	−3.9120	−3.9120	15.3039	15.3039	15.3039	0.02	0.02
27	−3.9120	−3.9120	15.3039	15.3039	15.3039	0.02	0.02
28	−3.9120	−3.9120	15.3039	15.3039	15.3039	0.02	0.02
29	−3.9120	−3.9120	15.3039	15.3039	15.3039	0.02	0.02
30	−3.9120	−3.9120	15.3039	15.3039	15.3039	0.02	0.02
31	−3.9120	−4.6052	21.2076	15.3039	18.0156	0.02	0.01
32	−3.9120	−3.9120	15.3039	15.3039	15.3039	0.02	0.02
33	−4.6052	−4.6052	21.2076	21.2076	21.2076	0.01	0.01
34	−4.6052	−4.6052	21.2076	21.2076	21.2076	0.01	0.01
35	−4.6052	−4.6052	21.2076	21.2076	21.2076	0.01	0.01

(continued)

TABLE 10.1   *(continued)*

SPECIES	$y = ln(\mu)$	$x = ln(p_o)$	x^2	y^2	xy	MEAN DENSITY OBSERVED	MEAN DENSITY PREDICTED $(e^{a+bln(x)})$
36	−4.6052	−4.6052	21.2076	21.2076	21.2076	0.01	0.01
37	−4.6052	−4.6052	21.2076	21.2076	21.2076	0.01	0.01
38	−4.6052	−4.6052	21.2076	21.2076	21.2076	0.01	0.01
39	−4.6052	−4.6052	21.2076	21.2076	21.2076	0.01	0.01
40	−4.6052	−4.6052	21.2076	21.2076	21.2076	0.01	0.01
41	−4.6052	−4.6052	21.2076	21.2076	21.2076	0.01	0.01
42	−4.6052	−4.6052	21.2076	21.2076	21.2076	0.01	0.01
43	−4.6052	−4.6052	21.2076	21.2076	21.2076	0.01	0.01
44	−4.6052	−4.6052	21.2076	21.2076	21.2076	0.01	0.01
45	−4.6052	−4.6052	21.2076	21.2076	21.2076	0.01	0.01
46	−4.6052	−4.6052	21.2076	21.2076	21.2076	0.01	0.01
47	−4.6052	−4.6052	21.2076	21.2076	21.2076	0.01	0.01
48	−4.6052	−4.6052	21.2076	21.2076	21.2076	0.01	0.01
49	−4.6052	−4.6052	21.2076	21.2076	21.2076	0.01	0.01
50	−4.6052	−4.6052	21.2076	21.2076	21.2076	0.01	0.01
51	−4.6052	−4.6052	21.2076	21.2076	21.2076	0.01	0.01
52	−4.6052	−4.6052	21.2076	21.2076	21.2076	0.01	0.01

NOTE: $y = ln$(mean density) and $x = ln$(proportion of occurrences) over 100 subplots of the Beni Reserve plot 01. The equation is ln(mean density) $= 0.4641 + 1.105 \cdot ln(p_o)$. The coefficient of determination (r^2) is 0.9859; SEE = 0.1623; standard error of x coefficient = 0.01872. Standard deviation of the y values is 1.8269.

rect. The estimates may be in error because other biological factors not included in this rule affect the relationship. We need some way to assess this *error of prediction*, or to assess the adequacy of our regression line as an explanation of the variation in density values.

We have already discussed in Chapter 6 the mechanism for defining this error. We want, on the average, to ensure that the difference between the predicted density value and the real one (that is, $Y_i - \hat{Y}_i$) is as small as possible. Remember, we cannot use just the average difference for each point on the graph because the sum of all such differences necessarily equals 0. The differences are taken both above and below the line, and therefore always add to 0 for any data set. That is, we know it is always true that $\Sigma(y_i - \hat{y}_i) = 0$. In regression, these differences are called *residual deviations*. The formula for the error that we minimize when we use the least squares method for regression is that given in Equation 6.6. When we use this equation for our regression, however, we give it another name. The measure of scatter of the data points about the regression line is called the *standard error of the estimate* (SEE) and is defined by the formula

$$S_{y \cdot x} = \sqrt{\frac{\Sigma_i (y_i - \hat{y}_i)^2}{N}}. \tag{10.3}$$

Equation 10.3 is equivalent to the familiar formula that we used in calculating the standard deviation about the (arithmetic) mean density (Equation 2.5). In Equation 10.3, the differences are taken not about a mean but about the predicted values; that is, about the regression line. Accordingly, this quantity is given a special name: the standard error of the estimate. It is not just called the standard deviation. However, the term *SEE* is misleading. It is decidedly not a standard error (the positive square root of the variance of the sampling distribution of a statistic), and it surely does not deal with a statistical parameter estimate in the usual sense. SEE, rather, is a standard deviation of the observed values spread about a regression line. It is an "estimate" in only a rather nontechnical sense in that it provides knowledge of the relative size of the variation likely to be encountered when using the regression line for making predictions. The subscript $(y \cdot x)$ used for the notation $S_{y \cdot x}$ indicates that this is the variation in the estimate for y when we use x for predicting y. It is a reminder that we cannot interchange the x and y variable and predict in reverse (that is, we cannot use y to predict x after our regression line was calculated with x as the independent variable) without some consequences.

Correlation Coefficient for Regression

A very helpful formula for conceptualizing the relationship of correlation to our regression needs is to rewrite Equation 10.3 in the form

$$S_{y \cdot x} = \sigma_y \sqrt{1 - r^2} \tag{10.4}$$

where σ_y represents the usual standard deviation of y; that is, from Equation 2.3, $\sigma_y = \sqrt{\frac{\Sigma_i (y_i - \mu_y)^2}{N}}$. We used the subscript on the mean density symbol merely to show that it is the mean of the y values, or density that is being subtracted. Of course, we can substitute the $\hat{\sigma}_y$ of Equation 2.5 into Equation 10.4 when we are using the observed data. The quantity r is the *correlation coefficient*. A correlation coefficient is a measure of the interdependence between the 2 variables. It is also a standardized covariance (see Chapter 9).

Equation 10.4 lets us see that with $S_{y \cdot x}$, the standard deviation, or spread, of the y values is corrected by multiplying, by using the relationship, or correlation, between the y and x values. For the Bolivian forest example we would use the values from Table 10.1 to obtain $S_{y \cdot x} = 1.8269 \cdot \sqrt{(1 - 0.9859)} = 0.2172$. It is usually true that by correcting the variance of the y values, we reduce it. That is, $S_{y \cdot x} < \hat{\sigma}_y$.

We could, of course, rewrite Equation 10.4 to solve for, or obtain a formula for, the *least squares method of computing the correlation coefficient*. That is,

$$r = \sqrt{1 - \frac{S_{y \cdot x}^2}{\sigma_y^2}}. \tag{10.5}$$

This is not the usual way that we find a formula written for r. This different re-write is helpful, however, in showing that the correlation coefficient is also the square root of 1 minus the relative amount of variation in y that is not explained by the x values. In this way of thinking, r is the square root of the *relative amount of explained variation*. The formula for the (*product moment*) correlation coefficient is most com-monly given by

$$r = \frac{\text{Cov}(x, y)}{\sqrt{\text{Var}(x) \cdot \text{Var}(y)}}$$

where the covariance of x and y (σ_{xy} in Chapter 9) is written $\text{Cov}(x, y)$. Then, using the defining formulas for the covariance (Equation 9.1) and the variance estimates (Equa-tion 2.5), we can write

$$r = \frac{\Sigma_i (x_i - \hat{\mu}_x)(y_i - \hat{\mu}_y)}{N \sigma_x \sigma_y}. \tag{10.6}$$

The least squares regression line was used as the basis for computing r in Equation 10.5, whereas Equation 10.6 is the formula used for computation when there is no rea-son to first obtain a regression line. These two methods give equivalent answers.

Interpretation of Predictive Error Measure

How do we interpret the SEE quantity $S_{y \cdot x}$ as a measure of scatter of our data points? It is, of course, a measure of the extent to which the residual error above and below the line when squared is large, and therefore a measure of the poorness of our linear rule for predicting our dependent variable (in this case, density). However, the SEE quan-tity is also interpreted in a way quite similar to the interpretation of the other standard deviations we calculated (for density and for proportions) in earlier chapters.

A very nice result from mathematical statistics says that we can expect a Normal distribution of these residual deviations if the y variable (mean density) is subject to *random observational error*. This means that any factors that could offer more informa-tion on density, but which are omitted from the regression equation, are assumed to be uncorrelated with, and of minor consequence to, the occurrences. That says we re-ally don't have to include these extraneous factors in the regression equation to gain knowledge of density, and their effect is expected to be inconsequential (merely ran-dom) on the residuals. So, the closer the fit, as judged by the measures r and $S_{y \cdot x}$, the better is the job that occurrences are doing for linear prediction. Then, when density can be assumed subject only to random observational error, we may expect, for example, that a Normal (distribution) amount, about 68%, of the points in the data set will be within the range of the regression line, plus-or-minus 1 standard error of estimate, and 95% will be within plus-or-minus 2 standard errors.

There is a relationship between Equation 10.3 or 10.4 and the number of constants used in the regression equation (two in our case: a and b), as well as the number of ob-servations. The more constants we put into the equation (if we had more x variables) relative to the size of N, the lower will be the SEE quantity. We can therefore use a *correction* to ensure that the SEE will not necessarily have this dependence on N and the other number of x variables. The *corrected SEE* is written

$$S^*_{y \cdot x} = S_{y \cdot x} \sqrt{\frac{N-1}{N-c}} \tag{10.7}$$

where c is the number of constants in the regression equation. For our example $c = 2$, with $N = 52$ species. The corrected value using Equation 10.7 would be $0.21724 \cdot \sqrt{\frac{51}{50}} =$ 0.21940. Clearly, for such a small number of constants and moderate sample size, there is hardly any difference in the corrected versus uncorrected forms of the SEE. In the case of *simple linear regression* for occurrences, the correction is needed only for small samples, as a *small sample correction*.

Because the quantity $S_{y \cdot x}$ is still affected by sampling error, this correction does not make, necessarily, $S^*_{y \cdot x}$ the true standard error of estimate in the actual population from which we obtained the sample. What we have done is correct a *known bias* in $S_{y \cdot x}$.

Coefficient of Determination

For Equation 10.4 we stated that r is the correlation coefficient between x and y. This correlation coefficient is often written $r_{y \cdot x}$, or equivalently, $r_{x \cdot y}$. However, in Equation 10.4 the square of this quantity, $r^2_{y \cdot x}$, or just r^2, is used, and a different name is given to it. Here r^2 is called the *coefficient of determination*. This coefficient expresses the proportion of variance in the data that is explained by use of the linear rule we chose. For our example, $r^2 = 0.9859$ indicates an extremely good fit with more than 98% of the variation in (log) density explained by (log) proportion of occurrences. Another way to make this clear is to rearrange and square the terms in Equation 10.5 to obtain

$$r^2 = 1 - \frac{S^2_{y \cdot x}}{\hat{\sigma}^2_y}$$

or, with a little algebra,

$$r^2 = \frac{\hat{\sigma}^2_y - S^2_{y \cdot x}}{\hat{\sigma}^2_y}. \tag{10.8}$$

Let us look at each term in this formula. First, $S^2_{y \cdot x}$ (*the variance of the estimate*) is the "error" variance we get when we use the linear regression equation; it is the variability that is not explained by the linear regression. Then, $\hat{\sigma}^2_y$ represents the total amount of variance in the y values, which are the mean density values in our case. Thus the numerator in Equation 10.8, $(\hat{\sigma}_y^2 - S_{y \cdot x}^2)$, must be the variance that is explained by the linear regression. The entire formula given by Equation 10.8 says that the coefficient of determination must be the *relative reduction in variance* achieved by using the regression equation, or, more specifically, by using the observed numerical values of *x*, as well as the linear predictive rule.

If we look at Equation 10.8 and pretend that the left side is 0, then we get $0 = \hat{\sigma}_y^2 - S_{y \cdot x}^2$. Alternatively, we can write this as $\hat{\sigma}_y^2 = S_{y \cdot x}^2$. This means that the variance of estimate for the *y* observations equals the total variance of the *y* observations themselves, when we do not even know *x*. Therefore, when the coefficient of determination, r^2, equals 0, the regression equation is of no help for prediction. This conclusion does not mean that we cannot predict *y* from *x*. It only means that if we can predict, the rule for prediction is not linear.

Now, looking at Equation 10.8 again, let us pretend that *r* equals +1 or −1. Then, $r^2 = 1$, and we get an equation of the form $\hat{\sigma}_y^2 - S_{y \cdot x}^2 = \sigma_y^2$. For this equation to work, the only value that $S_{y \cdot x}^2$ can be is 0. Then, if the *prediction error variance*, $S_{y \cdot x}^2$, equals 0, each prediction must be exactly correct, so *x* and *y* must be (functionally) related, and our linear rule must describe that relationship. Therefore, we can always think of our coefficient of determination as a measure of the *strength of the linear relationship*. From Equation 10.8, as well as Equation 10.4, we can see that the larger the *absolute value* of the coefficient (do not consider its sign, + or −, just the magnitude of the number; we write this as $|r|$), the smaller the standard error (and variance) of estimate will be.

We can correct the SEE for *N* and for *c* but, regardless, the value of the standard error of estimate, or the variance of estimate, is still biased. That is, on average, it is not equal to the true population value. An unbiased estimate of the true variance of estimate for predicting density from occurrences is given by

$$\hat{\sigma}_{y \cdot x}^2 = \frac{N}{N-2} \cdot S_{y \cdot x}^2. \tag{10.9}$$

However, unless the number *N* is quite small, the unbiased estimate of the true variance for predicting density from occurrences will be an unnoticeable correction for natural population study.

INFERENTIAL USES OF REGRESSION

In order to use occurrence data confidently for predicting density values within the sampled plots or habitats, it is important to ascertain if the predicted density value is reasonable. We can answer this question with a confidence interval. By so doing, we will no longer be dealing with just a descriptive use of regression; we will be entering into an inferential use. In order to make this transition, we must first make some assumptions and examine some limitations on the procedure.

For a regression problem, the target area in which we are interested is selected for the *x* value observation, but there are no assumptions on the distribution of the occurrences. We must, however, feel comfortable about making three important assumptions:

1. Within each of the habitats or plots the mean density is Normally distributed.
2. Within each of the plots the variances of the occurrences are about the same (called *homoscedasticity*).
3. The errors in the density measurements are independent; there are no systematic problems of measurement.

None of these assumptions present a problem; even the Normal distribution for mean density we have shown (in Chapter 3) to be close enough for our Bolivian data although certainly not for all natural population samples, especially of small size. One difference when we deal with inference, not just description, in regression is the assumption that if we can predict at all, prediction will best be done with our linear rule. Provided all three assumptions in our list are reasonable, we can form an interval estimate. However, the confidence interval that we shall construct has an unusual aspect.

For a specific occurrence, interest lies in the actual, not the predicted, mean density. Here we have a complicated situation. We showed in Figure 10.3 that there is always some error in prediction. This is true both for our sample and for the population regression. Thus we have a compounding of errors of prediction for the actual mean density.

There are three predominant sources of variability in the prediction of actual densities y_{ij} in the population. These sources of variability in prediction occur when (1) the value of $\hat{\mu}_r$, the estimated mean density, varies over different samples, depending upon fieldwork quality, the sampling plan, and the cooperation of the organisms; (2) the value of $\hat{b}$, the regression coefficient estimate, varies over different samples; and (3) the value of $\hat{y} - \hat{Y}$ varies as well. The mathematical notation $\hat{y} = \hat{a} + \hat{b}x$ represents the sample estimate of the true predicted value $\hat{Y} = a + bX$, not the true population density Y. We can see from the formula for the confidence interval estimate for 1 mean density value that these three sources cause variation in the final interval. The formula for a confidence interval about a true density, for which we use $\hat{Y}_j$ as the predicted density estimate for a particular (xj) occurrence, is

$$\hat{Y}_j - t_{(n-1),\alpha}\hat{\sigma}_{y\cdot x}S_j \leq Y_{ij} \leq \hat{Y}_j + t_{(n-1),\alpha}\hat{\sigma}_{y\cdot x}S_j. \tag{10.10}$$

In this formula there were so many terms that we wrote S_j to represent

$$\sqrt{\frac{1 + \dfrac{1}{N} + (x_j - \hat{\mu}_x)^2}{N\hat{\sigma}_x^2}}.$$

Then, from Equation 10.9 we obtain

$$\hat{\sigma}_{y\cdot x} = \sqrt{\hat{\sigma}_{y\cdot x}^2}.$$

The term $t_{(n-1),\alpha}$ indicates the Student's t value, with $n-1$ degrees of freedom at the α (say, $0.95 = \alpha$) level of confidence. The component $\hat{\mu}_x$ is our estimated mean for the

x values or occurrences. It is important to notice that this predicted interval is not an interval about the entire regression line. This is an interval for or about 1 point or 1 occurrence, y_{ij}, for which we predict a density value y. Two subscripts are used; the j for all the x values and the i to show it is a single point. That is, we used a *double subscript* to indicate that the true y value is of 1 occurrence in 1 sample. The result will be a surprisingly wide interval for small samples, especially if we unwittingly used extreme values of occurrences x_j (for example, because we misidentified or missed). A too-wide interval could also occur if we tried to pool information on unrelated localities or habitats. However, if the observed trend is based on at least 30 biological samples, the variability of $\hat{\mu}_x$ and $\hat{b}$ over the samples is minimal or absent (that is, the sampling distributions have very small standard deviations). Also, we have assumed that the variances for occurrences are equal, so we can replace Equation 10.10 with the *large-sample confidence interval for the actual density* by using the formula

$$\hat{y}_j - z\hat{\sigma}_{y \cdot x} \le y_{ij} \le \hat{y}_j + z\hat{\sigma}_{y \cdot x} \tag{10.11}$$

where z is a value from the Normal table (Table 3.2), y_{ij} is the actual density, and $\hat{y}_j$ is the predicted density for the selected x_j value. The use of the subscript j in Equation 10.11 denotes that this interval must be recomputed for each occurrence value on the x axis; an interval estimate will be given for each of the associated density values. This interval is not equally good for each of the occurrence values in the sample. The precision is closest for the particular x_j that equals the mean $\hat{\mu}_y$ and the interval gets wider elsewhere. This shows that if we use occurrence data quite disparate from the data in our original sample, we cannot expect to obtain a reasonable density estimate. In fact, any inference about the linear relationship, or its strength in the population, applies to only the particular occurrence values in the sample. There is no assurance that over any other range of occurrences the predictive relationship will be upheld at all.

PREDICTION WITH MEASUREMENT ERROR IN THE OCCURRENCE DATA

Let us now return to the Bolivian forest data graphed in Figure 10.2. This logarithmic plot of occurrences against mean density was approximately linear. One of the more complicated problems is inference from regression when there is measurement error in the x values, which are our species occurrences. Such errors can have serious consequences in regression, and may even necessitate the total avoidance of a regression approach. However, we are fortunate in that there is a situation in which such errors are not expected to have a major influence.

A condition on the x values that allows us to apply a regression analysis strategy, in light of measurement errors in occurrences, is that the errors are of a certain type. There must be no *systematic bias* in the recording of the occurrence data, and the errors that do exist in the occurrence data must not be related to those in the density values upon which we base our linear rule relationship. This condition will be sustained if we use reliable museum records, or if we are familiar with the reputation of the collector. Mathematically, statisticians state that the errors must be independently and

Normally distributed and have expected values of 0 with equal variance. An alternative approach, in other regression problems (for example, in morphological or observational studies), is to use set, not sampled, x values.

Relationship of Density and Occurrences Over the Target Area

The observed linear relationship between density and occurrences on a log–log scale is usually approximated by a power curve in the ecological literature. This approximation is written

$$\hat{\mu} = a p_o^b. \tag{10.12}$$

For the data set of Bolivian trees we determined that $\log(\hat{\mu}) = 0.4641 + 1.105 \cdot \log(p_o)$ from Equation 10.2. Transforming back from logarithms, as we did in Chapter 6, we obtain from Equation 10.12 an estimated density of $\hat{\mu} = 1.5902 p_o^{1.105}$, which is the same form as Equation 6.4.

Either form of the equation (that is, Equation 10.2 or 10.12) can be used to obtain density values. However, there is a minor complication. If we use Equation 10.2, we predict an estimate of log density (log μ), not the density itself. The usual procedure in ecology is to transform back to obtain a prediction for the raw density value. As we discussed in Chapter 9, when a transformation is used merely as a convenience, the estimate obtained, after transforming back again to the original units (that is, to density), does not have the smallest possible variance. It is an *inefficient estimate*. In many practical applications this is a serious deficiency. As mentioned in Chapter 6, computer packages routinely incorporate *nonlinear modeling* options with which the power curve could be estimated directly; in this way a density estimate, and a direct measure of fit, could be obtained without transformation. In our case, when predictions are based upon the occurrence data, the values for r and $S_{y \cdot x}$ show such strength of relationship for natural population data, regardless of taxa, that it matters very little whether one uses the power curve or the linear relationship formula for predicting an estimate of mean density.

Using the power curve calculated from the total set of Bolivian data, we can obtain the estimated mean density, $\hat{\mu}$, from the actual proportion of occurrences, p_o, for the most abundant species and for some of the rarer ones. The results are shown in Table 10.1. They indicate a predictable relationship between the list of species occurrences and the mean densities. The greatest deviation is evident with the more abundant species, which, based upon the number of times each occurs, yield higher densities than the equation would predict. Except for the most abundant species, however, the density estimates are not very far off. Thus we are on fairly certain ground when we use our information on species occurrences to predict species densities for the same localities, within some reasonable amount of error. For example, if a species we had not observed was found to have occurred in a prior inventory of that locality, we could use this methodology for incorporating those results. For example, if a particular species was represented by 25 occurrences recorded from a total of 50 sites in that area, we could estimate that species' density with the equation $1.5902 \cdot (0.5)^{1.105}$, where $p_o = \dfrac{25}{50} = 0.5$. That would give a mean density of 0.739 per quadrat, or an area-wide density of $0.739 \cdot 50$ sites = 36.95, or 37 individuals.

PREDICTION OF DENSITY FROM A
SAMPLE OF LOCALITY OCCURRENCES

The Bolivian Example: Terrestrial, Small Spatial Scale

A second question to ask when assessing the relationship of density and species occurrences is how occurrences could be used to predict densities in an area for which we have no density data. Let us reexamine the Bolivian data set for an answer. In the previous section we showed the relationship of density to occurrence by using known data on the entirety of Bolivian plot 01. Instead, let us first take a sample from the site for which we know both types of information, and use this sample to predict density in a similar target area for which we have only occurrence data. Bolivian plot 02 will be that similar target area. Both plots 01 and 02 are 10,000 m². Plot 02 is located near 01; it has similar environmental variable levels, but a slightly different habitat profile.

First, we select a random sample of size $n = 8$ (Table 10.2) from the 100 subplots of Bolivian plot 01. There are a total of 32 occurrences of 15 species represented by 47 individuals in these 8 quadrats. Table 10.3 shows the species mean density and proportion of occurrences for this sample from plot 01.

We then compute a predictive equation using this data, after a natural log (base e) transformation. From Equation 10.2 the appropriate equation is $ln(\hat{\mu}) = 0.459 + 1.171 \cdot ln(p_o)$. The coefficient of determination, r^2, is 0.89 for this sample. For comparison, $\hat{\sigma}_y = 0.90$. After transformation, the equation becomes $\hat{\mu}_y = 1.583 p_o^{1.171}$. Notice that the mechanics of this regression are quite simple but must be kept clear. If we use mean density and proportion occurrences to develop the predictive equation, then for our predictions ($\hat{y}$) we must use these same quantities for substitution into the regression equation. If we had used the log of the mean density and the log of the occurrences (not the proportion of occurrences), we would have obtained a different constant in our equation, but we would have obtained the same x coefficient (or $\hat{b}$), r^2, and $S_{y \cdot x}$ values for the equation. Finally, in order to predict, we would substitute occurrences, not proportion of occurrences. For example, had we used the occurrences, our linear equation would have been (from Table 10.3), $ln(\hat{\mu}) = -1.9761 + 1.171 \cdot ln(\text{occurrences})$, or the power curve would be $\hat{\mu} = 0.1386(\text{occurrences})^{1.171}$.

We divided plot 02 into 100 subplots also, and developed a species list for each of the 100 subplots. Summarizing this information over the entire hectare gave the occurrence data. Of the 15 species found in the plot 01 sample of 8 subplots, 11 were also found in plot 02. The question then is whether we can predict the species densities for these 11 species in plot 02, using the predictive equation for density from plot 01. This exercise could be construed as using community-level information from one or a series of localities or habitats to predict community-level characteristics of a second, but similar, locality. The final 3 columns in Table 10.3 show the results. If we compare the

TABLE 10.2 Occurrences of Tree Species in 8 Randomly Selected Quadrats

SPECIES No.	MEAN DENSITY (μ)	PRESENCE (+) OR ABSENCE IN QUADRAT								OCCURRENCES
		3	15	16	29	57	67	69	90	
1	2.52		+		+	+	+	+	+	6
2	1.31	+	+	+	+	+	+	+	+	8
4	0.25			+		+	+			3
5	0.19	+								1
6	0.17			+						1
7	0.17		+							1
8	0.17		+	+		+				3
14	0.06				+					1
17	0.04						+			1
18	0.03							+		1
19	0.03	+								1
20	0.03		+							1
21	0.03			+	+					2
22	0.03		+							1
32	0.02	+								1

NOTE: $N = 100$. The symbol + denotes occurrence.

SOURCE: Data drawn from Appendix 1.

plot 02 predicted density values ($\hat{y}$) with the actual figures from plot 02, it is clear that this use of regression methods and power curve predictions can be quite important for biodiversity work and for studying natural populations. The quantitative procedure for evaluating this comparison is found in the last column in Table 10.3, whose sum (0.33) gives the value we need for the numerator of the SEE. Equation 10.3 is used to obtain residual deviations. Then, the average of this last column yields our standard error of estimate, which is quite small in this case. It appears that we did remarkably well in estimating the mean density of a second locality, based on just a very small number of subplots. Had the number of samples drawn from plot 01 been larger (than 8), thus providing more representative ecological information on the relative abundance, we could have expected an even better prediction. However, these two 1-hectare plots and their species composition are very similar.

A Gulf of Mexico Example: Marine, Large Spatial Scale

It may not be possible to predict the mean density of a second locality so precisely when it is not a similar habitat or it is not geographically close to the original locality from which we are observing our trends. Let us consider an example of marine inver-

TABLE 10.3   Regression Results of Predicting Species Densities in Bolivian Forest Plot 02, Using Species Occurrences in Plot 01

| SPECIES No. | PLOT 01 | | | | PLOT 02 | | |
	MEAN DENSITY OF SAMPLE (y)	PROPORTION OF OCCURRENCES OF SAMPLE (x)	$ln(x)$	$ln(y)$	OBSERVED MEAN DENSITY (y)	PREDICTED MEAN DENSITY $(\hat{y})$	$y - \hat{y}$
1	1.875	0.75	−0.2877	0.6286	1.55	1.17	0.148
2	1.500	1.00	0.0000	0.4055	0.35	0.37	0.0004
4	0.375	0.375	−0.9808	−0.9808	0.07	0.06	0.0001
5	0.250	0.125	−2.0794	−1.3863			
6	0.125	0.125	−2.0794	−2.0794	0.09	0.09	0.0
7	0.125	0.125	−2.0794	−2.0794	1.68	1.25	0.18
8	0.375	0.375	−0.9808	−0.9808	0.59	0.56	0.0009
14	0.125	0.125	−2.0794	−2.0794	0.06	0.06	0.0
17	0.125	0.125	−2.0794	−2.0794	0.02	0.02	0.0
18	0.125	0.125	−2.0794	−2.0794			
19	0.250	0.125	−2.0794	−2.0794			
20	0.125	0.125	−2.0794	−2.0794	0.03	0.03	0.0
21	0.250	0.250	−1.3863	−1.3863	0.01	0.01	0.0
22	0.125	0.125	−2.0794	−2.0794	0.01	0.01	0.0
32	0.125	0.125	−2.0794	−2.0794			Total: 0.33

NOTE: In Plot 01 a sample of size 8 containing 47 individuals was used. Plot 01 samples selected are quadrat numbers 3, 15, 16, 29, 57, 67, 69, and 90. Regression equation is $ln(\hat{\mu}) = 0.459495 + 1.17128 \cdot ln$(proportion occurrences). $S_{y \cdot x}^2 = 1.1$ individuals (calculations to 2 places only).

tebrates from several off-shore traverses in the Gulf of Mexico. Buzas (1967) used the multivariate technique of canonical discriminant analysis to calculate that the data collected by the first 2 transects (0–30 meter depth) in Phleger (1956, traverses III and V) were similar. The stations along the traverses and the traverses themselves were spread over a distance of about 10 nautical miles.

There were 30 samples or localities sampled along the first transect (Table 10.4). Using the ln(occurrences) and ln(mean density) values, the predictive equation is ln(mean density) $= -3.356 + 1.4726 \cdot ln$(occurrences), with $r = 0.94$ and $S_{y \cdot x} = 29.8$. A comparison of the observed and the predicted density values for this transect in Table 10.4 shows that we are able to reproduce densities quite well, although in most cases the predictions for the most abundant species are too low. Figure 10.4 shows the correspondence of the observed and predicted densities.

Many times a researcher will find that the amount of work required to tabulate density information is excessive. Let us say that after a cost/benefit evaluation of the

TABLE 10.4 Gulf of Mexico Species Density Prediction from Occurrences over the Same Transect

Species No.	Observed Occurrences		Observed Mean Density		Predicted Mean Density	
	RAW	ln	RAW	ln	RAW	ln
1	2	0.6931	0.1	−2.3026	0.0968	−2.3353
2	2	0.6931	0.067	−2.7081	0.0968	−2.3353
3	15	2.7081	2.333	0.8473	1.8811	0.6319
5	24	3.1781	6.467	1.8667	3.7584	1.3240
6	1	0	0.033	−3.4012	0.0349	−3.3560
8	11	2.3979	1.000	0	1.1914	0.1751
9	1	0	0.033	−3.4012	0.0349	−3.3560
10	6	1.7918	0.500	−0.6932	0.4880	−0.7175
12	3	1.0986	0.333	−1.0986	0.1758	−1.7382
13	2	0.6931	0.067	−2.7081	0.0968	−2.3353
14	20	2.9957	2.367	0.8615	2.8734	1.0555
15	3	1.0986	0.233	−1.4553	0.1758	−1.7382
16	2	0.6931	0.067	−2.7081	0.0968	−2.3353
19	2	0.6931	0.133	−2.0149	0.0968	−2.3353
20	1	0	0.033	−3.4012	0.0349	−3.3560
21	8	2.0794	0.433	−0.8363	0.7454	−0.2938
22	15	2.7081	1.267	0.2364	1.8811	0.6319
23	1	0	0.033	−3.4012	0.0349	−3.3560
24	1	0	0.033	−3.4012	0.0349	−3.3560
25	3	1.0986	0.167	−1.7918	0.1758	−1.7382
26	14	2.6391	2.067	0.7259	1.6994	0.5303
27	1	0	0.033	−3.4012	0.0349	−3.3560
28	5	1.6094	0.200	−1.6094	0.3731	−0.9860
29	12	2.4849	1.100	0.0953	1.3543	0.3033
30	3	1.0986	0.333	−1.0986	0.1758	−1.7382
31	10	2.3026	0.700	−0.3567	1.0354	0.0348
32	13	2.5649	1.767	0.5691	1.5237	0.4211
33	13	2.5649	1.967	0.6763	1.5237	0.4211
34	8	2.0794	1.233	0.2097	0.7454	−0.2938
35	7	1.9459	0.333	−1.2040	0.6123	−0.4905
36	6	1.7918	0.533	−0.6286	0.4880	−0.7175
37	3	1.0986	0.333	−1.0986	0.1758	−1.7382
38	2	0.6931	0.100	−2.3026	0.0968	−2.3353
42	6	1.7918	0.267	−1.3218	0.4880	−0.7175
43	1	0	0.033	−3.4012	0.0349	−3.3560
44	27	3.2958	4.367	1.4740	4.4770	1.4974
45	29	3.3673	10.70	2.3702	4.9662	1.6027

(continued)

TABLE 10.4 *(continued)*

Species No.	Observed Occurrences		Observed Mean Density		Predicted Mean Density	
	RAW	*ln*	RAW	*ln*	RAW	*ln*
46	4	1.3863	0.233	−1.4553	0.2686	−1.3146
47	4	1.3863	0.367	−1.0033	0.2686	−1.3146
50	1	0	0.067	−2.7081	0.0349	−3.3560
52	20	2.9957	2.667	0.9808	2.8734	1.0555
54	7	1.9459	0.300	−1.2040	0.6123	−0.4905

NOTE: Prediction of species density on transect 1 by species occurrences on transect 1, using the equation ln(mean density) $= -3.356 + 1.4726 \cdot ln$(occurrences), for which $S_{y \cdot x} = 29.79$ and $r = 0.94$.

SOURCE: Data from Phleger 1956.

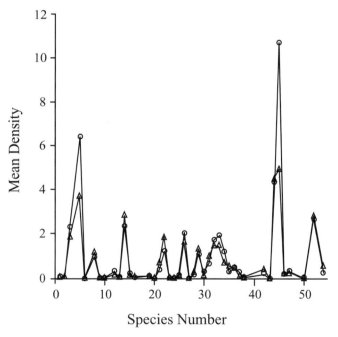

FIGURE 10.4 Correspondence between observed and predicted mean densities. Regression on occurrences for species of foraminifera in the Gulf of Mexico, transect 1 (Phleger 1956, traverse III). Mean densities are the values given in Table 10.4 (raw).

time it took for collecting, identifying to species, counting, and recording mean densities for the entire species list on transect 1, the decision was made to develop only a species list for transect 2. Only presence or absence (not counts) would be recorded. In such a case, how well would the information gathered from transect 1 serve as a basis for predicting mean density estimates for species in transect 2? For comparison we list in Table 10.5 the observed mean densities from transect 2 and note that the correlation between occurrences and mean density on transect 2 is 0.83; lower than the similar measure for transect 1. The last 2 columns in Table 10.5 are the predicted density values obtained by putting the ln(occurrences) data into the equation developed on transect 1, and given in Table 10.4.

Clearly, the predictions are quite accurate. However, we find again that the predictions for the most abundant species are too low. This problem is due to the errors of prediction discussed in the regression section of this chapter. The most precise predictions are always near the average value, while those at the ends, which pertain to the most abundant species, have wider confidence bands and are less accurate. The same caveat applies to the rare species, but they have the self-imposed limit of 0 to counteract this problem. Clearly, given the fact that predicted values will never be exact, and that these 2 transects are similar although not precisely equivalent and quite a distance apart, we are able to provide density information that is quite useful.

One way to provide better estimates for the most abundant species is to use the standard error of the coefficient. For example, we calculated the regression equation to be ln(mean density) $= -3.14196 + 1.262 \cdot ln$(occurrences) for observed data in the second transect. To obtain a density estimate for the most abundant species, number 45, we substituted the occurrence value and solved the equation ln(mean density) $= -3.14196 + 1.262 \cdot ln(30)$. Now the $ln(30) = 3.4012$. Multiplying this out yields ln(mean density for species 45) $= 0.872739$. Taking the inverse of the logarithm (for example, @exp in Excel or INVERSE-LN on a calculator) gives a mean density of 2.393457. In order to obtain an actual species density estimate over the 35 sites/localities along this transect we merely multiply the mean density by 35 sites/localities to obtain the estimate of 84. We can compare this value with the actual observed figure of 267 and see that it is a severe underestimate.

Had we remembered that the coefficient for the occurrences is itself a sample estimate, we could have used its standard error (available from any regression output) in order to get an upper bound for this imprecise estimate. That is, the equation to be solved for the most abundant species would be ln(mean density) $= -3.141957 + [(1.262 + 0.093297) \cdot ln(30)]$, or ln(mean density) $= -3.141957 + [1.355297 \cdot ln(30)]$. Following the same steps we would then estimate the species density to be 115. We could also have decided to use twice the standard error, and the estimate would then have been 158. The problem with imprecise estimates at the rare species end is solved by substituting an estimate of 1 if the calculated value becomes negative.

In this example, we had a reasonable number of localities upon which to base a prediction of the species densities; our estimates were quite good overall, but imprecise for the most abundant species. However, the severe underestimation of the most abundant species can be improved by a simple calculation.

It appears that the environmental or habitat similarity is as, if not more, important than sample size for the prediction of density from species occurrences. Therefore, the use of species occurrences for the prediction of species density across similar habitats can be achieved with limitations upon the researcher to define "similar." Certainly, ocean transects that are more than 10 nautical miles apart are in this case still

TABLE 10.5 Comparison of Predicted Versus Observed Foraminifera Density for the Second Transect from the Gulf of Mexico

Species No.	Observed Occurrences		Observed Mean Density		Predicted Mean Density	
	RAW	*ln*	RAW	*ln*	RAW	*ln*
1	1	0	0.0571	−2.8622	0.0349	−3.3560
2	1	0	0.0286	−3.5555	0.0349	−3.3560
3	21	3.0445	1.9143	0.6493	3.0874	1.1273
5	27	3.2958	3.4571	1.2404	4.4702	1.4974
8	10	2.3026	0.8286	−0.1881	1.0354	0.0348
9	11	2.3979	0.7714	−0.2595	1.1914	0.1751
12	3	1.0986	0.2000	−1.6094	0.1758	−1.7382
13	2	0.6931	0.0571	−2.8622	0.0968	−2.3353
14	14	2.6391	0.6286	−0.4643	1.6994	0.5303
15	6	1.7918	0.2857	−1.2528	0.4880	−0.7175
16	1	0	0.2857	−3.5554	0.0349	−3.3560
18	2					
19	44	3.7842	0.1429	−1.9459	9.1759	2.2166
20	4					
21	11	2.3979	0.5714	−0.5596	1.1914	0.1751
22	13	2.5649	0.7714	−0.2595	1.5237	0.4211
23	1	0	0.0286	−3.5554	0.0349	−3.3560
24	3	1.0986	0.1143	−2.1691	0.1758	−1.7382
25	5	1.6094	0.3429	−1.0704	0.3731	−0.9860
26	15	2.7081	2.3714	0.8635	1.8811	0.6319
28	2	0.6931	0.0571	−2.8622	0.0968	−2.3353
29	13	2.5649	0.8286	−0.1881	1.5237	0.4211
30	8	2.0794	0.5429	−0.6109	0.7454	−0.2938
31	18	2.8904	1.6286	0.4877	2.4604	0.9003
32	21	3.0445	3.1714	1.1542	3.0874	1.1273
33	18	2.8904	2.5143	0.9220	2.4604	0.9003
34	2	0.6931	0.0571	−2.8622	0.0968	−2.3353
35	9	2.1972	0.2857	−1.2528	0.8866	−0.1204
36	7	1.9459	0.2571	−1.3581	0.6123	−0.4905
37	1	0	0.0285	−3.5554	0.0349	−3.3560
38	2	0.6931	0.0571	−2.8622	0.0968	−2.3353
42	2	0.6931	0.0857	−2.4567	0.0968	−2.3353
43	1	0	0.0571	−2.8622	0.0349	−3.3560
44	22	3.0910	1.6000	0.4700	3.3064	1.1959
45	30	3.4012	7.6285	2.0319	5.2205	1.6526

(*continued*)

TABLE 10.5 (*continued*)

Species No.	Observed Occurrences		Observed Mean Density		Predicted Mean Density	
	RAW	*ln*	RAW	*ln*	RAW	*ln*
46	8	2.0794	0.5429	−0.6109	0.7454	−0.2938
47	7	1.9459	0.2857	−1.2528	0.6123	−0.4905
49	6					
52	18	2.8904	2.4000	0.8755	2.4604	0.9003
53	3					
54	1	0	0.0286	−3.5554	0.0349	−3.3560

NOTE: Second transect for recent foraminifera occurrence data used in predictive equation (Table 10.4) for transect 1 to obtain information on species mean densities. Observed mean densities listed for comparison with predictions. When there is a species number and an occurrence value only in a row of the table, that species did not occur in transect 1, so no prediction of its mean density could be made for transect 2. Notice the largest discrepancy is for species 19, for which there were 44 occurrences, while for transect 1 species 19 occurred only twice.

SOURCE: Data from Phleger 1956.

similar enough for prediction to be reasonable. The prediction of density can be exceptionally useful for many purposes, especially biodiversity study, in which abundance data are not easily obtained for certain taxa.

SUMMARY

1. We do not need any assumptions on either the density or the occurrence data in order to employ the descriptive statistics of regression for predicting density from occurrence data. We can do this from a species list, a museum search, or the results of an inventory.

2. The possible values that the correlation coefficient between occurrences and mean density can take on depends to some extent upon the forms of the *marginal distributions* of both variables. That is, unless both density and occurrences are free to take on all values in their range, the number we get when we calculate the correlation coefficient (r) will be limited. Not every r can range from +1 to −1, even though texts always state these values as the limits. If a very limited range of occurrences were observed, during a very short sampling period or over a very restricted habitat or number of localities, then the correlation coefficient can never actually take on values near +1 or −1. This leads to confusion in research results.

3. It is important that the occurrence data have no major systematic restrictions on the range of occurrences. Otherwise, we run the risk of a decided bias in the value of the

correlation coefficient, and its absolute value is lowered. An example of this bias would be when occurrence data were gathered during a poorly planned inventory, or by poorly trained personnel, or in which only selected habitats were examined.

4. The residual deviations about the regression line will be Normally distributed, since for natural populations density can usually be assumed subject only to random observation error. In this case about 68% (95%, 99%) of the observed data will be within 1 (2, 3) SEEs (the standard errors of estimate) of the regression line.

5. Statisticians consider the proportion of occurrences to be a linear transformation of the observed occurrences because the transformation is performed simply by dividing the number of occurrences for each species by the total number of samples, which is a constant number.

6. The proportion of occurrences is a linear transformation of occurrences. Therefore, where a is the slope and b is the intercept, we know that for 2 possible equations, $ln(\hat{\mu}) = \hat{a}_1 + \hat{b}_1 \cdot ln(p_o)$ and $ln(\hat{\mu}) = \hat{a}_2 + \hat{b}_2 \cdot ln(\text{occurrences})$, in which the only difference is the use of either the occurrences themselves or the proportion of occurrences, it must be true that $\hat{b}_1 = \hat{b}_2$. In addition, the correlation coefficient (r) and $S_{y \cdot x}$ will be the same for both equations. Only $\hat{a}_1$ will differ from $\hat{a}_2$.

PROBLEMS

10.1 Using the bat data of Problem 9.2:

   a. Calculate a regression on the occurrences for predicting density.

   b. What is the standard error of estimate?

   c. Make a table containing the observed and expected numbers from the regression.

11
SPECIES OCCURRENCES

As discussed in chapter 10, sometimes species lists may be the only available information on an area of interest. When using data from prior surveys, the investigator may find that the earlier researchers confined themselves to compiling lists of the species observed at localities that gave scant or no information on abundance. Some monographic works in systematics list the species of interest for the problem under investigation and where these species were observed, but the monographs may have ignored other members from different families in the same group. When using museum collections along with the literature and needed synonymy, researchers may be able to compile a file from numerous sources that lists how often the various species in a taxonomic or ecologic group occur at a number of localities within an area of interest. Although little or no information on species densities was included in many original observational studies, the compilation of species occurrences can often be used as a surrogate to approximate and replace species densities and other information.

In coping with an absence of actual counts, researchers can capitalize on the tight relationship between density and occurrences demonstrated in Chapter 10. The savings in time and money offered by such an approach, versus mounting a new expedition to obtain baseline data, is enormous. The occurrence data set obtained from older studies is often all that is required to constitute a pilot study, which allows a researcher to plan for more detailed sampling in the field.

Sometimes the object of a survey is to establish biofacies (faunal or floral zones) over a relatively large area. Such a survey may require an extensive sampling plan and the acquisition of many biological samples. Noting the presence of each taxon or species (a species list) in each of the biological samples is far less time-consuming than assigning each individual encountered to a specific taxon. By doing so, we create a conceptual situation equivalent to reducing the size of each quadrat so that it can contain a maximum of only 1 individual of each species. At this artificially induced low density, the observations are always on the lower part of the power curve relating the variance and the mean (Chapter 6). Recall that even species with highly aggregated spatial distributions will appear random at very low mean densities. By using species occurrences, the variability inherent in natural populations is greatly reduced. Rather than expending time and effort counting individuals from a few samples, we concentrate on obtaining species lists from many samples to obtain biodiversity information. In other words, we maximize n (the number of biological samples or quadrats; Chapter 5). This results in a better areal coverage of the study area and reduced variability. In most cases, the observations will be easier to interpret in terms of areal patterns, and the goal of establishing biofacies is easily met.

For all of these reasons, the best advice may be simply to ignore the alleged need to obtain counts of individuals. One could instead look to existing species occurrence information, as well as new species lists, and make use of these with no apologies. Occurrences can be an extremely reliable surrogate for the number of individuals, alone or in distribution form. Species ranking in a community as well as a variety of diversity measures can be calculated from occurrences.

Let us now investigate the usefulness of occurrence, or presence/absence, data. In this chapter we show that a species occurrence data set is merely a look at a natural population at a different scale, a scale on which the natural variability is greatly reduced.

SURROGATE RANK ORDER OF SPECIES

Occurrence data can provide insight into the interrelationships of species. Density data often vary greatly in an area from species to species and are somewhat unwieldy. In Bolivian plot 01, for example, the most abundant species is represented by 252 individuals, the third most abundant by 62 (Appendix 1). Most researchers would be comfortable concluding that the most abundant species makes up about 38% of the population and the third most abundant 9%. There is such a large amount of variation in natural populations, however, that for some researchers even percentages suggest too much certainty. They are most comfortable with a simple ranking of species. There is, then, a great interest in the ranking of species, or *ordering of species* and other groups, as the rank may be more reliable in many ways than the actual density or percentage figures.

The *rank* of an observation in a set is its ordinal number (or *place-value*: first, second, third, and so on). The set is ordered according to some criterion, such as most to least abundant. Appendix 1 shows the relation of *rank order* of the species abundances to that of the densities from Bolivian plot 01. Clearly, they are equivalent. When group or species densities are unknown and only species occurrences are available, it is sensible to assume that the rank order of species occurrences will provide a reasonable estimate of the rank order of species densities.

However, the Bolivian data are homogeneous and cover a relatively small area. Use of occurrence data to infer rank order of species is more problematic when (1) a much larger area is under study, (2) a small number of samples are collected, or (3) the organisms are much smaller relative to the study area or sample, and particular species might be clustered at a single or at a few localities. An example is a species with high densities at a single locality. In such a situation, consideration of occurrences alone could give an erroneous view of the density of such a species. When we are studying a large area with many sampled localities, the problem diminishes. Indeed, over a large area great abundance of a single species at only one locality would perk our interest. But such a phenomenon would surely be noted by any investigator making a field survey.

The size and number of localities must always be considered. For example, when we reduce the number of subplots in the Bolivian forest from 100 to 25 over the same area (1 hectare), we not only enlarge the size of each area to be sampled but the resultant number of occurrences (presence or absence) changes. This can, and in this case does, change some of the rank orders, especially for the more abundant species (Appendixes 1 and 2). One sign that a study design has divided the target area into too few localities is when the most abundant species are found to occur in all or nearly all of the quadrats. When this is the case, as in the Bolivian plot with $N = 25$ (Appendix 2), the species occurrences have no way to sort out into a rank order. That is, there will be many tied ranks. Consequently, the homogeneity of the distribution, the size of the target area, and the size of quadrats, plots, cores, or biological samples relative to the organism's size must always be kept in mind. Nevertheless, when only species occurrence data are available, they do afford us with a reasonable gauge of the rank order of density. The condition for success is that the statistical sample must be large enough. That is, the sample must include enough biological samples, or localities, to be representative of the target area.

USES OF SMALL SPECIES LISTS OR SMALL SAMPLES OF OCCURRENCES

Let us consider the rather common problem in which the researcher has information on only a relatively few localities or sites. How does one use this modest amount of information to develop a rank order of occurrences in the larger target area? A problem is that no matter how many individuals are recorded for the sampled area, there is no getting around the liability of using a small sample for quantitative analysis. This is because the individual is not the primary sampling unit; the locality itself serves that purpose (Chapter 7). Therefore, we must ask the question "Is a small sample, for which we have information only on occurrences, of any use for predicting abundance values?"

In order to attack this problem, let us further examine the distribution of occurrences. If we randomly choose 1 quadrat, core, subplot, or biological sample, then a particular species (say, species i) either occurs in the quadrat or does not. In this case, when we are interested in whether a particular species occurs in the sample, the quadrat is the sampling unit, not the individual or species. Within each quadrat or biological sample we do not have a count as our random variable; rather, we have a dichotomous (yes/no; occurrence/nonoccurrence) situation. The occurrence of species in the sample constitutes success for the quadrat, and its absence (nonoccurrence) is failure. Provided that the probabilities remain the same during the sampling process, which is quite reasonable, then each observation (occurrence/nonoccurrence) constitutes a Bernoulli trial (Chapter 8). We use a 1 in our data file if species occurs in our selected quadrat, and a 0 otherwise. If we are working with a total of N such quadrats or biological samples, the chance (or probability) of finding species in any of those randomly chosen quadrats is this ratio: the sum over all N quadrats of the number of occurrences of species in each quadrat (the 1's), divided by the total number of quadrats (N), or

$$p_o = \frac{1}{N} \Sigma_{i=1}^{N} O_i \tag{11.1}$$

where p_o is the probability of finding this species in any single quadrat, or the proportion of this species over the N. The numerator of this proportion in Equation 11.1 is a sum of 1's and 0's, and there are N of them; the denominator is always N. Thus the numerator can range only from 0, indicating that species was never found, to N, when that species i occurred in each quadrat. This shows that the fraction p_o is a number between 0 and 1, as a probability should be. The probability of not finding this species (q_o) in the quadrat is then $q_o = 1 - p_o$. The appropriate probability distribution can once again be the Binomial, and all the equations introduced in Chapter 8 for dealing with species proportions with the Binomial are applicable, provided three conditions can be met or assumed. First, as we saw in Chapter 8, we must have a dichotomous situation. Indeed, our setup of identifying an occurrence by presence or absence of the species allows this condition to be met. Second, we must have a constant chance (probability) of selection, which we have assumed to be reasonable. Third, we need independence of the contents of the quadrats over the area. This is a most important condition for the application of the Binomial. To the extent that such independence is not true, the Binomial fails to be robust and our inferential statements become problematic. Let us examine some situations in which the Binomial can help us with our fieldwork.

In previous chapters we selected 8 quadrats at random from 100 for our statistical sample. We shall use these same 8 quadrats here for an illustration. In these 8 quadrats 48 individuals belonging to 16 species were encountered. Table 10.2 shows in which quadrats each occurred, along with the total number of occurrences for each species. For this sample from this homogeneous area, the number of occurrences for each species does relate in a general way to species density, but far from perfectly. The second-ranking species, *Brosimum lactescens*, occurs in all 8 quadrats, whereas the most abundant species, *Scheelea princeps*, occurs in only 6. Those who are not statistically inclined may find it surprising that the third-ranking species, *Pouteria macrophylla*, does not occur at all. Among the rarer species, *Liana* and *Pera benensis* occur too frequently when compared to their actual densities.

The general equation for calculating the *estimated proportion of occurrence* (p_o) is

$$\hat{p}_o = \frac{1}{n} \sum_{i=1}^{n} o_i. \tag{11.2}$$

As usual, in Equation 11.2, n is the number of quadrats in the statistical sample, out of a possible number of quadrats N. We calculate p_o from the observed occurrences for each of the species found in the 8 quadrats (Table 11.1). The differences in rank order of species, already noted for Table 10.3, are also apparent in Table 11.1. Because $n = 8$, the minimum value of $p_o = \frac{1}{8} = 0.12$ (single occurrence), the next possible value is 0.25 $\left(\frac{2}{8}\right)$, and the next 0.38 $\left(\frac{3}{8}\right)$, and so on. Clearly, the proportions can, in this case, have only 8 values. Even so, except for *Pouteria macrophylla*, these very restricted estimated proportions do give us an idea of which species are the most abundant.

If you have read Chapter 8 carefully, these results should come as no surprise. What we have done here is equivalent to selecting 8 individuals at random from some large number N and attempted to estimate the species proportions by assuming that a Binomial is applicable. When individuals are involved, most researchers would say 8 is

TABLE 11.1 Species Rank Abundance, Using Occurrence Data

Species No.	True μ	True p_o	p_o Rank	Estimated $n = 8$	p_o Rank	Estimated $n = 25$	Rank
1	2.52	0.87	1	0.75	2	0.84	1
2	1.31	0.72	2	1.00	1	0.72	2
3	0.62	0.42	3			0.44	3
4	0.25	0.22	4	0.38	3	0.28	4
5	0.19	0.16	5	0.12	6	0.24	5
6	0.17	0.14	8	0.12	7	0.24	6
7	0.17	0.15	6	0.12	8	0.16	7
8	0.17	0.15	7	0.38	4	0.16	8
9	0.17	0.13	9			0.12	10
10	0.09	0.07	11			0.08	12
11	0.09	0.08	10			0.16	9
12	0.08	0.07	12				
13	0.06	0.06	13			0.04	15
14	0.06	0.06	14	0.12	9	0.12	11
15	0.04	0.04	15				
16	0.04	0.04	16				
17	0.04	0.04	17	0.12	10	0.08	13
18	0.03	0.03	18	0.12	11		
19	0.03	0.02	23	0.12	12		
20	0.03	0.03	19	0.12	13	0.04	16
21	0.03	0.03	20	0.25	5	0.04	17
22	0.03	0.03	21	0.12	14	0.04	18
23	0.03	0.03	22			0.04	19
24	0.02	0.02	24				
25	0.02	0.02	25			0.04	20
26	0.02	0.02	26				
27	0.02	0.02	27			0.04	21
28	0.02	0.02	28			0.04	22
29	0.02	0.02	29				
30	0.02	0.02	30			0.08	14
31	0.02	0.01	32				
32	0.02	0.02	31			0.12	15
33	0.01	0.01	33				
34	0.01	0.01	34			0.04	23
35	0.01	0.01	35			0.04	24
36	0.01	0.01	36				
37	0.01	0.01	37				
38	0.01	0.01	38			0.04	25

(continued)

TABLE II.I   (*continued*)

Species No.	True μ	True p_0	p_0 Rank	Estimated $n = 8$	p_0 Rank	Estimated $n = 25$	Rank
39	0.01	0.01	39			0.04	26
40	0.01	0.01	40				
41	0.01	0.01	41				
42	0.01	0.01	42				
43	0.01	0.01	43			0.04	27
44	0.01	0.01	44				
45	0.01	0.01	45				
46	0.01	0.01	46				
47	0.01	0.01	47				
48	0.01	0.01	48				
49	0.01	0.01	49				
50	0.01	0.01	50			0.04	28
51	0.01	0.01	51				
52	0.01	0.01	52			0.04	29

NOTE: True values of μ and p_0 are from all data in Bolivian plot 01 ($N = 100$). Estimated values $\hat{p}_0$ and occurrence rank are from $n = 8$ and $n = 25$ quadrats sampled at random.

not enough. Surprisingly, or maybe not, when biological samples are involved, researchers may not notice the small sample size as a problem. This is because of the failure to recognize that for occurrences the biological sample is the primary sampling unit, even though we are discussing the occurrence of a particular species.

A more realistic and trustworthy approach would be to sample a larger number of smaller-sized quadrats. To accomplish this we might divide the same area into 100 (rather than 25) quadrats. For this $N = 100$, we might sample $n = 25$ random quadrats (Appendix 1), and from this estimate p_0. The results shown in Table 11.1 are now much more reasonable, as is the Binomial condition of independence. Notice that when $n = 25$ (and $N = 100$) the p_0 and rank orders are quite similar to the actual values. Thus, our results depend not only on the proportion sampled but also upon other factors that contribute to the Binomial conditions being upheld.

CONFIDENCE LIMITS ON OCCURRENCES

By applying the Binomial distribution to occurrence data, we can obtain additional and valuable information. We know that for any given quadrat, species i is either in it or not. Then, if we can assume that the appropriate distribution is the Binomial, we know also that the variance of p_0 is given by Equation 8.13. The standard error of $\hat{p}_0$ is from Equation 8.15, $\hat{\sigma}_{po} = \sqrt{\dfrac{\hat{p}_0 \hat{q}_0}{(n - 1)}}$. With this information, let us calculate the confidence limits for some of the species encountered in the random sample of 8 quadrats. We do this to clarify how well our estimates are working. We are still assuming interest in 1 particular species, not simultaneous estimation of all relative abundances.

For the most abundant species, *Scheelea princeps*, we calculate $\hat{p}_o = 0.75$ and $\hat{\sigma}_{po} = \sqrt{\dfrac{(0.75)(0.25)}{7}} = 0.1637$. To obtain the 95% confidence limit we use Equation 8.15 and multiply $\hat{\sigma}_{po}$ by the appropriate value of t_p (which is the Student's t value) from Table 4.1. In this case $t_p = 2.36$ and the confidence limits are $(0.1637)(2.36) = p_o \pm 0.39$. For those species that appeared only once in the random sample $\hat{\sigma}_{po} = \sqrt{\dfrac{(0.12)(0.88)}{7}} = 0.1228$ and the confidence limits are $(0.1228)(2.36) = p_o \pm 0.29$. These are, of course, impossibly wide limits, and we should have expected them from our previous consideration of the Binomial distribution in the discussion of species proportions (Chapter 8). By introducing the lower limit of 0, and thereby reducing the effective range, we also introduce statistical problems that affect the final results. A sample of 8 quadrats obviously is much too small to use to obtain reliable rank order information. The problem is not inherently statistical; this size sample simply is not biologically representative of the species ordering in the target area.

On the other hand, with $n = 25$ we are in a much better situation. Table 11.1 shows that for a $\hat{p}_o = 0.84$, the value of $\hat{\sigma}_{po} = \sqrt{\dfrac{(0.84)(0.16)}{24}} = 0.0748$. From Table 4.1 for 95% confidence, we find a value of $t = 2.06$. Consequently, $(0.0748)(2.06) = p_o \pm 0.15$, which is much better than the interval calculated with only $n = 8$ ($p_o \pm 0.39$). For $n = 25$, the minimum occurrence is $\dfrac{1}{25} = 0.04$ and we have $\hat{\sigma}_{po} = \sqrt{\dfrac{(0.04)(0.96)}{24}} = 0.04$, so that $(0.04)(2.06) = p_o \pm 0.08$ (compared to $p_o \pm 0.29$ for $n = 8$).

SAMPLE SIZE FOR OCCURRENCES

What size of sample would be helpful in obtaining a useful estimate of rank order? In other words, how many quadrats, n, should we sample from a target population of N quadrats?

If we were willing to accept a precision value of, for example, $d = \pm 0.10$ (where d is width about the estimate, not the interval's width), then we would be able to calculate the number of required quadrats from Equation 8.24. For *Scheelea princeps*, with $\hat{p}_o = 0.75$, and from the $n = 8$ ($N = 25$) sample, we have $n = \dfrac{z^2 p_o q_o}{d^2}$, where, for 0.95 confidence, $z_{po} = 1.96$, $\hat{p}_o = 0.75$, $\hat{q}_o = 0.25$, and $d = 0.10$. This gives $\dfrac{(3.84)(0.75)(0.25)}{0.01} = 72$ as the value of n we would need to achieve our goal of being correct to within 10% of the mean proportion for *S. princeps*. However, note that for $\hat{p}_o = 0.84$, drawn from the $n = 25$ ($N = 100$) sample, we obtain a value for n of

$\dfrac{(3.84)(0.84)(0.16)}{0.01}$ = 52. Clearly, to achieve a reasonable degree of precision, we need a large number of quadrats or localities, just as we needed for calculating species proportions. Considering the equation for the variance of the Binomial (Equation 8.8), we would not expect to estimate species proportions with much precision on the basis of a very few individuals, nor should we expect an accurate estimate of the proportion of species occurrences on the basis of a few quadrats, biological samples, or localities. However, the number of quadrats or localities needed will depend on the size of the quadrat, plot, or locality. Larger sample plots can include more species, and thus fewer plots will be required to give adequate information on occurrences. However, this is true only if the plots are not so large that the more abundant taxa appear in all or nearly all of them, in which case sequentially ranking these abundant species will be hopeless (we will have many tied ranks).

It should now be apparent that a sampling scheme aimed at presence or absence should maximize the number of localities (quadrats). The advice given in Chapter 8 is still germane. As we pointed out there, the observations from each quadrat can constitute a Bernoulli trial, and we will need a large number of trials for the Binomial to be applicable and for the construction of suitable confidence limits.

OCCURRENCES AS SURROGATE FOR ABUNDANCE DISTRIBUTION

We have shown how species occurrences can be used as an alternative to recording the number of individuals for each species (Chapter 10). Consequently, for data sets containing many localities and species, it seems a reasonable alternative to consider the use of occurrences rather than individuals. Such an approach is the only one possible when the data consist of species lists, indicating which species were present at a particular locality. In addition, it gives us an optional view of the spread of the species over the localities. So far in this chapter we have been using proportions of species occurrence, p_o; now we can use just the number of occurrences themselves.

Once again, consideration of Bolivian plot 01 will help clarify understanding. Imagine dividing the hectare into very small quadrats so that each quadrat could fit around only 1 tree. As long as we are imagining, we can make all the trees the same size and stipulate that a quadrat must fit around a tree, not cut it into parts. What we have then is a carnival board (a ring toss) with pegs (the individual trees) onto which we toss a ring (the quadrat). Different color pegs can represent different species. The total number of different pegs the ring goes around after repeated tosses is the number of possible occurrences, $N_{occurrences}$. This number is equal to the total number of pegs, $N_{individuals}$ (for our special case), and there are S different colored pegs among the N. Of course, we could have obtained the same result (but a different statistical representation) by considering the entire board (hectare) as a unit, counting the number of pegs (individuals) and noting the different colors (species). In Chapter 9 we showed that given $N_{individuals}$ and S (species), we could fit various statistical distributions to the data. In our imaginary situation $N_{individuals} = N_{occurrences}$, so the results would be the same.

In the real world more than 1 individual is likely to be in a quadrat. Regardless of the size of the quadrat, if a species is represented by 1 individual, it can occur only once in our species list of occurrences.

Depending on the spatial configuration, however, a species with 2 or more individuals might occur in only 1 quadrat and hence be listed as a single occurrence. Furthermore, an abundant species, no matter how many individuals it has, can only occur as many times as there are numbers of quadrats. This is why in Chapter 10 the regression of density and occurrences did not predict the density of the most abundant species as well as it predicted the others. As quadrat size is increased, the number of occurrences of our abundant species ultimately reaches a value of only 1 (the entire hectare), and each species, no matter how abundant, can occur only once.

It is most common, then, to have a situation in which $N_{occurrences} < N_{individuals}$. In fact, the quadrat size and target population usually present us with a problem between the two extremes outlined above, so we find that $N_{occurrences} < N_{individuals}$. The relationship between occurrences and individuals, however, appears strong enough to warrant examining the distribution of occurrences and species in the same way as we examined the distribution of individuals and species in Chapter 9. We consider a change not only of the perceived distribution of the species but of the *scale* at which we study the natural population of interest. This scale of occurrences dampens the effect of very high densities of a species at a locality and hence removes some of the spatial aggregation encountered when individuals are observed. This is an area of statistical application that has been overlooked by both mathematical statisticians and biologists.

All varieties of distributions presented in Chapter 9 relied upon the number of individuals (N) and the number of species (S) for their fit. In this chapter, instead of employing the number of individuals represented by 1 species, 2 species, and so on, we shall use the number of species occurring at only 1 locality, 2 localities, and so on. In this way N becomes the number of occurrences rather than the number of individuals, and we will fit a distribution to the occurrence data. We shall see that this fitting is equivalent to the fitting of the abundance data. There is evidence of a dampening of the spatial variability, but the measures of the area's diversity are still appropriate. However, no formal statistical theory exists at this time for relating the two applications.

FITTING SPECIES ABUNDANCE DISTRIBUTIONS TO OCCURRENCES

In Appendix 1, the Bolivian data are presented with the study area divided into 100 plots. The number of occurrences for each species in the quadrats is also given. We saw, for example, that the most abundant species, *Scheelea princeps*, which is represented by 252 individuals, occurs in 87 of the 100 quadrats. In all, over the 100 quadrats, there are 396 occurrences (of any species) and $S = 52$ species. As an example of the application of a *surrogate abundance distribution* using these numbers, the Log series distribution from Chapter 9 can be fit to this data set.

For a Log series distribution, Fisher required that the total number of species, S, and the total number of individuals, N, for the sample be known. Because of the relationship between number of individuals of a species and the number of occurrences of that species, we can make a substitution. We fit the Log series distribution by using *S* as the number of species, but *N* as the number of occurrences. We selected this Log series distribution because, as discussed in Chapter 9, it is elegant, simple, and fits as well or better than other major contenders fit. The values so obtained from Equations 9.15 and 9.16 for the whole Bolivian assemblage are $\alpha = 16.0110$ and $x = 0.9611395$. Using these values in Equation 9.12, $S^* = 52.00$, and in Equation 9.13, $N^* = 396.00$, indicating the appropriateness of our estimates. The expected occurrence quantities from the Log series then can be calculated from Equation 9.10. The results, shown in Table 11.2, exhibit the same pattern as the results obtained when individuals rather than occurrences were used for *N* (Table 9.6). Namely, the observed rare species are relatively too rare and the abundant species a little too abundant when compared to the expected

TABLE 11.2 Fitting Total Occurrences with a Log Series

				IN CLASSES	
NUMBER OF SPECIES	OBSERVED OCCURRENCES	EXPECTED LOG SERIES	CLASS ENDPOINTS	OBSERVED OCCURRENCES	EXPECTED LOG SERIES
1	21	15.39	[1, 2)	21	15.39
2	9	7.40			
3	5	4.74	[2, 4)	14	12.14
4	3	3.42			
5	0	2.63			
6	2	2.10			
7	2	1.73	[4, 8)	7	9.88
8	1	1.46			
9	0	1.25			
10	0	1.08			
11	0	0.94			
12	0	0.83			
13	1	0.74			
14	1	0.66			
15	2	0.59	[8, 16)	5	7.55
.					
31	2		[16, 32)	2	4.74
.					
63	1		[32, 64)	1	1.99
.					
127	2		[64, 128)	2	0.41

NOTE: The observed number of species having $i = 1, 2, \ldots, n$ occurrences and the expected number for a Log series, where $N = 396$, $S = 52$, $\alpha = 16.0110$, and $x = 0.9611395$.

SOURCE: Data from Bolivian forest plot 01, Appendix 1.

numbers. As before, the expected number of species with 1 occurrence is very close to the value of α (15.39 vs 16.01), differing only in the first decimal place. Inspection of the tables for individuals (Table 9.6) and occurrences (Table 11.2) indicates that the Log series fit to the occurrences can give us as much information as the Log series fit to the individuals.

In general, the deviation $(1 - x)$ is an indication of the acceptability of fit of the Log series distribution. For $0 \leq (1 - x) \leq 0.05$, the fit is considered exceptional (we have 0.04 in the above example). When $(1 - x)$ is in the interval $0.10 \leq (1 - x) < 0.05$, the fit is good, and when $(1 - x)$ is greater than 0.10, the fit of the Log series is usually not close enough to the observed to please a naturalist. Remember we fit $\dfrac{N}{S}$ when we use the Log series. There are many sequences of observed values that sum to the same N and S values. For any 1 set of N and S values there is only 1 Log series; some sequences of observed values will be closely approximated by the expected values of that Log series and some will not. Because the first term in the series (Equation 9.10) is αx, the farther x departs from 1, the farther the expected number of species with 1 occurrence will be from α. However, used as a diversity measure, α will still provide intuitive information for biodiversity purposes.

From our earlier discussion, we would expect the number of species with 1 occurrence and the value of α to increase as quadrat size is increased. That is, as we use larger quadrat sizes, it will take fewer quadrats to cover the target area and it is less likely that a single occurrence will be represented by only 1 individual of a species. Continuing the logic, sometimes a quadrat will have 2, 3, or more individuals of the same species but still be listed as 1 occurrence because they all occur in one place. Using the data for $N = 25$ quadrats (Appendix 2), we can check this expectation by comparing this data set with our 100-quadrat example (Appendix 1). The data in Appendix 2 show that for $N = 25$ quadrats there are 233 occurrences ($N_{occurrences}$; N_{occ}). S is 52, as before. Using Equations 9.16 and 9.14, we obtain these values: $x = 0.9181214$ and $\alpha = 20.77907$. To check the accuracy of estimation, Equation 9.12 gives $S^* = 52.00$, and Equation 9.13 gives $N^* = 233.00$. While the value of x for this 25-quadrat example indicates a good fit, it is not as good as with $N = 100$ quadrats. As predicted, the value of α has increased from 16.01 to 20.78.

As a final exercise, let us reduce the $N = 100$ quadrats in Appendix 1 into $N = 10$ quadrats by combining into larger samples the quadrats numbered 1 to 10, 11 to 20, ..., 91 to 100. Enumerating the occurrences, we have $N_{occ} = 157$, while S remains 52 because we are considering the same hectare. Using the values 157 and 52, we obtain $\alpha = 27.17$ and $x = 0.85245188$. Equation 9.12 gives $S^* = 51.99$. Equation 9.13 gives $N^* = 156.97$. The expected value for the number of species with 1 occurrence is $\alpha x = 23.16$.

These exercises show that by increasing the quadrat size for the same set of data, we effectively change the scale of observations. As the quadrat size increases, so does α, while x decreases. This outcome does not necessarily diminish the usefulness of α as a measure or index for diversity, but it does mean that standardization of quadrat size is necessary if we wish to compare indices based on occurrences.

Sampling the Log Series: An Example

In the previous section, we calculated the Log series, or its parameter, for different-sized quadrats from Bolivian plot 01. Throughout this book we have also sampled the plot to see how well estimates calculated from samples of various configurations approached the true values determined by complete enumeration. We believe it will be instructive to do the same with the Log series for occurrences. That is, we will use a sample of species occurrences to examine how useful will be our actual prediction. From Appendix 1 ($N = 100$) we chose $n = 25$ quadrats at random and counted the number of times each species we encountered occurred in the 25 samples. In all, there were $N_{occ} = 108$ occurrences and $S = 29$ species. For the Log series, we obtained the values $\alpha = 12.9993$ and $x = 0.89256714$. Checking their accuracy with Equations 9.12 and 9.13, we obtain $S^* = 29.00$ and $N^* = 108.00$, respectively. We recall for $N = 100$, the value of α was 16.01. The standard error of α calculated from Equation 9.17 is $\sigma_\alpha = 1.98$. Multiplying by 2, which is the rounded value of 1.96 from Table 3.2, we obtain 95% confidence limits of about 13.00 ± 4. Consequently, the true α is well within the interval about the estimated value of 13, even though we had only a borderline good fit of the Log series.

Let us also examine the results for occurrences using 8 randomly chosen quadrats from a possible $N = 25$ (Appendix 2). The randomly chosen quadrats have a total of $N_{occ} = 72$ and $S = 31$ species (Table 11.3). In this case, the maximum number of occurrences for any species can be only 8, because our sample had only 8 quadrats. We

TABLE 11.3   Fitting a Sample of Occurrences with a Log Series

				IN CLASSES	
NUMBER OF SPECIES	OBSERVED OCCURRENCES	EXPECTED LOG SERIES	CLASS ENDPOINTS	OBSERVED OCCURRENCES	EXPECTED LOG SERIES
1	14	16.05	[1, 2)	14	16.05
2	9	6.24			
3	3	3.23	[2, 4)	12	9.47
4	0	1.88			
5	2	1.17			
6	1	0.76			
7	1	0.50	[4, 8)	4	4.32
8	1	0.34			
.					
15			[8, 16)	1	1.16

NOTE: The observed number of species having $i = 1, 2, \ldots, n$ occurrences and the expected number for a Log series, where $N = 72$, $S = 31$.

SOURCE: Data from Bolivian tree plot 01, 8 random quadrats (1, 3, 4, 10, 14, 18, 19, and 23), Appendix 2.

obtained the values $\alpha = 20.6528$ and $x = 0.77095$. Again we check accuracy by use of Equations 9.12 and 9.13, with the result $S^* = 31.00$ and $N^* = 72.00$, respectively. The standard error of α calculated from Equation 9.17 is $\sigma_\alpha = 3.98$. Multiplying by 2, we obtain 95% confidence limits of about 21 ± 8. The true value of α calculated from the entire 25 quadrats is 20.78, and once again we are well within the confidence limits.

Our samples, then, did quite well in estimating α for their respective quadrat sizes. Notice that the α for $n = 25$ is close to that for $N = 100$, from which these samples were selected, and not for $N = 25$. This is so because samples from a Log series distribution will themselves be a Log series.

Using the Log Series for Occurrence Data: A Cretaceous Example

The preceding examples of how species occurrences can be fit by statistical distributions were based on a small sample of forest trees provided by Bolivian plot 01. An example from a large data set with a wide geographic distribution is provided by Late Cretaceous molluscs from the Gulf Coastal Plain (Buzas and colleagues 1982). Instead of using the number of individuals recorded, the number of species with $1, 2, \ldots, n$ occurrences from 166 localities are fit by the Log series. The number of mollusc species is $S = 716$, and the number of occurrences is $N = 6{,}236$. From Equation 9.14 the value of $x = 0.967610$ was calculated, and from Equation 9.16 we have $\alpha = 208.7548$. Using these numbers in Equations 9.12 and 9.13, we obtain $S^* = 716.01$ and $N^* = 6{,}236.28$, respectively. The series calculated from Equation 9.10 is shown in Table 11.4. Considering the value of x, the expected number of species with 1 occurrence (αx) is close to α. As with the Log series calculated on individual trees, the observed rare species of molluscs are too rare, and the abundant species are too abundant, when evaluated against expectations. However, the fit of the observed occurrences with those expected from a Log series is strikingly good.

Because the Log series fits occurrences so well, it is reasonable to use α as a measure of diversity when only presence or absence data are available. When the same quadrat sizes are used in different areas, the biodiversity from each area can be characterized by α and other diversity indices. The use of indices with occurrences puts the biological scenario in a different scale. This alternative way of examining species distributions using only occurrence data has the advantage (and disadvantage) of dampening much of the spatial variability observed with density. We are thus justified in using occurrence data for obtaining surrogate abundance information by fitting the data with a probability distribution and using diversity indices so obtained. Buzas and colleagues (1982) and Buzas and Culver (1991) give examples of this methodology applied to very large data sets.

SUMMARY

1. Occurrence data from museum collections, research literature, species lists, or field inventories are much more valuable than generally believed.

2. It is especially important not to think of species occurrences (presence/absence data) as limited information. Instead, think of it as constituting information on a different scale, a scale on which much of the spatial variability is dampened. Recognition of

TABLE 11.4   Fitting a Sample of Occurrences of Cretaceous Molluscs with a Log Series

				IN CLASSES	
NUMBER OF SPECIES	OBSERVED OCCURRENCES	EXPECTED LOG SERIES	CLASS ENDPOINTS	OBSERVED OCCURRENCES	EXPECTED LOG SERIES
1	211	201.99	[1, 2)	211	201.99
2	82	97.72			
3	64	63.04	[2, 4)	146	160.76
4	47	45.75			
5	36	35.41			
6	27	28.55			
7	21	23.68	[4, 8)	131	133.39
8	17	20.05			
9	20	17.25			
10	21	15.02			
11	14	13.21			
12	15	11.72			
13	4	10.47			
14	12	9.40			
15	5	8.49	[8, 16)	108	105.61
.					
31			[16, 32)	75	75.59
.					
63			[32, 64)	38	34.03
.					
127			[64, 128)	7	4.60

NOTE: The observed number of species having $i = 1, 2, \ldots, n$ occurrences and the expected number for a Log series, where $N = 6,236$, $S = 716$, $\alpha = 208.7548$, and $x = 0.967610$. Data from Late Cretaceous molluscs of the Gulf Coastal Plain.

SOURCE: Buzas and colleagues 1982.

biofacies can easily and efficiently be accomplished through the use of presence/absence data.

3. As long as a suitable and constant quadrat size is chosen, occurrences provide very reliable information on the rank order of species and species proportions.

4. The statistical distributions that are used to fit species abundance observations can also be used for fitting species occurrences (Chapter 9). When N occurrences are substituted for N individuals, distributions like the Log series do fit the data very well. This is true for relatively small data sets as well as large ones.

5. Measures such as Fisher's α (as well as others, see Chapter 13) are suitable for measuring species diversity from occurrence data.

PROBLEMS

11.1 *Astrocaryum macrocalyx* is the seventh most abundant tree species in Bolivia plot 01. How many samples would be required in order for us to be 95% confident that the proportion of occurrences was within ±0.05 of the true value?

11.2 If you could sample individual trees at random, how many individuals would have to be sampled in order for us to be 95% confident that the species proportion of *A. macrocalyx* is within ±0.05 of the true value?

11.3 Use the bat data of Problem 9.2 (Appendix 5) to perform the following:

  a. Fit a Log series distribution to the occurrences (see Problem 10.1).

  b. Find the Log series parameter and put a confidence interval about it.

  c. Discuss the fit of the Log series to this bat occurrence data versus the fit to the bat abundance data (Problem 9.2).

12

SPECIES DIVERSITY:
THE NUMBER OF SPECIES

THOSE WHO GATHER SAMPLES of natural populations for biodiversity purposes always want to know how many species actually occur in their study area. For nearly all groups of organisms, it appears that fewer species occur in the Arctic than in the tropics. The classic increase in the number of species occurring with decreasing latitude is used as the rationale for scientific investigation. Such simple but intriguing observations have led to innumerable studies and papers concerned with the number of species occurring in different areas. On grand scales covering entire regions, the number of species encountered is often referred to as *gamma* (γ) *diversity* (Whittaker 1972). However, simple patterns often disappear as the number of studies increases and as the geographic area under consideration, as well as the methodology used to determine the number of species present, begin to vary greatly. When examining a single habitat, such as the Bolivian plot 01, the term *alpha* (α) *diversity* (Whittaker 1972) is frequently used to convey the notion that *within-habitat diversity* is being examined.

In the past, investigations examining the number of species were called studies of *species diversity*. Over the years a growing number of researchers included in such studies not only the number of species observed but also information on species abundances. This tendency has caused some anxiety among researchers. This is so because a number of different indices or measures are used to describe and measure species diversity when abundances are considered. A good deal of the literature consists of trying to determine which measure is "best" (see Magurran 1988; Hayek 1994, for a review). Consequently, in an attempt to avoid confusion, the number of species recorded in a study area is now often referred to as *species richness*, abbreviated *S*. In this chapter, we examine the role that the number of species, *S*, plays in sampling natural populations.

Measures using species abundances in their formulas are referred to under a variety of terms that often incorporate the name of the measure used. At the same time, the original term, *species diversity*, is still in use. Thus the interested reader must always ascertain what the author actually means by species diversity. To further complicate communication, the all-encompassing term *biodiversity*, or *biological diversity*, is now used when referring to any of the quantities just mentioned, as well as to a host of measures incorporating biomass, habitat diversity, phylogenetic diversity, and so on. While the terms used may present communication problems for some, those workers using a quantitative approach should experience little difficulty. If the author(s) state, in quantitative terms, precisely what was measured and how, the definition will lie in the quantification, and the terminology will become secondary.

MEAN NUMBER OF SPECIES

Because all of the individuals (trees) were counted in Bolivian plot 01, we know that $S = 52$ species are present. Obviously, a sample of a few quadrats from the plot will not contain all these species. Let us examine the division of this plot into 25 quadrats. Table 12.1 shows the number of individuals and species, along with accumulations, for the 25 quadrats (Appendix 2). One of the first questions we might ask is "What is the average number of species per quadrat?" A few strokes on a calculator or a computer reveal that the true population mean number of species per quadrat is $\mu = 9.320$ and $\alpha = 2.428$, using Equations 2.1 and 2.3, respectively.

Using the techniques illustrated in the density chapters, we will take a sample of 4 randomly chosen quadrats to examine how well we could estimate this true mean number of species per quadrat just from the sample. The randomly chosen quadrats we shall use are numbers 3, 4, 5, and 13 (Appendix 2). Using the formulas for sample estimators (Equations 2.2, 2.5, and 3.6), we obtain the estimates $\hat{\mu} = 10.25$, $\hat{\sigma} = 2.630$, and $\hat{\sigma}_{\mu} = 0.456$ (using the fpc), respectively. From Table 4.1 we obtain $t_{095,3} = 3.18$, so that the small-sample 95% confidence interval is $\mu \pm (t\hat{\sigma}_{\mu}) = \mu \pm (3.18)(0.456) = 10.25 \pm 1.45$, or $8.80 \leq \mu \leq 11.70$. We have, then, quickly and easily, obtained a reasonable interval estimate of the average number of species per quadrat in the entire plot.

If we wish to delineate the average number of species a particular habitat supports, the point estimate or the interval estimate will give useful information. We could, for example, say that a habitat supporting about 9 species per 400-m^2 quadrat within that target habitat has higher species diversity than does one supporting 3 species over an equivalent area. Samples that contain about the same number of individuals per quadrat (of a standard size) from a species-rich habitat will contain more species than will samples of the same size from a species-poor habitat. Knowledge of the average number of species per some standard-size biological sample or quadrat may be all that is necessary to distinguish 1 habitat from another in terms of species richness. However, this average S does not reveal how many different species we observed between samples, or how many species in total occur within the habitat. The researcher may also be interested in how species accumulate between samples within a habitat (hence, Table 12.1).

Another point of interest is the difference in the numbers of species encountered between 2 or more habitats, often called *beta diversity* (Whittaker 1972). The average number of species per sample or quadrat does not tell us how many different species we observed in the sum total of those samples. In addition, the researcher is interested in the number of different species that occur in some area of interest. For many biodiversity studies, the average number per quadrat may be all that is required. This estimate is quick, easy, and informative. Many different environments can be distinguished in this way.

TABLE 12.1   Basic Display of *N* and *S* for a Field Survey

Quadrat	N	Accum. N	S	S Added	Accum. S
1	30	30	11	11	11
2	20	50	6	2	13
3	24	74	9	4	17
4	24	98	14	8	25
5	29	127	10	2	27
6	30	157	8	3	30
7	28	185	8	1	31
8	32	217	9	1	32
9	29	246	9	0	32
10	24	270	5	1	33
11	31	301	13	4	37
12	24	325	11	2	39
13	21	346	8	1	40
14	20	366	11	0	40
15	26	392	10	1	41
16	28	420	13	2	43
17	17	437	8	2	45
18	23	460	6	2	47
19	37	497	9	0	47
20	37	534	13	1	48
21	22	556	7	0	48
22	28	584	12	2	50
23	19	603	7	1	51
24	32	635	8	1	52
25	28	663	8	0	52

NOTE: Bolivia tree plot 01, 25 quadrats. *N* is the number of individuals observed in each quadrat, and *Accum. N* is the cumulative sum over the hectare, quadrat by quadrat. *S* is the number of species observed in each quadrat; *S Added* is the number of previously unrecorded species added upon examination of the next quadrat; *Accum. S* is the number of species accumulated as each succeeding quadrat is examined.

NUMBER OF DIFFERENT SPECIES

Looking at Table 12.1, we see that 11 species occurred in sample 1, and 6 in sample 2. However, only 2 of the 6 species in sample 2 were not observed also in sample 1. Therefore, the total number of unique observed species in samples 1 and 2 together is $11 + 2 = 13$. This calculation appears in the rightmost column, as "Accum. *S*." As we examine additional samples, more previously unrecorded species appear. In sample 3, for example, 4 previously unrecorded species are added, so that after examining the first 3 samples, we have a record of 17 species. Not until we reach sample 24 (out of 25 possible) have we accounted for all 52 species in the plot.

Obviously, as we examine successive samples, we encounter more individuals; in turn, we necessarily must examine more individuals to record more species. Hence, the number of species observed (S) is related to, or is a function of, the number of individuals (N) examined. In mathematical notation this is written as $S = f(N)$, or S is a function of N.

Figure 12.1 shows a plot of the cumulative number of species versus the cumulative number of individuals. The curve is not at all smooth primarily because the mean number of individuals (N) per sample is only about 26, which is rather small. Had we chosen to plot this same total number of species by increments of approximately 100, using cumulative N of 98, 217, 301, 392, 497, 603, and 663 as our data points, the result would have been a smooth curve (Figure 12.2). Notice that at the outset species are added more rapidly than they are during later observations. After the first 157 individuals are counted (6 samples), species are added by 1's and 2's (Table 12.1), except for sample 11. This characteristic of fast initial accumulation, followed by a leveling-off, results in a curve rather than a straight line (Figure 12.1).

SPECIES EFFORT CURVE

A *species effort curve*, also termed a *species accumulation curve* (or even, occasionally, a *collector's curve*), is the cumulative number of species plotted against some measure of the effort it took to obtain that sample of species. The measure of effort can be the

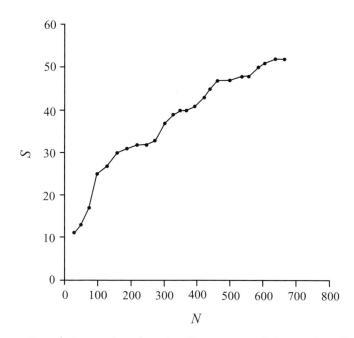

FIGURE 12.1   Cumulative number of species (S) versus cumulative number of individuals (N) for Bolivian plot 01 divided into 25 quadrats.

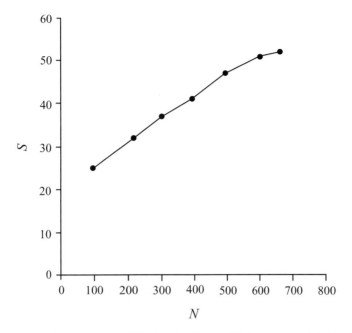

FIGURE 12.2   Another depiction of the data in Figure 12.1, but accumulated in increments of hundreds of individuals, rather than quadrat by quadrat, to produce a smoother curve.

number of individuals observed, as we already discussed. However, many times the researcher does not (or cannot) bother counting individuals of species previously discovered in the sample, and so records only new species as they occur. In this case (for example, see Lamas and colleagues 1991) the number of samples, traps, trap-days, or other areal or effort measurement may suffice.

Thus the curve in Figure 12.2 can also tell us something about collecting effort. Regardless of the measure of effort chosen, fieldwork will produce an initial period of rapid accumulation of information, as the first biological samples are examined. As each succeeding sample is observed, we will find fewer and fewer representatives of new species, until we reach the top portion of the curve, which may begin to level off. This leveling-off indicates that only *singletons* (species with 1 individual) are now being found. The question of interest to the fieldworker is "At what point is the next sample not worth taking, given the effort?"

It appears that in the temperate zone, this point for considering cutting off the fieldwork is probably in the range of 200 to 500 cumulative individuals for most taxa. However, in the tropics, across many taxa, we do not see a characteristic leveling-off (*asymptote*), and any discussion of expected unit effort is necessarily arbitrary and sample-dependent.

It is for this reason, among others, that there is interest in species-richness estimators. These are estimators of the richness of the target area or population. Because we never actually have a chance to obtain or observe some of the species in the target area, for reasons of scarcity and behavioral characteristics, among others, our sample estimate is almost always low (see for example Gotelli and Colwell 2001). Species-richness estimators have been developed to provide a larger richness value that is more representative of the total number of species in the area than the observations themselves show. However, at this time it appears that no one estimate is clearly the best, or most usable, for natural population work. In addition, although a few estimates, like the estimates of Chao (1984, 1987; and see Chao and colleagues 2009), have desirable properties, no one has shown that any one of these estimators actually comes close to providing a value with any consistency and that is an accurate enough approximation to the total number of species in a target area.

STANDARDIZING THE NUMBER OF INDIVIDUALS IN A SAMPLE

Standardizing in the Data Collection Stage: Standard Counts

The Bolivian example demonstrates the general fact that, for a species-rich habitat, it becomes very difficult to predict the total number of species present based on only a few samples from a relatively small area. After the first few samples or quadrats are examined, species keep accumulating by 1's and 2's within the target area, and usually will continue to do so even if the target area were to be expanded to several or many times the original areal dimensions. Resources, in terms of human life expectancy as well as funding, are insufficient for the task. Consequently, some compromise must be reached.

Many researchers are interested in the number of species occupying a habitat. A simple and effective way of determining a reasonable sample size (in terms of numbers of individuals) for estimating number of species is to plot S against N (in a cumulative distribution), as we did in Figure 12.1. Where the curve levels off, the researcher can assume that a level of diminishing returns is reached in terms of the number of species to be found. The researcher then declares, "enough." Alternatively, interest may be in the efficacy of estimates of species proportions for population information. Similarly, for this situation, plots of species proportions can be made for samples containing increasing numbers of individuals in order to determine when estimates of species percentages no longer vary by much (see, for example, Phleger 1960; Buzas, 1990). That is, the standard error is getting smaller, or stabilizes.

Over time it has been determined empirically that somewhere between 200 and 500 individuals are needed to obtain reasonable estimates of species proportions. In Chapter 6 we showed that if individuals are chosen at random, then at a species proportion of $p = 0.50$ and $n = 200$ the confidence interval about p would be of length $p \pm 0.07$ for $\hat{\mu} = p$; at $n = 300$ the length would be $p \pm 0.06$; at $n = 400$ $p \pm 0.05$ would be the length; and finally, at $n = 500$ we would have a confidence interval of length $p \pm 0.04$. That is, the use of the Binomial distribution confirms the conclusion reached empirically by researchers who drew a simple plot. Workers began using these cut-off values not just for determining species proportions but for counting the number of species to observe. For most taxa, when cumulative samples reach this range of 200 to 500 individuals, only rare species are encountered.

Using this primitive, but effective, methodology, researchers studying some groups of smaller organisms have *standardized* the number of individuals that they count and identify to species. For example, for micropaleontologists the number is

300 (Phleger 1960; Buzas 1990). For archeobotanists, among others, the number ranges up to 500.

To examine how well these cut-off numbers perform, let's look at Table 12.1. For a cumulative N of 217, we have $S = 32$ species; for $N = 301$, $S = 37$ species; for $N = 392$, $S = 41$ species; and for $N = 497$, $S = 47$. For this example, then, the difference between the micropaleontologists' value of $N = 300$ and the archeobotanists' value of $N = 500$ is about 10 species or types. This may seem like a lot, but we should keep in mind that these are all rare species. Indeed, any large incremental increase in N is likely to pick up merely 1 or more singleton (single individual) species during a particular sampling period.

In our Bolivian example, all of the 14 species that have a proportion or relative abundance greater than 0.009 are accounted for in the first 8 samples, for which $N = 217$ (Table 12.1). Within the Bolivian hectare that has been divided into 25 quadrats, 21 species occur in only 1 quadrat and 8 more in only 2. Consequently, any random sampling of a few quadrats from another survey of the same area might well produce approximately the same number of species for a given number of quadrats, but the list of rare species will almost certainly be different. Nevertheless, by standardizing for each of the various habitats the number of individuals observed from an area that has a large number of individuals (unlike many large mammals, for example), one may well be able to make a comparison of not only the species richness or mean number of species per quadrat but also of the component species of the habitat.

It should be apparent by now that there is no magical number of individuals that we can collect that will provide us with everything we want to know about a natural population. The more individuals that are counted and identified, the better will be the estimates for species proportions and the number of species. However, the advantage is balanced by a corresponding increase in money, time, and effort. The effort involved with field and laboratory time in going from 200 to, say, 500 individuals can be considerable. The amount of effort depends on the group and what procedures are involved.

Many groups of organisms are so abundant that each biological sample may contain hundreds or thousands of individuals. For such groups the question arises: "Should you process 2 smaller samples (say, of 200 or 250 each) rather than 1 larger sample (say, of 400 or 500), if the same amount of time is involved?" We would. Replication is vital to science. A few hundred individuals can easily be selected at random from the specimens in a single plot, quadrat, core, or a sample that has been homogenized in the laboratory. The Binomial confidence limits are applicable for the species proportions in this homogenized (randomized) type of sample. The assumptions present no problem for this approach. The variation in the natural population will, however, almost always be large, sometimes very large, and the estimation of this variance requires replicate samples. Because of the inherent variation in natural populations, estimates of variance that are based not on single biological samples but on a number of replicated biological samples become paramount.

In summary, for studies of species richness, because the number of species observed is a function of the number of individuals observed, the simplest way to assure compatibility between samples is to standardize N, the number of individuals observed. Some researchers customarily adhere to a fixed number of individuals; many do not. Standardization of sample size for a particular group or for a particular type of study promotes comparison of samples from various localities by different investigators.

Standardizing During the Analysis Stage: By Inference

Unfortunately, investigators working independently do not always agree with others working on the same taxa as to the procedure for collecting or counting fixed numbers of individuals. Even when they do, in many cases the organisms do not cooperate, or sampling gear may prevent the intended standardization. Consequently, if we wished to compare studies with differing numbers of individuals, in the analysis phase we might want to reduce or standardize the number of individuals to a common number. For example, if a comparison between Bolivian plot 01 ($N = 663$) and another plot containing only 200 individuals were desirable, we would wish to know how many species occur in plot 01 when only 200 individuals have been counted, or how many species the second study would have recorded if the researchers had continued working until 663 individuals were observed. Let us examine some ways to make this comparison.

Regression of N and S

Regardless of taxa, for data sets with observed values of N and S accumulated over biological samples, plots, cores, or quadrats, a picture similar to that found in Figure 12.1 always results. As we discussed in Chapters 6, 9, and 10, for any similarly shaped curve, a customary method for obtaining a linear relationship is to transform the arithmetic scale to a logarithmic scale.

When dealing with curves that plot N against S, it is usual to consider two alternative presentation methods. In Figure 12.3 the cumulative number of individuals is plotted on a log scale, while raw values for S are recorded; that is, S is on an arithmetic scale. In this instance, the appropriate equation is

$$S = a + bln(N) \tag{12.1}$$

where a and b are constants, defined in Equations 6.7 and 6.8, respectively. The relationship shown in Equation 12.1 between S and N is called *semilog* or *semilogarithmic*. Alternatively, in Figure 12.4, both scales are transformed to logarithms; that is, the axes are log–log. In this second instance, when both S and N are transformed to logs, we can write $ln(S) = ln(c) + dln(N)$. The appropriate equation relating N and S on a log–log plot is

$$S = cN^d \tag{12.2}$$

where c and d are constants, as defined in Equation 6.4. (Remember, the particular letters used are of no concern; they are arbitrarily chosen.) As noted in Chapter 6,

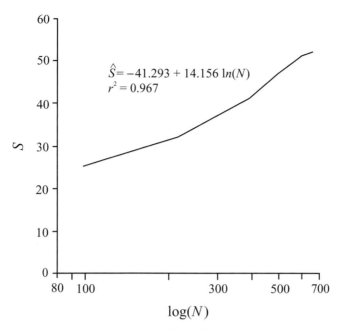

FIGURE 12.3 Semilog plot, S versus $\log(N)$, for Bolivian plot 01.

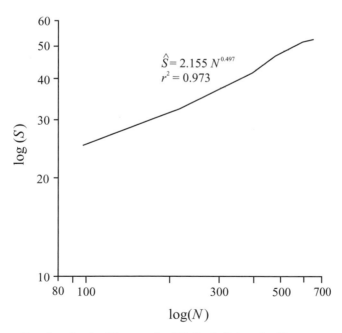

FIGURE 12.4 Log–log plot, $\log(S)$ versus $\log(N)$, for Bolivian plot 01.

TABLE 12.2　Species-Richness Estimation Based on Regressions of Individuals

QUADRAT	N	S	$\hat{S} = a + b\ln(N)$	$\hat{S} = cN^d$
1	98	25	23.61	21.04
8	217	32	34.86	31.24
11	301	37	39.50	36.75
15	392	41	43.24	41.91
19	497	47	46.60	47.16
23	603	51	49.33	51.91
25	663	52	50.67	54.42

NOTE: The number of species observed (S) and estimated ($\hat{S}$) for values of N. Here, $a = -41.293$, $b = 14.156$, $c = 2.144$, and $d = 0.497$.

when both variables have been placed on a logarithmic scale such as in Equation 12.2, this relationship is often called the power curve or power law.

For the semilogarithmic curve we obtained $\hat{S} = -41.293 + 14.156 \cdot \ln(N)$, with a coefficient of determination of $r^2 = 0.968$ (from Equation 10.8), and a standard error of estimate of $S_{y \cdot x} = 2.178$ (from Equation 10.4). We obtained $\hat{S} = 2.155N^{0.467}$ for the power curve. From the accompanying transformed linear equation (because r is not calculated for nonlinear equations) we get an $r^2 = 0.974$. In Table 12.2, we present the observed S and calculated log–log and semilog or estimated $\hat{S}$ for N in increments of about 100. As the values of r^2 indicate, both results fit the observed values from natural population data sets quite well. Choosing one over the other is simply a matter of preference.

REGRESSION OF AREA AND S: SPECIES–AREA REGRESSION

Clearly, species are accumulated as we observe more samples or quadrats. Because biological samples are portions of our area of interest, if we added just the areal measurements for each of the quadrats (1, 2, . . . , 25), we would expect a similar result to that obtained from accumulating values of N. That is, the relationship between the number of species, S, and the number of individuals (N) resembles the relationship between S and the area (A) in which the individuals are found or observed. The relationship holds, as well, for the time or effort it takes to collect or observe N and A.

Let us explain with an example. Each of the 25 quadrats in our Bolivian forest hectare has an area of $20\,m \cdot 20\,m = 400$ m^2. To illustrate the connection to S, we calculated the constants for the two predictive formulas (Equations 12.1 and 12.2), but with the accumulated numbers of individuals replaced with the related cumulative sample size or area. That is, we used the accumulated areas of $400\,m^2$, 800 m^2, . . . , $10,000$ m^2, corresponding to 30, 50, 74, . . . , 663 individuals, and accumulated S as shown in

Table 12.1. The results are $\hat{S} = -79.925 + 14.096 \cdot ln(A)$, with an $r^2 = 0.965$ and $S_{y \cdot x} = 2.30$. Then, $\hat{S} = 0.562A^{0.496}$, with an associated $r^2 = 0.976$ (based on the related semilog linear equation) and $S_{y \cdot x} = 0.066$ (this is in natural logs, or 1.07 untransformed). The slopes are 14.096 and 0.496 and are, as we might expect, very similar to the ones calculated using cumulative N, which were 14.156 and 0.467.

In Table 12.3, we compare the fit of the semilogarithmic and the power curve to the actual observations. Not surprisingly, the fit is quite good and we do as well with area as we do with individuals.

The *species–area relationship* was recognized by early researchers (see discussions in Arrhenius 1921 and Gleason 1922). The power curve for this relationship is usually written as

$$S = cA^z. \tag{12.3}$$

This is the common notation, but the letters chosen for the coefficients are arbitrary (we could have written, say, $S = aA^b$). The letter z is used arbitrarily in Equation 12.3 and is not to be confused with the standard Normal deviate. This relationship is widely used because of the popularity of the *equilibrium theory of island biogeography* (MacArthur and Wilson 1967). Extensive reviews are provided by Preston (1962), Connor and McCoy (1979), and Gilbert (1980).

Biological interpretations explaining the species–area relationship are numerous and often confusing. In this chapter, the relationship is used merely as a statistical tool. We are examining a single habitat (Bolivian plot 01) and showing the general result that species accumulate in an orderly fashion within a single habitat. Thus we show that there is a general phenomenon observed, which is that the use of the accumulation of the area (A) occupied by these species, rather than their abundances, can be used in a predictive manner.

TABLE 12.3   Species-Richness Estimation Based on Regression of Area

A	S	$\hat{S} = a + b ln(A)$	$\hat{S} = cA^d$
1,600	25	24.07	21.83
3,200	32	33.84	30.78
4,800	39	39.56	37.64
6,400	43	43.61	43.41
8,000	48	46.76	48.49
9,600	52	49.33	53.08
10,000	52	49.90	54.17

NOTE: The number of species observed (S) and estimated ($\hat{S}$) for value of A(in m^2). Here, $a = -79.924$, $b = 14.096$, $c = 0.562$, and $d = 0.496$.

RAREFACTION

Sample Size Standardization by Regression Methods

By summing the Bolivian data quadrat by quadrat for all 25 quadrats, we are able to calculate a regression based upon the entire plot. Substituting $N = 200$ individuals into the regression equation $\hat{S} = 2.144N^{0.497}$ that was estimated for plot 01 (Table 12.2), we obtain an estimated number of species of $\hat{S} = 29.84$, or about 30, for that value of N. Using the actual raw data in Table 12.1, we note that for 8 quadrats, which yielded an $N = 217$, there were $S = 32$ species observed. So now, if we use $N = 217$ in our equation, the regression yields $\hat{S} = 31.24\,\hat{S}$, which is a most reasonable estimate.

We can never expect any predictive equation to give us exact answers. However, we could compare such an estimate with the number of species observed when 200 individuals are sampled from another plot or locality. Then, for a given $N = 200$ from more than 1 habitat, we would be able to make statements concerning the comparative number of species, or comparative biodiversity.

Sanders's Method of Rarefaction

Unfortunately, for many of the observations made in the field, a sample-by-sample breakdown such as that shown in Table 12.1 may not be possible. The only information available may be the total number of individuals and the total number of species. This would be the case for a single inventory, for example.

For the Bolivian plot we found values for $N = 663$ and for $S = 52$. No regression is possible because, of course, we have only 2 values, 1 for each variable. With only this limited amount of information, then, how can we estimate the number of species at $N = 200$, say? Intuition, field experience (Table 12.1), and knowledge of the allocation of individuals among species discussed in Chapter 9 all indicate that had fewer individuals been collected, fewer species would have been recorded. Because so many of the species are represented by only a very few individuals, it is most likely that some of the rarer species would not be represented in a sample containing fewer individuals. Obviously, in any biological sample, a species must be represented by at least 1 individual in order to make the species list. Consequently, if the sample contains 200 individuals, the minimum species proportion that is possible is 1 individual out of the total of 200, or $\dfrac{1}{200} = 0.005$, which is 0.5%.

Using this line of thinking, Sanders (1968) developed a technique known as *rarefaction*, which is a *sample reduction method*. In Table 12.4, the 52 Bolivian species are listed in rank order showing the number of individuals of each, their species proportions (in percentages), and their cumulative percentages. Let's say we wish to reduce this group of 663 individuals to 200. The reasoning Sanders used to make this reduction goes as follows:

All the species with species proportions above $\dfrac{1}{200}$, or greater than 0.50%, would be included in the reduced sample. There are 17 of these non-singleton species with

TABLE 12.4 Data Needed for Sanders's Method of Computation for the Expected Number of Species, $E(S_r)$, in a Sample Size of r Individuals Chosen from n

Species Number	Number of Individuals	Cumulative Individuals	Percent	Cumulative Percent
1	252	252	38.01	38.01
2	131	383	19.76	57.77
3	62	445	9.35	67.12
4	25	470	3.77	70.89
5	19	489	2.86	73.76
6	17	506	2.56	76.32
7	17	523	2.56	78.88
8	17	540	2.56	81.45
9	17	557	2.56	84.01
10	9	566	1.36	85.37
11	9	575	1.36	86.73
12	8	583	1.21	87.93
13	6	589	0.90	88.84
14	6	595	0.90	89.74
15	4	599	0.60	90.35
16	4	603	0.60	90.95
17	4	607	0.60	91.55
18	3	610	0.45	92.00
19	3	613	0.45	92.46
20	3	616	0.45	92.91
21	3	619	0.45	93.36
22	3	622	0.45	93.82
23	3	625	0.45	94.27
24	2	627	0.30	94.57
25	2	629	0.30	94.87
.	.	.	.	.
32	2	643	0.30	96.98
33	1	644	0.15	97.13
34	1	645	0.15	97.28
.	.	.	.	.
52	1	663	0.15	99.99

SOURCE: Data from Bolivian plot 01.

$p > 0.50\%$ (Table 12.4). For our Bolivian data, when we look in the table at the level of 17 cumulative species, the corresponding cumulative percentage is 91.55%. That is, these 17 non-singleton species constitute 91.55% of the total individuals for this data set. The remaining $(52 - 17 = 35)$ rare species must be distributed in the last $(100 - 91.55)\% = 8.45\%$ of the observed frequency distribution. Therefore, the proportion of species that have at least 1 individual is $\dfrac{8.45}{0.50} = 16.9$ species. We have, then, the 17 observed species that made the original cut (in the top 91.55% of the data set), plus the number of rare species estimated to be in the remaining 8.45%. This yields $17 + 16.9 = 33.9$ species. Thus, using Sanders's methodology for rarefaction, we estimate that a sample size of 200 individuals recorded from the Bolivian plot of 663 individuals will produce about 34 species. How good is this estimate?

As a check, although we do not know for exactly $N = 200$, Table 12.1 shows that for $N = 217$, we observed 32 species. So, for this example, we see that Sanders's rarefaction is a reasonable estimation method.

Before we discuss the precision of the estimate, we shall compute an additional example. For this example, the first 19 quadrats in the Bolivian plot (of 25 total quadrats) contained 497 individuals (Table 12.1). Therefore, we will illustrate the procedure again using a target sample size of $N = 497$. For a sample size of 497, 1 individual would make up $\dfrac{1}{497} = 0.002 = 0.2\%$ of the data set. Table 12.4 shows that 32 species have a percentage larger than 0.2%, and that the cumulative percentage for 32 observed species is 96.98. Therefore, we will have $(100 - 96.98)\% = 3.02\%$ left for the remaining species. The division problem is the fraction $\dfrac{3.02}{0.20} = 15.10$. Thus we have $32 + 15.10 = 47.10$, or 47 species expected to be found in a plot similar to the Bolivian plot 01, if the observed number of individuals in the comparison was only 497 (and not the 663 we actually counted). Table 12.1 indicates that we actually observed 47, so, again, our estimate looks great.

The solution offered by Sanders to the problem of comparing areas of unequal richness, or of how many species would be present at a lower value of n (number of individuals), is ingenious, easily understood, and simple to calculate. Most important, Sanders's rarefaction methodology gives a fairly close, but inconsistent (sometimes bigger, sometimes smaller than actual), estimate. Probability theory, however, offers an even more elegant and more exact approach to the same problem.

The Hypergeometric Distribution for Rarefaction

Sanders's procedure actually can be better described by a statistical distribution called the Hypergeometric distribution. Before we present and explain the formula for this distribution, we shall provide an example to show why the Hypergeometric distribution is an alternative for rarefaction.

Suppose we have a sample of n individuals, of which n_1 belong to the bird species *Limosa fedoa* (marbled godwit) and $n_2 = n - n_1$ do not belong to species *L. fedoa*. Let us substitute some number to make this clear. Let us say we had a sample of 10 individuals,

4 of species *L. fedoa* and the rest of the 10 belonging to 3 more species in the same genus, *Limosa*, but not to the species *fedoa*. Then, $n = 10$, $n_1 = 4$, and $(n - n_1) = n_2 = 6$. We now choose a certain number of individuals (say, r) at random from the available n. What we require is the probability, denoted by q_k, that within the r number of chosen individuals exactly k are members of species *L. fedoa*. Following our example, we pick, say, 3 individuals from our collection of 10, and ask what the chance is that we get exactly 2 of *L. fedoa*. The value of k chosen must be an integer between 0 and n_1 (the total number of individuals of *L. fedoa* in the total sample of n individuals, where $0 \leq k \leq n$) or r (the number of individuals that we have selected to identify to species, where $0 \leq r \leq n_1$), depending on which is larger ($0 \leq k \leq r \leq n_1$).

It can be shown for our example that the probability q_k

$$q_k = \frac{{}_{n_1}C_k \cdot {}_{(n-n1)}C_{(n-k)}}{{}_nC_r} = \frac{\binom{n_1}{k}\binom{n - n_1}{n - k}}{\binom{n}{r}}. \tag{12.4}$$

Reading the right side of Equation 12.4 is really quite simple. The formula directs us to count the number of ways we could pick k individuals from the n_1 *L. fedoa* available, and then to multiply this by the number of ways for picking the remaining $(r - k)$ from the other $(n - n_1)$ species in the sample. We put the result of this multiplication in the numerator and divide (to achieve a number between 0 and 1) by the number of ways to get a total subsample of size r from the original sample of size n. Then, q_k is the answer, or the probability that we will obtain exactly k individuals of *L. fedoa* when we have selected a total of r individuals from the total n that are available to us.

The entire set of probabilities obtained from the above equation (one for each value of k) is called the Hypergeometric distribution. This combinatorial notation was defined in Equation 3.3. Recall that the name is apt because we look at the number of combinations; that is, not ordered, but unordered, selections.

To deal with the problem of rarefaction by using this distribution, we must find a formula for counting the expected number of individuals of the given species, *L. fedoa*, for different sample sizes. To begin, it is easiest to find the probability that no individuals of species *L. fedoa* were chosen in our sample of size r. That is, we shall calculate the value of q_0 for our example. Using Equation 12.4, with $k = 0$, we obtain

$$q_0 = \frac{\binom{n - n_1}{r}}{\binom{n}{r}} = \frac{{}_{(n-n1)}C_{(r)}}{{}_nC_r}. \tag{12.5}$$

Recall that these are equivalent ways to write each term. That is why there are 2 sets of formulas for q_0 in Equation 12.5. Using either notation, recall that when the value for r is 0, we define $_nC_r = {_nC_0} = 1$, regardless of the value we have for n. Now, q_0 is the probability that we did not get any individuals of *L. fedoa* when we separated out r individuals from our total sample of size n. Recall that for any probability distribution, the sum of all the individual probabilities must add to 1. Therefore, if q_0 can be taken as the probability of getting none, the quantity $(1 - q_0)$ must be the probability of obtaining at least 1 individual of the species *L. fedoa*. Now, if we calculate $(1 - q_0)$ for each of the $i = 1, 2, \ldots, s$ species in the sampled area, and sum these numbers, we will know the probability of obtaining, in our sample, at least 1 individual of each species. In turn, this will give us the expected number of species, $E(S_r)$, in the chosen group of r individuals. That is, we are adding up all those species for which we identified at least 1 specimen. Putting all this together, the formula that we need to describe the rarefaction situation is

$$E(S_r) = \Sigma_{i=1}^{s}\left[1 - \frac{_{(n-n_1)}C_{(r)}}{_nC_r}\right]. \tag{12.6}$$

What have we accomplished by deriving Equation 12.6? We can now take any field sample of size n and compare it with another collected sample of smaller size r. The comparison will be done on the basis of number of species observed in the smaller sample and the number estimated to be in an equivalent-size portion of the larger sample.

Although straightforward, this formulation (suggested in Hurlbert 1971) can, in practice, contain some large factorials. Recall from Chapter 3 that

$$_nC_r = \frac{n!}{r!(n-r)!}.$$

For example, our Bolivian case study contains a term ($n!$) that is 663!. This is a very large number, and in any practical application we would certainly encounter such terms. The size of such factorials can be troublesome, even for many computers. A simple solution to the problem of burdensome or impossible calculations is to factor; that is, reduce the combinatorial part of Equation 12.6. After suitable algebraic manipulation, an exact formula avoiding factorials for the combinatorial part of Equation 12.6 is

$$C_i = \frac{(n-r)(n-r-1)(n-r-2)\cdots(n-r-n_i+1)}{n(n-1)(n-2)\cdots(n-n_i+1)}. \tag{12.7}$$

We need the subscript i in the symbol C_i on the right side because we will need a separate calculation for each of the i species of interest. We can now rewrite Equation 12.6 as

$$E(S_r) = \Sigma_{i=1}^{s}(1 - C_i) \tag{12.8}$$

where $E(S_r)$ is the estimated number of species likely to be drawn in a sample of size r. Using Equation 12.8 greatly simplifies the calculation of the number of species expected to be found at a sample size smaller than that of the original sample (n).

While Equation 12.8 presents us with a calculation requiring less multiplication and smaller numbers to multiply, it is still a considerable computation, especially if done on a calculator. We will illustrate by calculating one of the 52 terms required in the summation of Equation 12.8 for our Bolivian example. Species number $i = 16$ (*Triplaris americana*) has $n_{i=16} = 4$ individuals. We will use $n = 663$, and we choose, for our example, $r = 301$ (11 quadrats; Table 12.1). The first term of the numerator $(n - r)$ for C_{16} is $(663 - 301) = 362$, and the last term $(n - r - n_{16} + 1)$ is $(663 - 301 - 4 + 1) = 359$. Then, the numerator consists of the factors $362 \cdot 361 \cdot 360 \cdot 359$. The first term of the denominator of C_{16} is $n = 663$, and the last term or factor is $(n - n_{16} + 1)$, or $(663 - 4 + 1) = 660$. The denominator, then, consists of the factors of $663 \cdot 662 \cdot 661 \cdot 660$. Note that in both the numerator and the denominator the number of terms is always equal to n_i ($n_{i=16} = 4$ for this example). Consequently, for the more abundant species, for which the number of individuals n_i is large, the numerator and denominator are correspondingly large. Multiplying out the numbers given above and dividing, we obtain $C_{16} = 0.0882$. The $i = 16^{\text{th}}$ term required is then $(1 - C_{16}) = (1 - 0.0882) = 0.9118$. Summing all of the $S = 52$ terms we would need to calculate for this reduction or rarefaction problem on the entirety of forest plot 01 gives us the expected number of species, $E(S_r)$, when r individuals are chosen at random from n.

A complication arises if $(n - n_i) < r$. For example, in the Bolivian plot 01, if we wanted to rarefy by choosing r to be 497, then for species $i = 1$ (*Scheelea princeps*), $(n - n_i) = (663 - 252) = 411$. In this case both Equations 12.6 and 12.7 present us with a problem. We cannot choose $r = 497$ individuals from 411 individuals in Equation 12.6, and the last term of the numerator in Equation 12.7 becomes negative. In this case, the solution is simply to award this term a value of 1. This merely says that the probability is equal to 1 because this most abundant species would show up in the sample of reduced size anyway.

The Binomial Distribution for Rarefaction

A simple alternative to the application of the Hypergeometric distribution is possible by considering a limiting distribution for the Hypergeometric distribution (Feller 1957). In mathematical statistics we state this as a *limit theorem*. When the sample size n is large, we can let the observed species proportion, n/n, equal the estimate of the proportion, p, for the Binomial distribution (Equation 8.3). Then, $p + (1 - p) = 1$, or $p + q = 1$, and a *Binomial approximation for the Hypergeometric probability q_k* (Equation 12.4) becomes appropriate for our natural population situation. Thus the probability of obtaining exactly k individuals of species i, Equation 12.5, can be rewritten in the familiar form of the Binomial distribution

$$q_k = {}_nC_r p^k q^{r-k}. \tag{12.9}$$

Because we are interested in q_0 for purposes of rarefying, as already explained, $k = 0$. Equation 12.9 reduces to

$$q_0 = {}_rC_0 p^0 q^{r-0} = 1 \cdot p^0 \cdot q^r = q^r. \tag{12.10}$$

In mathematical statistics we never calculate the quantity p^0. By definition, any number raised to the 0 power equals the quantity 1; that is, we define $p^0 = 1$, just as we define $_rC_0 = 1$. Then, the probability of finding at least 1 individual of species i is given by the formula $(1 - q^r)$. Equations 12.6 and 12.8 can now be expressed as

$$E(S_r) = \sum_{i=1}^{S}(1 - q_i^r).$$
(12.11)

Equation 12.11 is probably the easiest way to estimate the expected number of species from the species to be found in a sample of r individuals chosen at random from n individuals.

We again illustrate a single term of the required 52 terms for species $i = 16$ (*Triplaris americana*). There are 4 individuals of this species in the total population. So here $p_{i=16} = \dfrac{4}{663} = 0.0060$ (Table 12.4). Then, $q_{i=16} = 1 - 0.0060 = 0.9940$. If we choose $r = 301$ as before, then $q_{i=16}^r = 0.9940^{301} = 0.1634$. The $i = 16^{th}$ term in Equation 12.11 then becomes $(1 - 0.1634) = 0.8366$. This term is somewhat less than the 0.9118 calculated using the exact method in Equation 12.8. In general, we would expect that the $E(S_r)$ calculated by the Binomial would underestimate the exact value.

Equivalent Alpha Diversity for Rarefaction

In Chapter 9 we discussed a variety of probability distributions that have been fit to important biological relationships among species. One of the most viable of these distributions is Fisher's Log series. For a sample size of n from a Log series, we find from Equations 9.14 and 9.16, plus a little algebra, that the number of species represented by those n individuals is $S = \alpha ln\left(1 + \dfrac{n}{\alpha}\right)$. This formula can be used to rarefy, if we are willing to think that our sample results could indeed be described by a Log series distribution. As we saw, this is at best a mild restriction for any moderately large number of individuals. The Log series method will always work best if our data are exactly described by a Log series; otherwise, the rarefied estimate for S will be a slight overestimate.

To use Fisher's Log series for purposes of rarefaction, we can write

$$\hat{S}_1 = \alpha_2 ln\left(1 + \frac{n_1}{\alpha_2}\right).$$
(12.12)

In this *rarefaction equation* the subscripts 1 and 2 refer to the 2 samples being compared. The symbol α_2 represents the parameter of the Log series for N_2 and S_2. To illustrate, we will redo the second example we worked through earlier for Sanders's method, in which we compared the entire Bolivian plot, which had $N_2 = 663$ and $S_2 = 52$, with data from the first 19 quadrats of a total of 25 quadrats. For those 19 quadrats, $n_1 = 497$. We now ask how many species would be expected in the smaller sample if we wanted equivalent diversity. The calculations would be $\hat{S}_1 = 13.213 \cdot ln\left(1 + \dfrac{497}{13.213}\right) = 48.275$.

This resulting estimate of the number of species in the smaller sample can be evaluated by a look at Table 12.1. Here we see that we actually observed 47 species in a sample of that size.

Comparison of Rarefaction Methods

To evaluate the various methods for estimating numbers of species in populations smaller than that actually sampled, the results for the Bolivian plot 01 are shown in Table 12.5. The rarefaction method published by Sanders (1968) sometimes overestimates S, and sometimes underestimates S. The binomial rarefaction method is likely to underestimate S. The rarefied number of species, assuming equivalent diversity, will overestimate S to the extent that the Log series is not an exact fit. This inexactitude of fit of the Log series will be minimal for moderately large samples. The power curve based on all the quadrats and the Hypergeometric method both do very well in estimating S. In many instances, of course, the available data are not appropriate or extensive enough to allow a regression like the power curve to be calculated. Although the Sanders methodology, the Log series, and the Binomial do not give as accurate results as the Hypergeometric, their simplicity and their consistently small errors of prediction are appealing.

ABUNDIFACTION

Although statisticians warn about extending regression lines beyond the observations, it is, nevertheless, tempting to try to predict the number of species expected in an area that is larger than our actual target area (for an example, see Erwin 1982). As we expand the area of interest, we may, of course, also add supplementary and possibly dissimilar

TABLE 12.5 Comparison of Actual Numbers of Species, Using Five Different Rarefaction Methods

Quadrats	N	S	Sanders	$\hat{S}$ Hyper- geometric	$\hat{S}$ Binomial	$\hat{S}$ Equivalent α	$\hat{S} = 2.144 ln^{0.497}$
25	663	52	—	—	—	—	—
19	497	47	47.10	46.33	41.91	48.28	47.16
15	392	41	43.83	41.82	38.90	45.23	41.91
11	301	37	40.25	37.10	35.40	41.87	36.75
8	217	32	35.34	31.76	31.06	37.76	31.24

NOTE: Observed and predicted values of the number of species for values of N.

SOURCE: Data from Bolivian tree plot 01.

habitats, which compounds the problem by violating an assumption basic to many probabilistic methods that would have been of possible use. We note also that the primary variable is the number of individuals, while area is secondary. If we use area as the variable, instead of individuals, the presumption is that the individuals are distributed throughout it. Thus, even within a single habitat, the procedure is tenuous.

Regression and Power Curves for Extending Species–Area Predictions

We will now try three approaches for the analysis of observed species accumulation data, with the intent of finding an efficient method for predicting the number of species in an area larger than that of our observations. The three methods are the use of (1) the entire observed data set; (2) the observed data for only the initial area in which rapid accumulation of species occurs; and (3) the data from the end of the fieldwork in the observational area in which there is a fairly uniform, incremental addition of species data. We have used both semilogarithmic and power curves, and will now use these data sets to calculate three regressions for each.

To illustrate, we calculate semilogarithmic and power curve regressions for the area of only the first 12 of 25 quadrats in our Bolivian forest plot. This sample size constitutes approximately half of the total target area. If data for these 12 quadrats were the only observations available, how well could we predict the total number of species we would actually observe in the total plot of 25 quadrats?

Each of the 25 quadrats in the hectare is of dimension 20×20 m^2; therefore, the areas for these quadrats are the sequence 400 m^2, 800 m^2, . . . , 4,800 m^2. From Table 12.1 we note that 39 species were observed within this area, with a total cumulative number of individuals of 325. The semilogarithmic curve for these data yields a species-richness estimate of $\hat{S} = -62.546 + 11.727 \cdot ln(A)$, with an associated r^2 of 0.944 from the linear log equation (Equation 12.1). The power curve yields a species-richness estimate of $\hat{S} = 0.423A^{0.535}$ (Equation 12.2), with $r^2 = 0.952$. For most natural populations we will find such highly correlated situations with each of the fitting methods. For 10,000 m^2; that is, the entire plot, recall that $S = 52$ species. The semilog curve based on the sample of areas (400 m^2, 800 m^2, . . . , 4,800 m^2) predicts $S \sim 46$ for the entire hectare. The power curve estimates $\hat{S} \approx 58$ (Tables 12.6 and 12.7). These results compare with $\hat{S} = 50$ and $\hat{S} = 54$, respectively, which were our predictions based on the entire data set of areas over all 25 plots. These differences among the 4 predicted values for S appear to be relatively small. However, if we wish to project to a much larger area, the minor discrepancies become magnified greatly, as we shall demonstrate.

In Figures 12.1 and 12.2, and Table 12.1 (the S Added column), notice that after quadrat 12, where cumulative $N = 325$, the species are added by 1's and 2's. Perhaps a better strategy to predict the number of species expected in an area considerably larger than our observed hectare would be to calculate regressions for the end of the field periods; that is, the last part of the data set for which species are added in this orderly fashion. Let's try this approach. For the remaining 13 quadrats, the summed areas are 5,200 m^2, 5,600 m^2, . . . , 10,000 m^2. The logarithmic curve can then be computed just for this data. We obtained $\hat{S} = -135.400 + 20.3984 \cdot ln(A)$, with $r^2 = 0.974$.

TABLE 12.6 Three Sets of Species–Area Predictions, Using Semilog Regressions

A	S	400, . . . , 10,000 m² $\hat{S}$, ALL	400, . . . , 4,800 m² $\hat{S}$, BOTTOM	5,200, . . . , 10,000 m² $\hat{S}$, TOP
1,600	25	24	24	15
3,200	32	34	32	29
4,800	39	40	37	38
6,400	43	44	40	43
8,000	48	47	43	48
9,600	52	49	45	52
10,000	52	50	46	52
20,000	80	60	54	67
30,000	?	65	58	75
40,000	?	69	62	81
100,000	?	82	72	99

NOTE: The number of species observed (S) and estimated ($\hat{S}$) by Logarithmic curves for values of A. For the semilog equation $\hat{S} = a + b\ln(A)$, at 400, . . . , 10,000 m², $a = -79.924$, $b = 14.096$; for 400, . . . , 4,800 m², $a = -62.546$, $b = 11.727$; for 5,200, . . . , 10,000 m², $a = -135.4000$, and $b = 20.398$.

TABLE 12.7 Three Sets of Species–Area Predictions, Using Power Curves

QUADRATS	A	S	400, . . . , 10,000 m² $\hat{S}$, ALL	400, . . . , 4,800 m² $\hat{S}$, BOTTOM	5,200, . . . , 10,000 m² $\hat{S}$, TOP
4	1,600	25	22	22	23
8	3,200	32	31	32	32
12	4,800	39	38	39	38
16	6,400	43	43	46	43
20	8,000	48	48	52	48
24	9,600	52	53	57	52
25	10,000	52	54	58	53
50	20,000	80	76	85	72
75	30,000	?	93	105	86
100	40,000	?	108	123	98
1000	100,000	?	170	200	147

NOTE: The number of species observed (S) and estimated ($\hat{S}$) by power curves for values of A. For the log–log equation $\hat{S} = cA^d$ at 400, . . . , 10,000 m², $c = 0.562$, $d = 0.496$; for 400, . . . , 4,800 m², $c = 0.423$, $d = 0.535$; for 5,200, . . . , 10,000 m², $c = 0.865$, and $d = 0.446$.

The power curve yields $\hat{S} = 0.865A^{0.446}$, with $r^2 = 0.971$. At 10,000 m^2 for the semilogarithmic curve, we predict $\hat{S} \approx 52$, and for the power curve the prediction is $\hat{S} \approx 53$. As we might expect, the regressions for the top half (5,200 m^2, 5,600 m^2, . . . , 10,000 m^2) of the observed data set fit that portion (the top half) of the curve the best. In turn, the regressions for the bottom half (400 m^2, 800 m^2, . . . , 4,800 m^2) fit that section best. The fit using all 25 quadrats on the entire plot is best in the middle and throughout the observed range (Tables 12.6 and 12.7).

Table 12.6 shows the number of species expected, $\hat{S}$, when using each of the 3 exemplar data sets for computing semilogarithmic curves, each of which has data that cover 1,600 m^2. This table also includes our projections beyond those for the 1 observed plot. The third column, $\hat{S}$ all, based on all the data, gives a good overall fit, especially in the middle range, but it has the largest deviations at upper extremes of 9,600 m^2 and 10,000 m^2. The regression in the fourth column, based on the bottom (400 m^2, 800 m^2, . . . , 4,800 m^2) segment, fits well only in that beginning segment of the range. The last regression (fifth column), which is based on the final 5,200 m^2, 5,600 m^2, . . . , 10,000 m^2 segment, fits best only at that end. Each of these results is an expected consequence of the least squares criterion that we are using.

Table 12.7 shows the number of species expected for the 3 power curves when these predictions are also based upon intervals of 1,600 m^2. Again, the overall curve, based upon the totality of the observed data, fits best in the middle, near the observed mean value. The curve based on the initial (bottom) segment (400 m^2, 800 m^2, . . . , 4,800 m^2) fits best at the low end but is generally the overall poorest fit. The curve based on the top segment (5,200 m^2, 5,600 m^2, . . . , 10,000 m^2) fits best at the high end and is best overall, since we are here interested in predictions at this upper end. The best fit of the top segment is simply a consequence of using the least squares criterion, but it is not always the obvious conclusion, nor are the consequences of these choices clear for the ecologist.

In addition, Tables 12.6 and 12.7 show upward projections to 2, 3, 4, and 10 hectares, based upon our original observations in the full hectare, the first half of the hectare, and the second half of the hectare. As can be seen, the last row of values in these tables are projections to an area only 10 times that of our observed plot and yet the estimated value of S varies substantially, depending on the choice of the observed data set for the regression. As stated earlier in this chapter, predictions beyond the data, from which the regression was calculated, is risky business. The semilogarithmic and the power curve for the upper segment (5,200 m^2, 5,600 m^2, . . . , 10,000 m^2) fit the observed number of species very well at this high end of the areal values, but when an abundifaction to 10 hectares is undertaken, the semilog (based on the final segment of observed data) predicts 99 species. The power curve for the same observed data set predicts 147 species. Among the power curves, the estimate of the "slope" coefficient for the equation based upon the entire 25 quadrats is 0.496 and for the final half of the quadrats we calculated a value of 0.446. Figure 12.5 shows how this small difference (0.496 versus 0.446) magnifies the discrepancies between observed and expected values rapidly as the regression lines are extended on a log scale. The discrepancy grows so fast that at 10 hectares 1 power curve predicts 170 species and the other power curve predicts 147 species.

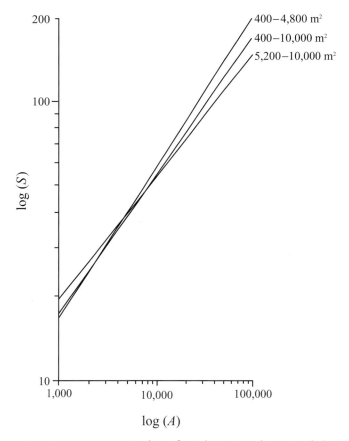

FIGURE 12.5   Species–area regression lines for Bolivian tree data extended to 100,000 m².

We have just shown that our differing methods provide for discrepant answers. Is any one of these procedures usable for predicting to a similar plot or target area? Researchers in natural populations expect, intuitively, that regressions based on the final samples, in which species are added in 1's and 2's, would best predict (have the smallest discrepancies between observed and expected numbers) the number of species found in a larger area with the same habitat. We have shown that the predicted and observed values are close when using this method. Now we will use a second Bolivian data set to check how well a prediction would work for a similar plot, beyond the original hectare plot 01.

Fortunately for us, Dallmeier and colleagues (1991) surveyed a second plot of 10,000 m² in the same Beni Reserve. The second plot is somewhat different, with slightly better drainage and a change in rank order of the abundant species. Nevertheless, many of the more abundant taxa are present in both areas.

To examine the problem of prediction to a similar but larger area, we shall use the regressions based on plot 01 to predict the additional number of species found in plot 02. The total area would then be 20,000 m², or 2 hectares. The predictions are presented in the 20,000 m² rows of Tables 12.6 (using the semilog methods of abundifaction) and Table 12.7 (using the power curve methods). In examining the species list from plot 02, we found 28 species not listed in plot 01. The total number of species found in both plots is then $52 + 28 = 80$. Table 12.6 shows that for the 20,000 m² row, all three estimates based upon semilog and log–log equations gave predictions that were considerably below the mark. The estimate (67), based upon the last half of the area sampled (5,200 m², . . . , 10,000 m²), was the closest to the value of 80 known species using the semilog regression (Table 12.6). For the power curve estimates (Table 12.7), two predictions were underestimates and one was an overestimate. The closest estimate using the power curve method was that based upon the entire data set from plot 01. Only slightly less accurate was the estimate based on the lower segment (400 m², . . . , 4,800 m²). The intuitive choice, that calculated from the upper segment of data, did more poorly.

If the amount of effort required is part of the choice for the extrapolated prediction to 20,000 m², we would choose the species predictions based upon the 400 m², . . . , 4,800 m² set because only 325 individuals need to be counted and identified to obtain it, and the result is almost as good as that obtained for the entire data set. However, as Figure 12.5 shows, the regressions diverge more and more as we attempt to predict for larger and larger areas. Thus the only safe basis for predicting numbers of species presumed to occur in similar habitats or larger areas of the same habitat is the use of the entire, biologically representative set of observations; that is, the sample from the entire target area. Even so, there is no method yet developed that will provide reasonable and trustworthy estimates when the projections are well beyond the range of the observed data.

Extending Species Predictions with a Ratio of Log Relationships

We discussed the fact that S plotted against $ln(N)$ will yield an approximately straight line, whether N represents the number of individuals, quadrats, biological samples, or the target population area itself. Since this is true, it tells us that the ratio of the sample values, $\dfrac{S}{ln(n)}$, is approximately linear over the area.

If we are interested in the number of species expected in the entire target area consisting of N individuals or samples, based upon a sample of n (where $n < N$), we could try to estimate merely by multiplying. To obtain an estimate (using any base for the logarithms) of the expected number of species $E(S)$ we could use

$$E(S) = \frac{S}{\log(n)} \cdot \log(N).$$

To be able to consider relatively small, but still biologically representative, numbers of random or systematically collected samples n, we could adjust this formula and use alternatively,

$$E(S) = \frac{S}{\log(n - 1)} \cdot \log(N + 1). \tag{12.13}$$

Using our Bolivian example, let us say we had taken a random sample of 12 quadrats, or approximately half the entire plot. Using Equation 12.13 and the data in Table 12.7, we predict that on the entire 25 quadrats we would find $E(S) = \dfrac{39}{\log(11)} \cdot \log(26) = 52.99$. This is a rather good estimate, considering the true value is 52 species. Let us try this calculation again, but with n and N representing individuals instead of quadrats. From the data in Table 12.1, we would get for these 12 quadrats $E(S) = \dfrac{39}{\log(324)} \cdot \log(664) = 43.84$.

For individuals instead of quadrats the species estimate is not so good. We might suspect that the reason for this discrepancy (44 instead of the actual 52 species) is the spatial distribution. When the value of N was the number of quadrats, the homogeneity of the spatial distribution provided for a constancy in this ratio method, because each of the additional quadrats not in the sample (the last 13) contained a similar average number of species to those quadrat counts that were observed. However, when the graph of the cumulative relationship of the individuals and species is examined, the first part of the curve is a period of rapid accumulation, whereas the unobserved part is picking up many singleton species. Comparing the two parts of the curve, we observe that a larger number of individuals must be collected in the last part to increase the cumulative tally of species just by one. That is, the part observed, or the sample, is not like the part that is not observed. However, for this plot, the homogeneity of the spatial distribution aids our prediction, whereas the accumulation of samples does not. A total regression, as in Table 12.6, works because it is over the total plot (it minimizes the error over the plot), but this ratio (Equation 12.13) uses just the final total size, not a range of sizes, for the basis of the prediction.

Now, let us try Equation 12.13 for a comparison procedure with the species–area relationship. That is, for the first example, let us say that 2 sets of samples are available and that we want to compare the habitats. After we have examined 1 habitat of 1 hectare area and obtained $S_1 = 52$ species from $N_1 = 25$ quadrats or samples, we want to ask how many more species will be found in a second habitat from a similar area nearby. We can adapt Equation 12.13 to accommodate $N_2 = 25 + 25 = 50$ quadrats,

$$\hat{S}_2 = \frac{S_1}{\log(N_1 - 1)} \cdot \log(N_2 + 1). \tag{12.14}$$

Plugging in the values, we obtain an estimate of the number of species in both hectares combined (plots 01 and 02 together): $\hat{S}_2 = \dfrac{52}{\log(24)} \cdot \log(51) = 64.33$. As can be seen from the tables, this estimate based on the ratio of logs approach for abundifaction is much lower than those predicted from the semilog and log–log power curve regressions, and it is quite deviant from the true number of 80 species in the 2 plots combined. By merely selecting the last accumulated value of 52 species from the entirety of plot 01, and projecting this ratio to the combined plot 01 plus plot 02, we are in effect ignoring the slope of the line from the regression. Thus our prediction is poor because our sampled data set is not actually linear, and because the combined plot is merely similar to, but not the same as, that from the first plot. These are the cautions that must be thoroughly understood when using a single ratio as an approach to estimate species number for an area larger than what we have sampled.

Clearly, these examples show that Equation 12.13 is not a reliable formula to use for abundifaction purposes when comparing sampled data from natural population work. Under certain very stringent conditions of application, it can work; but in general, it does not. Therefore, we need to continue our search for a trustworthy method of abundifaction.

Equivalent Alpha Diversity for Abundifaction

Rather than trying to extrapolate the number of species from a single small area to a much larger one, a better plan would be to spread samples out into all of the areas under consideration. This is the *equivalent alpha diversity method for abundifaction*, meaning we use the parameter of the Log series from a first locality or site to *abundify* and then compare to a second locality.

We can begin with the formula that employs equivalent alpha diversity for rarefaction (Equation 12.12) and adapt it to abundify a sample for comparative purposes. If our first sample contains S_1 species in N_1 individuals, units, biological samples (quadrats), or occurrences, and if we wish to compare this with a larger sample of size N_2 to find how many species the larger sample would contain, we could call this number of species S_2. We would then substitute into Equation 12.12 to obtain the *equivalent alpha diversity abundifaction equation*,

$$\hat{S}_2 = \alpha_1 ln\left(1 + \frac{N_2}{\alpha_1}\right) \tag{12.15}$$

where α_1 represents the parameter α determined for N_1 and S_1. To facilitate such comparisons we have calculated selected values of α in Appendix 4. This appendix gives values of the Log series parameter α for selected incremental values of N and S. For example, let us imagine that we wish to compare 2 habitats from which samples of $N_1 = 100$ and $N_2 = 1,100$ have been collected. In the first sample, 30 species were found. If we believed that the 2 areas were of equivalent diversity, how many species would we expect to find in the second population of 1,100 individuals? We need to abundify sample 1, so we calculate $\hat{S}_2 = 14.531 \cdot ln\left(1 + \frac{1100}{14.531}\right)$, which gives us the estimated number of species, $\hat{S}_2 = 63$. That is, by stating that we are interested in equivalent diversity, we use $\alpha_1 = 14.531$ (our value for $N_1 = 100$, $S_1 = 30$) as the diversity level for both localities. From Appendix 4, when $N = 1,100$ and $S = 63$, we find that $\alpha = 14.512$. If we had used just the rounded value of $\alpha = 15$, we would have obtained 64 species; certainly the difference of a single species in a large area is not a problem. Thus, if we had sampled 1,100 individuals from our first locality, we would expect to find about 64 species.

This equivalent alpha diversity method of abundifaction will work best when the data set conforms exactly to a Log series. When the observed data set departs from a

Log series, the abundified S_2 value will be an underestimate. Consider, once again, Bolivian plots 01 and 02. The value of $N = 663$ and $S = 52$ of plot 01 yields an $\alpha = 13.21$. In plot 02, $N = 664$ so that $N_2 = 1{,}327$. Using Equation 12.15, we obtain $\hat{S}_2 = 61.02$. This is considerably below the known value of 80 and is similar to the estimates obtained from the semilog regressions (Table 12.6). The low estimate is what we would expect if plot 02 is not distributed as the same Log series (Chapter 13).

Let us consider the α table in Appendix 4. It has always been noted in the literature on the Logarithmic series distribution that α is independent of sample size. However, a look at the values in this appendix would not confirm this, since the values for α increase as we read larger values of N and S. What is meant by this use of the word *independent* is that the formula for α does not have the total sample size written in it. That is, α is not a function of total n (it is not *functionally related to n*). This statement is often mistakenly interpreted to mean that α does not change with sample size. We can force an example to be without change, or we can have a data set that is an exact Log series, but in field practice it is usually not possible to see this. This point is discussed more thoroughly from another perspective in the next chapters.

The literature of most taxa is filled with examples of the use of α as a diversity index, and this use will be discussed. In Chapters 9 and 11 we show that the parameter of the Log series distribution is estimated by calculations from actual sets of biological data. What is not available is combination information on the fit of the distribution for diversity studies. It has been noted that α can be calculated and used as a diversity measure without checking the fit to the observed data. In the following chapters we will show that this is true, but only within certain limits. Consider that α is defined to be the upper limit that the first term (the number of species with 1 individual, or 1 occurrence [singletons]) can reach; that is, n_1. Because this is true, it should be clear that this meaning collapses for biological applications after α reaches a value equal to that of N. That is, if all N individuals were each in a single species, the value of n_1 would be N. This would be the limit, albeit unrealistic. Thus, $\alpha \leq N$ must be the limiting situation for natural population work. From Appendix 4 we can see that when α reaches the value of N, regardless of what this number is, x has a value of 0.50. Notice that it is always true that $\dfrac{N}{S} = 1.44$ when $\alpha = N$; that is, when $x = 0.50$. Beyond this point of $x = 0.50$, it would make no sense to use the Log series. In addition, when x approximates 0.63, the value of $\alpha = S$. At some point between $x = 0.63$ and 0.612 (depending on the N and S values), α becomes larger than S. We have therefore used 0.612 as a cutpoint for this appendix.

Also, as discussed in Chapter 9, when x reaches or exceeds 0.90, the Log series becomes statistically appealing to fit for the data set. However, when $x \geq 0.90$, it does not mean that the observations we collect are definitely well fit by a Log series, as we discussed. There can be a myriad of possible n_i values (the numbers of species with i individuals) that add to the same total values of N and S (an example is given in Chapter 13).

All this explanation taken together means that for fitting the Log series to a set of data, we use different constraints than when we want to use the fitted value of α as a diversity index. For the latter usage, the value of x may be quite far below $x = 0.90$; but in no case should x fall below the point $x = 0.50$ (or, $\dfrac{N}{S} = 1.44$). However, as we shall explain, if data are not distributed as Log series when α is used for a diversity measure, then especially when x is very low, interpretation of α can be problematical.

Finally, intermediate values of N and S that are not available in Appendix 4 may be obtained by linear interpolation. For example, assume that our fieldwork yielded 214 individuals, from which we recognized 81 species. Those values are not in Appendix 4, but we can approximate them by noting that for $N = 220$ and $S = 81$ we find $\alpha = 46.299$. Also, for $N = 210$ and $S = 81$ we obtain $\alpha = 48.318$. Using these values we can form the proportion problem of $46.299 \cdot 220 = 214a$, where a is the value of α we want to find for our sample. Solving, we obtain $\alpha = 47.597$. If we had directly calculated the value from Equation 9.16, we would have obtained $\alpha = 47.476$.

Actually, in most, if not all, natural population survey work, it is unnecessary to use exact decimal places for α. Rounding to whole numbers for α will suffice in many cases. In those cases, "guesstimation" of intermediate values for α will be the simplest procedure. For example, using the values of $N = 214$ and $S = 81$ for our field sample, and knowing that for $N = 220$ with $S = 81$ the value of $\alpha = 46.299$, while for $N = 210$ and $S = 81$ the value of $\alpha = 48.318$, there would be no problem with guessing that our sample would have a value of α of about 47. This procedure always works.

SUMMARY

1. Species diversity may mean merely species richness, S, or the number of species together with each of their abundances. Therefore, we must make the definition clear in our work.

2. When plotting the number of species obtained against the areal values used for our sampling effort, or against the individuals observed, or against any accumulated measure of effort, the resulting curves are similar in shape.

3. For most taxa, after approximately 200 to 500 individuals are collected, estimates of species proportions become reasonable, provided individuals are the primary sampling unit. This means that their standard errors become small, and all common species have been observed. Thus we will find only rare species after this cutpoint.

4. In many taxa, especially those containing microscopic organisms, it is usual to standardize the number of organisms counted in the sampling effort. However, this is a matter of preference or habit among different groups of researchers.

5. A log–log plot and a semilog plot of N versus S are equivalent for ecological purposes. Usually the choice is a matter of preference or habit among a group of researchers.

6. Some specialists agree to examine a standardized number of individuals, say 200 to 500. If this is done, comparison across habitats and of data sets collected by different researchers becomes easy. If this convention is not followed, several methods of rarefaction can be used to facilitate the comparison. All of these work to provide an estimate of a subsample of the larger sample that matches in area or numbers of individuals the size of the smaller sample.

7. For large numbers, rarefaction by use of Fisher's alpha (α) is as good or better than other methods. While computationally more difficult, Sanders's method, the

Hypergeometric, and the Binomial all yield reasonable results. Regressions can be the most accurate, but they require cumulative data.

8. It is possible to abundify, that is, to compare 2 habitats or samples at the larger of the 2 sizes by using the equivalent alpha diversity method. This method requires the assumption that both samples represent areas of equivalent alpha diversity. Under this assumption, the estimated S should be quite reasonable.

9. No statistical method yet devised will guarantee absolutely reasonable estimates of any biological quantity when predictions are beyond the range of the observed data. When cumulative data are available, regressions can be used to abundify. If only single values of N are available, Fisher's equivalent alpha diversity is the best choice.

PROBLEMS

12.1 Using the bat data from Appendix 5, calculate the power curve $\hat{S} = aN^b$. Recall from Table 12.1 that at $n = 301$, $S = 37$ trees. A colleague has stated that the bats have about the same species richness as the trees because at $n = 579$ there are $S = 38$ species. Comment.

12.2 Using the equation from Problem 12.1, rarefy to predict S for $n = 206, 344, 579, 707, 901, 1,002,$ and $1,253$.

12.3 We found for $N = 1,405$ and $S = 50$ that $\alpha = 10.1207$. Using Equation 12.12, rarefy for $n = 206, 344, 579, 707, 901, 1,002,$ and $1,253$. How do these answers compare with the answers for Problem 12.2?

12.4 Suppose that for the bat data (Appendix 5) you stopped observing at $n = 344$ and $S = 34$. For these numbers $\alpha = 9.3651$ (Appendix 4). Using Equation 12.14, abundify to $n = 579, 707, 901, 1,002,$ and $1,405$.

12.5 Using the proportions for bats given in Appendix 5, rarefy to $n = 344$ by the Binomial method.

12.6 Using the proportions in Appendix 5, use Sanders's method of rarefaction to predict S for $n = 344$.

12.7 Using the species occurrences in the bat data (Appendix 5), rarefy using Equation 12.12 to predict the number of species in 17 quadrats.

13

BIODIVERSITY:

DIVERSITY INDICES USING *N* AND *S*

I N THIS CHAPTER, we present an introduction to quantitative biodiversity. We organized this presentation in a stepwise fashion based upon the categories of data collected by the researcher. That is, we first discussed in Chapter 12 how the use of only
the total richness *S* could be used as a descriptor of the diversity of an assemblage
alone and in combination with total *N*. Indeed, species richness is at the core of biodiversity measurement.

Because the *S* that is observed depends upon how many individuals, *N*, are counted
or observed, we say that *S* is a function of *N* and write, $S = f(N)$. This statement means
that simply counting the number of species without regard to *N* is not sufficient for
quantitative analysis. Merely trying to count how many species are out there is analogous to recording the number of individuals without any reference to the amount of
space in which you find them. Thus, the next type of biodiversity measure available
incorporates both *N* and *S* for biodiversity assessment.

We shall begin by defining and discussing some of the simplest and most commonly used diversity measures that incorporate just *N* and *S*. The indices mentioned
in this chapter are those that are single measures based upon the observed data with
no reference to wider concepts or distributions.

More information often is collected in addition to the simple species count. However, samples covering a wide range of values for *N* and *S* are not always available because the investigator records only the total number of species and/or total individuals
from the field. The resultant data will consist of small values of total number of species, *S*, and individuals, *N*, primarily because of many features associated with field
work, for example, the low numbers of a rare taxon being studied or the lack of time,
resources, or other factors that limit the fieldwork.

Consequently, diversity measures, or diversity indices, were devised to aid in summarization of such data. However, an ecological community is multidimensional and
information or data on its many facets, of necessity, will be sacrificed when summarizing with a single numerical value. In fact, it is clear after reviewing the scientific
literature that users of such single values (indices or scalars) do not fully understand
that this reduction in dimension from community information to a single number
will result in inevitable inconsistencies in their results.

Since the number of indices formulated for species diversity purposes by now has
become quite extensive and seems to approach the number of workers, this area is

filled with problematic approaches. Many indices are merely variations of previously derived measures, or simply another way of looking at the same index, or algebraically equivalent to an already developed index (e.g., Peters 2004). Some indices work well for the particular data set on which the researcher was working but are not applicable or have not been tested on other data sets as they should have been and usually are not in general use. Throughout this book we have not attempted to catalog all the approaches used for sampling natural populations (for a review see, e.g., Hayek 1994). Instead, we have presented some that, in our judgment, have a sound statistical or mathematical foundation under certain conditions and, at the same time, provide us with enough of the best tools for sampling natural populations. We will follow that approach here.

Researchers use indices in most, if not all, scientific fields, each with slight variations in meaning. A fact that is not commonly known but is important to understand is that a diversity index is also a *statistic,* not merely a numerical value. That is, the diversity measure calculated from a sample can provide an estimate or approximation for the related population quantity if used correctly. In turn, the true value of the index for a community or assemblage of interest, which is always unknown, is a parameter. When used properly, the calculated diversity measure can tell us many interesting aspects and characteristics of our assemblage.

Diversity of species and communities has always been a fundamental question of concern in ecology. If the investigator counts both the species, S, and the total number of all individuals observed, N, this information can provide useful data from the sample. A measure that uses both S and N can result in a measure of S that is characteristic of an area, without the necessity of standardizing N.

Let us think about how to describe the diversity of an assemblage when the data recorded is total N and total S over some space or time. First, it is important to realize that whenever we consider 2 or more populations with distinctly different numbers of individuals within each species, there will be recognizable shortcomings when attempting to measure diversity with only total values. Any such measure devised, like the two we discuss below (d and ι), will not be sensitive to the proportional differences within the observed species categories. Consequently, the most serious shortcoming of any index or measure using only N and S is that such a measure incorrectly may treat the differing communities as if they were equivalent, despite their having ecologically meaningful differences in composition.

MARGALEF'S d

In Chapter 12 we demonstrate that a semilog plot of S against log N is most often a straight line through ecological data. This relationship was recognized by Gleason (1922) and prompted Margalef (1957) to propose the simple measure

$$d = \frac{S-1}{ln(N)} \tag{13.1}$$

in an attempt to consider species richness independently of sample size.

TABLE 13.1 Margalef Index: $d = \dfrac{(S-1)}{ln(N)}$

	SET 1				SET 2				SET 3		
N	S	d	$\dfrac{d}{d_{max}}$	N	S	d	$\dfrac{d}{d_{max}}$	N	S	d	$\dfrac{d}{d_{max}}$
217	32	5.77	0.057	217	22	3.90	0.038	2,170	32	4.04	0.005
301	37	6.31	0.062	301	27	4.56	0.045	3,010	37	4.49	0.006
392	41	6.70	0.066	392	31	5.02	0.049	3,920	41	4.83	0.006
497	47	7.41	0.073	497	37	5.80	0.057	4,970	47	5.40	0.007
663	52	7.85	0.077	663	42	6.31	0.062	6,630	52	5.80	0.008

NOTE: The maximum possible value of d is $d_{max} = \dfrac{(N-1)}{ln(N)}$. This is true because this maximum occurs when each time we observe a new individual it is a member of a new species. In this case, then, S would equal N. A standardized version of the index d is d_{max}. The first set of numbers are accumulated values from Appendix 2. The second set decreases all values of S by 10. The third set multiplies all values of N by 10.

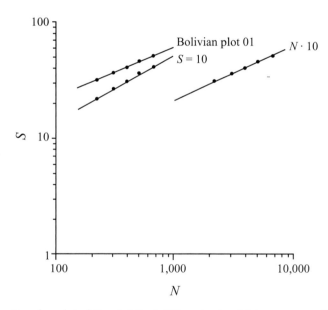

FIGURE 13.1 Log–log plot of N and S for 3 different sets of data, corresponding with data in Tables 13.1, 13.2, and 13.3.

Margalef's d is an attractive measure because it is easily calculated, as is evident from the formula in Equation 13.1. For the entire Bolivian plot 01 we have, $d = \dfrac{51}{ln(663)} = 7.85$. The random sample of 8 out of 25 quadrats from Appendix 2, as used in Table 8.8, has $d = \dfrac{30}{ln(201)} = 5.66$.

In Chapter 12 we accumulated the quadrats 1 through 25 (Appendix 2). The first 8 quadrats (Table 12.1) have $S = 32$ and $N = 217$, which gives us $d = \dfrac{31}{ln(217)} = 5.76$. The random quadrat selection, then, gives about the same value of d as do the first 8 quadrats. We note, however, that N is about the same for both sets selected, and simply using S would be sufficient to describe the diversity of these samples. That is, for 8 random quadrats we have $S = 31$ and $N = 201$; for the first 8 quadrats, $S = 32$ and $N = 217$.

Let us examine more intensively the properties of the ratio $\dfrac{(S-1)}{ln(N)}$ by using some of the data presented in Table 12.2. In Table 13.1 the first set of numbers contains each of the observed quadrat values of N and S added over the $N = 25$ quadrats (Table 12.2, Appendix 2). Because they are values that are accumulated over the entire plot, they get larger as we add in more quadrats to obtain the final total numbers for the entire plot. Thus, N is about 200, 300, 400, 500, and finally, equals the total. The second set decreases the number of species S by 10 in each case. That is, we subtracted 10 from each value of S in the first set. The third set multiplies N by 10 in each case. The 3 sets, 1 real and 2 created, are plotted in Figure 13.1. At first glance, d (Margalef's measure) seems to behave well (Table 13.1). For any particular data set, d increases as N gets larger. But when N remains constant and S decreases, d shows a decrease in diversity. For example, at $N = 301$ for $S = 37$, $d = 6.31$. At $N = 301$ for $S = 27$, $d = 4.56$. However, at $N = 3{,}010$ for $S = 37$, $d = 4.49$, which is very close to the value of $d(4.56)$ for $S = 27$ at $N = 301$. Clearly, we lack discrimination in this instance. This is so because we can multiply the denominator, $ln(N)$ by 10 while adding only 10 to the numerator to achieve the same ratio. In the first instance we have $\dfrac{26}{ln(301)} = \dfrac{26}{5.7071} = 4.56$. In the second, $\dfrac{36}{ln(3010)} = \dfrac{36}{8.0097} = 4.49$. The lack of discrimination with this index is revealed throughout any range of field results. For example, at $N = 30{,}100$ and $S = 47$, we have $\dfrac{46}{ln(30100)} = \dfrac{46}{10.3123} = 4.46$.

This problem of lack of differentiation with an index cannot be alleviated by standardizing. To accomplish standardization, we note that the maximum possible value of d occurs when each individual represents a different species. In this case, $d_{max} = \dfrac{(N-1)}{ln(N)}$. To standardize each set of data we divide each value of d in the set by the maximum value of d; that is, we use $\dfrac{d}{d_{max}}$ as our index, which is standardized to range only between values of 0 and 1. For the first 2 data sets in Table 13.1 (the base case that was observed in the field, and the data with $S-10$) we have $d_{max} = \dfrac{662}{ln(663)} = \dfrac{662}{6.4968} = 101.8967$. We now standardize by dividing each d entry by this value.

Thus at $N = 663$ for $S = 52$ and $d = 7.85$, we have $\dfrac{7.85}{101.8967} = 0.0770$. For another example, in the third data set, we have $\dfrac{6629}{ln(6630)} = \dfrac{6629}{8.7994} = 753.3502$, so that at

$N = 6630$ for $S = 52$ and $d = 5.80$, we have $\dfrac{5.80}{753.3502} = 0.0077$, or 0.008. The index

$\dfrac{d}{d_{max}}$ is a standardized index. However, in Table 3.1 we show that the values of the standardized index are still not distinct for differing values of N and S.

A problem with this standardization approach for the Margalef diversity index is that for field data we usually do not know the value of the total N and S (although we may be able to get a good estimate of total N from the methodology of Chapter 4 and an estimate of total S from the methodology of Chapter 12).

However, for the bottom row of values of $\dfrac{d}{d_{max}}$ in Table 13.1 that we have just

calculated, we note that $\dfrac{d}{d_{max}} = \dfrac{\left[\dfrac{(S-1)}{ln(N)}\right]}{\left[\dfrac{(N-1)}{ln(N)}\right]}$, which equals $\dfrac{(S-1)}{(N-1)}$. For the base case in

Table 13.1, this gives $\dfrac{51}{662} = 0.0770$. Thus, if we have standardized the area or space

under consideration, then $\dfrac{(S-1)}{(N-1)}$ or, of course, $\dfrac{S}{N}$ becomes a reasonable measure of diversity. For example, let us say that for 3 field sampling periods the same amount of area or the same number of quadrats yielded $N = 301$ and $S = 37$, $N = 301$ and $S = 27$, and $N = 3,010$ and $S = 37$ (second row of Table 13.1). Our values for $\dfrac{S}{N}$ will be 0.1229,

0.0897, and 0.0123, respectively. In this case, the index $\dfrac{S}{N}$ differentiates among the

areas. However, the tricky problem with the use of indices has not disappeared. Suppose 1 local area has $N = 392$ and $S = 41$; another $N = 217$ and $S = 22$ (Table 13.1). Now

d separates them quite nicely (6.70 vs 3.90); however, $\dfrac{S}{N} = 0.10$ for both. Unless a

range of data like that shown in Table 13.1 and Figure 13.1 is available, we just can't tell which is the correct index to use for our particular set of samples.

LOG–LOG INDEX, IOTA

In Chapter 12 we also relate S and N, not just by a semilog plot but a log–log plot, which is the power curve given by $S = aN^b$ (Equation 12.2). On log–log axes this curve plots as a straight line and can be expressed as $ln(S) = ln(a) + b \cdot ln(N)$, where $ln(a)$ is the y-intercept and b is the slope of the straight line. If we ignore the intercept, or set

$ln(a) = 0$, then we can solve this equation to get $b = \dfrac{ln(S)}{ln(N)}$. The ratio $\dfrac{ln(S)}{ln(N)}$ is then the "slope" of the line, and $S = N^b$. We have not seen this measure in the diversity literature, nor are we advocating herein its use as yet another diversity index. It is, however, a logical extension when one considers a semilog versus a log–log plot (Chapter 12) and adds another iota of complexity to the entire subject. We here define this ratio as iota, ι, so that

$$\iota = \frac{ln(S)}{ln(N)}. \tag{13.2}$$

In Table 13.2, we see that the iota's are similar not across the rows, but down the columns. This is desirable because each column represents the same data set but with different values of N. From the original Bolivian data set, we would say that a natural population in which $N = 301$, $S = 37$, and $\iota = 0.6327$ is similar to a population in which $N = 497$, $S = 47$, and $\iota = 0.6201$. We would predict from the ι calculated by Equation 13.2 for $N = 301$ and $S = 37$ that at $N = 497$, $S = 497^{0.6327} = 50.81$, which is not too far off from 47. We would not, of course, use this crude measure of slope if we had the entire data set for predictions. Instead we would calculate a power curve with two parameters; our expected S would be 47.16 (Table 12.2).

When only an N and S are available, ι does provide us with a crude measure. We must keep in mind, however, that because ι is the "slope" of a power series, very small changes may produce sizable differences in possible values for richness or S. For example, $497^{0.6300} = 49.97$; a difference of 0.01 in the slope b value in this equation produces either $497^{0.6200} = 46.96$ or $497^{0.6400} = 53.17$. A difference of 0.05 in the b value gives $497^{0.5800} = 36.63$ or $497^{0.6800} = 68.16$. Consequently, if we wish to discriminate among situations such as those given in Table 13.2, we must look for small differences in the slope coefficient, or b value, on the order of less than 0.05. If we are examining several

TABLE 13.2   Log–Log Index: $\iota = \dfrac{ln(S)}{ln(N)}$

	SET 1			SET 2			SET 3	
N	S	ι	N	S	ι	N	S	ι
217	32	0.6442	217	22	0.5746	2,170	32	0.4511
301	37	0.6327	301	27	0.5775	3,010	37	0.4508
392	41	0.6219	392	31	0.5751	3,920	41	0.4488
497	47	0.6201	497	37	0.5816	4,970	47	0.4524
663	52	0.6082	663	42	0.5753	6,630	52	0.4490

NOTE: The first set of numbers are the accumulated values from Appendix 2. The second set decreases all values of S by 10. The third set multiplies all values of N by 10.

areas and necessarily must confront values of N and S that are widely different, the ratio provided by this log–log index could prove useful.

At this point we have exhausted the discussion of the most commonly advanced diversity measures using N and S only. We consider now a bit more sophisticated and quite popular index. This index, called Fisher's α, while still relying upon N and S, introduces the abundance of each of the species though not their actual observed values, and their differences. We shall discuss this measure first without any references to a statistical distribution for these relative abundances or any statistical properties. We then continue by showing how such a reference distribution can prove advantageous for this and other diversity measures.

FISHER'S ALPHA

For the two diversity indices discussed thus far (d and iota) we used only N and S in the calculations. No assumption was made concerning the variability or spread of individuals within species in the assemblage. With Fisher's alpha (α), we have a departure.

Let us begin once more with the ratio of $\dfrac{N}{S}$. Using Fisher's method, this ratio alone can predict the number of species represented by 1 individual, 2 individuals, and so on. Recall from Equation 9.10 that by using just N and S we can generate an entire series, given by the expression $\alpha x + \dfrac{\alpha x^2}{2} + \cdots + \dfrac{\alpha x^n}{n} = -\alpha ln(1-x)$. Here we solve the equation for α, a constant depending upon N and S only, while we determine x, which is also a constant. Then the value for αx is the expected number of species represented by 1 individual, $\dfrac{\alpha x^2}{2}$ the number of species with 2 individuals, and so on. To calculate the series, we need to estimate the values for α and x. Given N and S, we can calculate α using Equation 9.16, which is Fisher's relationship, or,

$$\frac{N}{S} = \frac{\left(e^{\frac{S}{\alpha}} - 1\right)}{\left(\dfrac{S}{\alpha}\right)}. \tag{13.3}$$

Some researchers prefer to use the relationship that solves for S given in Equation 12.15 rather than Equation 13.3. However, either way, the solution is not simple, but iterative. We provide in Appendix 4 the solution for α for a great many combinations of N and S. Although computers and calculators are readily available, this summary of values in Appendix 4 is helpful because the iterative solution demands more than spreadsheet or calculator capability.

Because α is a well-known measure of diversity, many researchers are content to use the value of α to express diversity, but they often have no idea how it relates to their observations. Consequently, we will proceed just a little farther (as we did in Chapter 9) to clarify the meaning of α.

Once α is calculated, we can easily obtain $x = \dfrac{N}{(N + \alpha)}$ (from Equation 9.14). For Bolivian plot 01, $\alpha = 13.21$. Therefore, $x = \dfrac{663}{(663 + 13.21)} = 0.98$ (when actually calculating the series, x should be carried out to at least 6 or 7 decimal places). Then, for our first term, αx, we have $(13.21) \cdot 0.98 = 12.95$. Because x is most often a number that is close to 1 for natural populations, Fisher's α is a number close to the number of species we expect to be represented by 1 individual (12.95 vs 13.21). By thinking of α in this way, we can more easily visualize the meaning of the numerical value and take it out of the realm of the mysterious. Most experienced naturalists would agree that in any large collection with many species, quite a few species are represented by just 1 individual. Because of the logarithmic relationship of N and S, as both N and S increase, the number of singletons becomes, intuitively, a good measure of diversity.

When $x < 0.50$, the value of α loses its meaningfulness for biological work. Notice in Appendix 4 that if the ratio $\dfrac{N}{S} < 1.44$, we obtain $x < 0.50$. Solving backwards for the computed value of α will always give an α that is greater than or equal to N; our total number of individuals. Thus α can no longer be the limit of our number of species with 1 individual. As a reminder, when x is less than about 0.61 (this depends upon N; the value varies between 0.63 and 0.612), the value of α is greater than S, which is also unacceptable. Consequently, Appendix 4 does not contain values of α below the cut-point at which $x < 0.61$.

FISHER'S ALPHA AND THE LOG SERIES DISTRIBUTION

The quantity α, as Williams observed in Fisher and colleagues (1943), can be used as a measure of diversity, even when the Log series is not a good statistical fit to the data. Indeed, many subsequent authors including Magurran (1988) confirm this recommendation. However, such usage is not always profitable for clear interpretation. If we can determine that the data actually are distributed as, or are close to, a Log series, then alpha becomes an extremely good and more adaptable diversity measure. We shall highlight some of the advantages and disadvantages.

Table 13.3 shows the values of α for the Bolivian trees, using the same observed base case and 2 other sets as in Tables 13.1 and 13.2. Notice that α increases along with N and S in all the examples. This correspondence causes a great deal of confusion for some researchers because of the true statement that α is a constant that is independent of sample size. Many researchers believe that, regardless of N, α will remain constant; a perfect measure. Looking at Table 13.3, one could conclude that α is not independent of sample size because it increases with N and S, as discussed in Chapter 9. Consequently, an argument ensues. This apparent paradox is easily understood by considering the equation for the Log series that relates S, N, and α. This relationship is

$S = \alpha ln\left(1 + \dfrac{N}{\alpha}\right)$. In this equation, α is a constant as N and S vary.

TABLE 13.3   Fisher's α

SET 1			SET 2			SET 3		
N	S	α	N	S	α	N	S	α
217	32	10.36	217	22	6.12	2,170	32	5.32
301	37	11.08	301	27	7.18	3,010	37	5.94
392	41	11.53	392	31	7.90	3,920	41	6.38
497	47	12.74	497	37	9.24	4,970	47	7.19
663	52	13.21	663	42	9.97	6,630	52	7.69

NOTE: The first set of numbers are the accumulated values from Appendix 2. The second set decreases all values of S by 10. The third set multiplies all values of N by 10.

Recall that for the Bolivian plot 01, with $S = 52$ and $N = 663$, we obtained $\alpha = 13.21$. For $N = 217$, $S = 13.21 \cdot ln\left(1 + \dfrac{217}{13.21}\right) = 37.75$. In words, for a Log series with $\alpha = 13.21$ and $N = 217$, we would thus expect about 38 species. In fact (Table 13.3), we observe 32. There is nothing wrong with the equation itself; α is a constant and any value of S can be estimated for any selected N. The problem is with the data. We observe 37 or 38 species at about 300 individuals, not 200; the data do not constitute a perfect Log series.

Notice that we began in the last section with the ratio $\dfrac{N}{S}$ in which we made no assumptions and did not even consider a reference distribution. In this section, we add that the value for this ratio actually can be generated from knowledge of a statistical distribution for the data, the Log series distribution, discussed in Chapter 9. In other words, the problem with interpreting or using Fisher's α as a diversity measure is that we may not know if the observations adhere perfectly or even closely to a Log series. If that is the true case, and alpha is actually the parameter of a distribution, then we fail to understand some necessary properties of this measure and run the risk of misinterpreting the ecological scenario. For a true Log series in this current example, we would have, at $N = 217$ and $S = 37.75$, a value for α of 13.21 and for x of 0.9426705 (see Chapter 9 for the procedures). Note that x has changed, not α. Equation 9.12 is $S^* = \alpha ln(1 + x) =$

$- 13.21 \cdot ln(1 - 0.9426705) = 37.77$. Then, Equation 9.13 is $N^* = \dfrac{\alpha x}{(1 - x)} = \dfrac{13.21 \cdot 0.9426705}{1 - 0.9426705} =$

217.21. The constant is α, and the change in x ensured the correct values for S^* and N^*;

but at $N = 663$, we obtain $S = \alpha ln\left(1 + \dfrac{N}{\alpha}\right) = 13.21 \cdot ln\left(1 + \dfrac{663}{13.21}\right) = 51.99$, which is cor-

rect. What we have demonstrated is that if the data set can be fit exactly by a Log series, then α is a constant that is independent of N. Mother Nature, however, rarely cooperates exactly. As N and S from natural populations accumulate or increase with more sampling or fieldwork, an increase in α (like that in Table 13.3) usually is observed. However, if we use Fisher's α as a simple diversity measure, with no attempt to see how the data could be distributed, most likely we won't be able to interpret our resultant value for this measure.

The α's in the first, or base, case shown in Table 13.3 are clearly distinguished from the α's obtained for the last two manufactured cases in the table. The middle data set was constructed by subtracting 10 from all the values of S. The last set increases the first values of N by a multiple of 10 while preserving the same S values from set 1. For these two manufactured data sets there is overlap in the values of α.

Let us consider two entries. At $N = 301$ and $S = 27$, $\alpha = 7.18$; at $N = 4,970$ and $S = 47$, $\alpha = 7.19$ (Table 13.3). If α is our measure of diversity, we would conclude that these two observations represent equivalent diversity of $\alpha = 7.2$. At $N = 1,000$, we would expect

$$S = \alpha \cdot ln\left(1 + \frac{N}{\alpha}\right) = 7.2 \cdot ln\left(1 + \frac{1000}{7.2}\right) = 35.57 \text{ species. Despite the congruence in } \alpha$$

values in these two cases, Figure 13.1 indicates that we are looking at 2 quite different N and S plots. By extending the plots of the middle data set to $N = 1,000$, we would expect about 53 species. For the last data set, we would expect about 23 species at $N = 1,000$ (Figure 13.1). If we did not have the N and S plots and their regressions and instead had only the second point in the middle data set and the second-to-last point in the last data set, then we most likely would assume a constant $\alpha = 7.2$ and a regression line between the two. We have demonstrated in this text, as we all know, that a regression line based on only 2 points is an extremely dangerous procedure. It will only work, in this case, if the relationship is exactly Fisher's Log series. To avoid this difficulty, we need N versus S plots at several points (as we have done here) or else a standard size N. If we had an $N = 1,000$ with $S = 23$, and an $N = 1,000$ and $S = 53$, then the α values would be 4.20 and 11.94, respectively. The 2 sets of data are now clearly distinguished by α. Of course, they would be distinguished by S as well.

VARIANCE AND STANDARD ERROR OF FISHER'S α INDEX

Recall from Chapter 9 that Fisher (1943) gave a formula (our Equation 9.18) for the variance of α, and Anscombe (1950) also gave an estimate (our Equation 9.19). The square root of either of these variances is the standard error of α, denoted by σ_α. Consequently, confidence limits can be calculated for α. Unfortunately, the two formulas do not give exactly the same answers for σ_α. Table 13.4 shows the estimates for σ_α, and the 95% confidence limits (Chapter 4) for α or the same 3 data sets shown in Table 13.3.

Let us examine Fisher's confidence limits first. Recall that this quantity σ_α is called an *asymptotic standard error*, and, therefore, will be of use only for large samples. As usual, the problem arises as to how large is large enough. For the base case data set, the *Fisher confidence interval* would include all the estimates of α except for the first two (third column of Table 13.3). For the second and third data sets, Fisher's confidence interval would include estimates of α for only the last two rows (Table 13.3).

TABLE 13.4   Standard Errors, σ_α, and Confidence Limits for Fisher's and Anscombe's Estimates

$N = 663,\ S = 52,\ \alpha = 13.21$	$N = 663,\ S = 42,\ \alpha = 9.97$	$N = 6{,}630,\ S = 52,\ \alpha = 7.69$

Fisher's estimate:

$\sigma_\alpha = 1.0013$	$\sigma_\alpha = 0.80171$	$\sigma_\alpha = 0.4006$
$11.25 \leq 13.21 \leq 15.18$	$8.40 \leq 9.97 \leq 11.54$	$6.91 < 7.69 \leq 8.48$

Anscombe's estimate:

$\sigma_\alpha = 2.1289$	$\sigma_\alpha = 1.7661$	$\sigma_\alpha = 1.1557$
$9.04 \leq 13.21 \leq 17.39$	$6.51 \leq 9.97 \leq 13.43$	$5.43 \leq 7.69 \leq 9.96$

NOTE: The first data set is the Bolivian tree study plot 01 (Appendix 2). The second set has $S - 10$. The third has $N \cdot 10$.

Looking across the data sets (Table 13.4), we observe that Fisher's confidence intervals overlap for data sets 1 and 2 as well as for 2 and 3. Only data sets 1 and 3 lack overlapping values.

The confidence limits with Anscombe's estimate are always larger than for Fisher's. Anscombe's formula also is for large samples (*asymptotic*), but in addition it has the disadvantage of being an estimate with an unknown amount of bias. For the base case data set (in Table 13.3), all of the estimates of α fall within the confidence interval. For the second and third data sets, all the estimates of α, except for the first one, fall within the interval. The larger estimate for *Anscombe's standard error* allows us to include most of the estimates for α on the same data sets within the confidence interval. We must, of course, pay a premium as we look across the data sets. As Table 13.4 shows, the confidence intervals from the Anscombe formula overlap for all 3 sets of data.

Consideration of either formula for the variance of α suggests that we can be much more confident of the true value of α when N is large (in the thousands) and x is close to 1. In general, we prefer Fisher's formula to that of Anscombe, although Fisher's has the slight disadvantage that calculation is more cumbersome. Neither formula is useful for very small values of N. When N is greater than about 30 to 50, Fisher's formula gives a smaller, more reassuring value of the standard error. As N and S increase, Fisher's formula performs in a dependable manner, although it does increase. On the other hand, Anscombe's estimator is of unknown bias and disintegrates for $x < \dfrac{e}{(1+e)}$, or $x < 0.7311$, approximately. Thus we can use α as a diversity measure throughout the range of $1 > x > 0.61$, but Anscombe's standard error is defined only within the range $1 > x > 0.7311$. In addition, as x decreases from 0.90 to 0.7311, Anscombe's estimate becomes unrealistically large in comparison to the calculated α value being used for diversity purposes.

EVALUATION OF BIODIVERSITY MEASURES ON OBSERVATIONS WITH *N* AND *S* ONLY

A review of the ecological literature indicates to us that many researchers seek a single measure or index of diversity that will characterize the "species diversity" of their natural populations. No single measure has, however, emerged as that index. Instead, various indices appear to be in vogue at any particular time, and some researchers have their favorites.

Let us consider the implications of using an index that incorporates (1) just *S*, (2) both total *N* and total *S*, or (3) *N*, *S*, and individuals within species, without assumptions on any of the numbers or properties of these observed summary values. Recall from Chapter 2 that the mean density by itself is not a particularly informative measure of abundance unless it is combined with another measure of the variability. In a similar fashion, any one measure of diversity may not be sufficient for diversity purposes. As measures of diversity, we saw that richness, Margalef's *d*, and iota each suffer from oversimplification of the ecological situation being studied. That is, an assemblage merely described or explained by a single value, say *S* = 10, certainly is not maximally informative by itself. Let us say we add to this *S* = 10 that *N* = 20; we still know little from the data even after calculating an index. With these two pieces of information, we could propose several hypothetical scenarios but with no firm or quantitative basis for selecting one ecological truth over the other. Certainly, the numbers of individuals within species could provide a more solid basis for knowledge of the overall assemblages and for inference. However, one measure we presented, α, does take into account the number of individuals, their expected but not observed abundances, and the number of species, but it works best and its values are more interpretable when abundances are compared to a Log series distribution. When *N* and *S* are fairly large, this comparison with a reference Log series distribution appears reasonable though. However, as we saw, even the single value of α is not particularly informative alone. Consider that even a casual perusal of natural populations clearly indicates that species abundances vary for any observed number of total species, *S*. Should we consider the diversity of a local habitat with *S* = 5 species consisting of 46, 26, 16, 8, and 4 individuals equivalent to the diversity of a habitat with 90, 6, 2, 1, and 1 individuals? Most researchers would say no. Consequently, indices are preferable that consider not only the number of species but also species abundances, which we shall discuss further in the next chapter. Such indices are, of course, more complicated than a simple species richness value but also more informative, as we shall see.

SUMMARY

1. A single diversity index based only on total *N* or total *N* plus total *S*, though simple and informative, will not be sufficient to describe within-habitat diversity.

2. If a single index is of paramount importance, α may be a good choice. This index can be used with or without reference to a statistical distribution for the data according to some authors, but we do not recommend application without knowledge of the distribution fit to the set of relative abundances, or the relative species abundance

vector (RSAV), **p**, unless the sample size is quite large. Provided that the underlying distribution used to calculate α is a reasonable Log series (which is not uncommon in nature), α will be a constant, and its value is easily interpreted as approximately the number of species expected to be represented by 1 individual. In addition, α has the further advantage of being easier to use yet as reliable as any other method for rarefaction or abundifaction.

Problem

13.1 You have 2 biological observations or samples. In the first you observed $S = 8$ species, with the following number of individuals: 50, 25, 15, 5, 2, 1, 1, 1. In the second sample you observed $S = 11$ species, with the following number of individuals: 150, 100, 75, 40, 20, 10, 1, 1, 1, 1, 1. Calculate d, standardized d, ι, and α for each. Comment on the result.

14

BIODIVERSITY: DIVERSITY MEASURES
USING RELATIVE ABUNDANCES

In CHAPTER 12 WE SAW that when more individuals, N, are added to the sample or accumulated by observation or collection, the number of species, S, encountered also gets larger. This phenomenon, when plotted on Cartesian coordinates, often is called the *collectors or effort curve*. The number of species is a function of (depends on) the number of individuals, N, and we write $S = f(N)$ as done in Chapter 13.

As shown in Chapters 12 and 13, simple measures, either S alone, or those composed of N and S, can be calculated easily, but no single such number will describe the complexity of biological and ecological information inherent in a multispecies community. In this chapter, we first discuss the importance of the numbers of individuals within each species, or their relative abundances, and then examine how these numbers can provide important data for diversity measurement. We relate these observed relative abundances to basic statements on probability so that we can show how relevant reference statistical distributions can be identified simply and used to provide vital added dimensions to diversity study.

Species Abundances

The number of individuals in each species $\{n_i\}$ and, hence, the total number of individuals N can both vary greatly as every investigator knows. Therefore, for simplicity's sake, as well as because theoretical distributions often are fit to proportions, researchers use species proportions instead of the actual number of individuals (density). We treat this subject extensively in Chapter 8. We can simply write for each of the $i = 1$, $2, \ldots, S$ species that the abundance of species i is n_i. Adding each of these abundances for all S species, we write

$$N = n_1 + n_2 + \cdots + n_S = \sum_{i=1} n_i \tag{14.1}$$

to show that all of these abundances together add to the total number of individuals N. From these abundances, for the proportion of each species we write

$$p_i = \frac{n_i}{N}, \tag{14.2}$$

which expresses not just a proportion, relative frequency, or relative abundance but also a *probability estimate*. The frequency $\frac{n_i}{N}$, denoted by p_i, is (an estimate of) the probability that, if an individual is chosen at random from N individuals, it will belong to species i. Researchers often multiply the p_i by 100 and state the result as *relative percent abundance, species percent*, and a variety of other names.

Using this notation, we can write

$$\sum_{i=1}^{S} \frac{n_i}{N} = \sum_{i=1}^{S} p_i \qquad (14.3)$$

to represent the sum of each of the species proportions without having to write out the entire expression with addition signs. Clearly since we showed that the numerator is actually equal to N, then what we have in Equation 14.3 is $\frac{N}{N} = 1$. This equality is convenient for using probabilities in any selected probability space, since the sum of the probabilities must equal 1. The *probability space* is the space in which we have the $\{p_i\}$ and on which we can have a *frequency distribution*. A distribution can be defined as the apportionment of the individuals into species or merely a list of the values recorded in the sample. In addition, a probability or statistical distribution is often fit to these listed values. Therefore, confusion is rampant because of the different uses of the word *distribution*. This apparent confusion is clarified in Chapter 15 where we distinguish between the measurement of diversity and that of evenness.

First, the observed data set for the individual species obtained in the field or museum collection provides us with relative abundances. Next, these relative abundances are considered the frequencies with which individuals were found in particular species. Finally, these frequencies are also the proportions of each species, as well as estimates of the probabilities with which an individual can be found within a particular species. So for our purposes, it is quite important to realize that the set itself of observed species relative abundances constitutes a distribution. This set of values $\{p_i\}$ will then be "fit" with some statistical distribution for further insight into the underlying ecological processes. One of the major problems in ecology is determining which known statistical distribution (Chapter 9) actually can be best fit to the set of $\{p_i\}$ values.

In the following sections of the biodiversity chapters, we describe diversity measures using the set of $\{p_i\}$ values as both single points and as members of an observed partitioning into species to which we try and fit a known statistical distribution.

The Relative Species Abundance Vector, RSAV, and Notation

Equation 14.3 describes the species proportion of each of the i species. Clearly, if we add all the p_i's together, they will equal 1. As an equation, we write

$$\sum_{i=1}^{S} p_i = 1. \qquad (14.4)$$

If we rank all the species from the most abundant to the least abundant, we form a rank ordering of species abundances such as in Appendix 2. In this rank ordering the most abundant species is labeled with the numeral 1, the second most abundant is labeled or ranked 2, and so on until all are taken into account. The most abundant species in Bolivia plot 01 has $p_1 = 0.38$. We infer from this value that that if we pick individuals at random, we can expect that about every third individual, or more specifically, about every 2 out of 5 (38%) individuals, will belong to the most abundant species.

SIMPLIFYING SUMMARY NOTATION

The relative abundance of each species can be listed in a column as we do in a spreadsheet. Although these relative abundances can also be written in a row, it is more usual in the study of natural populations, and more helpful for quantitative analysis, to form columnar lists. Any single column or single row of numbers is called a *vector* and is designated by a lowercase bolded letter. A *matrix* is merely a name given to simplified notation for describing data that consists of more than 1 row and more than 1 column. A matrix is designated by a capitalized, bolded letter. The entire ensemble of species proportions shown in Appendix 2 from Bolivia plot 01 is a column of numbers (proportions) and can be expressed as a vector we shall designate by **p**. We have then,

$$p_{(S \times 1)} = \begin{bmatrix} p_1 \\ p_2 \\ \vdots \\ p_S \end{bmatrix}. \tag{14.5}$$

This "shorthand" notation for the relative species abundance vector is designated as **p**, while the parentheses, or subscript, indicates the *dimension* of the vector; that is, number of rows (S) and the number of columns (1). We always write the dimension $(r \times c)$, so that the number of rows is written first and then the number of columns. For example, if the data set has 5 rows and 7 columns we write (5×7) as the dimension. A vector is always of dimension $(1 \times c)$ or $(r \times 1)$. The number of actual entries in a vector is the *rank* of that vector; in biodiversity study the rank is always S; the species richness. For Bolivia plot 01 the rank is $S = 52$ or $\mathbf{p}_{(52 \times 1)}$. This vector **p** of probability estimates constitutes a probability distribution sometimes referred to as the *relative abundance distribution*, or *RAD* (Wagner and colleagues 2006). Much of the remainder of this book is concerned with examining the properties of this distribution or the vector **p**.

It may be confusing to those new to quantitative development that the vector of relative species abundances is sometimes referred to as the relative abundance distribution, or RAD. The RAD in vector format is also called the *relative species abundance vector,* or *RSAV*. Now we have said that this same vector is also a probability distribution, so it appears we are adding to the confusion. In addition, the literature is filled with many attempts to "fit" a distribution to the data in the RSAV/RAD. However, we hope to make clear in the following chapters exactly how this observed set of abundances is a distribution of values to be fit by some selected probability or statistical distribution and exactly how and why this is of interest in biodiversity.

PLOTTING THE DATA

Before any numerical manipulation or analysis, most statisticians recommend simply plotting the data. In Chapter 9 (Figures 9.2 and 9.3), we examine the so-called hollow curves that are plots of the number of individuals versus the number of species (Fisher et al. 1943; Preston 1948). Two alternative plotting methods are in vogue at the present time. The first is often called the *Whittaker plot*, after Whittaker (1975); see Krebs (1999) and Magurran (2004). This graph visually presents species proportions on a log scale on the *y* axis and the species rank ordering on the *x* axis. Figure 14.1 shows the plot for the Bolivia plot 01 tree data. The Whittaker curve is a graphical representation of all the data that can easily be compared with other data sets; hence, its popularity.

Another representation suggested by Pielou (1975), and see Tokeshi (1993) for a review, involves placing the cumulative proportions on the *y* axis and the species in rank order on the *x* axis. This is also called the *k-dominance plot* (Lambshead and colleagues 1983; Platt and colleagues 1984; Magurran 2004). In Figure 14.2 we plot the Bolivian plot 01 tree data in this manner but with rank order of species on a log scale (Hayek and Buzas 2006). This *k*-dominance presentation is a bit more satisfying in the visual sense and it too can be used to compare all the data in the community with other communities. The graphical representations provide an ordering to the diversity of areas to be compared and are particularly useful when dealing with communities with a relatively small number of species. However, as the number of species increases, they become somewhat unwieldy.

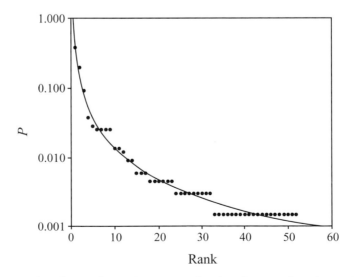

FIGURE 14.1   Plot of p_i on a log scale versus rank order of species for Bolivian plot 01.

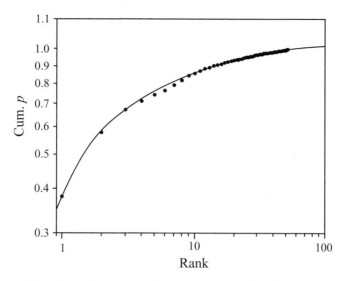

FIGURE 14.2   Plot of cumulative p_i on a log scale versus rank order on a log scale for Bolivia plot 01.

RICHNESS MEASURES AND THE RSAV

At this point in our diversity investigations, it should be clear that for a diversity measure to be a meaningful representation of the ecological complexity in a community, we must somehow incorporate the vector **p**, that is, the information in the RSAV. In Chapters 12 and 13 we examine measures that use just richness or both richness, S, and total number of individuals, N. Such measures were shown to lack any information on the multiplicity of ways in which richness can be concentrated or spread into abundance or rarity categories. In the technical sense then, such measures are not diversity measures at all; they cannot provide comparative data across assemblages or communities. To know N and S, or estimates of them, is not sufficient for ecological comparison and differentiation of assemblages, although indices with only these quantities are both simple and intuitively appealing. It seems appropriate then to have some rules or characteristics for quantitative measures of diversity.

EFFICACY AND RULES FOR DIVERSITY MEASURES

Huston (1994) stated that the patterns of latitudinal gradients first attracted scientific attention for the problem of observed species diversity. In 1993, Latham and Ricklefs made the case that in 1807, von Humboldt and Bonpland first compared tropical and temperate diversity patterns. In these examples, we understand that the concept of biological diversity relates to the variety of life as represented by species richness.

With the introduction and use of *diversity measures*, which are quantitative measures of that variety, sets of communities were ranked and compared in various ways.

As we have seen, inconsistencies are inevitable when reducing to a single value a multidimensional concept or complex situation like a species assemblage. However, if that single value incorporates the RSAV, then a potentially more fruitful description is possible because a set of probabilities is also a *discrete probability distribution*. Then, this set of values $\{p_i\}$ or **p** can be modeled by some statistical distribution with known characteristics. There are solid reasons to increase knowledge by using the statistical parameters that define such a fitted distribution to judge and compare biodiversity among communities. For those readers who are mathematically inclined, this incorporation of **p** into a quantitative descriptor of diversity allows us to define a real-valued function on the set of real numbers and we all know how terrific that is!

What are some essential characteristics of a quantitative diversity measure? First, let us use a generic symbol I to indicate and discuss a diversity index or measure. Since we want our index I to use the entries in the RSAV, which are probabilities, we can also call I a "function." Then, for I, which we shall use to represent any index of diversity that provides a measure of the concentration of the classification into the species or taxon groups, we require that:

1. I must be nonnegative.
2. I must be symmetric. One way to look at what this property for biodiversity study is that if we were to change the order of any of the entries in **p**, we will get the same value for I. Another way to say this is that I is constant over all permutations of the species proportions or frequencies in **p**.
3. I has a coherence property. That is, I reaches its minimum for ecological work when all except 1 of the p_i values equals 0 and the remaining entry equals 1. In addition, I is largest or it reaches its maximum when all proportions p_i coincide.
4. I must be order-preserving. This says that the nearer the set of entries in **p** are to the maximum or equalizing point value (in our case, $\{1/S, 1/S, \ldots 1/S\}$), the higher, or at least not lower, is the diversity represented by **p**.
5. I must be zero-indifferent. This statement means that if some $n_i = 0$, then the value of I is the same when calculated on only the observed entries that are greater than 0 or when calculated on all values including the 0 entries. Of course this situation will occur only if, for example, we tabulate a species list and record in the field whether we find individuals within each species or not. So the value of I would be exactly the same for the data on 6 species $\{15, 22, 37, 0, 0, 0\}$ and for data on the same area in which we recorded only those abundances that were not 0, or $\{15, 22, 37\}$. Alternatively, using only the relative abundances, in some sense the vector **p** $= (0.6, 0.4)$ is different from the vector $(0.6, 0.4, 0, 0, 0)$. The second vector with zeros is sensible only if 3 species, which could be present in the community, and/or could be found in a complete census, have not been observed. If this is the case, the vector $(0.6, 0.4, 0, 0, 0)$ is much more informative than the vector $(0.6, 0.4)$ for comparison purposes in biodiversity.

Using these rules for diversity indices or measures and attempting to compare communities that share the same number of species S, but differ in the apportionment of N (or biomass) into the categories of taxa, is of vital importance in biodiversity study. Below we list and discuss some of the most popular diversity measures, which adhere to the above set of important rules.

Using the information on richness, total individuals, and the set of apportioned relative species abundances, we now present and evaluate for biodiversity use, one simple and two quite important diversity measures that each adhere to the above rules.

DIVERSITY MEASURES WITH NO ASSUMPTION ABOUT STATISTICAL DISTRIBUTION

One of the great advantages of having a data set fit by a statistical distribution is that the entire natural population data ensemble can then be succinctly summarized by at most one or two (occasionally three) parameters of the fitted distribution. The calculated values of such a measure will be interpretable in light of the behavior of the known reference distribution. If we can estimate the parameters, we can generate the fitted distribution. Then, if we wish, we can make comparisons with the fitted parameters from distribution fit to other sets of observations on other assemblages in time or space. However, such comparison is not often achievable or possible with small samples. In addition, distinguishing among several different statistical distributions is not simple, even with large samples. Therefore, an ability to characterize or summarize a fauna or flora, taking into account both the number of species and the species abundances, while making no assumptions about an underlying distribution, is most attractive to researchers. To achieve this end, ecologists have used a variety of measures. Below we shall discuss a simple measure as well as two of the most common measures (Simpson's λ and Shannon's H) with a short development of their history, usage, and interpretation for biodiversity purposes.

SIMPSON'S LAMBDA

In 1912, Gini, and then later Simpson (1949) for biological data, showed that if N is "large enough," selection of an individual and then its replacement does not affect the assemblage's structure. Each individual will belong to one S group (species) with proportions or relative abundances $\{p_1, p_2, \ldots, p_s\} = \mathbf{p}$ (where $\Sigma p = 1$), and a measure of concentration of the classification into species is

$$\lambda = \Sigma p_i^2. \tag{14.6}$$

This formula in Equation 14.6 says that to calculate this measure, we take each entry in the RSAV, $\mathbf{p}$, square it, and then add them all. Since λ has the entries p_i in its formula, we know that λ must be a function of the vector $\mathbf{p}$.

When all species have an equal number of individuals, then each species has N/S individuals. The calculated value of Simpson's index can be any number between 1 and $\frac{1}{S}$. When the diversity is the largest possible with S groups (that is, each species has an equal number of individuals), $\lambda = \Sigma p_i^2 = \frac{1}{S}$ as a maximum. When all individuals are in a single species, then $\lambda = \sum p_i^2 = 1$, since $S = 1$ and $\frac{1}{1} = 1$. When S is large and most individuals are concentrated in a single species, λ will approach or get close to 1.

Simpson derived an unbiased estimate of λ, which is

$$\hat{\lambda} = \sum \frac{n_i(n_i - 1)}{N(N - 1)}. \tag{14.7}$$

The reduction by 1 in Equation 14.7; that is, the introduction of the terms $(n_i - 1)$ and $(N - 1)$, make clear that we select the second member of the pair after, or without replacing, the first individual. This verifies our earlier statement; namely, that λ can be considered as the probability that 2 individuals chosen at random from N belong to the same i^{th} species.

Let us examine the characteristics of λ with an illustration of its calculation. Table 14.1 presents the calculation of Simpson's index λ using an accumulation from the first 11 quadrats of the 25 from Appendix 2.

Looking at the $\sum p_i^2$ column in Table 14.1, notice that the first 3 species account for $\frac{0.1929}{0.1991} \cdot 100 = 96.89\%$ of the λ value. Because the more abundant species contribute so heavily to the value of λ, Simpson's index is often thought of as a *measure of dominance*. Also note that the unbiased estimate of λ differs from λ by only $0.1991 - 0.1958 = 0.0033$ in this example. This shows that because most researchers calculate a column of p_i's as part of their data analysis, the calculation of an unbiased estimate is hardly warranted. Let us examine the variance of λ to confirm this conclusion. Simpson (1949) indicated that if N is very large, the variance of λ can be approximated by

$$\sigma_\lambda^2 = \frac{4}{N}\left\{\sum p_i^2 - \sum(p_i^2)^2\right\}. \tag{14.8}$$

For our Bolivian example we have $\sigma_\lambda^2 = 4/301 \cdot (0.1991 - 0.0396) = 0.0021$. Hence, $\sigma_\lambda = 0.0460$. Multiplying by 1.96 because Normality can be assumed for this index, we obtain for the confidence interval a value of 0.1991 ± 0.0902. This value is larger than the difference between the biased and unbiased estimates substantiating our earlier conclusion based on the results of calculations from Table 14.1 that estimating Simpson's index requires only the squares of the p_i's. We can calculate the variance from the exact formula given by Simpson and we will obtain the same result. Thus 300 is certainly a large enough N for this formula to work.

In Table 14.2 values of λ have been calculated for accumulated quadrats, individuals, and species. The value of λ changes very little with increasing N and S. We would expect this to be the case because the addition of more rare species adds very little to the sum of λ values as we discussed. Provided that the most abundant species do not decrease their proportions as N increases (they don't in the Bolivian plot), then λ will remain approximately constant and a stable measure.

It may not be apparent that for the same N, if we observe or obtain more or less S, that we can obtain the same value of lambda. In Table 14.3, we show a Broken Stick model

N	$N(N-1)$	p_i	p_i^2	$\sum p_i^2$
111	12,210	0.3688	0.1360	0.1360
68	4,556	0.2259	0.0510	0.1870
23	506	0.0764	0.0058	0.1929
6	30	0.0199	0.0004	0.1933
8	56	0.0266	0.0007	0.1940
10	90	0.0332	0.0011	0.1951
8	56	0.0266	0.0007	0.1958
9	72	0.0299	0.0009	0.1967
10	90	0.0332	0.0011	0.1978
1	0	0.0033	0.00001	0.1978
1	0	0.0033	0.00001	0.1978
6	30	0.0199	0.0004	0.1982
1	0	0.0033	0.00001	0.1982
4	12	0.0133	0.0002	0.1984
3	6	0.0100	0.0001	0.1985
1	0	0.0033	0.00001	0.1985
1	0	0.0033	0.00001	0.1985
3	6	0.0100	0.0001	0.1986
2	2	0.0066	0.00004	0.1987
3	6	0.0100	0.0001	0.1988
2	2	0.0066	0.00004	0.1988
2	2	0.0066	0.00004	0.1988
2	2	0.0066	0.00004	0.1989
1	0	0.0033	0.00001	0.1989
2	2	0.0066	0.00004	0.1989
1	0	0.0033	0.00001	0.1989
1	0	0.0033	0.00001	0.1989
1	0	0.0033	0.00001	0.1990
1	0	0.0033	0.00001	0.1990
1	0	0.0033	0.00001	0.1990
1	0	0.0033	0.00001	0.1990
1	0	0.0033	0.00001	0.1990
1	0	0.0033	0.00001	0.1990
1	0	0.0033	0.00001	0.1990
1	0	0.0033	0.00001	0.1990
2	2	0.0066	0.00004	0.1991
1	0	0.0033	0.00001	0.1991
301	17,738	1.0094	0.1991	$\hat{\lambda} = \dfrac{17738}{90300} = 0.1958$

NOTE: Data based on the first 11 (of 25) quadrats of Bolivia plot 01, as presented in Appendix 2. $N = 301$ and $S = 37$. Simpson's lambda and an unbiased estimate of Simpson's lambda are calculated.

TABLE 14.2   Simpson's Lambda for Accumulated Samples of Bolivian Plot 01

N	S	λ	p_{max}
217	32	0.1829	0.3318
301	37	0.1991	0.3668
392	41	0.2012	0.3827
497	47	0.2037	0.3843
663	52	0.1982	0.3801

NOTE: Data from the $N = 25$ quadrats, Appendix 2.

(Chapter 9) with $S = 10$ and a manufactured, but not unreasonable, set of proportions with $S = 20$. Both yield about the same value for λ (0.1702, 0.1707). Consequently, if we are looking for a single index to characterize species diversity, λ must be viewed with caution.

Because λ is the probability that 2 individuals will belong to the same species when selected at random, $1 - \lambda$ is the probability that the 2 individuals selected will not be from the same species. In other words, it is the probability that if 2 individuals are picked at random, they will belong to different species. Ecologists believe this probability, called the *probability of interspecific encounter* (PIE), $1 - \lambda$, is an important concept because it allows us to look at interactions in populations (Hurlbert 1971). An evaluation of the intended usefulness of $1 - \lambda$ for ecology is given by Olszewski (2004). We return to PIE in the next chapter when we examine evenness. Hill (1973), May (1975), and Hayek and Buzas (1997) prefer $\dfrac{1}{\lambda}$ as a diversity measure, and later on in this chapter we examine why (although, see Gadagkar [1989] for one of the undesirable properties of this inverse measure). In subsequent chapters we also strengthen the interpretive and inferential use of Simpson's index for diversity evaluation by describing the way in which we can make use of the vector **p** as a probability distribution.

BERGER AND PARKER INDEX

Berger and Parker (1970) suggested the proportion of the most abundant species (p_1 or p_{max}) as an extremely simple but effective index to account for the diversity of planktonic foraminifera. We write

$$D_{BP} = p_1 = p_{max}. \tag{14.9}$$

The last column in Table 14.2 shows the value for the most abundant species in Bolivia plot 01. As N increases, the proportion of the most abundant species reaches the terminal value of about 0.38 rather quickly at 300 or 400 individuals. This closely parallels the pattern of the value of λ that reaches 0.20 at about 300 individuals. Berger

TABLE 14.3   Simpson's λ for a Broken Stick Distribution and a Hypothetical Distribution

BROKEN STICK DISTRIBUTION $S = 10$			HYPOTHETICAL DISTRIBUTION $S = 20$		
p_i	p_i^2	Σp_i^2	p_i	p_i^2	Σp_i^2
0.2929	0.0858	0.0858	0.3500	0.1225	0.1225
0.1929	0.0372	0.1230	0.1480	0.0219	0.1444
0.1429	0.0204	0.1434	0.0950	0.0090	0.1534
0.1096	0.0120	0.1554	0.0750	0.0056	0.1591
0.0846	0.0072	0.1626	0.0550	0.0030	0.1621
0.0646	0.0042	0.1667	0.0550	0.0030	0.1651
0.0479	0.0023	0.1690	0.0400	0.0016	0.1667
0.0336	0.0011	0.1702	0.0300	0.0009	0.1676
0.0211	0.0004	0.1706	0.0250	0.0006	0.1682
0.0100	0.0001	0.1707	0.0250	0.0006	0.1689
Total:	1.0001	0.1707	0.0200	0.0004	0.1693
			0.0200	0.0004	0.1697
			0.0150	0.0002	0.1699
			0.0150	0.0002	0.1701
			0.0050	0.00002	0.1701
			0.0050	0.00002	0.1702
			0.0050	0.00002	0.1702
			0.0050	0.00002	0.1702
			0.0050	0.00002	0.1702
			0.0050	0.00002	0.1702
			Total:	0.9980	0.1702

NOTE: The index λ equals 0.1707 for the Broken Stick distribution and 0.1702 for the Hypothetical, yet the first distribution has $S = 10$ and the second distribution has $S = 20$.

and Parker (1970) pointed out an inverse relationship between their index and $1 - \lambda$, which would, of course, turn into a positive relationship had they used λ. As May (1975) indicated, this index is a very effective measure and is attractive because of its utter simplicity. It also reinforces the notion that λ is a measure of dominance. We visit the Berger and Parker measure again in the last chapter of this book.

INFORMATION

When computers came into popular use, researchers in information studies wanted a quantitative measure for the number of digits required for the unique expression of a particular message. For computers a message had to be in a *binary* (0 or 1) *format*. The 0 and 1 binary digits are called *bits*. Let us explore how this binary approach applies to characterizing natural populations of organisms.

Consider a case in which there are 4 species or categories that we wish to represent with a binary "alphabet." The $S = 4$ species could be represented uniquely in the English alphabet by A, B, C, and D. In binary form, we could represent them uniquely as 11, 10, 01, and 00, respectively. In each case for the binary expression, there are $n = 2$ digits representing each of the species. For each digit then there can be only 2 choices; that is, the digits 1 or 0. For the entire group of interest of size n, there are a total of 2^n categories, since we have a binary choice (2 possibilities) to make each time. That is, for our example, it took $2^n = 2^2 = 4$ categories for a unique representation. For our needs, these categories are $2^n = S$, the number of species represented.

Now, recalling that a logarithm is defined as the power to which a base must be raised to equal a given number, we can take the logarithm of each side of this equation in order to simplify it and to solve for n. That is, $2^n = S$ can be written in log format as $n = \log_2(S)$, after we take the log of each side. We used $\log_2$ (that is, log to the base 2) because we wanted to solve for n and eliminate the 2 in the equation. That is, we knew that $\log_2(2) = 1$. Now, for our example we have $2 = \log_2(4)$. We had to use 8 bits (4 groups of 2 bits each) in all to represent our 4 species. If we now divide $8/4 = 2$, this is the average number of bits required per species. Another way to say this is that, on the average, we need 2 bits to represent each species in a unique manner.

INFORMATION FROM NATURAL POPULATIONS

In natural populations, the representation just presented must be modified because individuals are not distributed evenly among the species. In the species proportions discussion (Chapter 8), and in this chapter, we noted that the proportion of the ith species, p_i, can be considered as the probability of a randomly chosen individual belonging to species i. For simplicity, let us start with a situation in which each p_i must be an *integral power* (the exponent is an integer and the base is the proportion p_i) of the base $1/2$. Because there are two choices, the probability each time is 1 out of 2. This may sound complicated, but it means merely that because each species in the grouping or locality of interest has a different relative abundance, we will make an assumption that each of these relative abundances must be one of the values 0.5, $0.5^2 = 0.25$, $0.5^3 = 0.125$, . . . for species in decreasing order of abundance. A simple set of values was chosen for our calculations only to simplify the presentation of the concepts, not necessarily to present a realistic example. The possibilities for the proportions then would be 0.5, 0.5^2, 0.5^3, and so on. The notation we use for the example is that each species proportion must be $p_i = 0.5^{d_i}$, where d_i is a positive integer ($d_i = 1, 2, 3, . . .$).

Let us use as an example an assemblage of 5 species, with species proportions of 0.25, 0.25, 0.25, 0.125, and 0.125. We wish to represent this set of proportions uniquely in a binary format. Table 14.4 shows a possible binary code and p_i for each species. This table shows the p_i's as well as their values of 0.5^2 and 0.5^3, so that we would then know that each d had to be either a 2 or a 3. Knowing this exponent means that we would be able to represent each of these 5 species uniquely in either a 2-bit or a 3-bit binary format. That is, our exponents (d_i's) are the number of digits needed to represent each species in the code. The average number of digits or bits for these species is

TABLE 14.4   A Unique Binary Representation of 5 Species

SPECIES	BINARY CODE	d_i	0.5^{d_i}	OR	p_i
A	11	2	0.5^2		0.25
B	10	2	0.5^2		0.25
C	01	2	0.5^2		0.25
D	001	3	0.5^3		0.125
E	000	3	0.5^3		0.125

NOTE: A binary code, positive integer d_i, and proportions p_i for 5 hypothetical species with differing abundances.

the sum of the d_i's weighted by the probabilities, p_i's, which algebraically is written as average $= \Sigma p_i d_i$.

In our example we have $(0.25 \cdot 2) + (0.25 \cdot 2) + (0.25 \cdot 2) + (0.125 \cdot 3) + (0.125 \cdot 3) = 0.50 + 0.50 + 0.50 + 0.375 + 0.375 = 2.25$. On average, it takes about 2.25 bits to represent uniquely both the identity and the abundance of any 1 species in our data set of 5 species.

Recalling that $p_i = 0.5^{d_i}$ (since we want this example to work out), we take $\log_2$ of both sides of the equation and obtain $\log_2(p_i) = d_i \cdot \log_2(0.5)$. Now, using negative exponents, we can write $1/2 = 2^{-1} = 0.5$. Then, $\log_2(0.5)$ must equal -1 because the logarithm base 2 is defined as the exponent on the base of 2. Then we have $\log_2(p_i) = d_i(-1)$ or since $d_i(-1) = -d_i$, solving for d_i we get $d_i = -\log_2(p_i)$.

SHANNON'S INFORMATION FUNCTION

The formula above can now be substituted into the expression for the average. That is, the average is $\Sigma p_i d_i$, so after substituting we obtain as the average $-\Sigma p_i \log_2(p_i)$. This average is given the symbol H and called *Shannon's information index* or *measure*. Then H is usually written as

$$H = -\sum p_i \log_2 p_i. \tag{14.10}$$

This equation 14.10 is also known as the *information function* (Shannon 1948). It is related to entropy in statistical mechanics, as well as to likelihood and information in mathematical statistics. If the true values of proportions p_i are not integral powers of 0.5, then Equation 14.10 will be an approximation, but a very good one.

The first researchers to use the information function chose $\log_2$ for their computations. This is somewhat inconvenient, so most workers now use either $\log_{10}$ or $\log_e (= \ln$, or the natural log). Recall that multiplication by a constant is all that is required to move from 1 logarithmic base to another. For example, for any value of n, $\log_{10}(n) \cdot 3.3219 = \log_2(n)$, and $\ln(n) = 1.4427 \cdot \log_2(n)$. Comparing the results from computations using different bases then should present no difficulty. Unfortunately, some researchers do not indicate which base they used, and so a subsequent researcher must use some ingenuity to ferret it out. Henceforth we will use the natural log, *ln*, for all further discussion of natural populations in this book. For our purposes then Equation 14.10 becomes

$$H = -\sum p_i ln p_i. \tag{14.11}$$

Table 14.5 provides values of H for proportions from 0.000 to 1.000.

The amount each species contributes to the value of H depends on its proportion in the assemblage, p_i. We will draw an example from Table 14.4 to illustrate this point. A value of $p_i = 0.25$ yields a value for $p_i ln(p_i) = -0.3466$ from Table 14.5, whereas a value of $p_i = 0.125$ yields 0.2599. Figure 14.3 shows a plot of $p_i ln(p_i)$. The curve has a maximum at about 0.4, or, more precisely, at $p = \dfrac{1}{e} = 0.3679$. Thus, values of $p_i = 0.25$ and 0.50 are just about the same, although this curve is not quite symmetric. Notice also that for a $p_i = 0.90$, the value of $p_i ln(p_i)$ is about 0.09, whereas for $p_i = 0.10$ the value is 0.23. Rare species ($p_i < 0.01$), which constitute the majority of those found in biodiversity studies, individually contribute little to the value of $H(p_i = 0.01$ has a value of only -0.0461, and 0.001 of -0.0069). Clearly, those species with proportions in the middle range influence the value of this measure most heavily. While examining Figure 14.3 to ascertain the contribution various values of p_i make to H, one should keep in mind that $\sum p_i = 1$, and that each p_i will be proportional to all others because they are probability estimates.

To offer more insight as to how H works as an index of diversity, we examine 3 hypothetical populations, each containing 5 species. Introducing a new notation that lets us list the proportions contributing to the value of H, the first example in Table 14.6 is $H(p_1, p_2, p_3, p_4, p_5) = H(0.90, 0.04, 0.03, 0.02, 0.01$, respectively).

In the first example, the relative abundance values show a strong dominance by 1 species (A = 0.90). This most abundant species (A), however, contributes less to the value of $H(0.0948)$ than does either species B(0.1288) or C(0.1052), which together contribute to over half the value of $H(0.2340$ of a total of 0.4530). In the second example, $H(0.46, 0.26, 0.16, 0.09, 0.03)$, most of the species contribute substantially, and in the same rank order as their proportions. Consequently, their sum, the value of H, is much larger than that for the first example. In the final example, $H(0.20, 0.20, 0.20, 0.20, 0.20)$, each species contributes the same amount, and H has its maximum possible value, that is, the largest value that could be achieved by any example with exactly 5 species. This maximum value, which will occur when all relative abundances are equal, is $H = ln(S) = ln(5) = 1.6094$. Alternatively, from the definition of natural logarithms we may write $e^H = S$, or $e^{1.6095} = 5$. This result is in keeping with our first example of 4 species in which each was represented by 2 digits and $2^n = S$ or $2^2 = 4$.

H AS A DIVERSITY MEASURE

As should be evident from the above explanation and example, Shannon (1948) considered the amount of statistical information associated with a given RSAV, or vector **p**, to be measurable with the quantity H. Another way of looking at this information

TABLE 14.5 Values of $-p_i \ln(p_i)$ for p_i, Where Values of $i = 0.0001$ to 0.99

	0.00	0.01	0.02	0.03	0.04	0.05	0.06	0.07	0.08	0.09
0.0001	0.0000	0.0009	0.0017	0.0024	0.0031	0.0038	0.0045	0.0051	0.0057	0.0063
0.001	0.0000	0.0069	0.0124	0.0174	0.0221	0.0265	0.0307	0.0347	0.0386	0.0424
0.01	0.0000	0.0461	0.0782	0.1052	0.1288	0.1498	0.1688	0.1862	0.2021	0.2167
0.1	0.2303	0.2428	0.2544	0.2652	0.2753	0.2846	0.2932	0.3012	0.3087	0.3155
0.2	0.3219	0.3277	0.3331	0.3380	0.3425	0.3466	0.3502	0.3535	0.3564	0.3590
0.3	0.3612	0.3631	0.3646	0.3659	0.3668	0.3674	0.3678	0.3679	0.3677	0.3672
0.4	0.3665	0.3656	0.3644	0.3629	0.3612	0.3593	0.3572	0.3549	0.3523	0.3495
0.5	0.3466	0.3434	0.3400	0.3365	0.3327	0.3288	0.3247	0.3204	0.3159	0.3113
0.6	0.3065	0.3015	0.2964	0.2911	0.2856	0.2800	0.2742	0.2683	0.2623	0.2560
0.7	0.2497	0.2432	0.2365	0.2297	0.2228	0.2158	0.2086	0.2013	0.1938	0.1862
0.8	0.1785	0.1707	0.1627	0.1547	0.1465	0.1381	0.1297	0.1212	0.1125	0.1037
0.9	0.0948	0.0858	0.0767	0.0675	0.0582	0.0487	0.0392	0.0296	0.0198	0.0100

NOTE: To read for $p_i = 0.0002$, look at the first entry on the left, 0.0001. Now, go across to the column headed 0.02 and read $-p_i \ln(p_i) = 0.0017$. To read for $p_i = 0.38$, find 0.3 on the left and go across to 0.08, where you read $-p_i \ln(p_i) = 0.3677$.

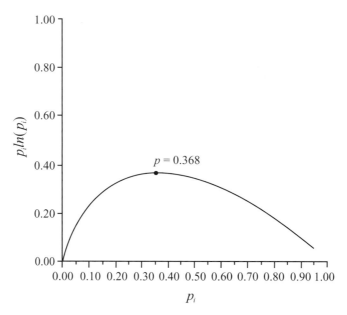

FIGURE 14.3   Plot of p_i and $p_i ln p_i$.

function, H, is as a *measure of uncertainty* in predicting the abundances of the species that compose the relative abundance vector **p**. An individual chosen at random from the first example, which is our high-dominance example, would have about 0.90 or a 90% chance of belonging to species A. Thus, we are relatively certain of the outcome of our selection. The uncertainty value is low. In the third example, where the individuals are distributed evenly among the 5 species, an individual chosen at random would have the same estimated probability of belonging to any 1 of the species. When picking an individual at random, we would have an equal chance of getting a representative individual from any 1 of the 5 species, so this equiprobability case represents the maximum uncertainty possible. Overall, for any given number of species, S, the higher the value of H, the more uncertain we are of the outcome of our selection process. In summary, the amount of information we receive from observing the result of our sampling experiment or fieldwork, depending upon chance, is seen as numerically equal to the amount of uncertainty in the outcome before we go into the field or carry out this field experiment. So, Shannon's information function can also be seen either as a measure of uncertainty or of information about **p**.

As already stated, the maximum uncertainty will occur when each of the species is equally represented as in our third example in Table 14.6. That is, $H_{max} = ln\ S$ in biodiversity studies. The more species there are, and the more evenly the individuals are spread across these species, then the higher will be the value of H, because we will be more unsure which species we will be most likely to observe next time in the field. For example, we might think that if all species were equally sized, then a sample with 5 species would be about $\frac{1}{3}$ as diverse as a sample with 15 species. However, $H = ln(5) = 1.6095$ for the first case and $H = ln(15) = 2.7081$ for the second case. Even

TABLE 14.6   Shannon's Index, H, for
Three Different Levels of Dominance

SPECIES	p_i	$p_i ln(p_i)$

Example 1: Dominance of One Species

A	0.90	−0.0948
B	0.04	−0.1288
C	0.03	−0.1052
D	0.02	−0.0782
E	0.01	−0.0461

$\Sigma p_i = 1.00$; $-\Sigma p_i ln(p_i) = 0.4531 = H$

Example 2: Broken Stick Distribution

A	0.46	−0.3572
B	0.26	−0.3502
C	0.16	−0.2932
D	0.09	−0.2167
E	0.03	−0.1052

$\Sigma p_i = 1.00$; $-\Sigma p_i ln(p_i) = 1.3225 = H$

Example 3: All Abundances Equal or Even

A	0.20	−0.3219
B	0.20	−0.3219
C	0.20	−0.3219
D	0.20	−0.3219
E	0.20	−0.3219

$\Sigma p_i = 1.00$; $-\Sigma p_i ln(p_i) = 1.6095 = H$

NOTE: $H = -\Sigma p_i ln(p_i)$.

though it appears ecologically reasonable when all species are equally common, Shannon's measure does not indicate that diversity is proportional to the number of species, or the richness. Shannon's index provides the uncertainty we have in the identity of the species in the sample, not the number of species in the assemblage. When we are interested in picking individuals, or sampling in this way, the value of H cannot be considered as an average, nor as a measure that is an absolute description. The reason is that our sample size is always finite; whatever species are included in the sample, we usually do not expect them to cover the entirety of the population. Therefore, this estimated value of H may not be reasonable as a population estimate because the population of organisms may contain many more species than are in the sample. Of

TABLE 14.7   Calculation of the Variance of H

p_i	$p_i(ln(p_i))^2$	
Example 1: Dominance of One Species		
0.90	0.0100	
0.04	0.4144	
0.03	0.3689	$\sigma_H^2 = [1.3115 - (0.4531)^2]/300 = 0.0037$
0.02	0.3061	
0.01	0.2121	
1.00	1.3115	
Example 2: Broken Stick Distribution		
0.46	0.2774	
0.26	0.4718	
0.16	0.8732	$\sigma_H^2 = [2.5131 - (1.3225)^2]/300 = 0.0025$
0.09	0.5218	
0.03	0.3689	
1.00	2.5131	
Example 3: Equal or Even		
0.20	0.5181	
0.20	0.5181	
0.20	0.5181	$\sigma_H^2 = [2.5903 - (1.6095)^2]/300 = 0.0$
0.20	0.5181	
1.00	2.5903	

NOTE: Calculations use Equation 14.10 r σ_H^2 and data from Table 14.6 with $N = 301$.

course, if many more rare species were to be included, the value of H would not change appreciably (this is our property of zero-indifference of a diversity measure). We can thus conclude that using H as a diversity measure is not without problems, but the relative size of H is a sound indication of the diversity in terms of the number of species plus the spread of the individuals into those species.

Some confusion exists as to what constitutes the variance of H. Various texts and authors use different forms of this quantity. The reason appears to be that the quantity in the variance formula is a series of additive terms, with no single formula for the sum. We chose first to derive and then to use the formula we derived that approximates the sampling variance of a difference function (Samuel-Cahn 1975). A reasonable expression is obtained from the first terms of an asymptotic formula. The formula we use to calculate the variance of the diversity index H is

$$\sigma_H^2 = \frac{\{\sum p_i (lnp_i)^2\} - H}{N - 1}.$$
(14.12)

In Table 14.7 we show the calculation of Equation 14.12 for the data presented in Table 14.6, using an arbitrary $N = 301$. In the first example we obtain $\sigma_H^2 = 0.0037$; the standard error is $\sigma_H = 0.0607$. Multiplying by 1.96, we have confidence limits of $H \pm 0.1190$. In the second example $\sigma_H^2 = 0.0025$, so that $\sigma_H = 0.0505$ and the confidence limits are $H \pm 0.0989$. In the third example $\sum p_i ln(p_i)^2 = H^2$, and so the variance is 0. This makes sense because in the last case $H = ln(S)$.

GENERALIZATION OF DIVERSITY MEASUREMENT

As shown above by example, when both Shannon's and Simpson's indices are calculated for any given data set, the answers or resultant values are not equal. Since we usually study the diversity of a species assemblage for its connection with and illumination of ecosystem functioning and internal organization, obtaining these differing values for multiple diversity indices can make interpretation quite problematical. We do not agree with those who are convinced that diversity is a "meaningless concept" (Hurlbert 1971), but rather we see the concept confused by researchers and agree with Hill (1973) that diversity can be as unequivocal as any other ecological parameter. Therefore, we derive a unified treatment to eliminate confusion.

Although we presented each of these indices separately, and although in the current biodiversity literature they appear to be independent, we showed that they rely upon the RSAV, $\mathbf{p}$, in their formulae. In addition, both Simpson's and Shannon's are information-based indices that rely for their quantification upon the concept of entropy, which we discuss more in depth in subsequent chapters. Therefore, it is possible to write one general formula that will yield either index. That is, these 2 indices are members of a general *family of diversity measures*. Many mathematicians have derived such families of formulae, which are in fact simple to devise, and here we choose to present two of the more important for our purposes. All such families contain parametric or nonparametric diversity indices that use the vector $\mathbf{p}$.

Renyi (1965) developed a family of parametric measures by extending Shannon's concept of entropy. He did this by defining what he called the *parametric entropy of order a* given by the equation

$$H_a = \frac{1}{(1 - a) ln(\sum p_i^a)}.$$
(14.13)

Of course, once again, p_i is the relative abundance of species i in the RSAV, $\mathbf{p}$. Renyi (1965) also showed that Equation 14.13 constitutes a *general measure of information*, as discussed above. We already know that for these quantities p_i, we must have that $0 \geq p_i \geq 1$ and $\sum p_i = 1$. The parameter for this family of Renyi's is also called the "order" a. Looking at the formula it is clear that we could set a equal to any real number, small or large. However, a restriction that confines a only to be $a \geq 0$ must be used for biodiversity work so that H_a can be adequate for ecological research (Patil and Taillie 1982).

In 1973, Hill cleverly showed that H_a can be used quite successfully to summarize and measure the diversity of an ecological community. As an exponential alternative (to get rid of the use of logarithms) to the family of Renyi, Hill developed his family of measures, which are likewise dependent upon the amount of emphasis given to the rare species just as is Renyi's formulation. We elaborate on this family of Hill's measures in the next chapter.

An especially convenient aspect of the use of Renyi's H_a is that the important entropy or information-based indices we have discussed, Shannon's and Simpson's, as well as Berger–Parker's and species richness, are each members of this family. Since Renyi's (1965) formula depends only upon the parameter a, we obtain members of the family by substituting values for a into Equation 14.13. For example, when $a = 0$, then

$H_0 = ln(S)$; the richness. When $a = 2$ then $H_2 = ln\left(\dfrac{1}{\sum p_i^2}\right) = ln\left(\dfrac{1}{\lambda}\right)$, where λ is Simp-

son's index. When $a = 1$ we can see from Equation 14.13 that the denominator of the

multiplier $\dfrac{1}{(1-a)}$ equals 0. Therefore, since division by 0 is impossible, we have to

examine what happens when a gets closer and closer to 1, not when it equals 1. Interestingly, we obtain Shannon's index. The Berger–Parker index is identified by this formula when a gets very large. It should also be clear that the maximum of H_a occurs when $a = 0$, or the value of species richness, which of course includes equal weighting to all species.

Let us illustrate the way these families of information-based measures work by using the data from the first quadrat of Bolivia plot 01. The vector **p** consists of all the relative abundances, or p_i's, found in the first quadrat and is listed in Table 14.8.

In Table 14.8, we observe that the column in which we raised each p_i to the 0 power is simply a column of 1's and adds to the observed values of S. The frequencies raised

to the 0.90 power give a sum for the $\{p_i\}$ of 1.197 and $\dfrac{1}{(1-0.90)} = 10$, so that the

$ln(1.197)$ raised to the 10th power yields 1.798. Calculating Shannon's information by the traditional method from our Equation 14.8 yields 1.769. Consequently, we have a very nice approximation by simply raising p_i's to a number close to 1; namely, 0.90. This approximation gets closer to the actual result we would obtain when we actually use Equation 14.13 as we try values even closer to 1 (e.g., 0.91, 0.92). In the last case, when $a = 2$ we simply square each p_i, just as we would when calculating Simpson's γ

from Equation 14.5. Substituting 2 for a into our formula, the ratio $\dfrac{1}{(1-2)}$ is equal to

-1 so that $ln\left(\dfrac{1}{\lambda}\right)$ becomes therefore our measure of information H_2. Table 14.8 shows

how for a fixed RSAV, Renyi's general formula is a decreasing function. Because

$ln\left(\dfrac{1}{\lambda}\right)$ is the measure of information, Hill (1973) and May (1975), as well as Hayek and

Buzas (1997), preferred it over both γ and $1 - \gamma$ as a diversity measure.

TABLE 14.8   Illustration of Renyi's Generalized Information Family of Measures for Data from Quadrat 1, Bolivia Plot 01 (Appendix 2)

SPECIES NO.	p_i	$p_i^{a=0}$	$p_i^{a=0.90}$	$p_i^{a=2}$
2	0.500	1	0.536	0.250
3	0.033	1	0.046	0.001
5	0.133	1	0.163	0.018
9	0.033	1	0.046	0.001
19	0.067	1	0.088	0.004
27	0.033	1	0.046	0.001
32	0.067	1	0.088	0.004
35	0.033	1	0.046	0.001
38	0.033	1	0.046	0.001
41	0.033	1	0.046	0.001
43	0.033	1	0.046	0.001
	0.998	11	1.197	0.283

$a = 0 H_0 \quad ln(\sum p_i^0)^1 = ln(11) = 2.398$

$a = 0.90 \quad H_1 = ln(\sum p_i^{0.90})^{10} = ln(1.197)^{10} = ln(6.039) = 1.798$

$a = 2 H_2 \quad ln(\sum p_i^2)^{-1} = ln(0.283)^{-1} = ln(1/\gamma) = ln(3.534) = 1.262$

NOTE: $H = -\sum p_i ln p_i = 1.769$. Calculated values are seen to decrease as the values of the order a increase.

Finally, summarizing our measures of information using Renyi's generalized notation, we have

$$H_0 = lnS, \tag{14.14}$$

$$H_1 = -\sum p_i ln p_i, \tag{14.15}$$

$$H_2 = ln\left(\frac{1}{\gamma}\right). \tag{14.16}$$

Now we can link Shannon's and Simpson's measure together with the following equations and can therefore write

$$H_1 = H_2 + \text{constant}, \tag{14.17}$$

$$H_2 = H_1 - \text{constant}. \tag{14.18}$$

We examine the exact nature of the constant in the next chapter after we undergo more development of the information approach to species diversity. At this point we have one formula for generating a set of three separate, commonly used diversity indices. We continue in this unified treatment and develop notation for a set of evenness or equitability measures. We then continue by expanding the development of our fundamental overall unification of all measures as well as our suggestions for usage with all quantitative biodiversity study projects.

SUMMARY

1. The relative species abundance vector (RSAV), **p**, is a set of species proportions and at the same time a set of probabilities (estimates) adding to 1. This vector, **p**, forms the basis for examining the relative abundance distribution (RAD). In summary, any vector of relative abundances is also a vector of probabilities.

2. Importantly, when we have a vector containing entries that are probabilities, such a vector and its entries constitute a *discrete probability distribution*. That is, the entries of **p** constitute a distribution that identifies the probability of each value of a random variable, or the number of individuals within the taxon groupings.

3. Simpson's index gives the probability that 2 individuals selected at random from the population will be from the same species.

4. The unbiased estimate of Simpson's index given by Equation 14.7 is so close to the index itself that its use is unwarranted in large samples. However, this estimate does have a Normal distribution and therefore can prove useful for inclusion in statistical testing.

5. The quantity $-ln(p_i)$ in the formula for Shannon's measure H is often referred to as the *surprise* associated with species i. If p_i is small, then we would be surprised if we found species i. Accordingly, $-ln(p_i)$ will be large for small relative abundances p_i. On the other hand, if p_i is large, then there would be considerable surprise if we did not find that species. Viewing Shannon's index in this way reinforces the fact that H is an uncertainty measure associated with **p**.

6. Both Shannon's and Simpson's measures are special cases of formulae for a generalized family of diversity measures. We chose to present both Renyi's and Hill's formulae, where one is equivalent to the other (with different parameterization only). Therefore, while other diversity measures we have discussed provide point estimates of diversity and community structure, a general family of measures provides a continuum of measures that differ in their sensitivity to abundant or rare species. The general formula is written with a single parameter a, called its order, and this family of measures becomes increasingly dominated by the commonest species as we consider larger and larger values of a. When a is 0, the resultant measure is completely insensitive to species relative abundances contained in **p** and yields only S, the species richness.

7. For any specific species assemblage, the general formula given in Equation 14.13, H_a for the family of diversity measures, always decreases as we showed in Table 14.8. More mathematically we can say that the different diversity measures we obtain by using differing values of a are also different moments of the same basic general function describing **p**. In this way when we change a we are performing a scaling operation and we can plot each of the results of H_a against a as a *diversity profile* to examine the entirety of the biodiversity of the assemblage.

8. Using a generalized information approach developed by Renyi, species proportions $\{p_i\}$, S, Simpson's index, λ, Shannon's H and Berger–Parker are all measures of information: In Renyi's notation these become $H_0 = lnS$, $H_1 =$ Shannon information; $H_2 =$ Simpson's information.

9. H_1 (Shannon's measure) differs from H_2 (Simpson's measure) only by the addition of a constant.

Problems

14.1 When using the information function H, we can write out the species proportions/ relative abundances, or we can use the notation $H(p_{1;} \, p_2, \, p_{1;} \ldots, \, p_n)$. Using Table 14.5 find the following values for the information function, H.

 a. $H(0.90, 0.10)$

 b. $H(0.90, 0.09, 0.01)$

 c. $H(0.90, 0.09, 0.009, 0.001)$

 d. Look at the values of H that you just calculated. What does this demonstrate?

14.2 You have two biological observations or samples. In the first you observed $S = 8$ species, with the following number of individuals: 50, 25, 15, 5, 2, 1, 1, 1. In the second sample you observed $S = 11$ species, with the following number of individuals: 150, 100, 75, 40, 20, 10, 1, 1, 1, 1, 1. Calculate H_1 and H_2 for each. Comment on the result.

15

BIODIVERSITY:

DOMINANCE AND EVENNESS

C LEARLY, RELATIVE ABUNDANCES from different communities form distinct patterns when they respond differently to external environmental, compositional, or other changes that alter both absolute and relative abundances of the constituent species. Therefore, the most useful quantitative diversity measures used in biodiversity studies, as we have seen in the previous chapters (Chapters 12 through 14), should incorporate information on species richness, S, as well as N and especially $\mathbf{p}$.

Thus far in our treatment of biodiversity measurement we have learned that the set of values of the species proportions that are contained in $\mathbf{p}$; that is, $\{p_i\}$, compose what is termed both an $RSAV$, the list or vector, and an RAD, a relative species abundance distribution. For each set $\{p_i\}$, we first examined the apportionment of an assemblage into the species categories and called this "diversity." We saw in Chapter 14 that the diversity measures (as members of the family of indices developed by Renyi), Shannon's H_1 and Simpson's H_2, can each provide us with information on the allocation of individuals within the S species that compose $\mathbf{p}$. We shall now consider the second aspect of the term *distribution* and examine how those $\{p_i\}$ values are spread over the set of S fixed-species categories. This property is called *evenness* or *equitability*. A measure of evenness is also often defined as a summary statistic of the equality (e.g., Pielou 1966; Routledge 1983) of the entries $\{p_i\}$, while Hill defined an evenness index as a ratio of his diversity numbers. We hope to clarify by showing that for quantitative biodiversity assessment, an evenness index is a measure of how an assemblage is spread or dispersed into its species categories and, at the same time, such an index can quantitatively be related to species richness and information. In this sense, evenness is comparable or analogous to a variance, or range of a variable.

Let us consider how the individuals have spread themselves into the observed species after our field sampling or museum investigation. If we plot the data, the visual pattern of the observed values becomes clear. On the one hand, each and every species could have the same number of individuals as was the case with our evenness example shown in Table 14.6. This pattern is called *complete evenness* or *equitability*. On the other extreme, occurring in very few ecological situations, all individuals could belong to a single species except for 1 individual's being found in each of the remaining species. This other extreme pattern describes *complete dominance*, or maximal unevenness. In this sense, evenness and dominance are merely two ends of a continuum of possible apportionment into the set of observed species.

An important aspect of this pattern or description of the relative abundances within species is that the total number of species S is fixed. That is, in trying to develop a quantitative measure of evenness we must have only a single value for S. Note that this is not the case for our diversity measures. The quantitative measurement of evenness can be thought of as obtaining a measure of the degree to which the $\{p_i\}$ are equitably distributed among the fixed set of observed species. As we shall see, there are 2 basic methods by which to achieve this: Either we standardize our measures (say by dividing by S) or we somehow obtain equal values of S (e.g., by rarefaction or abundifaction as in Chapter 12).

While these end or extreme conditions are quite clearly defined and described, they are not usually observed in nature. Hence, researchers desire a measure for identifying more subtle differences among the values in **p**. We shall continue by presenting a unification of the available evenness measures that shows how each suggested evenness measure is a member of a general family.

DEVELOPMENT OF GENERALIZED NOTATION FOR EVENNESS AND DOMINANCE

The development of measures of diversity, evenness, and dominance has been seemingly everlasting, and these concepts and the descriptive, single-value indices are now well cemented yet still misunderstood in the ecological literature. In Chapter 14 we discussed the fact that each of the diversity indices is often presented and treated as a distinct entity. Commonly, we find lists of measures calculated on a single data set and never compared or discussed except as individual numerical values. Although each of the indices for a research study is calculated from the same set of $\{p_i\}$ values, N and S, no comparisons are usually involved. However, by using a general formula for a family of diversity measures in the last chapter, we show the interrelationships among the most common diversity measures and provide a unification. In this chapter, we continue our discussion of evenness and dominance as two ends of a spectrum and then present a unified treatment of evenness measures and a general formula for a family of evenness measures for encapsulating these concepts quantitatively.

There have been very few parametric families of measures developed that can be used for quantifying the concepts of ecological evenness or dominance. However, in 1973, Hill devised a family based upon that of Renyi's (see Chapter 14). In Hill's general notation, which transformed Renyi's logarithmic family to the arithmetic scale, he combined diversity measures into effective measures of dominance or evenness. The conceptual basis for this family of measures was Shannon's information theory. Hill's numbers of order a, N_a, for orders 0, 1, and 2 coincided with the popular information-based diversity measures that Renyi combined in his diversity family. Hill defined these numbers as

$$N_a = e^{H_a} \tag{15.1a}$$

where H_a is Renyi's formula in Equation 14.13. Likewise, we could write

$$H_a = ln(N_a)$$ (15.1b)

to express the relationship between Renyi's and Hill's. Equations 15.1a and 15.1b merely show what we emphasize in Chapter 14 that Hill's family uses a different way to write the parameters of Renyi's family; both families are equivalent.

In this way N_a in Equation 15.1a is the equation to find the *numbers equivalent* expressions for each of the diversity indices (Ricotta 2005). For example, e^H is the equivalent number of species needed to be evenly distributed (all p_i's equal) to obtain the value for H that was observed with the unequal p_i values (MacArthur 1965). Jost (2006) recommends the transformation of information indices to obtain the effective number of species rather than the calculated value of the index itself. A second explanation is that N_a in Equation 15.1a represents the number of species we would need to have to yield the same information measure H_a if all species were equally represented (Patil and Taillie 1982; Ludwig and Reynolds 1988). This statement seems intuitively appealing as a description of the ecological concept of diversity, and Hill's numbers have a firm basis in information theory, so we shall continue using Hill's formulation.

Hill's family uses a quantity with order a, like that of Renyi, and we can write for Hill's number N_a:

$$N_a = (\Sigma p_i^a)^{\frac{1}{1-a}} = \frac{1}{(a-1)\sqrt{(\Sigma p_i^a)}}$$ (15.2)

where

$N_{-\infty}$ = the reciprocal of $\min\{p_i\}$, for $i = 1, \dots, S$, or the rarest species
$N_0 = S$, species richness (all species both rare and common are included)
$N_1 = e^{H_1}$, for H_1, which is Shannon's index (more emphasis on rare than common and abundant)
$N_2 = \dfrac{1}{\text{Simpson's } \lambda}$, or the multiplicative inverse of Simpson's index (more emphasis on rare and less on abundant than Shannon's)
$N_\infty = \dfrac{1}{\text{Berger–Parker index}}$, or the reciprocal of $p_{i=1}$, the most abundant species (so that only the most dominant species is included).

As we use larger values for a, each resultant index will be smaller than the prior one; that is, N_0 is the largest and N_∞ will have the smallest value. As we can see, both formulations depend only upon the value of a in Equation 15.2 to which we raise the species proportions. Since this is true, and because of the characteristics of each of these measures discussed in the last chapter, we know that when we change a, going from 0 up to larger values, we are thereby selecting different members of this family of measures and making decisions concerning to what extent we want rare or dominant species to provide information on the diversity of the assemblage at issue in the research plan. Changing values for *a* is a scaling process from total species richness to the dominance concentration in the assemblage as we increase *a* (Podani 1992). Merely selecting a particular diversity or evenness measure because it worked in a particular situation, as is commonly done, or because a colleague suggested it, or for any poorly thought out reason will leave the research open to problematical interpretation and non-replicability.

We shall use Hill's notation to see how he arrived at some of his evenness measures. Recalling Equation 15.2 we expand it here to obtain

$$N_a = (p_1^a + p_2^a + \cdots + p_S^a)^{\frac{1}{1-a}} \tag{15.3}$$

where each p_i value is the relative abundance of one of the S species and *a* is the parameter. All orders *a* of the points of N_a are continuous and for our purposes can be summarized as a *diversity profile* (e.g., Patil and Taillie 1982), which many authors prefer rather than single-point estimates. The *profile* of any assemblage whose RSAV is **p** is plotted by changing the values of *a*, while plotting all graphs on the same axes. The data for Bolivia plot 01 is shown in Figure 15.1. If more than 1 community were plotted, this ordering would show the graph of the more diverse community above the graphs of those less diverse communities. The *k*-dominance plots of Lambshead and colleagues (1983) from Chapter 14 are also diversity profiles but with the ordering of the graphs reversed, with the less diverse above those plots of the more diverse communities. It is important to understand that when we consider 2 communities, even though a diversity profile plot such as this one in Figure 15.1 can show the more diverse community lying above the less diverse, this is not always the case and plots can intersect. That is because diversity plots show only a *partial ordering* and that ordering can be ambiguous as well. The ordering is only

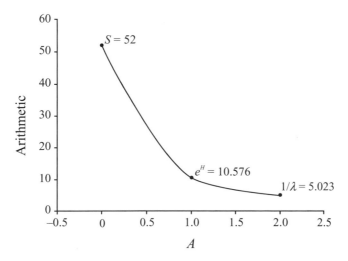

FIGURE. 15.1   Hill's continuum for Bolivia plot 01.

complete and unambiguous for the equiprobable case or distribution when all p_i values are equal. For the equiprobable case, $\mathbf{p} = \{1/S, 1/S, \ldots, 1/S\}$ and we always have

$$S \geq e^H \geq \frac{1}{\lambda} \geq \frac{1}{max\{p_i\}}. \tag{15.4}$$

With many but not all data sets these inequalities will hold as well.

Hill's family N_a expresses its members, which are the diversity measures, in the same relative abundance units. The members of this family as we have noted get increasingly sensitive to rare or abundant species as the parameter a changes. Therefore, Hill proposed to measure the evenness/dominance continuum with a continuum of ratios of diversity numbers.

Evenness Measures

Hill defined his family of measures using his numbers written as ratios:

$$E_{a,b} = \frac{N_a}{N_b}$$

where the a and b correspond to any 2 contiguous integers or pairs (a, b). However, Hill conceived of an evenness measure as a *standardized* measurement, and since his N_0 is equivalent to the total species richness S by which we need to standardize, he then defined his evenness family equivalently as

$$E_{a,0} = \frac{N_a}{N_0} \quad \text{for} \quad a > 0. \tag{15.5}$$

A look back at the last chapter on diversity measures shows that an information-based diversity measure when standardized is a measure of evenness. Taillie (1979) showed that this family met basic important mathematical rules for defining a quantitative measure of ecological diversity. In Figure 15.1 we plot this family of measures for the trees in Bolivia plot 01.

Evenness Measures Related as Special Cases in a Single Family

As we have discussed, when all the species are evenly distributed, that is when $p_1 = p_2 = \cdots = p_S$, we know that $H_{max} = H = lnS$. If we now get rid of the logarithm by raising each side to the power e, then we have $e^H = S$. Both Buzas and Gibson (1969) as well as Sheldon (1969) suggested the ratio of these 2 quantities as a measure of evenness because they saw that when $\frac{e^H}{S}$ is less than 1, this ratio measures the departure from complete evenness. Therefore, since $\dfrac{e^H}{S} = \dfrac{N_1}{N_0}$, in Hill's notation, we shall use this as our first evenness measure and write

$$E_1 = \frac{N_1}{N_0} = \frac{e^H}{S}. \tag{15.6}$$

For our Bolivian trees from plot 01, $E_1 = \dfrac{10.576}{52} = 0.203$. However, at the moment we have no way of evaluating what this value for E_1 suggests in terms of evenness, yet we usually see such single calculated evenness values presented in the peer literature as meaningful.

Continuing to use Equation 15.5 we now consider the measure corresponding to the next of Hill's measures: $\dfrac{N_2}{N_0}$ is

$$E_2 = \frac{N_2}{N_0} = \frac{\left(\dfrac{1}{\lambda}\right)}{S} = \frac{1}{\lambda S} \tag{15.7}$$

where λ is Simpson's diversity measure discussed in Chapter 14.

For our Bolivian example, we calculated $E_2 = \dfrac{5.045}{52} = 0.097$, but again, we have no idea of how the two measures E_1 and E_2 compare or how they can describe the evenness of the Bolivian plot. All that we can say about these calculated evenness values at this point is that the more the $\{p_i\}$ values differ from each other, the lower must be the calculated value of each of the ratios (Alatalo 1981).

Use of Hill's general evenness family in Equation 15.5 lets us see that both Shannon's H and Simpson's λ index, described as measures of diversity in the previous chapter, can also be considered to be measures of evenness when comparing populations with the same richness S. Curiously, Shannon's H is almost never used as a measure of evenness, but Simpson's λ is often cited as a measure of dominance because the first couple of p_i's account for most of the measure (Smith and Wilson 1996). As we saw in the Chapter 14 the complement $(1 - \lambda)$, which in ecology is called the probability of interspecific encounter, or PIE (Hurlbert 1971; Olszewski 2004), is also often used incorrectly as a measure of evenness. From the previous chapter and the beginning of this paragraph, however, it should be clear that both Shannon's H and Simpson's λ $\left(\text{including functions of Simpson's: } (1 - \lambda) \text{ and } \dfrac{1}{\lambda} \right)$ are measures of evenness *only* if comparing populations *with the same* S. Note also the example in Chapter 14 where the same value of λ was obtained with different values of S and clearly different values of evenness (Table 14.3). Use of $(1 - \lambda)$ as a measure of evenness, believing it to be independent of S, is incorrect and unwarranted (see, for example, Alroy and colleagues, 2008, who make this error).

In both Equations 15.6 and 15.7 the denominator is N_0, or S. However, Hill also suggested that other ratios of his diversity numbers might be of interest. So we also consider the ratio $\dfrac{N_2}{N_1}$ (Figure 15.1) as our third evenness possibility. We shall take this measure into account for completeness' sake and to show the consequences of use of a nonstandardized measure that does not adhere to the set given in Equation 15.5

$$E_3 = \frac{N_2}{N_1} = \frac{\left(\dfrac{1}{\lambda}\right)}{e^H} = \frac{1}{\lambda e^H}. \tag{15.8}$$

For our Bolivian tree plot 01 example (Figure 15.1), we have $E_3 = 5.045/10.576 = 0.477$.

AN ALTERNATIVE FAMILY OF EVENNESS MEASURES

Recalling from our discussion of E_1 that when all species are equally distributed $e^H = S$, or, $H = lnS$, we can form the ratio $\dfrac{H}{lnS}$. Then, just as $\dfrac{e^H}{S}$ when less than 1 is a measure of evenness, so also the same condition is true for $\dfrac{H}{lnS}$. This set of facts prompted Pielou (1966) to suggest the measure

$$J_1 = \frac{lnN_1}{lnN_0} = \frac{H}{lnS} \tag{15.9}$$

as an evenness measure; that is, the standardization of H with respect to its maximum, $H_{max} = lnS$. Although the two families are equivalent, the logarithmic notation for the family of measures is Renyi's, but we shall continue to expand these measures with Hill's formulation. For our Bolivian example, $J_1 = \dfrac{2.359}{3.951} = 0.597$.

Continuing to emulate the ordering we used for our E measures, the next measure in this J group is

$$J_2 = \frac{lnN_2}{lnN_0} = \frac{ln\left(\dfrac{1}{\lambda}\right)}{lnS}. \tag{15.10}$$

For Bolivian tree plot 01 we have $J_2 = \dfrac{1.618}{3.951} = 0.410$. Now if we recall the methods for manipulating logarithms, we know that since the logarithm of a fraction; that is, $ln\left(\dfrac{1}{\lambda}\right)$ is equivalent to the log of the denominator subtracted from the log of the numerator, then equivalently we have $J_2 = \dfrac{-ln\lambda}{lnS}$. This was a measure suggested by Smith and Wilson in 1996.

To complete our logarithmic-based ratios, we use

$$J_3 = \frac{\ln N_2}{\ln N_1} = \frac{\ln\left(\dfrac{1}{\lambda}\right)}{H}. \tag{15.11}$$

For the Bolivian trees, we calculated $J_3 = \dfrac{1.618}{2.359} = 0.686.$

We now have 6 measures of evenness and have calculated each of these for a single set of data (Bolivian tree plot 01). The question now becomes how does one evaluate these 6 different values and how well does each measure the evenness on data sets with clearly different evenness distributions?

COMPARING EVENNESS VALUES

Let us now examine 2 hypothetical data sets with clearly distinct evenness conditions to see whether we can detect how well each of the 6 measures actually provides a quantitative measure of this concept of evenness. In Table 15.1, the first set of 5 hypothetical observations clearly shows dominance by the most abundant species p_i. The second set of observations is a Broken Stick distribution (Chapter 9). This distribution has arguably the most even or equitable distribution observed in nature (Lloyd and Ghelardi 1964). What we wish to do is compare the columns of measures next to the observations for each of the two distributions.

For the distribution exhibiting dominance, all of the information-based diversity measures in the first column, except for λ, are smaller than for the more equitable Broken Stick distribution. Note, however, that N is the same for both distributions. Simpson's lambda, of course, is larger for the dominant distribution because the first value for calculation of λ is $p_i^2 = (0.90)^2 = 0.810$, or 99.5% of the total value of λ. On the other hand, for the calculation of the Broken Stick distribution, $p_i = 0.46$ so $p_i^2 = 0.212$, or 67.2% of the total λ. Note that in Table 15.1, λ is larger for the more dominant case, while $(1 - \lambda)$ is smaller. One of the reasons that many researchers prefer PIE or $(1 - \lambda)$ is because it is going the "right way"; that is, its values range from 0 to 1 as the assemblage changes from more to less dominant.

The most important point to keep in mind is that when we are comparing 2 populations with the same number of species in each, say $S = S_1 = S_2$, all our information-based measures of evenness in Table 15.1 adequately evaluate what researchers refer to as dominance, evenness, or equitability regardless of the spread of the $\{p_i\}$ values.

Let us now examine the second column containing 6 evenness measures. Recall that E_1, E_2, J_1, and J_2 are each standardized by S. The values for all measures of evenness are greater for the Broken Stick distribution, which is the more even. We can conclude correctly from each of these evenness measures that indeed the Broken Stick distribution is more even. Examining Figure 15.1 we see that $S > e^H > \dfrac{1}{\lambda}$, as we saw in Equation 15.2, even though we do not have complete evenness, this inequality may be

TABLE 15.1   A Dominant and Broken Stick Distribution with $S = 5$

Distribution with Dominance

n_i	p_i	p_i^2	$p_i ln p_i$	THE MEASURES	
90	0.90	0.810	−0.095	$\lambda = 0.814$	$E_1 = 0.315$
4	0.04	0.002	−0.129	$1 - \lambda = 0.186$	$E_2 = 0.246$
3	0.03	0.001	−0.105	$1/\lambda = 1.229$	$E_3 = 0.781$
2	0.02	0.0004	−0.078	$H_2 = ln(1/\lambda) = 0.206$	$J_1 = 0.281$
1	0.01	0.0001	−0.046	$H_1 = 0.453$	$J_2 = 0.128$
——	——	——	——	$e^H = 1.573$	$J_3 = 0.763$
100	1.000	0.814	−0.453		

Equitable Broken Stick Distribution

n_i	p_i	p_i^2	$p_i ln p_i$	THE MEASURES	
46	0.46	0.212	−0.357	$\lambda = 0.315$	$E_1 = 0.750$
26	0.26	0.068	−0.350	$1 - \lambda = 0.685$	$E_2 = 0.635$
16	0.16	0.026	−0.293	$1/\lambda = 3.175$	$E_3 = 0.846$
9	0.09	0.008	−0.217	$H_2 = ln(1/\lambda) = 1.155$	$J_1 = 0.821$
3	0.03	0.001	−0.105	$H_1 = 1.322$	$J_2 = 0.718$
——	——	——	——	$e^H = 3.751$	$J_3 = 0.874$
100	1.000	0.315	−1.322		

NOTE: The values of H_1, H_2, and $1 - \lambda$ all increase in value as we go from a dominant distribution to the more equitable Broken Stick distribution.

true in most cases but not necessarily. Evenness measures with the subscript 3 will be the largest because the values of e^H and $\frac{1}{\lambda}$ are much closer to one another than when an index is standardized or divided by S. Our examination of two distributions in which $S_1 = S_2$ indicates that all the measures we considered will discriminate the differences in evenness quite adequately. From this cursory examination both the E's and J's appear easy to understand, desirable, and perfectly acceptable measures of evenness. Are they?

The Repeat, or Mixtures of Distributions

Hill (1973) and Taillie (1979) and Jost (2007) indicated that a fundamental property of an evenness measure is that it should remain the same if the distribution is simply "repeated," or "doubled." Hill used a male and female split as a relatively uninformative and unclear example. By *repeat* it is meant that we start with some number of assemblages that are replicas of one another, but have no species in common. Then, the repeat is a *mixture* of these assemblages wherein these authors want the evenness of the final result or mixture to be identical to the evenness of each of the component assemblages. *Passing the repeat test* means that this equality of evenness measures between the mixture and the original component assemblages holds true despite the increase in richness and the consequent increase in diversity indices. Smith and Wilson (1996) regarded passing a repeat test as a sign that the evenness measure is independent of S, which we shall examine as well.

The best way to understand the repeat is with an example. We will repeat the distribution for the dominant and Broken Stick models used in Table 15.1 and present the results in Table 15.2.

From Table 15.2, it is obvious that if we repeat exactly the number of individuals within each species from Table 15.1, we double the total number of individuals; we have $N = 200$. Similarly, we also have doubled the number of species to obtain $S = 10$ species. The result of this composite in Table 15.2 for $S = 10$ is that with the increase in richness and relative abundances, there is a consequent increase in all diversity measures.

However, the evenness measures E_1, E_2, and E_3 remain the same in both tables, within rounding error, while those values of J_1, J_2, and J_3 do not. The set of evenness measures we denoted as the J's do not pass the repeat test.

Looking at the values across Tables 15.1 and 15.2 we see that all information-based diversity measures for each distribution have increased for the mixture except for Simpson's λ. All these measures, except for λ, are higher for the more even distribution; the Broken Stick. We also saw this in Table 15.1. Looking at our evenness measures, we observe that E_1, E_2, and E_3 remain the same within rounding error as they do in the $S = 5$ example. The J's, however, do not. Like the information measures and unlike the evenness measures E, all the J's increase.

In summary, if we compare the results in Table 15.1 with those in Table 15.2, clearly the original species richness for the distribution of values has doubled, or increased from 5 to 10, when we mix the two assemblages. There are 2 sets of identical absolute abundances for $S = 5$, while in the composite the relative abundances for $S = 10$ have changed. Thus, the diversity measures must change as well. It should be clear that when we speak of a mixture of identically distributed vectors, or a "repeat" distribution, we mean to refer only to the absolute abundance data, not that the values of $\{p_i\}$ remain the same. In addition, the set of evenness measures we called E_1, E_2, and E_3 pass the repeat test, and remain constant under mixing, while those designated by J_1, J_2, and J_3 do not.

Changes in Simpson's and Shannon's Measures Under Repeat

Using these tables once more we can see the rule for the change in the diversity measures. In Table 15.2, Simpson's index using $S = 10$ is one-half of that same calculation when $S = 5$. Thus for a mixture of 2 populations in which the original Simpson's

TABLE 15.2 A Dominant Distribution Exactly Repeated

Dominant Distribution

n_i	p_i	p_i^2	$p_i ln p_i$	THE MEASURES FOR MIXTURE DISTRIBUTION WITH $N = 200$ AND $S = 10$	
90	0.450	0.203	−0.359	$\lambda = 0.407$	$E_1 = 0.314$
4	0.020	0.0004	−0.078	$1 - \lambda = 0.593$	$E_2 = 0.246$
3	0.015	0.00022	−0.063	$1/\lambda = 2.457$	$E_3 = 0.783$
2	0.010	0.0001	−0.046	$H_2 = ln(1/\lambda) = 0.899$	$J_1 = 0.497$
1	0.005	0.000025	−0.026	$H_1 = 1.144$	$J_2 = 0.390$
90	0.045	0.203	−0.359	$e^H = 3.139$	$J_3 = 0.786$
4	0.020	0.0004	−0.078		
3	0.015	0.00022	−0.063		
2	0.010	0.0001	−0.046		
1	0.005	0.000025	−0.026		
200	1.000	0.407	−1.144		

Broken Stick Distribution

n_i	p_i	p_i^2	$p_i ln p_i$	THE MEASURES	
46	0.23	0.053	−0.338	$\lambda = 0.156$	$E_1 = 0.751$
26	0.13	0.017	−0.265	$1 - \lambda = 0.844$	$E_2 = 0.641$
16	0.08	0.006	−0.202	$1/\lambda = 6.410$	$E_3 = 0.851$
9	0.045	0.002	−0.140	$ln(1/\lambda) = 1.858$	$J_1 = 0.876$
3	0.015	0.00022	−0.063	$H_1 = 2.016$	$J_2 = 0.807$
46	0.23	0.053	−0.338	$e^H = 7.508$	$J_3 = 0.922$
26	0.13	0.017	−0.265		
16	0.08	0.006	−0.202		
9	0.045	0.002	−0.140		
3	0.015	0.00022	−0.063		
200	1.000	0.156	−2.016		

NOTE: Instead of $S = 5$, we now have $S = 10$ and a Broken Stick distribution exactly repeated so that instead of $S = 5$, we now have $S = 10$.

index is λ, then the value of Simpson's index for the mixture is $\dfrac{\lambda}{2}$. If we mix more than 2 populations, say m populations, then the value would be $\dfrac{\lambda}{m}$. In the repeat case, Simpson's diversity index decreases for the mixture.

Examining the values of Shannon's index for the mixture and original populations, a formula is not so easily discerned. However, when $S = 5$ we have $H_1 = 0.453$ while for $S = 10$, $H_1 = 1.144$. The rule for the change in Shannon's formula with mixing identical populations is that the value of H_1 for the mixture of 2 populations is $H(\text{original populations}) \cdot (ln2)$. For our example, we have that $1.144 = 0.453 + 0.693$. In general then for m populations, the value of Shannon's index for the original distribution is equal to

$$-\Sigma\left(\frac{p_i}{m}\right)ln\left(\frac{p_i}{m}\right) = -\Sigma\, p_i lnp_i + ln(m). \tag{15.12}$$

The diversity computed on the mixture is greater than the diversity computed on the original distribution. This statement is not true when we are considering evenness measures since it is desirable for them to remain constant under repeat.

The Extremes of Evenness Indices

The final property of evenness measures that has given rise to controversy in the biodiversity literature concerns the desired minimum and maximum of a measure. Although preferences are hotly debated, whether an index goes from 0 to 1, 1 to 0, or between two other limits is irrelevant to correct quantitative application. The maximum value that the evenness measures of E_1, E_2, E_3, J_1, J_2, and J_3 can take on is 1 when all species abundances are equal, or $\{n_1 = n_2, \ldots, = n_S\}$.

Ideally, most ecologists would like a measure of evenness to vary from this maximum of 1 to a minimum of 0 and at the same time show an independence from S. While examination at the maximum is straightforward, the minimum requires a little more explanation. The *true minimum* occurs when all but 1 species contain 0 individuals; this would be *true dominance*. However, this is not possible for biodiversity purposes because, simply put, if a species is not represented by at least 1 individual, we would not know it is there. Instead, in sampling natural populations we say that except for 1 species, all the species in the observed sample contain 1 individual for the minimum to occur in biodiversity work. Achieving the true minimum is simply not possible in biodiversity research.

Not realizing the difference between the 2 cases for the minimum above, Heip (1974) pointed out with an unrealistic example that the value of the E_1 index of evenness does not reach 0 if $S = 2$ and one species contains 999,999 individuals and the other 1. In this highly improbable situation, $H = 0.000001$ and $e^H = 1$ so that $E_1 = \dfrac{1}{2} = 0.500$. Consequently, he proposed modifying E_1 by subtracting 1 from the numerator and denominator to "correct" the measure, even though 1 is not the true minimum possible with this situation. Let us call this adjusted index E_1 – prime (E_1') and write

$$E_1' = \frac{(e^H - 1)}{(S - 1)} \tag{15.13}$$

so we can compare it to Equation 15.6 for illustrative purposes. Later, Smith and Wilson (1996) also believed reaching 1 was essential. These authors offered the same modifications to E_2 and E_3 as we used in Equation 15.8. We have then

$$E_2' = \frac{\left(\frac{1}{\lambda} - 1\right)}{(S - 1)} \tag{15.14}$$

and

$$E_3' = \frac{\left(\frac{1}{\lambda} - 1\right)}{(e^H - 1)}. \tag{15.15}$$

With $N = 1,000$ and $S = 10, 20, 40,$ and 80, let us examine how these E-prime values compare to the original evenness measures E_1, E_2, and E_3. While finding the maximum is straightforward, as S and N vary, finding the actual minimum for our not-quite-total-dominant case is not. For this minimum we would have to assign a single individual to all species except 1 and the remaining individuals to the 1 abundant species and see what value results, which is quite time-consuming. Fortunately, Fager (1972) provided simple solutions. The minimum for Simpson's measure he derived is

$$\lambda_{min} = 1 - \frac{(S - 1)(2N - S)}{N(N - 1)}. \tag{15.16}$$

And for Shannon's measure, we have

$$H_{min} = lnN - \frac{(N - S + 1)}{N} lnN(N - S + 1). \tag{15.17}$$

Using these equations we prepared Table 15.3 to offer a comparison of the values.

In Table 15.3, we have included the information measures so that the reader can easily see how the evenness indices are calculated. Note that for both Simpson's and Shannon's measures the values for E_1 and E_2 come close to 0 as we increase the richness. This shows that the value of each of these measures depends correctly on the fact that we are not using a case of true total dominance, only the dominance that is

TABLE 15.3   Comparing Minimum E and E′ Values at $N = 1,000$ and $S = 10$, 20, 40, and 80

MEASURE	$S = 10$	$S = 20$	$S = 40$	$S = 80$
λ	0.982	0.962	0.923	0.848
$1 - \lambda$	0.018	0.038	0.077	0.152
$\dfrac{1}{\lambda}$	1.018	1.040	1.083	1.179
$ln\left(\dfrac{1}{\lambda}\right)$	0.018	0.039	0.080	0.165
H_1	0.071	0.150	0.308	0.622
e^H	1.074	1.162	1.361	1.863
$E_1 = \dfrac{e^H}{S}$	0.107	0.058	0.034	0.023
$E'_1 = \dfrac{e^H - 1}{S - 1}$	0.008	0.009	0.009	0.011
$E_2 = \dfrac{1}{\lambda S}$	0.102	0.052	0.027	0.015
$E'_2 = \dfrac{\frac{1}{\lambda} - 1}{S - 1}$	0.002	0.002	0.002	0.002
$E_3 = \dfrac{1}{\lambda e^H}$	0.948	0.895	0.796	0.633
$E'_3 = \dfrac{\frac{1}{\lambda} - 1}{e^H - 1}$	0.243	0.247	0.230	0.208

observable in a field sample as we discussed. In addition, the resultant value of those evenness measures' actual closeness to 0 depends upon the sample size, or S, so a correction other than 1 is possible; simple to obtain but not particularly useful. All the Simpson-based measures $\left((1 - \lambda), \text{ and } ln\left(\dfrac{1}{\lambda}\right)\right)$ increase, except for λ itself, which decreases. The Shannon measure, H_1, and e^{H_1}, also increase.

Comparing the E and E' values we also see changes. Both E_1 and E_2 have minima of about 0.1 at $S = 10$, while E'_1 and E'_2 are closer to 0 as desired. However, for $S > 20$, the original E_1 and E_2 values are reasonably close to 0.

While $E'_3 < E_3$, both measures remain far above 0. The reason is obvious when looking at values of $\dfrac{1}{\lambda}$ and e^H; they are both close to one another but their ratio is not standardized by S or a function of S as was the case for the E_1 and E_2 measures. In Figure 15.2, we illustrate what is happening. As we move from $S = 10$ to $S = 80$, the denominator in E_1 and E_2 gets larger and so the values of the measures get closer to 0. This does not happen for E_3. Consequently, E_3 does not have the same range as the other 2

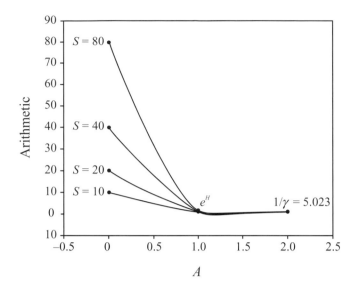

FIGURE 15.2   Minimum values of e^H and $1/\lambda$ at $S = 10$, 20, 40, and 80.

measures of E and is not desirable for an evenness index. Nevertheless, as we see in Chapter 17, it will have other useful properties.

E′ and the Repeat, or Mixtures

Let us now examine in Table 15.4 how the E' series that was designed to have a minimum of 0 behaves using the same repeat set of a dominant distribution and a Broken Stick distribution at $S = 5$, 10, 20, and 40 species.

The E' series clearly does not pass the repeat test, but the values get closer to E as S increases, as we would expect by examining the equations for their calculation. We also conclude that after S is greater than about 20, the E series is effectively at 0 for its minimum.

Therefore, because only the E series passes the repeat, only the E series is acceptable as the family of evenness measures. Derivation of decomposition equations in the next chapter reconfirms this position. In addition, since the diversity of a mixture of assemblages should always be greater than or at least not less than that for the original assemblage, Shannon's measure is preferred for diversity measurement.

SUMMARY

1. Using Hill's continuum, the family of established evenness measures contains

$$E_1 = \frac{e^H}{S}, E_2 = \frac{1}{\lambda S}, \text{ and } E_1 = \frac{1}{\lambda e^H}.$$

2. Other popular evenness measures can be represented as members of Renyi's family of log-based measures; namely, $J_1 = \dfrac{H}{\ln S}$, $J_2 = \dfrac{\ln\left(\dfrac{1}{\lambda}\right)}{\ln S}$, and $J_3 = \dfrac{\ln\dfrac{1}{\lambda}}{H}$. However,

TABLE 15.4   Values of E and E' for a Dominant Distribution Shown in Table 15.1 for the Repeat $S = 5$, 10, 20, and 40

Dominant Distribution

	$S = 5$	$S = 10$	$S = 20$	$S = 40$
$E_1 = \dfrac{e^H}{S}$	0.315	0.315	0.315	0.315
$E_1' = \dfrac{e^H - 1}{S - 1}$	0.143	0.238	0.278	0.297
$E_2 = \dfrac{1}{\lambda S}$	0.246	0.246	0.246	0.246
$E_2' = \dfrac{\dfrac{1}{\lambda} - 1}{S - 1}$	0.057	0.162	0.206	0.226
$E_3 = \dfrac{1}{\lambda e^H}$	0.781	0.781	0.781	0.781
$E_3' = \dfrac{\dfrac{1}{\lambda} - 1}{e^H - 1}$	0.400	0.681	0.740	0.762

none of these J measures pass the repeat test and cannot be recommended for general use in biodiversity research.

3. E' measures, designed to have minimum values close to 0 or at 0, can be written as members of a family of measures; namely, $E' = \dfrac{(e^H - 1)}{S - 1}$, $E_2' = \dfrac{\left(\dfrac{1}{\lambda} - 1\right)}{S - 1}$, and $E_3' = \dfrac{\left(\dfrac{1}{\lambda} - 1\right)}{(e^H - 1)}$. When S is less than about 20, they are as effective as the E values, but at greater values of S little is gained through their use. In addition, the E' measures do not pass the repeat test. We do not recommend these measures for general use in biodiversity research.

4. Like all the J indices, E_3' does not pass the repeat test.

5. There is no fundamental distinction between diversity and evenness when species richness is held constant for comparisons.

PROBLEMS

15.1 Using the values for a dominant distribution given in Table 15.1, evaluate the J's and E'(prime)'s at $S = 20$ and $S = 25$.

15.2 Smith and Wilson (1996) suggested a value of $E = \dfrac{(1 - \lambda)}{1 - \dfrac{1}{S}}$. Evaluate the repeat at $S = 5$, $S = 10$, and $S = 15$.

15.3 Evaluate the minimum values for the above evenness measure at $S = 10$ and $S = 80$.

15.4 Peters (2004) thought he had devised a new evenness measure, which had, however, been developed and discarded many years prior. This measure was $E = 1 - \left(\lambda - \dfrac{1}{S} \right)$. Evaluate the repeat of this measure at $S = 5$, $S = 10$, and $S = 15$.

15.5 Evaluate the minimum values for the above evenness measure at $S = 10$ and $S = 80$.

16

BIODIVERSITY:
UNIFYING DIVERSITY AND EVENNESS
MEASURES WITH CANONICAL EQUATIONS

IN CHAPTERS 12 THROUGH 15 we present the development of a family of measures that adequately summarize the attributes or characteristics of the RSAV, **p**. We showed that the most useful of these measures were based upon information theory as developed by Shannon, Renyi, Hill, and others. In addition, we tied together most of the widely accepted measures, showed how they could be viewed in sets with common characteristics. We then examined characteristics of each member of each set in light of usage in the biodiversity literature and finally discussed rules and properties of each set that made certain of the measures unacceptable and other measures the most useful for our purposes. In this chapter we show how each of our information-based evenness and diversity measures can be unified within a single canonical equation. Our canonical equation decomposes richness, information diversity measures, and evenness into their component parts.

The decomposition equation solves a problem that eluded solution for nearly 50 years and was declared unsolvable. The problem: How much of any information diversity measure is due to richness and how much to evenness? Alternatively, can richness and evenness be related quantitatively? An amazingly simple solution was derived by Hayek and Buzas (1997, 1998). We give the essence of their derivation here.

HISTORICAL DERIVATION OF DECOMPOSITION

Recall that when all species are equally distributed $(p_1 = p_2 = \cdots = p_S)$, then $e^{H_1} = S$ where e is, as everywhere in this book, the base of the natural logarithms, $H_1 = -\Sigma p_i ln p_i$, and S is the number of species. Therefore, in this case the ratio $\dfrac{e^{H_1}}{S} = 1$ at complete evenness and less than 1 in all other situations. Buzas and Gibson (1969) defined this ratio as a measure of equitability or evenness, and in this book we call it E_1. We have then $E_1 = \dfrac{e^{H_1}}{S} = 1$ or $e^{H_1} = S \cdot E_1$. By taking the natural log (ln) of each side of the equation $e_1^H = S \cdot E_1$, we obtain the desired decomposition

$$H_1 = lnS + lnE_1. \tag{16.1}$$

This is an incisive equation. Our decomposition equation is very important for all biodiversity work because researchers now have a simple way to examine evenness separately from richness within the *same biological assemblage or ecological system*. This equation is the mechanism by which to unify all biodiversity measurement. It is important to notice that in Equation 16.1 H_1 can be used as a vehicle for the decomposition of evenness and richness, as well as a simple diversity index. Remember also that we show in Chapter 15 that there is no fundamental distinction between information diversity and evenness when species richness, S, is held constant for comparisons.

To see what Equation 16.1 reveals, first consider that it is always true that $0 \le E_1 \le 1$. Therefore, the value of lnE must always be negative (*ln* of all numbers <1 is negative). Consequently, Equation 16.1 says that Shannon's H_1 is made up of *lnS* minus the amount of evenness exhibited by the abundances. Because we know the maximum value of H_1 is achieved when all species are equally distributed; that is, when $lnS = H_{1max}$ we can also write for Equation 16.1 $H_1 = H_{1max} + lnE_1$. In this last case, $E_1 = 1$ so that $lnE_1 = 0$.

In a similar fashion recall that when all species are equally distributed $\dfrac{1}{\lambda} = S$. Once again, the ratio $\dfrac{\frac{1}{\lambda}}{S} = \dfrac{1}{\lambda S} = 1$ in this case. We can define a measure of evenness as

$$E_2 = \dfrac{\frac{1}{\lambda}}{S} = \dfrac{1}{\lambda S} \text{ (Smith and Wilson 1996). As above } \dfrac{1}{\lambda} = S \cdot E_2. \text{ Taking the natural log } (ln)$$

of both sides we obtain

$$H_2 = lnS + lnE_2 \tag{16.2}$$

recalling that $H_2 = ln\left(\dfrac{1}{\lambda}\right)$. We have then the same decomposition with Simpson's index as with Shannon's.

DECOMPOSITION AND THE LOG CONTINUUM

In Chapter 15, we use Hill's continuum (illustrated in Figure 15.1) to establish the E measures for evenness; E_1, E_2, and E_3. These evenness measures were all ratios of the primary diversity measures S, e^{H_1}, and $\dfrac{1}{\lambda}$ obtained from the family (Equation 14.13) of measures by varying the order a. In Figure 16.1 we illustrate this family for the Bolivian trees of plot 01 on a log scale instead of the arithmetic scale used in Figure 15.1. The simple division used to form ratios of the E set of measures on the arithmetic scale no longer applies. Because we are using logs the quantities must be subtracted from one another to achieve the same result. By using the definitions in Chapter 15 we have

$$lnE_1 = H_1 - lnS \tag{16.3}$$

and also

$$lnE_2 = H_2 - lnS. \tag{16.4}$$

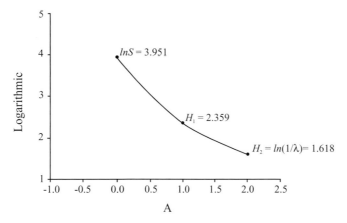

FIGURE 16.1   Hill's continuum for Bolivia plot 01 on a log scale.

These equations can be obtained directly from consideration of Figure 16.1 or by algebraically manipulating Equations 16.1 and 16.2. On Figure 16.1, we see the 3 points lnS, H_1, and H_2. The first 2 measures of evenness were obtained by subtracting the measure of lesser value (H) from the measure of higher value (lnS). Using the same procedure on 1 more measure is ($H_2 - H_1$). Hence, we have

$$lnE_3 = H_2 - H_1. \tag{16.5}$$

In Chapter 14, we point out via Equations 14.17 and 14.18 that Shannon's information function differed from Simpson's only by a constant. Now, we see that constant is lnE_3.

Finally, by subtracting Equation 16.3 from Equation 16.4 and using the equality in Equation 16.5, we have

$$lnE_3 = lnE_2 - lnE_1 \tag{16.6}$$

or

$$lnE_1 = lnE_2 - lnE_3. \tag{16.7}$$

We have now successfully decomposed or unified information measures, species richness, and evenness into a single set of canonical equations.

CHECKING THE UNIFIED MEASURES WITH A NUMERICAL EXAMPLE

We recall from Bolivia plot 01 that we have $S = 52$ and $H_1 = 2.359$, so that $E_1 = \dfrac{e^{2.359}}{52} = \dfrac{10.580}{52} = 0.203$. Using Equation 16.1 for decomposition we have $H_1 = lnS + lnE_1$, so that $2.359 = 3.951 + (-1.952) = 2.359$.

For Bolivia plot 01, we also have $\lambda = 0.198$ and $\frac{1}{\lambda} = 5.043$. $E_2 = \frac{\frac{1}{\lambda}}{S} = \frac{5.043}{52} = 0.097$. Using Equation 16.2 for decomposition we have $H_2 = lnS + lnE_2$ so that $1.618 = 3.951 + (-2.333) = 1.618$. For $E_3 = \frac{\frac{1}{\lambda}}{e^{H_1}} = \frac{5.043}{10.580} = 0.477$. From Equation 16.5 $lnE_3 = H_2 - H_1$ so that $-0.741 = 1.618 - 2.359 = -0.741$.

From Equation 16.6 we also have $lnE_3 = lnE_2 - lnE_1$, and from above $-0.741 = -2.333 - (-1.952) = -0.741$.

DECOMPOSITION WITH E' MEASURES

As an alternative to E_1, in Chapter 15 Equation 15.8, $E'_1 = \frac{e^{H_1} - 1}{S - 1}$ was suggested as an alternative because it could reach 0 at low values of S. In Chapter 15 we rejected Equation 15.8 because it would not pass the repeat test. Let us now examine how well it decomposes. $E'_1 = \frac{10.576 - 1}{52 - 1} = \frac{9.576}{51} = 0.188$ and $ln(0.188) = -1.673$. From Equation 16.1, substituting E'_1 for E_1, $2.359 = 3.951 + (-1.673) = 2.278$. Thus the E'_1 measure not only will not pass the repeat test, it will also not decompose. With a smaller S, the failure will become worse.

A suggested alternative for the measure E_2 was Equation 15.9, $E'_2 = \frac{\left(\frac{1}{\lambda} - 1\right)}{S - 1} = \frac{(5.043 - 1)}{(52 - 1)} = \frac{4.043}{51} = 0.079$ and $ln(0.079) = -2.535$. From Equation 16.2, substituting E'_2 for E_2, $1.618 = 3.951 + (-2.535) = 1.416$. Like E'_1, E'_2 also fails to decompose.

The alternative for E_3 was Equation 15.10, $E'_3 = \frac{\left(\frac{1}{\lambda} - 1\right)}{\left(e^{H_1} - 1\right)} = \frac{(5.043 - 1)}{(10.576 - 1)} = \frac{4.043}{9.576} = 0.422$. The $ln(0.422) = -0.862$. From Equation 16.6, substituting E'_3 for E_3, $lnE_3 = lnE_2 - lnE_1$, $-0.862 = -2.333 - (-1.952) = -0.381$. Like the other two alternatives E'_3 fails to decompose and repeat.

SUMMARY

1. Decomposition of the Shannon information function, H_1, into its component parts consisting of species richness and evenness is accomplished by $H_1 = lnS + lnE_1$.

2. Decomposition of the Simpson information function, H_2, into its component parts consisting of species richness and evenness is accomplished by $H_2 = lnS + lnE_2$.

3. Shannon and Simpson measures differ only by a constant and are related by

$$lnE_3 = H_2 - H_1 = ln\left(\frac{1}{\lambda e_1^H}\right).$$

4. All the E measures can also be related by $ln E_3 = ln E_2 - ln E_1$.
5. E' values for evenness will not decompose.

PROBLEMS

16.1 A count of foraminifera in a sample from Nueces Bay, Texas (Buzas-Stephens and Buzas 2005), had the following counts: 144, 4, 6, 25, 1, 3, 10, 1, 5, 163, and 1. Calculate the Shannon decomposition.

16.2 Using the data from Problem 16.1, calculate the Simpson decomposition.

16.3 Using the same data set again, calculate decomposition using E'_1 and E'_2. Comment on the result.

16.4 Comment on the efficacy of the measures E and E' for measuring evenness. Delete the value for 144, recalculate, and comment again.

17

BIODIVERSITY: SHE ANALYSIS AS THE ULTIMATE UNIFICATION THEORY OF BIODIVERSITY WITH THE COMPLETE BIODIVERSITYGRAM

T HIS CHAPTER IN COMBINATION WITH Chapter 18 composes the culmination of the chapters on quantitative biodiversity assessment. Here we introduce SHE analysis, the information–theoretic approach to obtaining a comprehensive biodiversity analysis. We show how SHE allows us to dissect our families of measures into components useful for powerful inferential statements; we elucidate the cogent aspects and give the fundamental properties of this technique. In addition, we explain how this new synthesis in biodiversity analysis can be used as a distribution-free methodology and considerably more. Strong inferences with SHE are enhanced by the addition of knowledge of an appropriate statistical reference distribution (Chapters 3 and 9). In addition, we provide a new, complete and all-inclusive plot for the totality of data from any biodiversity analysis called the Biodiversitygram, or BDG. Then, in the next chapter, we provide examples of the way in which SHE unifies the search for quantitative assessment of an assemblage or a total biological or ecological system.

Let us review what we have learned in the biodiversity chapters. Recall that we first defined the vector **p** of observed species proportions, and considered the single, commonly used diversity and evenness measures that rely on the set of relative abundances $\{p_i\}$ in **p**. We evaluated properties of each measure and showed how many should be discarded, and some were well-suited for quantitative biodiversity evaluation because they adhered to a set of standard rules. We then developed equations for encompassing these single measures into families of both diversity (in Chapter 14) and evenness measures (in Chapter 15). Finally, we showed how both of these families could be incorporated into one unified description (in Chapter 16) with our canonical equations. After all the measures were revealed to be related in a single family, we used the canonical equations to show how to decompose any family member or single diversity measure. We separated this unified development of diversity measures and each individual measure itself described by the canonical equations into species richness and evenness components. Throughout this treatment and discussion, we tried to make clear that in arriving at our canonical equations for the unification of biodiversity measurement and in the decomposition, we did not need to make any assumptions about the observed $\{p_i\}$. Despite the fact that we speak of the distribution of the p_i values, and that we also discussed how the $\{p_i\}$, or RSAV, could constitute a discrete probability distribution (Chapter 14), there was no mention of any further involvement of a statistical distribution. Each of the measures in our families can be calculated merely by using the observed sample values for the relative species abundances.

In this chapter we now describe SHE and show the additional enhancements and inferences for biodiversity study that can be obtained from SHE analysis. We describe the increase in knowledge of the complexities and the structure of species communities that results when we use the methodology of SHE to fit a particular statistical distribution that best describes the RSAV (or RAD).

SHE Analysis—An Overview

SHE (an acronym for species richness, diversity measure H, and evenness) is first a distribution-free methodology that allows the researcher to understand the observed data in a holistic manner. In addition, SHE is an inductive, analytical method by which we can hypothesize, fit, and verify a statistical distribution for the abundances based upon the unique one-to-one relationship of entropy as a distributional parameter.

Here we elucidate the all-encompassing and incisive SHE analysis (Buzas and Hayek 1996, 1998, 2005; Hayek and Buzas 1997, 1998, 2006) that was derived to provide a solid theoretical and unified basis for quantitative biodiversity measurement and analysis. SHE relies upon the fact that each distribution has a unique relationship with statistical entropy. There is a unique formula for the entropy of a statistical distribution and we show that this entropy can also be the limiting value for the information-based diversity measures, as the number of species gets larger with increased sampling or accumulation. Using this fact, we can describe deviations from this approach route of information H to the entropy, computed on successively larger samples, as changes in community structure indicative of assemblage change or biofacies, biome, or zonal end points.

In classification or clustering techniques, we use pairwise addition of individual species to group assemblages so that no overall view of the structure of the community as a whole exists. With SHE analysis the development is over multiple successive samples from a biome, ecosystem, or ecological community that gives us a snapshot of the total assemblage over space or time and provides a complete synthesis of the observed data.

SHE as Distribution-Free Methodology

When we derived the decomposition equation and then actually separated each diversity measure included in our canonical equation's description of its separate components for richness and evenness, we evaluated these components without regard to any particular conceptual or statistical distribution. Thus, this information-based SHE methodology can provide a total community synthesis that is distribution free.

In Chapter 12, we showed how the "collectors curve" indicates that as we accumulate individuals, we also accumulate species. In that case we wrote S as a function of the number of individuals, or $S = f(N)$. All the data in each sample, over the range of the sampling, can be accumulated either over space or time depending upon the scientific question. This accumulation can be done for both diversity and evenness evaluation as well as for collection effort, as we shall explain.

Consider our decomposition equation, which is $H = lnS + lnE$ (Equations 16.1 and 16.2). We can all agree by this point in the book that lnS is linearly related to lnN. Using the decomposition equation that shows for the diversity family that every lnS equals a diversity measure and an evenness measure, then it should be clear that both H and lnE must also have a linear relationship with the number of individuals. Therein lies the key to SHE.

ACCUMULATING OBSERVED SAMPLES

As we point out in Chapter 16, for field data there are two choices to obtaining a representative or large enough sample size for accurate, reliable biodiversity assessment: single or accumulated samples. In the accumulation of samples, we add together the new species that are found from each separate, successive sample and calculate a new diversity and a new evenness measure for each successive sample at each step in the accumulation. This method yields an ordered list or sequence of increasing or decreasing measures and some final value of N, S, and H and E that will be approached by the time all samples are so accumulated. When we come to the end of the accumulation, and have added in all the samples, we find that we have accumulated all values of lnN until the final sample size is reached. At this end point we will have obtained a final set of values for H, S, and E. Let us illustrate.

THE BIODIVERSITYGRAM: BDG

A biodiversitygram, or BDG, is a simple pictorial we developed (Hayek and Buzas 2006) for any sample data set of the interrelationships among all 3 variables in the decomposition equation. This plot combines lnS, H, and lnE, which are each plotted on the y axis against the same set of increasing values of lnN on the x axis. A BDG not only reminds us that each of our biodiversity measures, richness, evenness, and information-based diversity, is a function of the number of individuals, but gives us a clear way to examine and compare the biodiversity assessment of an assemblage from any area, biome/biofacies, or over any time span. This we do on Figure 17.1. In this BDG, the relationships of the relevant variables are clear without regression equations. For Bolivia plot 01 we see that the calculated values of Shannon's information function, H_1, and Simpson's H_2 remain constant across all values of lnN. The amount of separation between these values is given by lnE_3 (see Equations 16.5 and 16.6), which because the 2 lines are parallel to lnN, we can see this pictured in Figure 17.1 as a constant difference.

In Table 17.1 we can see entries for both lnN and lnS. On Figure 17.1, we then can observe just how closely lnS and lnN are associated. The regression is nearly perfect. The sign of the coefficient for the regression of lnS with lnN shows a positive slope, and the linear regression itself is highly significant with an R^2 of nearly 1 (see Chapter 10 for a summary of regression techniques).

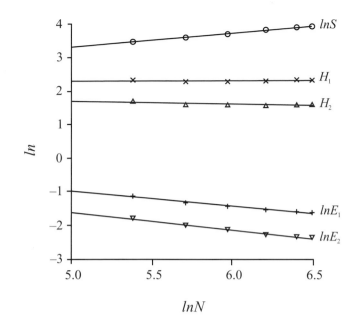

$$lnS = 1.057 + 0.447lnN, p = 0.000, R^2 = 0.996$$
$$H_1 = 2.110 + 0.037lnN, p = 0.201, R^2 = 0.368$$
$$lnE_1 = 1.054 - 0.410lnN, p = 0.0000, R^2 = 0.985$$
$$lnS = 2.225 - 1.077lnE_1, p = 0.000, R^2 = 0.986$$
$$H_2 = 1.978 - 0.059lnN, p = 0.150, R^2 = 0.441$$
$$lnE_2 = 0.921 - 0.506lnN, p = 0.000, R^2 = 0.980$$
$$lnS = 1.902 - 0.869lnE_2, p = 0.000, R^2 = 0.983$$

FIGURE 17.1   Biodiversitygram (BDG) for Bolivia plot 01, observed.

TABLE 17.1   Bolivian Tree Plot 01

N	S	lnN	lnS	H_1	lnE_1	H_2	lnE_2	lnE_3
217	37	5.380	3.466	2.337	−1.129	1.698	−1.768	−0.639
301	42	5.707	3.611	2.298	−1.313	1.614	−1.997	−0.684
392	45	5.971	3.714	2.314	−1.400	1.604	−2.110	−0.710
497	48	6.209	3.850	2.330	−1.520	1.590	−2.260	−0.740
603	51	6.402	3.922	2.367	−1.565	1.619	−2.313	−0.748
663	52	6.497	3.951	2.359	−1.592	1.619	−2.332	−0.740

NOTE: Beginning with $N = 217$, the variables are given in increments of about $N = 100$.

As the regressions show, each equation, H_1 against lnN and H_2 against lnN, has a slope near 0, a regression that is not significant, and a low value for R^2. The fact that the regression plots as a straight, almost horizontal line indicates that for our observed plot 01 the species proportions for the more abundant species change very little as we accumulate individuals and species.

From the decomposition equations we can also see the interesting fact that the slope or regression coefficient of H_1 equals the slope of lnS plus the slope of lnE_1. As an interesting aside, both lnE_1 and lnE_2, which have slopes about equal but of opposite sign to the slope of lnS, can be used in this example and each is equivalent to an inverse measure of richness since these slopes are equal to but of opposite sign to the slope of lnS against lnN in the Bolivia plot 01.

The relationships above for our Bolivian plot 01 actually are indicative of some general truths. If any comparison of our observed data with the biodiversity data from another plot is desired, then the values for lnE_1 and lnE_2 from each data set must be compared at a constant value of lnN. This is so because, as we have pointed out through use of the decomposition equation, not only lnS but also H and lnE are functions of N. We stressed the importance of a constant lnN value for comparison (Chapter 12) by rarefaction or sampling design, just as we now pointedly stress comparison only at standardized or constant lnN values.

In summary, because of the decomposition relationship, it is always true that any two of the regression equations fixes the third. This means that we could present our synthesis of the community with just two of the three (richness, evenness, information-based diversity measure) quantities, but for completeness and ease of understanding, we use all three at this point in our presentation.

We emphasize that a BDG will provide the most comprehensive snapshot of the population from which the data was sampled. From a BDG, which summarizes all relevant aspects in one plot, we can obtain enhanced insight for the interpretation of the observed data over multiple plots. In addition, with the development of the BDG, biodiversity researchers now have an established, standardized pictorial method by which to compare assemblages, species populations, or ecosystems over time and space.

FAMILY OF INFORMATION-BASED MEASURES AND ENTROPY

In the numerical examples in Chapter 16 we used our data from the Bolivian plot to give credence to the universality of the general canonical equations in Equations 16.6 and 16.7. In those examples we used the observed species values; no assumptions were necessary for the calculations to hold true. We obtain individual values for any member of our family of diversity and evenness measures merely by using the $\{p_i\}$ values. Now we explain the enrichment resulting from using the fact that we can invoke information–theoretic statistical properties and fit the $\{p_i\}$ with a statistical distribution.

In Chapter 9 we described the most widely accepted statistical distributions for ecological practice. A characteristic of each of these as well as of other statistical distributions is that they have certain *parameters* or *moments* that define them. The parameters most readers will recognize are the mean or average and the variance as discussed in Chapters 2 and 3. This average of a set of numbers also is denoted the *expected value* of the distribution of the data in statistical terms. There are other lesser known but equally important characteristics that are also related uniquely to a statistical distribution, and one such parameter is its *entropy*. This is the same entropy that we mentioned in the section on information theory (Chapter 14). For purposes of biodiversity study, we define the entropy of a distribution as the expected value of the information function based on our $\{p_i\}$: that is, the expected value of H. We write this as $E(H)$. Entropy, when defined as the expected value of information, is a measure related to the randomness within the community, not information itself nor uncertainty. Formulas for entropies of some of the statistical distributions in Chapter 9 are known (Bulmer 1974; May 1975) and discussed at length in the next chapter, while the entropy for other distributions can be derived from an information-generating function.

We saw in Chapter 14 that in application to communication theory, we can evaluate information H both before and after signal reception, or H from its source and H from the receiver. In ecology, using Shannon's probabilistic derivations we can evaluate the expectation, or expected value of H; that is, $E(H)$. We evaluate this population expected value based upon our sample, which uses in its calculation the vector **p**, and compare this to the observed value of H. This is essentially comparable to taking a sample from a population and calculating a statistical estimate from the observed data. Thus, we have a method for comparing our calculated value of H with the formula for $E(H)$ derived for any distribution to provide us with a wealth of added information for biodiversity purposes.

To make use of entropy in our underlying theoretical development it should be clear since $E(H)$ is the expected value of H that $E(H)$ can be decomposed just as H can be decomposed. Therefore, in a manner similar to the decomposition of H, likewise, we can decompose $E(H)$ into separate parts for richness and evenness. That is, just as we decompose the information in our sample we can also decompose the population values and we can write $E(H) = lnS + lnE$ for any particular statistical distribution relevant for biodiversity purposes. Since this is true, before we examine our Bolivia plot 01, we must first look at some alternative interpretations by which we can understand this factoring into richness and evenness for a statistical distribution.

ALTERNATIVE CONCEPTUALIZATIONS OF EVENNESS

Evenness as a Residual

By means of our decomposition Equations 16.1 and 16.2, any information-based diversity measure in our unified system can be decomposed into components of species richness and evenness. This decomposition is exact regardless of the observed values in **p**. However, if we fit a statistical distribution to the observed values in **p**, then we can make inferences that are considerably more comprehensive because the result is a succinct mathematical description with known properties to assess and compare with our actual experiential and sample data.

Using conditional probabilities and statistical marginal–distributional theory for the discrete entropy case, Hayek and Buzas (1997) presented the alternative scenarios for S, H, and E and the general solution that decomposed H into its maximum with an additive residual related to the distribution's dispersion or variance. That is, residual $lnE = H_{max} - H_{observed}$. Shannon understood this distinction and called the subtracted uncertainty "equivocation." For the application to quantitative biodiversity assessment, this residual defined as distributional evenness E allows for or describes the loss of information from $E(H)$, the entropy defined for a particular distribution $\mathbf{p}$. This loss of information is from the maximum evenness for the sample from that particular distribution, or $lnS(= H_{max})$. Clearly, the allocation of individuals within the species, its evenness or spread, is the commonality across any multispecies communities that keeps such a community from attaining its maximum $E(H)$.

For example, we derived that $E(H) = ln(S)$ is the expected value or entropy for an Equiprobable or Uniform statistical distribution. That is, when all species have the same number of individuals, the entropy, or expected value of Shannon's H, is equal to $ln(S)$. We also know that $ln(S)$ is the maximum value that Shannon's H can obtain when calculated from a data set containing S species in which each species has the same value for all p_i.

Therefore, if we decompose Shannon's H into its components of $ln(S)$ and $ln(E)$, and also know that the p_i are each equal, then it must be true that $H = ln(S) + 0$ from Equation 16.1. That is, H is at its maximum of lnS and lnE is at its minimum of 0 when we have a Uniform distribution. We see in Chapter 15 that evenness also is 0 when all species contain the same number of individuals. Therefore, this decomposition provides us with an alternative explanation for evenness.

If we fit any particular statistical distribution to the observed data, then evenness is the *residual* or difference between the maximum diversity that could be obtained in the population or assemblage that adheres to that distribution and the observed or sample diversity. Each entropy is uniquely related to its distribution. Therefore, if we know that $H = H_{max} = lnS$ we would know that we had a Uniform distribution. For any value of evenness we know exactly how different the RSAV is from complete evenness or from a Uniform distribution.

Evenness as Redundancy

Evenness can as well be considered to be *redundancy* as it can be a residual from distributional uniformity or equiprobability. In a physical sense, if matter or energy is distributed within some system, entropy is the measure of how much has been distributed toward (high entropy) or away from (low entropy) uniformity or equality. So if, for example, we consider that $E(H) = lnS$, then from the decomposition equation $E(H) = lnS + 0$, and the redundancy or amount of variability over uniformity or equality of the $\{p_i\}$ is 0. Therefore, redundancy or evenness is a measure that will increase as the values in $\mathbf{p}$ move away from being equal across all the species toward more inequality, order, or increased structure.

Evenness as Uncertainty

Because evenness is a measure over and above perfect equality of the values $\{p_i\}$, we can use evenness as an inequality measure to express the difference between random and ordered or structured assemblage behavior. In Chapter 14 we discuss Shannon's H as an uncertainty measure. H increases as both S increases and the relative abundances in $\mathbf{p}$ get close to uniformity or equiprobability. H decreases to its maximal certainty when all individuals are members of a single species. Thus, decrease in uncertainty is gain in information. Actually, the use of the term *uncertainty* or *uncertain* in this context is distinct from the general English usage. The characteristic feature of uncertainty is that in a "certain" situation, there is still an element of randomness or risk, and that particular part comes from a known probability distribution. Alternatively, in an uncertain situation, there likewise is a random aspect, but this risk comes from an unknown probability distribution.

In summary, regardless of the term used, evenness, residual, redundancy, or uncertainty, the conceptualization is equivalent. In essence, we are talking about evenness as a transformation of distributional deviation. Evenness is a correlate or measure of dispersion of the distribution of individuals within species: the less even, the more dominant, or the more rare species. We capitalize on this set of facts with SHE analysis.

A PERCENTAGE MEASURE OF DISTANCE FROM UNIFORM ABUNDANCES

Since $ln(E)$ is in essence the amount by which the distribution of the values in $\mathbf{p}$ differ from a Uniform or Equiprobable statistical distribution, then the ratio lnE/lnS expressed as a percent provides a measure of exactly how far the observed relative abundances are from a Uniform distribution. For example, in our Bolivian plot data set we have $\dfrac{lnE_1}{lnS} = \dfrac{-1.592}{3.951}$. In absolute value this shows that we are about $0.403 \cdot (100) = 40\%$ off from Equiprobability, or from a Uniform distribution. An interesting aspect of this measure is its relationship to Pielou's evenness measure, or J_1. From Equation 15.10 we know that $J_1 = \dfrac{H_1}{lnS}$. But from our decomposition equation we know that this must equal $\dfrac{lnS - lnE_1}{lnS}$. By rearranging terms this becomes $J_1 = 1 - \dfrac{lnE_1}{lnS}$. Clearly then $1 - J_1$ equals our measure of $\dfrac{lnE_1}{lnS}$. As we saw for the Bolivian data, Pielou's measure $J_1 = 0.597$, or about 60%, so that $1 - J_1$ is 40% when multiplied by 100.

SHE AS COMPLETE INFERENTIAL METHOD FOR BIODIVERSITY ANALYSIS AND EVALUATION

SHE is derived from and based upon fundamental ideas in probabilistic information theory and uses the theorem that entropy is related uniquely (a 1-to-1 relationship) to a distribution and vice versa. That is, if we know, for example, that the entropy of a distribution is $E(H) = lnS$, then we know that distribution must be Uniform.

Alternatively, when the distribution of the **p** is Uniform, which means that all the p_i are equal, then $E(H) = lnS$ must be true. We begin our detailed explanation by showing how this fundamental theorem can be used to identify the fit of a statistical or probability distribution to observed data. Since this result is mathematical and not probabilistic we do not require a statistical test or the calculation of confidence intervals; we know with certainty that the entropy is uniquely related to the distribution.

In Table 15.1 we calculated Shannon's H for 2 distinct distributions, each with $S = 5$. The values of H we obtained were 0.4531 for our example of a dominant distribution, 1.3225 for our Broken Stick, and $lnS = 1.6095 = H = 1.6095$ for our completely even or Uniform distribution. From Chapter 14 the most we can say is that these were increasing values and that $H_{max} = 1.6095 = ln(5)$ for the set of even abundances.

Using the results above on the entropy of a distribution, we now know not only that the maximum value of H was obtained but that we did not even have to use Equation 14.11 to calculate H. Because the p_i were all equal for the 5 species we could foresee that H was going to be $ln(5)$.

In addition, we know that evenness is 0 for a Uniform distribution. Now for the Broken Stick distribution example on 5 species we have that $H = 1.3225$. By the decomposition equation, we know $1.3225 = 1.6095 - lnE$, or $lnE = 0.2879$. Therefore, this Broken Stick, although purportedly the most even in nature (Lloyd and Ghelardi 1964), has 0.2879/1.6095, which says that it is 0.1783, or approximately 18% deviant from uniformity. Decomposing the H for the dominant distribution gives

$lnE = 1.6095 - 0.4531 = 1.1564$, so this distribution is approximately $\dfrac{1.1564}{1.6095} = 0.7185$, or 72% off from uniformity.

At this point, we could continue by providing the entropy for the other commonly fitted distributions for comparison with our observed data. However, two questions now arise: (1) How do we actually obtain a best-fitting distribution to our sample data? and (2) How large a sample do we need for our observed H to be "close to," or a good estimate of, entropy, $E(H)$? Clearly, we must compare observed H with expected H (entropy) in some way. Below we first discuss the usual approach to obtaining a fit to **p**, and then illuminate a new and different method that relies upon SHE that can provide for clearer decisions.

FITTING STATISTICAL DISTRIBUTIONS: GOODNESS OF FIT WITH FREQUENCY VERSUS DISTRIBUTION FUNCTION

Frequency Function Fit

It is quite popular to perform Goodness-of-Fit tests with conventional statistical tools. Such tests have many disadvantages, noteworthy among which is that no such approach will be able to provide an answer for the researcher concerning the optimal distribution to fit to the observed data unless under extreme conditions of disparity. When statistical distributions whose shapes are similar are being compared and tested for fit to a data set, there may be no definitive answer to the question of best-fit.

Deviation from a hypothesized frequency or density distribution for the $\{p_i\}$ is what most envision when discussing Goodness of Fit. We have discussed that the RSAV contains frequencies p_i and in Chapters 2 and 3 and elsewhere that the set of values $\{p_i\}$ is itself a distribution. That is, that the set of values $\{p_i\}$ or RSAV is a specification of the pattern of these frequencies for each species and how they are distributed according to the value of some variable of interest to the researcher. For the observed data, this specification could be a table, a histogram, or a list with the categories, the groupings, or in bins. After we form this summary plot, we try to fit some statistical or conceptual distribution or model.

We usually measure the Goodness of Fit of this distribution or model to the observed $\{p_i\}$ data by a criterion that depends upon the squares of the differences between the observed and the theoretical values in each bin or category. When this criterion is minimal, then we call this fit the "best." There are a variety of such tests; for example, the Chi-square, Kolmogorov–Smirnov, and the Anderson–Darling tests. However, each test criterion is devised under different constraints so that multiple tests might lead to multiple results that may be neither comparable nor consistent. In addition, there is no one absolute choice of method by which to make the ultimate decision between two distributions; say, for example, the fit of a Log series versus a Log normal (see Chapter 9). With field data on a number of different organisms, distinguishing between a Log series and a Log normal has proved impossible (for example, see Buzas and colleagues 1982; Magurran 1988; Buzas and Hayek 2005).

Distribution Function Fit

Alternatively, we can use the unique relationship of entropy with distribution to find the closest-fitting statistical distribution to the observed data. As we show above, we can approximate the amount of deviation from uniformity or equality that our observed distribution contains. However, if we combine our information on this deviation from equality with the best-fitting statistical distribution for describing our observations, this combined knowledge provides the foundation for making the strongest inferential statements on changes in assemblages and community structure.

As Ulrich (2001) and Chapters 13 through 16 demonstrate, the observed relative abundances $\{p_i\}$ must be available and known before evenness and diversity assessment are possible in quantitative biodiversity. If data from a natural population conformed exactly to some statistical distribution like one of those in Chapter 9, or even adhered to multiple such distributions, we would be able easily to use theoretical description. However, each data set we obtain provides an error-prone snapshot of a multispecies community, because a sample is never a perfect reflection of the population or assemblage.

Statistical theory can be of help in this situation. Uniquely related mathematically to each statistical density or frequency function is the *distribution function*. The distribution function is the graph of the accumulated frequencies used as the ordinate or y axis against the variate's increasing values plotted on the abscissa. For biodiversity evaluation, this distribution function is the total frequency of species when the variate is the number of individuals that are less than or equal to some x value. When standardized to unit area, this distribution function describes the proportion of spe-cies with values of observed individuals less than or equal to some number.

Our final value of H approaches the value of entropy $E(H)$ with large samples or over accumulations of the observed samples. Since each distribution has unique en-tropy, then it makes sense that we should be able to examine each entropy formula and see which is the closest to our last H. The distribution whose entropy is the clos-est to our observed value will be associated uniquely with a distribution that becomes our best-fitting model.

This distribution function approach (Hayek and Buzas 1997; Buzas and Hayek 2005) combined with the theorems from information theory we discuss here is used for fitting statistical distributions to the data in a precise way. We see in the next chap-ter that although this is a relatively simple procedure in the case of the Uniform distri-bution, it is not so straightforward for selecting among the most commonly used models in biodiversity study.

As we discuss in Chapter 18, despite the precision of the SHE methodology and the existence of the 1-to-1 relationship of entropy and distribution, there are times when more than 1 statistical distribution can be fit satisfactorily with a given or determined N and S. Especially with small sample sizes, we will be unsure which will be the most appropriate. Accumulating or summing sample information over some measure of space or time, like biome or faunal zone, ecosystem, locality, or season, will alleviate much of the inherent problem of fit and allow a single distribution to be selected and best-fit. Even with accumulation of the data, which is essential to examine the totality of the space or time component of our study, there are still many times that the sam-ple data are not sufficient for making a unique decision among a variety of models. We then need to consider additional features of the data. This is elaborated in the next chapter.

Summary

1. SHE analysis is a distribution-free mathematical methodology that uses two main relationships: (1) the decomposition equation $H = lnS + lnE$, and (2) the accumulative relationship in which the three quantities lnS, H, and lnE are seen as functions of the sample size N; that is, we can write each as $f(lnN)$, or $lnS = f(lnN)$, $H = f(lnN)$ and $E = f(lnN)$.

2. The Biodiversitygram (BDG) is a plot of lnS, H, and lnE on a single x axis of lnN, along with regression estimates. This plot shows the complete set of relationships for a biodiversity study.

3. The quantity E, or the form lnE, can be conceptualized as a measure of evenness, a residual, redundancy, and uncertainty.

4. The ratio $\dfrac{lnE}{lnS}$ calculated as a percent is a measure of how deviant observations are from a Uniform distribution.

5. When comparing different assemblages it is vital to do so only with standardized values of lnS, that is, with the same sample size used for the comparison.

PROBLEMS

17.1 The results of a biodiversity study of 3 communities over which we accumulated the observed data are as follows:

COMMUNITY	$\ln N$	$\ln S$	H	$\ln E$
1	5.30	4.09	3.69	−0.40
1	5.99	4.38	3.98	−0.40
1	6.68	4.60	4.20	−0.40
2	5.30	4.09	3.98	−0.11
2	5.99	4.38	3.98	−0.40
2	6.68	4.60	3.98	−0.62
3	5.30	4.09	3.72	−0.37
3	5.99	4.38	3.98	−0.40
3	6.68	4.60	4.18	−0.42

Plot these data on a BDG.

17.2 If the data were available only at $\ln N = 5.99$, how would you interpret the result? Discuss the difference between the frequency function fit at $\ln N = 5.99$ and the distribution function fit using all the accumulated data from Problem 17.1.

BIODIVERSITY:
SHE ANALYSIS FOR COMMUNITY STRUCTURE
IDENTIFICATION, SHECSI

I N CHAPTER 16, we introduce the reader to the decomposition into richness and evenness of information or information-based diversity via the canonical equations. We connected all biodiversity measures mathematically. In Chapter 17, we explain how information measures calculated from the RSAV, **p**, actually have an associated distribution, which has a parameter called entropy. Entropy can be designated as the expected value of H, and written as $E(H)$. In this chapter we introduce equations for $E(H)$, unique for each of the major and most commonly used distributions in ecological analysis. We then illustrate further depths of SHE analysis with the data from Bolivia plot 01 by calculating entropy values of these major distributions and comparing their plots with the observed data on BDG's. We also examine the efficacy of using regression equations for comparing distributional properties. Finally, we show why the Log series distribution serves well as a *Null Model* for comparison with any observed distributions.

SHE POSSIBILITIES

We begin by considering the possibilities for values of S, H, and E when comparing any 2 populations or samples; we will label the 2 populations to be compared as a and b. The possible scenarios for any diversity measure H and evenness measure E are

1. $S_a = S_b$, $H_a = H_b$, $E_a = E_b$
2. $S_a = S_b$, $H_a \neq H_b$, $E_a \neq E_b$
3. $S_a \neq S_b$, $H_a = H_b$, $E_a \neq E_b$
4. $S_a \neq S_b$, $H_a \neq H_b$, $E_a = E_b$
5. $S_a \neq S_b$, $H_a \neq H_b$, $E_a \neq E_b$

The first scenario indicates that the 2 sets of observations have an identical number of species (species richness) and species proportions (relative abundances) are also equal. Recall that the species proportions are used in calculating both the diversity measures H and evenness measures E. Since we have in scenario 1 that both $S_a = S_b$ and that $\mathbf{p}_a = \mathbf{p}_b$, then both the resultant two diversity values of H_a and H_b and the 2 evenness values E_a and E_b will be equal. In other words, assemblages a and b are identical, so the quantitative evaluation of them is identical and we have no game.

In the second case, the same number of species is observed in each sample, but the species proportions are different, resulting in different values of diversity measure H. We illustrate this scenario in Table 15.1 with a fabricated distribution showing strong dominance and a Broken Stick distribution. Looking back at Table 15.1, note that values of H_1, H_2 and E_1, E_2 all clearly indicate that there is a difference in the relative abundance of the 2 populations.

The third possibility has unequal richness but equal diversity or H values. From our decomposition equation we show that because the different S's and E's can cause a resultant offset to each other, the differences in effect are canceled and diversity measures can be equal. That is, one data set has more species and lower evenness than does the other, but in a very precise way that is seen to satisfy the decomposition equation.

In the fourth scenario there is differing richness in the 2 assemblages considered, and the RSAV's are also not equal. In this case then the diversity measures H_a and H_b will not be equal since they use these distinctly different sets of values, $\mathbf{p}_a$ and $\mathbf{p}_b$. However, evenness measures can sometimes be the same, a scenario we discuss below.

In the last or fifth case, the S values are unequal, the $\mathbf{p}_a$ and $\mathbf{p}_b$ are not equal, and this results in different diversity values, or $H_a \neq H_b$. Even though this scenario appears similar to our third possibility, in this situation, the unequal values do not offset one another and the evenness values therefore are also different from one another.

Entropy Used to Fit Distributions with SHE: Theoretical Comparison

In Chapter 9, we discuss at some length the Broken Stick distribution. As will become apparent, we show that the Log series and this Broken Stick distribution are at the opposite ends of the possible ecological scenarios listed above. The Broken Stick is related to scenario 4, in which the evenness is constant, while the Log series is represented by scenario 3 with constant diversity (H) values. While we could easily use a generating function to provide us with an exact formula for $E(H)$ for each of our distributions (Broken Stick, Canonical Log normal, Log normal, and Log series), just as we did for the Uniform distribution in the last chapter, there would be one important difficulty. The Uniform distribution is quite an elementary probability distribution, while the distributions given in Chapter 9 are complicated. Therefore, the generation or derivation of their exact entropy formulas would not, in general, be useful for our purposes here.

Because of these complicating factors, we shall defer to estimators of this parameter, or formulas for estimates of the exact entropy $E(H)$ for each of the distributions. An advantage of this approach is that our estimators will be given as functions, or related to, our information-based diversity measures H_1 and H_2. Then we shall show the efficacy of this approach by example. One word about notation: Ordinarily we would write the exact formula as $E(H)$ and its estimator as $\widehat{E(H)}$ as we did for our mean and variance parameters in Chapters 2 and 3. However, since we are using only the estimators that have our information-based family of diversity measures in their formulas, we shall revert to the simpler notation and call the estimators $E(H)$. Hopefully, no confusion will arise. Once we have derived the $E(H)$ associated with each distribution we can calculate our diversity measures H_1 and H_2 and compare these values with the $E(H)$ values for each distribution to find the closest to describe the relative abundances, which will be uniquely associated with our best-fitting theoretical distribution.

ENTROPY FOR THE BROKEN STICK DISTRIBUTION

The entropy or the expected value for Shannon's information measure is given by May (1975) as

$$E(H_1) = lnS - 0.42. \tag{18.1}$$

We checked the formula given above against a calculation using the formula for the Broken Stick given in Chapter 9 and found quite close agreement for $S > 100$. However, for $S < 100$, our results show that a constant of 0.40 rather than the 0.42 in Equation 18.1 gives an even closer approximation (Hayek and Buzas 1997). As Table 18.1 shows, either constant in the equation, 0.40 or 0.42, gives a very close and apparently unbiased estimate for Shannon's H_1.

When we compare Equation 18.1 with our decomposition equation, $H_1 = lnS - lnE_1$, it becomes clear that for a true Broken Stick distribution the constant subtracted from lnS is lnE_1. If $lnE_1 = -0.40$, then $E_1 = 0.67$ while if $lnE_1 = -0.42$, then $E_1 = 0.66$. So E is a constant for a Broken Stick, and clearly this is our scenario 4.

The entropy estimate for the Broken Stick can also be given in terms of H_2 or Simpson's λ as shown in Table 18.2. May (1975) devised a form of this estimate by first noting that

$$\lambda = \frac{2}{S} \tag{18.2}$$

or, by rearranging terms, he got

TABLE 18.1   Calculation from Broken Stick Distribution Equation and Approximations of Shannon's Information Measure H_1

S	H_1 CALCULATED	$H_1 = lnS - 0.40$	$H_1 = lnS - 0.42$
10	1.955	1.903	1.893
50	3.512	3.512	3.492
100	4.194	4.205	4.185
1,000	6.486	6.508	6.488

TABLE 18.2   Calculation from Broken Stick Distribution Equation and Approximations of Simpson's Information Measure H_2

S	λ CALCULATED	2/S	$H_2 = ln(1/\lambda)$	$H_2 = ln(S/2)$
10	0.163	0.200	1.814	1.609
50	0.038	0.040	3.270	3.219
100	0.019	0.020	3.963	3.912
1,000	0.002	0.002	6.215	6.215

$$\frac{1}{\lambda} = \frac{S}{2}. \tag{18.3}$$

Using these formulas, the estimate of entropy or expected value for the Broken Stick we obtain using Simpson's information-based measure is

$$E(H_2) = ln\left(\frac{1}{\lambda}\right) = ln\left(\frac{S}{2}\right). \tag{18.4}$$

Comparison with our decomposition equation shows that we can calculate a formula that is in the same format as Equation 18.1. That is, we obtain the relationship

$$E(H_2) = lnS - 0.693. \tag{18.5}$$

The values of $E(H_2)$ in Table 18.2 show that these simplified estimates for $E(H_2)$ for the Broken Stick distribution are quite reasonable.

The difference between the diversity measures H_1 and H_2 is a constant. Recall from Equation 16.5 that this difference $(H_2 - H_1)$ was shown to equal lnE_3. We can now use this difference between the formulas based on both Shannon's and Simpson's measures to estimate values of lnE_3 for the Broken Stick distribution if we choose. As Equation 18.6 shows, the difference between the estimates, that is, the value of lnE_3, is always the constant -0.273 for the Broken Stick distribution.

$$lnE_3 = ln\left(\frac{S}{2}\right) - (lnS - 0.42) = -0.273. \tag{18.6}$$

Table 18.3 indicates the values of lnE_3 for values of S as low as about 50 reach approximately -0.27, or -0.29 if we use -0.40 as the constant in the equation. However, the actual observed values of this quantity reach -0.27 only at about 1,000 species in our example. This is an example of why mathematical statisticians do not define "large" or "small" sample sizes—each problem can require different values or definitions of these words.

TABLE 18.3 Differences (lnE_3) Between Shannon's and Simpson's Measures for Calculated and Theoretical Values

S	$ln\left(\dfrac{1}{\lambda}\right) - H_1$	CALCULATED $\left(ln\left(\dfrac{S}{2}\right)\right) - (lnS - 0.42)$
10	−0.141	−0.284
50	−0.242	−0.273
100	−0.231	−0.273
1,000	−0.271	−0.273

TABLE 18.4 Calculated Theoretical Most Abundant Species Proportion Versus Estimated for lnS/S

S	CALCULATED	lnS/S
10	0.230	0.292
50	0.078	0.090
100	0.046	0.052
1,000	0.007	0.007

Finally, let us consider May's (1975) theoretical estimate for the expected value of the species proportion of the most abundant species from a Broken Stick distribution. This estimate is

$$D_{BP} = p_i = \frac{lnS}{S}.\tag{18.7}$$

Once again, we have actual calculations for this proportion for the most abundant species using the Broken Stick equation of Chapter 9 to compare with the simple theoretical estimate given in Equation 18.7. These are presented in Table 18.4. As with the earlier theoretical estimates, the estimate lnS/S approaches the actual value of p_1 as S becomes "large." Overall, we conclude that the simplified estimates of entropy shown above are remarkably good estimates for the exact values calculated from the related Broken Stick distribution.

ENTROPY FOR THE LOG NORMAL DISTRIBUTION

The complexities of the Log normal are discussed in Chapter 9. Unlike the Broken Stick distribution discussed above, the Log normal is not as succinctly defined. Because Preston (1948) first suggested the Log normal distribution to fit species abundance distributions, his notation, although somewhat awkward, is often followed. The development of estimators for the expected values for H were derived by May (1975), who used Preston's log base-2 classes, which he called octaves, R, after musical octaves. We present the estimators for both the Canonical Log normal and the non-Canonical Log normal as is the usual presentation (Magurran 1988).

Canonical Log Normal

Following May (1975) we consider a three-parameter (two of which are independent) Log normal. The first parameter is the number of species in the modal octave, S_0. A second parameter, called a, is a proportionality constant that necessarily always has a value close to 0.20. The third parameter is γ. For the particular relationship Preston (1962) referred to as the Canonical Log normal, the modal octave for the number of species per octave curve (R_{max}) and the modal octave for the number of individuals per octave curve (R_N) are a fixed distance apart and the modal octave of the individuals curve coincides with the modal octave of the species curve. In this situation the ratio of the modal octaves is 1 and thus, for the Canonical Log normal, the parameter discussed in Chapter 9 is $\gamma = 1 = R_{max}/R_N$. For this distribution, May (1975) gave the estimator or expected value for the Canonical Log normal based upon Shannon's information measure as

$$E(H_1) = 2\pi^{-0.5}\sqrt{\ln S}. \tag{18.8}$$

In addition, May (1975) gave an estimate for the most abundant species from this distribution as

$$D_{BP} = p_i = \frac{\ln 2}{\sqrt{\pi \ln S}}. \tag{18.9}$$

The estimate for entropy of this distribution, based upon Simpson's measure H_2, is too complicated to be useful, and our alternative derivations did not provide for any more useful version.

Non-Canonical Log Normal

When $\gamma < 1$, so that the Log normal is no longer Canonical, then May's (1975) entropy estimate based upon Shannon's H_1 measure is

$$E(H_1) = (1 - \gamma^2)\ln S. \tag{18.10}$$

Although Equation 18.10 may look unfamiliar, recall from Chapter 15, Equation 15.10, that $J_1 = H_1/\ln S$, which is the measure of evenness proposed by Pielou. Clearly, by comparison with Equation 18.10, we know that $J_1 = (1 - \gamma^2)$. We also recall from Chapter 17 that $J_1 = (1 + \ln E_1/\ln S)$. Because the parameter γ is a constant for the Log normal, the value of J_1 and the value of lnE_1/lnS are also constant. In addition, it should be evident that $\gamma^2 = lnE/lnS$. Estimators for Simpson's H_2 are complicated and not realistically useful for most biodiversity purposes. We will not use the values here.

ENTROPY FOR THE LOG SERIES

We discuss Fisher's Log series in Chapters 9 and 14 (Fisher and colleagues 1943). Recall that this distribution has one parameter, α, which is widely used as a diversity measure. Fortunately, α also is the key to estimating entropy for the Log series. Bulmer (1974) provided estimates for the entropy of the Log series using both Shannon's and Simpson's information measures. The entropy or expected value of H_1 is given by

$$E(H_1) = ln\alpha + 0.577 \qquad (18.11)$$

where α is the parameter of the Log series and 0.577 is Euler's constant. Actually, Euler's constant is an infinite or non-repeating decimal that we rounded to three places, which should prove to be of little consequence for our purposes here.

Then we also can show that

$$\alpha = \frac{1}{\lambda} \qquad (18.12)$$

where λ is Simpson's index. So, we put the above two results together and we obtain for the Log series that

$$E(H_2) = ln\frac{1}{\lambda} = ln\alpha. \qquad (18.13)$$

Thus, for the Log series distribution, estimates based on either H_1 and H_2 are tight and tidy. Because α is a constant and is equal to $1/\lambda$, the values of the information-based diversity measures are also constant as is their difference. The difference between the 2 measures, which we saw equaled lnE_3 in the last section, gives the constant 0.577.

That is, from Equation 16.5 we have

$$lnE_3 = (H_2 - H_1) = \left[\left(ln\frac{1}{\lambda}\right) - (ln\alpha + 0.577)\right] = -0.577. \qquad (18.14)$$

Further, to satisfy our decomposition Equations 16.1 and 16.2 for the Log series, as lnS increases with lnN, we shall see that lnE must decrease by the same amount as lnS increases. This is scenario 3 in our list of possibilities.

Finally, for this Log series distribution, the proportion for the most abundant species (May 1975) or the Berger and Parker index (1970) can be estimated by

$$D_{BP} = p_1 = \frac{\ln\alpha}{\alpha}. \qquad (18.15)$$

PATTERN AND VALUE

SHE Analysis lets us more deeply examine community structure for biodiversity purposes by allowing for the consideration of both the *pattern* and the *value* of a distribution. It is unlikely that any observed sample data set will be exactly described by any 1 statistical distribution; a sample is always an imperfect reflection of the population assemblage as we have said, and distributional best-fit will be relative or imperfect. An equation for the estimator of the entropy for a given S and N (actually, only N is required for the Log series) for each distribution provides us with a single point estimate for true or population entropy. We do not realistically expect the observed value to be exactly equal to the associated entropy, just as we do not expect our sample mean to exactly equal our population mean. However, the entropy values for each distribution are compared with the observed diversity measure to determine the appropriate best-fit distribution. We show this in Chapter 17 for the Uniform distribution; a very simplified example. However, the estimates we calculate from the formulas in Chapter 9 for the parameters α for the Log series and γ for the Log normal can result in equivalent H values at some value of S. It would then be impossible to determine the best-fitting choice for a distribution at that particular value of S. In Table 18.5, we tabulate a hypothetical situation in which three distributions have yielded the same value of H_1 (and hence $\ln E_1$) at $S = 80$. Since this overlap of values and distributions can occur with frequency in ecological applications, what can be done to make the correct determination?

To answer this question, instead of merely considering the closeness of the point estimate to the entropy for a distribution, we can examine both the value and the pattern of the distribution. An assessment of a series of diversity H values as both $\ln S$ and $\ln N$ get larger is possible only with an accumulation procedure (Chapter 17). For example, in Table 18.5 we can compare the pattern of what is happening to H at values of S both below and above the species richness at which we obtain these equivalent values for *observed entropy*. Just as we saw in Chapter 2, examining the pattern displayed by the three distributions allows for easy recognition of the correct distribution to fit. As Table 18.5 makes clear for an exact Broken Stick, $\ln E_1$ is constant at all values of increasing S; for a theoretical or exact Log series, H_1 will be constant, while for the Log normal the ratio $\ln E_1/\ln S$ (or J_1) will be constant. These patterns are always found for the exact distributions.

By examining not only the values but also the pattern, we can usually distinguish among the distributions for purposes of fit. This two-fold SHE procedure will also allow for further insight into community structure, as we discover a little later on, because the possibility exists that we may observe a particular pattern but not the values for a

TABLE 18.5   SHE Analysis for Three Levels of S for Broken Stick, Log Series, and Log Normal Distributions

DISTRIBUTIONS	S	$\ln S$	$E(H_1)$	$\ln E_1$	$\ln E_1 / \ln S$
Broken Stick	60	4.09	3.69	−0.40	−0.098
	80	4.38	3.98	−0.40	−0.091
	100	4.60	4.20	−0.40	−0.087
Log Series	60	4.09	3.98	−0.11	−0.028
	80	4.38	3.98	−0.40	−0.091
	100	4.60	3.98	−0.62	−0.135
Log Normal	60	4.09	3.72	−0.37	−0.091
	80	4.38	3.98	−0.40	−0.091
	100	4.60	4.18	−0.42	−0.091

NOTE: For the Log series $\alpha = 30$. For the Log normal $(1 - \gamma^2) = 0.9087$. At $S = 80$, all values are identical. Consideration of $S = 60, 80,$ and 100 shows that $\ln E_1$ is constant (−0.40) for the Broken Stick, H_1 is constant (3.98) for the Log series, and $\ln E_1 / \ln S$ is constant (−0.091) for the Log normal.

specific distribution. Consider the example in Table 18.5 when $N = 400$ and $S = 80$, for which $\alpha = 30$ (Appendix 2). From Equation 18.11 the value of H_1 is constant at 3.98 as shown. However, the actual observations could show more dominance than that expected from a Log series and, hence, a lower value of the observed H_1 at $S = 80$. At the same time, the dominant species could exhibit, for example, constant proportions with increasing sample size as N is accumulated, and so the value of H_1 would remain constant. We would, then, have a Log series pattern but with lower values than expected. We continue by taking the Bolivian plot 01 data and using this SHE methodology to attempt to find the most appropriate distribution to fit to this set of observations.

USING ENTROPY TO FIT DISTRIBUTIONS WITH SHE: EXAMPLES

In the following example we illustrate how to find the most appropriate fitted distribution for our observed data from Bolivia plot 01. The procedure involves steps in which we make the calculations for each proposed distribution we want to try and fit and compare both the theoretical pattern of each distribution and its expected value to the observed. The procedure is that we first use the observed richness and number of individuals; in our case for the Bolivian data, we use $N = 663$ and $S = 52$. We then calculate the entropy estimates for each theoretical distribution, whose formulas are provided in this chapter. Our aim is to find the closest fit to the observed **p**. We discussed the unique relationship of the entropy parameter to each distribution and also showed that our estimates based upon our information-based diversity measures were both adequate and useful as estimates.

Then, for this same $N = 663$ and $S = 52$, we use the complete set of observed data in **p** to construct a BDG for our observations, along with the calculated values of H_1 and H_2 and regressions based upon this **p**. The regression coefficients and results are compared with the known or expected values from each theoretical distribution that are also plotted. Finally, we examine both pattern and values expected from the theoretical-based BDG with the actual data-based BDG. The observed is compared with each theoretical and a decision made.

Broken Stick and Bolivia Plot 01

Let us consider whether the Broken Stick distribution could be the best distribution to fit our Bolivian plot 01 data. For the Broken Stick distribution, we use Equation 18.1 with a constant of -0.40, which is based on Shannon's information function to compare with observed Shannon's H_1 from our Bolivian plot. We also compare the observed Simpson's measure H_2 calculated from the RSAV, **p**, for Bolivian plot 01 with the expected value given in Equation 18.4. In both cases, because we are evaluating the fit of a Broken Stick distribution, the value of evenness, or the residual, we know must be constant.

Results for comparisons are shown in Table 18.6 and the BDG shown in Figure 18.1, along with slopes from the regression equations. Clearly, H_1 and H_2 are parallel to lnS as we expect for a Broken Stick. The values for lnE_1 and lnE_2 are constant for this distribution so that the slopes are 0 and regression is not useful. Similarly, the regression for lnS versus lnE is not meaningful because lnE is constant with an infinite or vertical slope.

In Figure 18.2, we present the BDG for the Bolivian plot's observed data; that is, the relative abundances. We use these observations to calculate lnS, H_1, and lnE_1 on the actual values of p_i in the observed RSAV, **p**. As is quite apparent when viewing Figure 18.2, neither the slopes of H_1 and lnE_1 nor the values are close to the actual Broken Stick pattern for $N = 663$ and $S = 52$ shown in Figure 18.1, or to the results given in Table 18.6.

TABLE 18.6   Broken Stick Values for Bolivia Plot 01

N	S	lnN	lnS	H_1	H_2	lnE_1	lnE_2	lnE_3
217	37	5.380	3.466	3.066	2.773	−0.40	−0.693	−0.293
301	42	5.707	3.611	3.211	2.918	−0.40	−0.693	−0.293
392	45	5.971	3.714	3.314	3.020	−0.40	−0.693	−0.293
497	48	6.209	3.850	3.450	3.157	−0.40	−0.693	−0.293
503	51	6.402	3.932	3.532	3.239	−0.40	−0.693	−0.293
663	52	6.497	3.951	3.551	3.258	−0.40	−0.693	−0.293

NOTE: N and S observed. $H_1 = lnS - 0.40$ and $H_2 = ln\left(\dfrac{S}{2}\right)$.

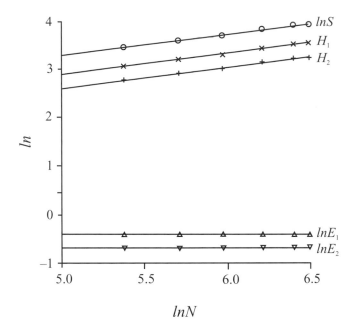

$$lnS = 1.057 + 0.447lnN, p = 0.000, R^2 = 0.996$$
$$H_1 = 0.657 + 0.447lnN, p = 0.000, R^2 = 0.996$$
$$lnE_1 = -0.40 + 0.000lnN \quad \text{or} \quad \text{No Regression}$$
$$lnS = -0.40 + 0.000lnN \quad \text{or} \quad \text{No Regression}$$
$$H_2 = 0.363 + 0.448lnN, p = 0.000, R^2 = 0.996$$
$$lnE_2 = -0.693 + 0.000lnN \quad \text{or} \quad \text{No Regression}$$
$$lnS = -0.693 + 0.000lnN, \text{No Regression}$$

FIGURE 18.1   Biodiversitygram (BDG) for Broken Stick distribution using lnS values from Bolivia plot 01.

The BDG for H_2 in Figure 18.3 is similar to that of Figure 18.2, which presents the calculations for H_1. So we see in this BDG that neither the actual Broken Stick pattern of results nor the expected values for this distribution are close to our observations.

In keeping with the high value for evenness we see in this plot, the Broken Stick estimate for the most abundant species (the estimate of D_{bp} diversity measure), with a total $S = 52$ from Equation 18.7, is $ln(52) / 52 = 0.076$. On the other hand, the calculated value using the original Bolivian data in Equation 8.2 is $p_1 = 0.380$.

Log Normal and Bolivia Plot 01

Canonical Log Normal   We now consider the possibility that our observed data could be best fit by a Canonical Log normal. The estimated value of entropy for the Canonical Log normal based upon H_1 is given by Equation 18.8. Using this equation we calculate the theoretical Canonical Log normal that would be most appropriate for our Bolivian data; that is, the Canonical Log normal with the same $S = 52$ and $N = 663$ as our ob-

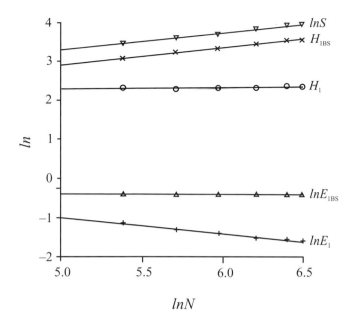

Observed: $lnS = 2.225 - 1.077lnE_1$
Broken Stick: $H = lnS - 0.40$, No regression

FIGURE 18.2   BDG for Broken Stick distribution and observed values for Shannon's measure for Bolivia plot 01.

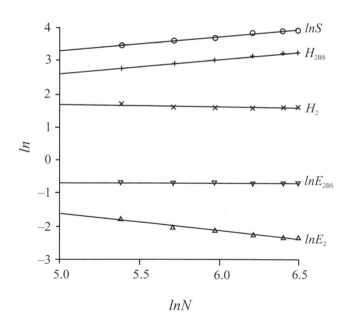

Observed: $lnS = 1.902 - 0.869lnE_2$, $p = 0.000$, $R^2 = 0.983$
Broken Stick: $lnS = -0.693$, No regression possible

FIGURE 18.3   BDG for Broken Stick distribution and observed values for Simpson's measure for Bolivia plot 01.

served data. The constant in this estimator is $2\pi^{-0.5}$, which is 1.128 for the observed data. So we know that $E(H_1) = 1.128\sqrt{\ln S}$. This equation tells us that for each successively larger value of $\ln S$, the increase in H_1 is close to the square root of $\ln S$. The results of the calculations using the observed N and S are shown in Table 18.7 and illustrated on Figure 18.4 along with the regression equations. Examining the figure, notice that H_1 has a slight increase (slope 0.131) as $\ln S$ increases. The slope of $\ln S$ versus $\ln E_1$ is −1.414 (Figure 18.4).

Figure 18.5 shows a comparison of the Canonical Log normal results with the values calculated from the observed vector **p**. Both pattern and value appear similar. However, the slopes for $\ln S$ versus $\ln E_1$ for the Canonical Log normal and for the observed differ by 0.337 (−1.414 and −1.077, respectively).

In addition, we know that the Canonical Log normal is appropriate when the value of γ equals 1. As an added test of appropriateness for our data, we shall estimate the value of γ. We use the methodological example described in Magurran (1988) that actually uses Pielou's (1975) method for fitting a truncated Log normal. We obtained a value of $\gamma = 0.255$; clearly not close to 1. Therefore, we shall consider a non-Canonical Log normal for potential fit to our observed data.

Non-Canonical Log Normal For Equation 18.10 for the non-Canonical Log normal we require an estimate of $(1 - \gamma^2)$. For this example, since $\gamma = 0.225$, then $(0.255)^2 = 0.065$ and $1 - 0.065 = 0.935$. Then by substitution, for this Log normal we have $E(H_1) = 0.935$ $\ln S$. In Table 18.8 we give the data required to plot the BDG and do the regressions shown in Figure 18.6.

Because the value of $(1 - \gamma^2)$ is so high, namely 0.935, the BDG shown in Figure 18.6 is similar to the one for the Broken Stick distribution shown in Figure 18.1. We have found this to be the case whenever we have a non-Canonical Log normal parameter ($\gamma = 0.255 \ J_1$) close to 1.

For the Broken Stick distribution above we calculated the slope of H_1 versus $\ln N$ as 0.447. This value was exactly the same as the slope for $\ln S$ versus $\ln N$. For the Log normal given here the slope is a very close 0.418 (Figure 18.6). Not surprisingly, when we plot the observed values and the Log normal values, neither the values nor the pattern is close to the observed, with Figure 18.7 similar to Figure 18.3.

TABLE 18.7 Canonical Log Normal Values for Bolivia Plot 01

N	S	$\ln N$	$\ln S$	H_1	$\ln E_1$
217	37	5.380	3.466	2.100	−1.366
301	42	5.707	3.611	2.143	−1.468
392	45	5.971	3.714	2.174	−1.540
497	48	6.209	3.850	2.213	−1.637
503	51	6.402	3.932	2.237	−1.695
663	52	6.497	3.951	2.242	−1.709

NOTE: N and S observed. $H_1 = 2\pi^{-0.5}\sqrt{\ln S}$.

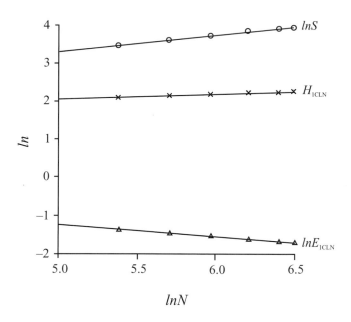

$$lnS = 1.057 + 0.447lnN, p = 0.000, R^2 = 0.996$$
$$H_1 = 1.395 + 0.131lnN, p = 0.000, R^2 = 0.996$$
$$lnE_1 = 0.338 - 0.316lnN, p = 0.0000, R^2 = 0.996$$
$$lnS = 1.535 - 1.414lnE_1, p = 0.000, R^2 = 1.000$$

FIGURE 18.4   BDG for Canonical Log normal distribution for Bolivia plot 01.

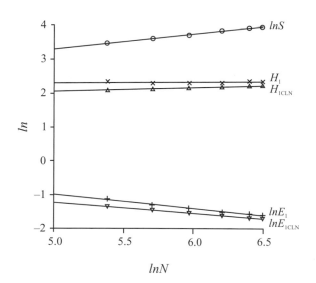

Observed: $lnS = 2.225 - 1.077lnE_1, p = 0.000, R^2 = 0.986$

Canonical Log Normal: $lnS = 1.535 - 1.414lnE, p = 0.000, R^2 = 1.000$

FIGURE 18.5   BDG for Canonical Log normal distribution and observed values for Bolivia plot 01.

TABLE 18.8 Log Normal Distribution Calculated from $E(H_1) = (1 - \gamma^2) \ln S$ Where $\gamma = 0.255$; $(1 - \gamma^2) = 0.935$

N	S	lnN	lnS	H_1	lnE_1
217	37	5.380	3.466	3.241	−0.225
301	42	5.707	3.611	3.376	−0.235
392	45	5.971	3.714	3.473	−0.241
497	48	6.209	3.850	3.600	−0.250
503	51	6.402	3.932	3.676	−0.256
663	52	6.497	3.951	3.694	−0.257

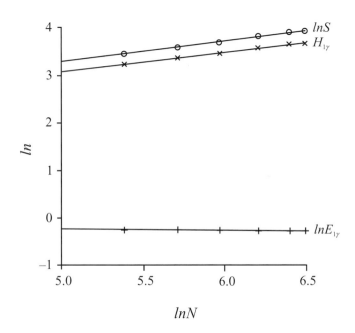

$lnS = 1.057 + 0.447lnN, p = 0.000, R^2 = 0.996$
$H_1 = 0.990 + 0.418lnN, p = 0.000, R^2 = 0.996$
$lnE_1 = -0.067 - 0.029lnN, p = 0.0000, R^2 = 0.995$
$lnS = 0.048 - 15.190lnE_1, p = 0.000, R^2 = 0.999$

FIGURE 18.6 BDG for Log normal distribution with $\gamma = 0.255$ for Bolivia plot 01.

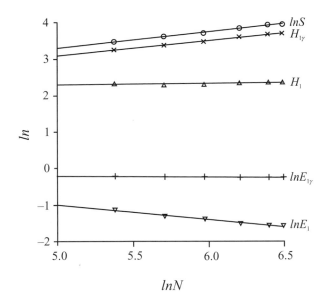

Observed: $lnS = 2.225 - 1.077 lnE$, $p = 0.000$, $R^2 = 0.986$
Log Normal: $lnS = 0.048 - 15.190 lnE_1$, $p = 0.000$, $R^2 = 0.999$

FIGURE 18.7   BDG for Log normal distribution with $\gamma = 0.255$ and observed values for Bolivia plot 01.

Log Normal Pattern from Last H_1

The Log normal plotted in Figure 18.7 is clearly not at all close in pattern or value to the calculations for H_1 and lnE_1 for Bolivia plot 01. We can easily make the values closer by using the last value of H_1. To examine the pattern, we use the last H_1 and lnS as values for an estimate of $\dfrac{H_1}{lnS} = J_1 = (1 - \gamma^2)$. Thus, we examine a Log normal pattern without actually estimating γ through a solution for the truncated Log normal. In our data $\dfrac{H_1}{lnS} = \dfrac{2.359}{3.951} = 0.597 = J_1 = (1 - \gamma^2)$. Solving for gamma we have $\gamma = 0.635$. This value is not near the 0.255 estimate we made from the original counts, although it is much closer to 1, the value for the Canonical Log normal. The values are given in Table 18.9 wherein all H_1 values are 0.597 lnS. This means that the ratio $J_1 = (1 - \gamma^2)$ is a constant equal to 0.597 for all entries. The ratio lnE_1/lnS is also a constant and it equals $\gamma^2 = 0.403$.

In this example we plot the Log normal distribution whose values are obtained by using the last calculations for the quantity H_1/lnS shown in the BDG in Figure 18.8. This figure appears similar to Figure 18.4, the BDG for the Canonical Log normal. To examine exactly why this is true, we devised Figure 18.9, in which this BDG shows the plot of the Canonical Log normal, the Log normal based upon calculations using the last $\dfrac{H_1}{lnS}$, and the observed Bolivian values.

TABLE 18.9   Log Normal Pattern Calculated from Last Observed $H_1/lnS = 0.597$

N	S	lnN	lnS	H_1	lnE_1
217	37	5.380	3.466	2.069	−1.397
301	42	5.707	3.611	2.156	−1.455
392	45	5.971	3.714	2.217	−1.497
497	48	6.209	3.850	2.298	−1.552
503	51	6.402	3.932	2.347	−1.585
663	52	6.497	3.951	2.359	−1.592

NOTE: For each entry $H_1 = 0.597\ lnS$.

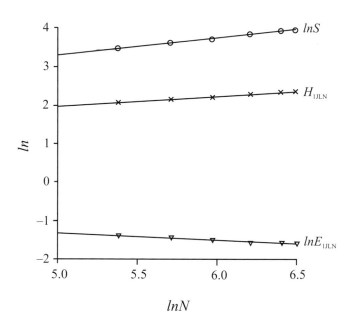

$$lnS = 1.057 + 0.447lnN, p = 0.000, R^2 = 0.996$$
$$H_1 = 1.395 + 0.131lnN, p = 0.000, R^2 = 0.996$$
$$lnE_1 = 0.338 - 0.316lnN, p = 0.0000, R^2 = 0.996$$
$$lnS = 1.535 - 1.414lnE_1, p = 0.000, R^2 = 1.000$$

FIGURE 18.8   BDG Log normal distribution pattern from last H_1 of Bolivia plot 01, $J_1 = 0.597$.

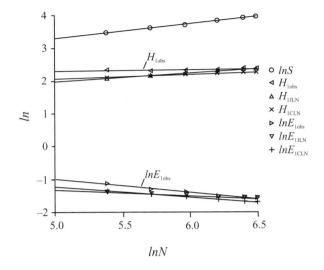

Observed: $lnS = 2.225 - 1.077lnE_1, p = 0.000, R^2 = 0.986$
Canonical Log Normal: $lnS = 1.535 - 1.414lnE_1, p = 0.000, R^2 = 0.996$
Log Normal: $lnS = 0.001 - 2.480lnE_1, p = 0.000, R^2 = 1.000$

FIGURE 18.9   BDG for observed, Canonical Log normal, and Log normal from last $H_1 (\gamma = 0.635)$ for Bolivia plot 01.

The result for the H_1 calculation that uses the Bolivian RSAV, **p**, is more horizontal than that predicted by the Log normals; the observed Bolivian values for lnE_1 decline more rapidly than those for the Log normal. We are approaching uniformity of the assemblage more rapidly. In addition, although the slope for lnS versus lnE_1 for the Canonical Log normal is –1.414 and the slope for the Log normal based upon the last $\dfrac{H_1}{lnS}$ is –2.480; nevertheless, they both have quite similar plots as seen in Figure 18.9, with values given in Tables 18.7 and 18.9.

Log Series and Bolivia Plot 01

To obtain a value of the Log series parameter α for $N = 663$ and $S = 52$ we may use Appendix 4 with some interpolation. Alternatively, we can substitute N and S into the equation $S = \alpha ln\left(1 + \dfrac{N}{\alpha}\right)$, given in Chapter 9. For this equation we find by substitution that we can solve iteratively for S to obtain $52.003 = 13.215 ln\left(1 + \dfrac{663}{13.215}\right)$. Using Equation 18.11 we calculate the Log series entropy or expected value based upon Shannon's information measure to be $E(H_1) = ln(13.215) + 0.577 = 2.581 + 0.577 = 3.158$. We can also use Equation 18.13 to find the entropy estimate based upon Simpson's measure. Application of this equation using the above calculated values yields $E(H_2) = ln(13.215) = 2.581$. We have already showed in the last Log series section that α is a

TABLE 18.10   Log Series Values for Bolivia Plot 01

N	S	$\ln N$	$\ln S$	H_1	H_2	$\ln E_1$	$\ln E_2$	$\ln E_3$
217	37	5.380	3.631	3.158	2.581	−0.473	−1.050	−0.577
301	42	5.707	3.734	3.158	2.581	−0.576	−1.153	−0.577
392	45	5.971	3.812	3.158	2.581	−0.654	−1.231	−0.577
497	48	6.209	3.877	3.158	2.581	−0.719	−1.296	−0.577
503	51	6.402	3.927	3.158	2.581	−0.769	−1.345	−0.577
663	52	6.497	3.951	3.158	2.581	−0.793	−1.370	−0.577

NOTE: N is observed. S is calculated from $S = \alpha ln\left(1 + \dfrac{N}{\alpha}\right)$ where $\alpha = 13.215$.

constant for the Log series distribution as N and S accumulate or get larger. Therefore, since the Log series entropy formulas (Equations 18.11 and 18.13) both are functions of α (that is, they have α in their formulas), then both $E(H_1)$ and $E(H_2)$ also remain constant as N and S increase. In Table 18.10, we present the results of the calculations for illustration.

Because we know from the decomposition Equations 16.1 and 16.2, respectively, that H_1 and H_2 are constant as $\ln S$ increases, then $\ln E$ must decrease by exactly the same amount to satisfy the equations when H is constant. We illustrate this on the BDG shown in Figure 18.10.

As the regression equations accompanying the figure show, the slopes for $\ln S$ versus $\ln N$ and $\ln E_{1,2}$ versus $\ln N$ are exactly the same (0.285) but of opposite sign. Given this situation, the regression for $\ln S$ versus $\ln E$ has a slope of exactly −1 and the equation for $\ln S$ versus $\ln E_{1,2}$; namely, $\ln S = $ constant $+ \ln E_{1,2}$ is the decomposition equation with the constant equal to H. To avoid clutter we did not plot $\ln E_3$, which is a constant −0.577, that is, the difference between H_1 and H_2.

In Figure 18.11 we show the estimated Shannon values for a calculated Log series alongside of the values calculated from the observational vector **p**. While the expected values (except for $\ln S$) for a Log series are much larger than the observed, the regression slopes are nearly identical. The slope for $\ln S$ versus $\ln E_1$ observed is −1.077 while for the theoretical we obtain −1.000. We have a situation then where the pattern is a Log series but the values are not those that were expected if this were really a true Log series.

Figure 18.12 is a BDG of the observed and estimated values for Simpson's information values. It is, as we would expect, much like Figure 18.11. The slope for the observed $\ln S$ versus $\ln E_2$ is not as close to −1 based on the Shannon information, but, nevertheless, we would interpret the pattern as the same. Also, the estimated values for the Log series are clearly larger than those expected for a theoretical or exact Log series. The higher values result because the theoretical Log series distribution is more even than the observed distribution of the $\{p_i\}$ from Bolivia plot 01. For example, the observed value at $N = 663$ is $E_1 = 0.204$, while the theoretical Log series has $E_1 = 0.452$.

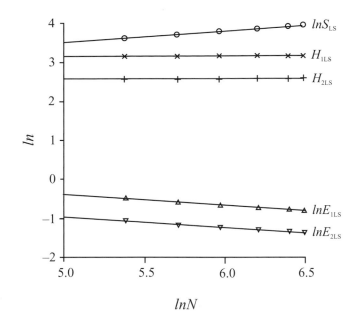

$lnS_{LS} = 2.101 + 0.285lnN, p = 0.000, R^2 = 0.999$
$H_1 = 3.158 + 0.000lnN$
$lnE_1 = 1.057 - 0.285lnN, p = 0.000, R^2 = 0.999$
$H_2 = 2.581 + 0.000lnN$
$lnE_2 = 0.480 - 0.285lnN, p = 0.000, R^2 = 0.999$
$lnS = 3.158 - 1.000lnE_1$
$lnS = 2.581 - 1.000lnE_2$

FIGURE 18.10 BDG for Log series distribution from Bolivia plot 01, $\alpha = 13.215$.

In addition, the estimated proportion of the most abundant species from the Log series is $D_{bp} = p_1 = \dfrac{ln\alpha}{\alpha} = \dfrac{2.581}{13.215} = 0.195$ while the observed most abundant species in the Bolivian data set is actually 0.380 (Appendix 2).

Log Series Pattern from Last H_1

Just as we used the last values of H_1 and lnS to examine the Log normal pattern, now we take the value of the last H_1 (2.359) calculated from the Bolivian data and use it as a constant for all entries of N and S shown in Table 18.11. On Figure 18.13, we plot the values from Table 18.11 along with the observed data calculations. The result is a remarkable fit.

The difference in the slopes for lnS versus lnE_1 for the observed and theoretical is 0.077 (Figure 18.13). The pattern for the Bolivia plot 01 data is apparently a Log series. However, looking at Figure 18.11, we see the values are not what they should be

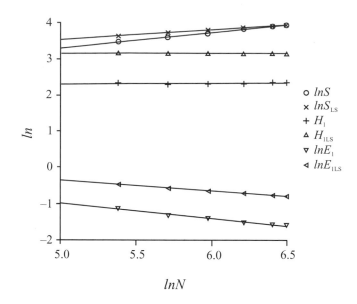

Observed: $lnS = 2.225 - 1.077lnE_1$

Log Series: $lnS = 3.158 - 1.000lnE_1$

FIGURE 18.11 BDG for observed values and Log series for Shannon's information for Bolivia plot 01.

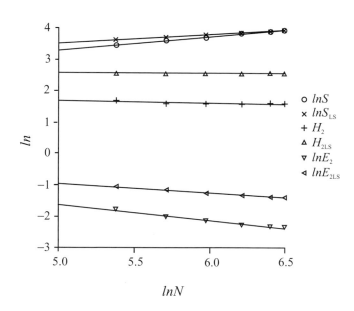

Observed: $lnS = 1.902 - 0.869lnE_2$

Log Series: $lnS = 2.581 - 1.000lnE_2$

FIGURE 18.12 BDG for observed values and Log series for Simpson's information for Bolivia plot 01.

TABLE 18.11   Log Series Pattern Calculated from Last Observed $H_1 = 2.359$

N	S	lnN	lnS	H_1	lnE_1
217	37	5.380	3.466	2.359	−1.107
301	42	5.707	3.611	2.359	−1.252
392	45	5.971	3.714	2.359	−1.355
497	48	6.209	3.850	2.359	−1.491
503	51	6.402	3.932	2.359	−1.573
663	52	6.497	3.951	2.359	−1.592

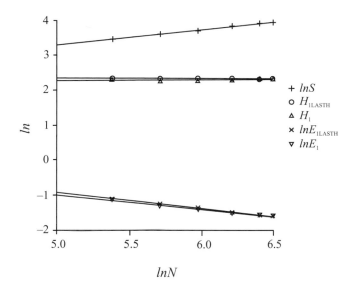

Observed: $lnS = 2.225 − 1.077lnE_{21}$, $p = 0.000$, $R^2 = 0.986$

Log Series Pattern: $lnS = 2.359 − 1.000lnE_1$, $p = 0.000$, $R^2 = 1.000$

FIGURE 18.13   BDG for observed values and Log series using last H_1 for Bolivia plot 01.

for a Log series. The last value of H_1 ($= 2.359$), however, is close to the estimate for a Canonical Log normal where $H_1 = 2.242$ (Table 18.7). If we calculate α from $H_1 = 2.359 = ln\alpha + 0.577$, then $ln\alpha = 1.782$ and $\alpha = 5.942$, or about 6. Hence, the Bolivia plot 01 data has a measure of H_1 close to what we would obtain for an exact Log series with a much lower value of α. However, a Log series with about this value would be difficult to discriminate from a Canonical Log normal.

We also note that higher values of H_1 near 3 or in the 3's; for example, those that are common in nature, would be impossible to attain for most biodiversity studies because the formula that uses $\sqrt{lnS}$ requires over 1,000 species to result in such a value. At any rate, in Bolivia plot 01, we have a situation where the pattern is Log series but the values are not, yet we have learned a considerable amount more about the community structure of our Bolivian tree data than has ever been possible by any other method by using this SHE approach.

Null Models and Neutral Models

Null Model

We define a *null model* as a baseline or benchmark against which observed data can be compared. All three of our distribution models discussed above have been suggested as null models at one time or another. For example, Lloyd and Ghelardi (1964) suggested the Broken Stick, McGill (2003) suggested use of the Log normal, while Buzas and Hayek (2005) offered evidence for the Log series.

Existing Models   In the ecological literature, there is a bewildering number of models suggested to explain relative species abundance distributions (RSAD's) (for example, May 1975; Tokeshi 1993). McGill and colleagues (2007) list 5 "families" of distributions with 27 members and state there are actually over 40 members in their system. The definitions they used are not the strict statistical definitions we use to define our cohesive groups of member distributions or measures. So, although they list both the Log series and Log normal as "purely statistical," they place the Broken Stick in the family called "niche partitioning." As you read on, it will become clear that such a bewildering number of families and non-exclusive categories with little mathematical rigor in their definitions may prove impossible for purposes of discrimination. Nevertheless, the entropy for many of these distributions has not been worked out. Obviously, there is still much work to do.

Using our SHE analysis, we have presented a tripartite scheme to describe observed patterns in which a plot of the accumulation of H with lnN can (1) exhibit H parallel to lnS, which indicates a Broken Stick distribution; (2) be horizontal or parallel to lnN, which indicates the distribution is a Log series; and (3) have a positive slope that is less than lnS, which indicates a Log normal distribution. The Log series and the Broken Stick represent the two extremes in the list of possibilities for fitting observed ecological data. Theoretical values for all other intermediate distributions will have slopes of H versus lnN between these two extremes. Many of the plots of H_1 versus lnN for numerous other models not explicitly included here lie in the area between the Broken Stick and the Log series as well. As we point out, by changing the value of the Log normal parameter γ, a wide variety of slopes are possible.

Distinguishing the pattern of one model from another in this area of overlapping and contiguous slopes on a plot can prove to be extremely difficult or likely impossible. For example, even with accumulation, Figure 18.9 shows a Canonical Log normal that is difficult to distinguish from a non-Canonical Log normal, while Figures 18.1 and 18.6 demonstrate that a Log normal can be difficult to distinguish from a Broken Stick distribution under certain conditions. We have already explained that a Canonical Log normal can be tricky to distinguish from a Log series that has an α of about 6. Thus, although the BDG plot yields a complete visual representation of biodiversity data along with regression equations, in some cases it may not be a simple matter to assign a specific distributional pattern based upon just our entropy or its estimate, even when we also use our tripartite scheme. However, in most instances the additional consideration of the tripartite scenario does lead to meaningful analysis of species abundances (Murray 2003; Buzas 2004; Hayek and colleagues 2007; Buzas and colleagues 2007a; Wilson 2008).

As we mentioned above, on a BDG plot, the lines for plots of H measures in our tripartite scheme will have slopes between those of lnS and lnN. If we envision this space as the area of a windshield where the wipers move from near vertical to horizontal when we consider or plot the differing possible distributions, we see that all possible lines are defined by the numerous positions in which the wipers may stop. What is being displayed here is that many of these possible distributions are members of a single larger family of distributions, as we discuss in Chapter 9.

Neutral Models

The development of neutral models (listed by McGill and colleagues [2007] under the "population dynamics" family) has sparked a renewed interest in species abundance models (Gotelli and McGill 2006). One neutral model proposed by Caswell (1976) has a complex development through population genetics and assumes no species interactions (see for example Platt and Lambshead 1985; Lambshead and Platt 1988). Table 18.12 shows, however, that calculated values of Shannon's H_1 for Caswell's model cannot be distinguished from those of a Log series. The zero sum multinomial development of the neutral model proposed by Hubbell (1997), also interpreted as a process-oriented model (Gotelli and McGill 2006), leads to the development of a "fundamental diversity number" that is asymptotically (for large and not-so-large samples) identical to Fisher's α (Hayek and Buzas 1997; Hubbell 2001). Thus, at least two models listed by McGill and colleagues (2007) under "population dynamics" cannot be distinguished from a model in the "purely statistical" category. Based on these evaluations, we consider the Caswell and Hubbell neutral models as equivalent to a Log series.

TABLE 18.12   Expected Values for Caswell's (1976) Neutral Model and a Log Series

S	N 200		N 400		N 800	
	$E(H_{Cas})$	$E(H_{LS})$	$E(H_{Cas})$	$E(H_{LS})$	$E(H_{Cas})$	$E(H_{LS})$
10	1.520	1.372	1.402	1.198	1.301	1.053
20	2.285	2.288	2.127	2.066	1.992	1.891
30	2.774	2.858	2.594	2.594	2.440	2.394
40	3.135	3.287	2.940	2.981	2.773	2.759
50	3.423	3.640	3.216	3.291	3.038	3.047

Log Series as a Null Model

Neutral models are particularly attractive as null models because, theoretically at least, they can be simple, and make no assumptions about the myriad of biological variables or ecological traits that enter into an organism's ability to survive. While assumptions such as a zero sum model with a constant N may be unrealistic for the majority of organisms whose densities fluctuate greatly, observations concerning *simple* models do provide a yardstick against which abundances can be evaluated.

An important set of criteria for a null model is (1) simplicity of calculation, (2) neutrality, and (3) easy comparison with observation. Based on the evaluation of properties shown in Table 18.13, we propose here the Log series as a null model as did Buzas and Hayek (2005). Especially important considerations are (1) the simplicity of calculation for the single-parameter Log series; (2) the simplicity and consistency of an information-based family of measures such as Simpson's λ and Shannon's H; and (3) for a Log series, a regression on the observations that equates the equation $lnS = \beta_0 + \beta_1 lnE_1$ to the decomposition equation $lnS = H - lnE_1$ so that $\beta_1 = -1$ and $\beta_0 = H$ must be true for any Log series. In addition, as we discuss above, the Log series is both indistinguishable from and equivalent (asymptotically) to some of the most popular neutral models.

Quantitative Assessment of Null Model

We have just shown that SHE analysis for community structure identification (SHECSI) compares both characteristics of pattern and value from a theoretical or exact distribution with an observed RSAV for identifying the most appropriate fit to a set of observed biodiversity data. This means that when we propose a null model as the baseline for our observations, then both pattern and value of these observations must be close to the theoretical values we obtain from the proposed model. It is also important that the BDG should look similar to the plot in Figure 18.13. Let us provide an example to consider value at the outset.

From the Bolivian plot 01 data we obtained a value for Fisher's alpha of $\alpha = 13.215$ so that $E(H_1) = ln(13.215) + 0.577 = 2.581 + 0.577 = 3.158$. The last value of H_1 calculated from the Bolivian data is 2.359, so $2.359 - 0.577 = 1.782 = ln\alpha$. Then, using this number, $e^{1.782} = 5.942 = \alpha$. We shall designate the first value of 13.215 as α_a and the second value of 5.942 as α_b. The ratio $\dfrac{\alpha_b}{\alpha_a} = \dfrac{5.942}{13.215} = 0.45$. This says that the last value is only 45% as large as the first.

Alternatively, we could use Equation 16.5 to compute the difference between the diversity H measures as $H_b - H_a$, or $2.359 - 3.158 = -0.799$, giving us $e^{-0.799} = 0.45$. In addition, we can also use all lnE's (remember that we can use any letters we want to distinguish quantities, so we shall use c to indicate the difference between quantities designated with a and b) recalling Equation 16.6 and calculate $lnE_c = lnE_b - lnE_a$, or $-1.592 - (-0.793) = -0.799$. Note in Equations 16.5 and 16.6 we defined lnE_3 as the difference between Simpson's and Shannon's information measures. Here this difference of the two values, or lnE_c, is used to define the difference in estimates based on Shannon's measure. We have then three alternative ways to obtain the same answer. We have calculated a measure of how much the observed and theoretical values differ.

TABLE 18.13 Population Evaluation with Accumulation of lnN and lnS

Distribution	Entropy	Observed H	Fisher's α	Simpson's λ	lnE_1	J_1	Relation of N and S	p_1 Estimate	Null Model Comparison $lnS = \beta_0 + \beta_1 lnE_1$	
Log series	$ln\alpha + 0.577$	constant	constant	constant	large decrease	small decrease	semilog	$\dfrac{ln\alpha}{\alpha}$	$\beta_1 = -1$	$\beta_0 = H$
Log normal $\gamma = 1$	$2\pi^{-0.5}\sqrt{lnS}$	small increase	increase	decrease	small decrease	small decrease	logarithm	$\dfrac{ln2}{\sqrt{\pi lnS}}$	$\beta_1 < -1$	
Log normal $\gamma < 1$	$(1 - \gamma^2)lnS$	small increase	increase	decrease	small decrease	constant	logarithm		$\beta_1 < -1$	
Broken Stick	$lnS - 0.40$	large increase	increase	decrease	constant	small increase	arithmetic	$\dfrac{lnS}{S}$	not significant	

For both simplicity and clarity we recommend the difference between the observed and Log series estimate of Shannon's H as a measure of similarity between the observed and theoretical Log series values.

For an evaluation of the pattern, recall for a Log series that our decomposition equation and the regression equation of lnS versus lnE are analogous. We have for a Log series that $lnS = \beta_0 + \beta_1 lnE$ where $\beta_0 = E(H_{Log\ series})$ and $\beta_1 = -1.000$. The difference in the β_1's between the Log series value of -1.000 and the observed value is the measure of how far the patterns differ. We have then

$$\beta_{1\ Log\ series} - \beta_{1\ observed}$$

as a measure. In the Bolivia plot 01 for the slope or pattern evaluation we have $-1.000 - (-1.077) = 0.077$. The slopes are very close.

The quantitative differences calculated above are illustrated by the BDG shown in Figure 18.11 where the slopes or patterns are similar but the values of H and lnE are quite distinct. We therefore reject the null hypothesis that the relative abundance data from Bolivia plot 01 is distributed as a Log series, concluding that a Log series pattern is present but not the appropriate values. Bolivia plot 01 shows too much dominance; that is, the last value for a Log series is $lnE_1 = -0.793$ or $E = 0.452$ (Table 18.10) while the observed is $lnE_1 = -1.592$, $E = 0.204$ (Table 18.11).

Had the slopes proved to be different from those we obtained above, we could then, of course, have pursued other patterns and values for the appropriateness of fit of different distributions. The ability of SHE analysis to recognize both pattern and value allows for a more comprehensive and sophisticated approach to community structure evaluation than simply accepting or rejecting a neutral model on the basis of a single set of numerical values from a single observed data set without accumulation. The possibility of accounting for ecological traits that may adjust abundances over space and time is an exciting new frontier (Shipley and colleagues 2006). SHE analysis for the identification of community structure (SHECSI) can and should play an important role in the renaissance of the study of species abundances (McGill 2006).

SUMMARY

1. In SHE analysis the number of individuals are accumulated and the values H, lnS, and lnE are all related to and calculated as functions of lnN.

2. Shannon's entropy estimate for a Broken Stick model is $E(H_1) = lnS - 0.42$. Simpson's entropy estimate for a Broken Stick is $E(H_2) = ln\left(\dfrac{1}{\lambda}\right) = ln\left(\dfrac{S}{2}\right)$, $\lambda = \dfrac{2}{S}$.

3. The entropy estimate based on Shannon's measure for a Canonical Log normal where the parameter $\gamma = 1$ is $E(H_1) = 2\pi^{-0.5}\sqrt{lnS}$.

4. Shannon's entropy for a Log normal where the parameter $\gamma < 1$ is $E(H_1) = (1 - \gamma^2)lnS$.

5. Shannon's entropy for a Log series where α is the parameter is $E(H_1) = ln\alpha + 0.577$. Simpson's entropy for a Log series is $E(H) = ln\left(\dfrac{1}{\lambda}\right) = ln(\alpha)$, $\alpha = \dfrac{1}{\lambda}$.

6. Similarity and confusion of interpretation of values of H_1 for given values of N and S are overcome by SHE analysis through accumulation of lnS, H, and lnE with lnN. As N and S change, or increase, then for a Broken Stick, lnE will remain constant; for a Log normal with $\gamma < 1$, $\dfrac{lnE}{lnS} = \gamma^2$ will remain constant; while for a Log series, Shannon's H_1 remains constant. These patterns can be observed even if the values of the diversity measures as components are not those expected.

7. We strongly suggest the Log series as a null model for biodiversity assessment. For the Log series, the decomposition equation $lnS = H + lnE$ becomes for the observations $lnS = \beta_0 + \beta_1 lnE$ where $\beta_0 = H$ and $\beta_1 = -1.000$.

8. When evaluating observed species abundances using SHE analysis for community structure identification (SHECSI), both pattern and value of the distribution must be considered. Knowledge of these aspects of any biodiversity analysis provide the most comprehensive in-depth evaluation available of the biodiversity from any ecological and biological systems. The system provides us with a succinct and, at the same time, dynamic summarization of community structure.

PROBLEMS

18.1 Why are the following relationships impossible?

a) $S_a \neq S_b$, $H_a = H_b$, $E_a = E_b$

b) $S_a = S_b$, $H_a = H_b$, $E_a \neq E_b$

c) $S_a = S_b$, $H_a \neq H_b$, $E_a = E_b$

18.2 Perform a SHE analysis on the bat data in Appendix 5 for $n = 206, 344, 579, 707, 901, 1,002, 1,255$, and $1,405$. How does the result of this analysis compare with the one for Bolivian trees given in Table 18.10?

18.3 Off the west coast of New Zealand at mid-bathyal depths, Buzas and colleagues (2007) made the observations shown in the table below:

N	S	H_1	E_1
200	41	3.22	0.62
400	51	3.28	0.43
1,000	67	3.35	0.43
1,300	72	3.37	0.41

Evaluate these results using the Log series as a null model.

18.4 Why are the values for H_1 generated by the entropy estimate for the Canonical Log normal impossible for the New Zealand data given above?

APPENDIX 1

Number of Individuals per 100 m² Quadrats of the
Beni Biosphere Reserve Plot 01, $N = 100$

Sp. No.	Species Name	Individuals per Quadrat No.									
		1	2	3	4	5	6	7	8	9	10
1	Scheelea princeps	0	0	0	0	1	2	0	2	0	3
2	Brosimum lactescens	1	5	4	5	1	2	4	3	1	1
3	Pouteria macrophylla	0	1	0	0	0	0	0	0	2	0
4	Calycophyllum spruceanum	0	0	0	0	0	0	0	0	0	0
5	Heisteria nitida	2	0	1	1	0	1	0	0	0	1
6	Ampelocera ruizii	0	0	0	0	0	0	0	0	0	0
7	Astrocaryum macrocalyx	0	0	0	0	0	0	0	0	0	0
8	LIANA (Bejuco)	0	0	0	0	0	0	0	0	1	2
9	Salacia	1	0	0	0	0	0	2	0	0	0
10	Sorocea saxicola	0	0	0	0	0	0	0	0	0	0
11	Trichilia pleena	0	0	0	0	0	0	0	0	0	0
12	Aspidosperma rigidum	0	0	0	0	0	0	0	0	0	0
13	Pithecellobium latifolium	0	0	0	0	1	0	0	0	0	0
14	Swartzia jorori	0	0	0	0	0	0	0	0	0	0
15	Guapira	0	0	0	0	0	0	0	0	1	1
16	Triplaris americana	0	0	0	0	0	0	0	0	0	0
17	LEGUM (Leguminosae-Mim)	0	0	0	0	0	0	0	0	0	0
18	Acacia loretensis	0	0	0	0	0	0	0	0	0	0
19	Cupania cinerea	0	0	2	0	1	0	0	0	0	0
20	Gallesia integrifolia	0	0	0	0	0	0	0	0	0	0
21	Pera benensis	0	0	0	0	0	0	0	0	0	0
22	Spondius mombin	0	0	0	0	0	0	0	0	0	0
23	Swartzia	0	0	0	0	0	0	0	0	0	0
24	Erythroxyl (Erythroxylaceae)	0	0	0	0	0	0	0	0	0	0

		Individuals per Quadrat No.									
Sp. No.	Species Name	1	2	3	4	5	6	7	8	9	10
25	Eugenia	0	0	0	0	0	0	0	0	0	0
26	Garcinia brasiliensis	0	0	0	0	0	0	0	0	0	0
27	Inga cinnamomea	1	0	0	0	0	0	0	0	0	0
28	Inga edulis	0	0	0	0	0	0	0	0	0	0
29	Pithecellobium angustifolium	0	0	0	0	0	0	0	0	0	0
30	Simaba graicile	0	0	0	0	0	0	0	0	0	0
31	Virola sebifera	0	0	0	0	0	0	0	0	0	0
32	Xylopia ligustrifolia	1	0	1	0	0	0	0	0	0	0
33	Byrsonima spicata	0	0	0	0	0	0	0	0	0	1
34	Copaifera reticulata	0	0	0	0	0	0	0	0	0	0
35	Cordia	0	0	0	1	0	0	0	0	0	0
36	Drypetes	0	0	0	0	0	0	0	0	0	0
37	Faramea	0	0	0	0	0	0	0	0	0	0
38	Ficus trigona	0	0	0	1	0	0	0	0	0	0
39	Genipa americana	0	0	0	0	0	0	0	0	0	0
40	Guazuma ulmifolia	0	0	0	0	0	0	0	0	0	0
41	Inga	1	0	0	0	0	0	0	0	0	0
42	Luehea cymulosa	0	0	0	0	0	0	0	0	0	0
43	Maytenus ebenifolia	0	1	0	0	0	0	0	0	0	0
44	Miconia	0	0	0	0	0	0	0	0	0	0
45	Platymiscium	0	0	0	0	0	0	0	0	0	0
46	Randia armata	0	0	0	0	0	0	0	0	0	0
47	Salacia	0	0	0	0	0	0	0	0	0	0
48	Toulicia reticulata	0	0	0	0	0	0	0	0	0	0
49	Unonopsis mathewsii	0	0	0	0	0	0	0	0	0	0
50	Vochysia mapirensis	0	0	0	0	0	0	0	0	0	0
51	Unidentifiable genus	0	0	0	0	0	0	0	0	0	0
52	Unidentifiable genus	0	0	0	0	0	0	0	0	0	0
Total:		7	7	8	8	4	5	6	5	5	9
Occurrences:		6	3	4	4	4	3	2	2	4	6

SOURCE: Dallmeier and colleagues 1992.

Number of Individuals per 100 m² Quadrats of the

Beni Biosphere Reserve Plot 01, *N* = 100

Individuals per Quadrat No.

Sp. No.	11	12	13	14	15	16	17	18	19	20	21	22	23	24	25
1	2	3	2	1	3	0	3	0	2	4	1	2	4	3	3
2	1	1	0	0	2	1	2	0	2	3	3	1	3	0	0
3	1	0	1	0	0	0	0	1	0	0	1	1	2	3	0
4	0	0	0	0	0	1	0	1	0	0	0	1	0	0	0
5	0	0	0	0	0	0	0	0	0	0	0	0	0	0	0
6	0	0	1	0	0	1	0	0	2	0	0	0	0	0	0
7	0	0	0	0	1	0	0	2	1	1	0	0	0	0	0
8	0	0	0	1	1	1	0	1	0	0	0	0	0	0	0
9	1	0	1	0	0	0	0	0	0	0	1	0	0	0	0
10	0	0	0	1	0	0	0	0	0	0	0	0	0	0	0
11	0	0	0	0	0	0	0	0	0	0	0	0	0	0	1
12	0	0	1	0	0	0	0	2	0	0	0	0	0	0	0
13	0	0	0	0	0	0	0	0	0	0	0	0	0	0	0
14	0	0	1	0	0	0	0	0	0	0	0	0	0	0	0
15	0	0	0	0	0	0	0	0	0	0	0	0	0	0	0
16	0	0	0	0	0	0	0	0	0	0	0	0	0	0	0
17	0	0	0	0	0	0	0	0	0	0	0	0	0	0	0
18	0	0	0	0	0	0	0	0	0	0	0	0	0	0	0
19	0	0	0	0	0	0	0	0	0	0	0	0	0	0	0
20	1	0	0	0	1	0	0	0	0	0	0	0	0	0	0
21	0	0	0	0	0	1	0	0	0	0	0	0	0	0	0
22	0	0	0	0	1	0	0	0	0	0	0	0	0	0	0
23	0	0	0	0	0	0	0	0	0	0	0	0	0	0	0
24	0	0	0	0	0	0	0	0	0	0	0	0	0	0	0

Individuals per Quadrat No.

Sp. No.	11	12	13	14	15	16	17	18	19	20	21	22	23	24	25
25	0	0	0	0	0	0	0	0	0	0	1	1	0	0	0
26	0	0	0	0	0	0	0	0	0	1	0	0	0	0	0
27	0	0	0	0	0	0	0	0	0	0	0	0	0	0	0
28	0	0	0	0	0	0	0	0	0	0	0	0	0	0	0
29	0	0	0	0	0	0	0	0	0	0	0	0	0	1	0
30	0	0	0	0	0	0	0	0	0	0	0	0	0	0	0
31	0	0	0	0	0	0	0	0	0	0	0	0	0	0	0
32	0	0	0	0	0	0	0	0	0	0	0	0	0	0	0
33	0	0	0	0	0	0	0	0	0	0	0	0	0	0	0
34	0	0	0	0	0	0	0	0	0	0	0	0	0	0	0
35	0	0	0	0	0	0	0	0	0	0	0	0	0	0	0
36	0	0	0	0	0	0	0	1	0	0	0	0	0	0	0
37	0	0	0	0	0	0	0	0	0	0	0	0	0	0	0
38	0	0	0	0	0	0	0	0	0	0	0	0	0	0	0
39	0	0	0	0	0	0	0	0	0	0	0	0	0	0	0
40	0	0	0	0	0	0	0	0	0	0	0	0	0	0	0
41	0	0	0	0	0	0	0	0	0	0	0	0	0	0	0
42	0	0	0	0	0	0	0	0	0	0	0	0	0	1	0
43	0	0	0	0	0	0	0	0	0	0	0	0	0	0	0
44	0	0	0	0	0	0	0	0	0	0	0	0	0	0	0
45	0	0	0	0	0	0	0	0	0	0	0	0	0	0	0
46	0	0	0	0	0	0	0	0	0	0	0	0	0	0	0
47	0	0	0	0	0	0	0	0	0	0	0	0	0	0	0
48	0	0	0	0	0	0	0	0	0	0	0	0	0	0	0
49	0	0	0	0	0	0	0	0	0	0	0	0	0	0	0
50	0	0	0	0	0	0	0	0	0	0	0	0	0	0	0
51	0	0	0	0	0	0	0	0	0	0	0	0	0	0	0
52	0	0	0	0	0	0	0	0	0	0	0	0	0	0	0
Total:	6	4	7	3	9	5	5	8	7	9	7	6	9	8	4
Occurrences:	5	2	6	3	6	5	2	6	4	4	5	5	3	4	2

Number of Individuals per 100 m² Quadrats of the
Beni Biosphere Reserve Plot 01, $N = 100$

Individuals per Quadrat No.

Sp. No.	26	27	28	29	30	31	32	33	34	35	36	37	38	39	40
1	4	3	7	6	5	4	2	4	4	4	1	4	2	6	2
2	0	0	2	1	0	3	1	0	2	2	2	1	2	0	3
3	2	1	1	0	0	0	0	0	1	0	0	0	0	0	0
4	0	0	0	0	2	0	1	0	0	0	0	0	0	0	0
5	1	0	0	0	0	1	0	0	0	0	0	0	0	0	0
6	0	0	1	0	0	1	1	0	2	0	0	0	0	0	0
7	0	0	0	0	0	0	0	1	1	0	0	0	1	0	0
8	0	0	0	0	0	0	0	0	0	0	1	0	0	0	0
9	0	0	0	0	0	0	0	0	0	0	3	1	0	0	0
10	0	0	0	0	0	0	0	0	0	0	0	0	0	0	0
11	0	0	0	0	0	0	0	0	0	0	0	0	0	0	0
12	0	1	0	0	0	0	0	0	0	0	0	0	0	0	0
13	0	0	0	0	0	0	0	0	0	0	0	0	0	0	0
14	0	0	0	1	0	0	0	0	0	0	1	0	0	0	0
15	0	0	0	0	0	0	0	0	0	0	0	0	0	0	0
16	0	0	0	0	0	0	0	0	0	0	0	0	0	0	0
17	0	0	0	0	0	0	0	0	0	0	0	0	0	0	0
18	0	0	0	0	0	0	0	0	0	0	0	0	0	0	0
19	0	0	0	0	0	0	0	0	0	0	0	0	0	0	0
20	0	0	0	0	0	0	0	0	0	0	0	0	0	0	0
21	0	1	0	1	0	0	0	0	0	0	0	0	0	0	0
22	0	0	0	0	0	0	0	0	0	0	0	0	1	0	0
23	0	0	0	0	0	0	0	0	0	0	0	0	0	0	0
24	0	0	0	0	0	0	0	0	0	0	0	0	0	0	0

Individuals per Quadrat No.

Sp. No.	26	27	28	29	30	31	32	33	34	35	36	37	38	39	40
25	0	0	0	0	0	0	0	0	0	0	0	0	0	0	0
26	0	0	0	0	0	0	0	0	0	0	0	0	0	0	0
27	0	0	0	0	0	0	0	0	0	0	0	0	0	0	0
28	0	0	0	0	0	0	0	0	0	0	0	0	0	0	0
29	0	0	0	0	0	0	1	0	0	0	0	0	0	0	0
30	0	0	0	0	0	0	0	0	0	0	0	0	0	0	0
31	0	0	0	0	0	0	0	0	0	0	0	0	0	0	0
32	0	0	0	0	0	0	0	0	0	0	0	0	0	0	0
33	0	0	0	0	0	0	0	0	0	0	0	0	0	0	0
34	0	0	0	0	0	0	0	0	0	0	0	0	0	0	0
35	0	0	0	0	0	0	0	0	0	0	0	0	0	0	0
36	0	0	0	0	0	0	0	0	0	0	0	0	0	0	0
37	0	0	0	0	0	0	0	0	0	0	0	0	0	0	0
38	0	0	0	0	0	0	0	0	0	0	0	0	0	0	0
39	0	0	0	0	0	0	0	0	0	0	0	0	0	0	0
40	0	0	0	0	0	0	0	0	0	0	0	0	0	0	0
41	0	0	0	0	0	0	0	0	0	0	0	0	0	0	0
42	0	0	0	0	0	0	0	0	0	0	0	0	0	0	0
43	0	0	0	0	0	0	0	0	0	0	0	0	0	0	0
44	0	0	0	0	0	0	0	0	0	0	0	0	0	0	0
45	0	0	0	0	0	0	0	0	0	0	0	0	0	0	0
46	0	0	0	0	0	0	0	0	0	0	0	0	0	0	0
47	0	0	0	0	0	1	0	0	0	0	0	0	0	0	0
48	0	0	0	0	0	0	0	0	0	0	0	0	0	0	0
49	0	0	0	0	0	0	0	0	0	0	0	0	0	0	0
50	0	0	0	0	0	0	0	0	0	0	0	0	0	0	0
51	0	0	0	0	0	0	0	0	0	0	0	1	0	0	0
52	0	0	0	0	0	0	0	0	0	0	0	0	0	0	0
Total:	7	6	11	9	7	10	6	5	10	6	8	7	6	6	5
Occurrences:	3	4	4	4	2	5	5	2	5	2	5	4	4	1	2

	Individuals per Quadrat No.														
Sp. No.	41	42	43	44	45	46	47	48	49	50	51	52	53	54	55
1	1	2	3	6	4	3	4	2	3	2	1	2	2	0	1
2	0	2	0	1	0	0	0	1	2	0	2	1	0	1	0
3	4	0	1	0	0	0	0	1	0	0	0	1	0	0	0
4	0	0	0	0	1	0	0	0	0	0	0	0	0	1	1
5	0	0	0	0	0	0	1	0	0	0	0	0	0	0	0
6	0	1	0	0	0	0	2	0	1	0	0	1	0	0	0
7	0	0	0	0	1	0	0	0	0	0	0	0	1	1	0
8	0	1	0	0	0	0	0	1	1	0	0	0	1	0	0
9	0	0	0	0	0	0	0	0	0	0	0	0	0	0	0
10	0	0	0	0	0	0	0	0	0	0	0	0	0	0	0
11	0	0	0	0	0	0	0	0	0	0	0	2	0	1	0
12	0	1	1	0	0	0	0	0	0	0	0	0	1	0	0
13	0	0	0	0	0	0	0	0	0	0	0	0	0	0	0
14	0	0	0	1	0	0	0	0	0	0	0	1	1	0	0
15	1	0	0	0	1	0	0	0	0	0	0	0	0	0	0
16	0	0	0	1	0	0	0	0	0	0	0	0	0	0	0
17	0	0	0	0	0	0	0	0	0	0	0	0	0	0	0
18	0	1	0	0	0	0	0	0	0	0	0	0	0	0	0
19	0	0	0	0	0	0	0	0	0	0	0	0	0	0	0
20	0	0	0	0	0	0	0	0	0	0	0	0	0	0	0
21	0	0	0	0	0	0	0	0	0	0	0	0	0	0	0
22	0	0	0	0	0	0	0	0	0	0	0	0	0	0	0
23	0	0	0	0	0	0	0	0	0	0	0	0	0	0	0
24	0	0	0	0	0	0	0	0	0	0	0	0	0	0	0

Individuals per Quadrat No.

Sp. No.	41	42	43	44	45	46	47	48	49	50	51	52	53	54	55
25	0	0	0	0	0	0	0	0	0	0	0	0	0	0	0
26	1	0	0	0	0	0	0	0	0	0	0	0	0	0	0
27	0	0	0	0	0	0	0	0	0	0	0	0	0	0	0
28	0	0	0	0	1	0	0	0	0	0	0	0	0	0	0
29	0	0	0	0	0	0	0	0	0	0	0	0	0	0	0
30	0	0	0	0	0	0	1	0	0	0	0	0	0	0	0
31	0	0	0	0	0	0	0	0	0	0	0	0	0	0	0
32	0	0	0	0	0	0	0	0	0	0	0	0	0	0	0
33	0	0	0	0	0	0	0	0	0	0	0	0	0	0	0
34	0	0	0	0	0	0	0	0	0	0	0	0	0	0	0
35	0	0	0	0	0	0	0	0	0	0	0	0	0	0	0
36	0	0	0	0	0	0	0	0	0	0	0	0	0	0	0
37	1	0	0	0	0	0	0	0	0	0	0	0	0	0	0
38	0	0	0	0	0	0	0	0	0	0	0	0	0	0	0
39	0	0	0	0	0	0	0	0	0	0	0	0	0	0	0
40	0	0	1	0	0	0	0	0	0	0	0	0	0	0	0
41	0	0	0	0	0	0	0	0	0	0	0	0	0	0	0
42	0	0	0	0	0	0	0	0	0	0	0	0	0	0	0
43	0	0	0	0	0	0	0	0	0	0	0	0	0	0	0
44	0	0	0	0	0	0	0	0	0	0	0	0	0	0	0
45	0	0	0	0	0	0	0	0	0	0	0	0	0	0	0
46	0	0	0	0	0	0	0	0	0	0	0	0	0	0	0
47	0	0	0	0	0	0	0	0	0	0	0	0	0	0	0
48	0	0	0	0	0	0	0	0	0	0	0	0	0	0	0
49	0	0	0	0	0	0	0	0	0	0	0	0	0	0	0
50	0	0	0	0	0	0	0	0	1	0	0	0	0	0	0
51	0	0	0	0	0	0	0	0	0	0	0	0	0	0	0
52	0	0	0	0	0	0	0	0	0	0	0	0	0	0	0
Total:	8	8	6	9	8	3	8	5	8	2	3	8	6	4	2
Occurrences:	5	6	4	4	5	1	4	4	5	1	2	6	5	4	2

Number of Individuals per 100 m² Quadrats of the

Beni Biosphere Reserve Plot 01, $N = 100$

Individuals per Quadrat No.

Sp. No.	56	57	58	59	60	61	62	63	64	65	66	67	68	69	70
1	3	2	1	4	5	4	2	2	3	2	0	1	0	2	4
2	2	1	2	0	1	0	2	1	0	0	3	1	2	1	2
3	1	0	1	0	1	0	1	2	0	0	0	0	0	0	0
4	0	1	0	0	0	0	0	0	0	0	0	1	0	0	0
5	0	0	1	1	1	0	1	0	0	0	0	0	0	0	0
6	1	0	0	0	0	0	0	1	0	0	0	0	0	0	0
7	0	0	0	0	0	0	0	0	0	1	0	0	0	0	0
8	0	0	0	0	1	0	0	1	0	0	0	2	0	0	0
9	1	0	0	0	0	0	0	0	1	0	0	0	0	0	0
10	0	0	0	1	0	0	0	0	0	0	0	0	0	0	0
11	0	0	0	0	0	0	0	0	0	0	0	0	1	0	0
12	0	0	0	0	0	1	0	0	0	0	0	0	0	0	0
13	0	0	0	0	0	0	0	1	0	0	0	0	0	0	0
14	0	0	0	0	0	0	0	0	0	0	0	0	0	0	0
15	0	0	0	0	0	0	0	0	0	0	0	0	0	0	0
16	0	0	0	0	1	0	0	0	0	0	0	0	0	0	0
17	0	0	0	0	0	0	0	0	0	0	0	0	0	1	0
18	0	0	0	0	0	0	0	0	0	0	0	0	0	0	0
19	0	0	0	0	0	0	0	0	0	0	0	0	0	0	0
20	0	0	0	0	0	0	0	0	0	0	0	0	0	0	0
21	0	0	0	0	0	0	0	0	0	0	0	0	0	0	0
22	0	0	0	0	0	0	0	1	0	0	0	0	0	0	0
23	0	0	0	1	0	0	0	0	0	0	0	0	0	0	0
24	0	0	0	0	0	0	0	0	0	0	0	0	0	0	0

Individuals per Quadrat No.

Sp. No.	56	57	58	59	60	61	62	63	64	65	66	67	68	69	70
25	0	0	0	0	0	0	0	0	0	0	0	0	0	0	0
26	0	0	0	0	0	0	0	0	0	0	0	0	0	0	0
27	0	0	0	0	0	0	1	0	0	0	0	0	0	0	0
28	0	0	0	0	0	0	1	0	0	0	0	0	0	0	0
29	0	0	0	0	0	0	0	0	0	0	0	0	0	0	0
30	0	0	0	0	0	0	0	0	0	0	0	0	0	0	0
31	0	0	0	0	0	0	0	0	0	0	0	0	2	0	0
32	0	0	0	0	0	0	0	0	0	0	0	0	0	0	0
33	0	0	0	0	0	0	0	0	0	0	0	0	0	0	0
34	0	0	0	0	0	0	0	0	0	0	0	0	0	0	0
35	0	0	0	0	0	0	0	0	0	0	0	0	0	0	0
36	0	0	0	0	0	0	0	0	0	0	0	0	0	0	0
37	0	0	0	0	0	0	0	0	0	0	0	0	0	0	0
38	0	0	0	0	0	0	0	0	0	0	0	0	0	0	0
39	0	0	0	0	0	0	0	0	0	0	1	0	0	0	0
40	0	0	0	0	0	0	0	0	0	0	0	0	0	0	0
41	0	0	0	0	0	0	0	0	0	0	0	0	0	0	0
42	0	0	0	0	0	0	0	0	0	0	0	0	0	0	0
43	0	0	0	0	0	0	0	0	0	0	0	0	0	0	0
44	0	0	0	0	0	0	0	0	0	0	0	0	0	0	0
45	0	0	0	0	0	0	0	0	0	0	0	0	0	0	0
46	0	0	0	0	0	1	0	0	0	0	0	0	0	0	0
47	0	0	0	0	0	0	0	0	0	0	0	0	0	0	0
48	0	0	0	0	0	0	0	0	0	0	0	0	0	0	0
49	0	0	0	0	0	0	0	0	0	0	0	0	0	0	0
50	0	0	0	0	0	0	0	0	0	0	0	0	0	0	0
51	0	0	0	0	0	0	0	0	0	0	0	0	0	0	0
52	0	0	0	0	0	0	1	0	0	0	0	0	0	0	0
Total:	8	4	5	7	10	6	9	9	4	3	4	5	5	4	6
Occurrences:	5	3	4	4	6	3	7	7	2	2	2	4	3	3	2

APPENDIX 1 (CONTINUED)

Number of Individuals per 100 m² Quadrats of the
Beni Biosphere Reserve Plot 01, *N* = 100

Individuals per Quadrat No.

Sp. No.	71	72	73	74	75	76	77	78	79	80	81	82	83	84	85
1	5	2	2	3	3	6	3	3	4	5	2	1	2	0	3
2	1	2	1	4	2	1	0	1	1	1	2	2	1	2	1
3	1	0	3	2	0	3	1	3	0	1	2	1	1	1	0
4	1	0	0	1	1	0	1	1	1	2	0	0	0	0	0
5	0	0	0	0	0	0	1	0	0	0	0	0	0	0	0
6	0	0	0	0	0	0	0	0	0	0	0	0	0	0	0
7	0	0	0	0	0	0	0	0	0	0	1	0	0	1	1
8	0	0	0	0	0	0	0	0	0	0	0	0	0	0	0
9	0	0	0	0	0	0	0	0	1	0	0	0	1	0	0
10	0	0	0	1	0	0	0	0	0	0	0	0	0	0	0
11	0	0	1	0	0	0	0	0	1	0	0	0	0	0	0
12	0	0	0	0	0	0	0	0	0	0	0	0	0	0	0
13	0	0	0	1	0	0	0	1	0	0	1	0	0	0	0
14	0	0	0	0	0	0	0	0	0	0	0	0	0	0	0
15	0	0	0	0	0	0	0	0	0	0	0	0	0	0	0
16	0	0	0	0	0	0	0	1	0	0	0	1	0	0	0
17	0	0	1	0	0	0	0	0	1	0	0	0	0	0	1
18	0	0	0	1	0	0	0	0	0	0	0	0	0	0	0
19	0	0	0	0	0	0	0	0	0	0	0	0	0	0	0
20	0	0	0	0	0	0	1	0	0	0	0	0	0	0	0
21	0	0	0	0	0	0	0	0	0	0	0	0	0	0	0
22	0	0	0	0	0	0	0	0	0	0	0	0	0	0	0
23	0	0	0	0	0	0	1	0	0	0	0	0	0	0	0
24	0	1	0	0	0	0	0	0	0	0	0	0	0	0	0

Individuals per Quadrat No.

Sp. No.	71	72	73	74	75	76	77	78	79	80	81	82	83	84	85
25	0	0	0	0	0	0	0	0	0	0	0	0	0	0	0
26	0	0	0	0	0	0	0	0	0	0	0	0	0	0	0
27	0	0	0	0	0	0	0	0	0	0	0	0	0	0	0
28	0	0	0	0	0	0	0	0	0	0	0	0	0	0	0
29	0	0	0	0	0	0	0	0	0	0	0	0	0	0	0
30	0	0	0	0	0	0	0	0	0	0	0	0	0	0	0
31	0	0	0	0	0	0	0	0	0	0	0	0	0	0	0
32	0	0	0	0	0	0	0	0	0	0	0	0	0	0	0
33	0	0	0	0	0	0	0	0	0	0	0	0	0	0	0
34	0	0	0	0	0	0	0	0	0	0	0	0	0	0	0
35	0	0	0	0	0	0	0	0	0	0	0	0	0	0	0
36	0	0	0	0	0	0	0	0	0	0	0	0	0	0	0
37	0	0	0	0	0	0	0	0	0	0	0	0	0	0	0
38	0	0	0	0	0	0	0	0	0	0	0	0	0	0	0
39	0	0	0	0	0	0	0	0	0	0	0	0	0	0	0
40	0	0	0	0	0	0	0	0	0	0	0	0	0	0	0
41	0	0	0	0	0	0	0	0	0	0	0	0	0	0	0
42	0	0	0	0	0	0	0	0	0	0	0	0	0	0	0
43	0	0	0	0	0	0	0	0	0	0	0	0	0	0	0
44	0	0	0	0	0	0	0	1	0	0	0	0	0	0	0
45	0	0	0	0	0	0	0	0	0	0	0	0	0	0	0
46	0	0	0	0	0	0	0	0	0	0	0	0	0	0	0
47	0	0	0	0	0	0	0	0	0	0	0	0	0	0	0
48	0	0	0	0	0	0	0	0	0	0	0	0	0	0	1
49	0	0	0	0	0	0	0	0	0	0	0	0	0	0	1
50	0	0	0	0	0	0	0	0	0	0	0	0	0	0	0
51	0	0	0	0	0	0	0	0	0	0	0	0	0	0	0
52	0	0	0	0	0	0	0	0	0	0	0	0	0	0	0
Total:	8	5	8	13	6	10	8	11	9	9	8	5	5	4	8
Occurrences:	4	3	5	7	3	3	6	7	6	4	5	4	4	3	6

Individuals per Quadrat No.

Sp. No.	86	87	88	89	90	91	92	93	94	95	96	97	98	99	100
1	2	3	3	1	1	0	5	5	4	4	1	3	2	2	2
2	0	0	0	0	1	3	1	2	1	1	2	3	1	0	1
3	0	0	0	1	0	1	0	0	1	2	1	1	2	2	1
4	0	0	0	0	0	2	0	0	0	0	1	0	1	1	1
5	0	1	3	0	1	0	0	0	0	0	0	0	0	0	0
6	0	0	1	0	0	0	0	0	0	0	0	0	0	0	0
7	0	2	0	0	0	0	0	0	0	0	0	0	0	0	0
8	0	0	0	0	0	0	0	0	0	0	0	0	0	1	0
9	2	0	0	0	0	0	0	0	0	0	0	1	0	0	0
10	0	0	0	0	0	0	0	0	0	1	3	0	0	1	1
11	0	0	1	0	0	0	0	0	1	0	0	0	0	0	0
12	0	0	0	0	0	0	0	0	0	0	0	0	0	0	0
13	0	1	0	0	0	0	0	0	0	0	0	0	0	0	0
14	0	0	0	0	0	0	0	0	0	0	0	0	0	0	0
15	0	0	0	0	0	0	0	0	0	0	0	0	0	0	0
16	0	0	0	0	0	0	0	0	0	0	0	0	0	0	0
17	0	0	0	0	0	0	0	0	0	0	0	0	0	0	0
18	0	0	0	0	1	0	0	0	0	0	0	0	0	0	0
19	0	0	0	0	0	0	0	0	0	0	0	0	0	0	0
20	0	0	0	0	0	0	0	0	0	0	0	0	0	0	0
21	0	0	0	0	0	0	0	0	0	0	0	0	0	0	0
22	0	0	0	0	0	0	0	0	0	0	0	0	0	0	0
23	0	0	0	0	0	0	0	0	0	0	0	1	0	0	0
24	0	0	0	0	0	0	0	1	0	0	0	0	0	0	0

Individuals per Quadrat No.

Sp. No.	86	87	88	89	90	91	92	93	94	95	96	97	98	99	100
25	0	0	0	0	0	0	0	0	0	0	0	0	0	0	0
26	0	0	0	0	0	0	0	0	0	0	0	0	0	0	0
27	0	0	0	0	0	0	0	0	0	0	0	0	0	0	0
28	0	0	0	0	0	0	0	0	0	0	0	0	0	0	0
29	0	0	0	0	0	0	0	0	0	0	0	0	0	0	0
30	0	1	0	0	0	0	0	0	0	0	0	0	0	0	0
31	0	0	0	0	0	0	0	0	0	0	0	0	0	0	0
32	0	0	0	0	0	0	0	0	0	0	0	0	0	0	0
33	0	0	0	0	0	0	0	0	0	0	0	0	0	0	0
34	0	0	0	0	0	1	0	0	0	0	0	0	0	0	0
35	0	0	0	0	0	0	0	0	0	0	0	0	0	0	0
36	0	0	0	0	0	0	0	0	0	0	0	0	0	0	0
37	0	0	0	0	0	0	0	0	0	0	0	0	0	0	0
38	0	0	0	0	0	0	0	0	0	0	0	0	0	0	0
39	0	0	0	0	0	0	0	0	0	0	0	0	0	0	0
40	0	0	0	0	0	0	0	0	0	0	0	0	0	0	0
41	0	0	0	0	0	0	0	0	0	0	0	0	0	0	0
42	0	0	0	0	0	0	0	0	0	0	0	0	0	0	0
43	0	0	0	0	0	0	0	0	0	0	0	0	0	0	0
44	0	0	0	0	0	0	0	0	0	0	0	0	0	0	0
45	0	0	0	0	0	0	0	0	0	1	0	0	0	0	0
46	0	0	0	0	0	0	0	0	0	0	0	0	0	0	0
47	0	0	0	0	0	0	0	0	0	0	0	0	0	0	0
48	0	0	0	0	0	0	0	0	0	0	0	0	0	0	0
49	0	0	0	0	0	0	0	0	0	0	0	0	0	0	0
50	0	0	0	0	0	0	0	0	0	0	0	0	0	0	0
51	0	0	0	0	0	0	0	0	0	0	0	0	0	0	0
52	0	0	0	0	0	0	0	0	0	0	0	0	0	0	0
Total:	4	8	8	2	4	7	6	8	7	9	8	9	6	7	6
Occurrences:	2	5	4	2	4	4	2	3	4	5	5	5	4	5	5

Sp. No.	Total	Mean (μ)	Variance (σ^2)	Standard Deviation (σ)	Proportion	Occurrences
1	252	2.52	2.6496	1.6278	0.3801	87
2	131	1.31	1.3739	1.1721	0.1976	72
3	62	0.62	0.7756	0.8807	0.0935	42
4	25	0.25	0.2475	0.4975	0.0377	22
5	19	0.19	0.2339	0.4836	0.0287	16
6	17	0.17	0.2011	0.4484	0.0256	14
7	17	0.17	0.1811	0.4256	0.0256	15
8	17	0.17	0.1811	0.4256	0.0256	15
9	17	0.17	0.2411	0.4910	0.0256	13
10	9	0.09	0.1419	0.3767	0.0136	7
11	9	0.09	0.1019	0.3192	0.0136	8
12	8	0.08	0.0936	0.3059	0.0121	7
13	6	0.06	0.0564	0.2375	0.0090	6
14	6	0.06	0.0564	0.2375	0.0090	6
15	4	0.04	0.0384	0.1960	0.0060	4
16	4	0.04	0.0384	0.1960	0.0060	4
17	4	0.04	0.0384	0.1960	0.0060	4
18	3	0.03	0.0291	0.1706	0.0045	3
19	3	0.03	0.0491	0.2216	0.0045	2
20	3	0.03	0.0291	0.1706	0.0045	3
21	3	0.03	0.0291	0.1706	0.0045	3
22	3	0.03	0.0291	0.1706	0.0045	3
23	3	0.03	0.0291	0.1706	0.0045	3
24	2	0.02	0.0196	0.1400	0.0030	2

Sp. No.	Total	Mean (μ)	Variance (σ^2)	Standard Deviation (σ)	Proportion	Occurrences
25	2	0.02	0.0196	0.1400	0.0030	2
26	2	0.02	0.0196	0.1400	0.0030	2
27	2	0.02	0.0196	0.1400	0.0030	2
28	2	0.02	0.0196	0.1400	0.0030	2
29	2	0.02	0.0196	0.1400	0.0030	2
30	2	0.02	0.0196	0.1400	0.0030	2
31	2	0.02	0.0396	0.1990	0.0030	1
32	2	0.02	0.0196	0.1400	0.0030	2
33	1	0.01	0.0099	0.0995	0.0015	1
34	1	0.01	0.0099	0.0995	0.0015	1
35	1	0.01	0.0099	0.0995	0.0015	1
36	1	0.01	0.0099	0.0995	0.0015	1
37	1	0.01	0.0099	0.0995	0.0015	1
38	1	0.01	0.0099	0.0995	0.0015	1
39	1	0.01	0.0099	0.0995	0.0015	1
40	1	0.01	0.0099	0.0995	0.0015	1
41	1	0.01	0.0099	0.0995	0.0015	1
42	1	0.01	0.0099	0.0995	0.0015	1
43	1	0.01	0.0099	0.0995	0.0015	1
44	1	0.01	0.0099	0.0995	0.0015	1
45	1	0.01	0.0099	0.0995	0.0015	1
46	1	0.01	0.0099	0.0995	0.0015	1
47	1	0.01	0.0099	0.0995	0.0015	1
48	1	0.01	0.0099	0.0995	0.0015	1
49	1	0.01	0.0099	0.0995	0.0015	1
50	1	0.01	0.0099	0.0995	0.0015	1
51	1	0.01	0.0099	0.0995	0.0015	1
52	1	0.01	0.0099	0.0995	0.0015	1
Totals:	663				1.0000	396

$\mu = 6.63$

$\sigma^2 = 4.6931$

$\sigma = 2.1664$

APPENDIX 2

Number of Individuals per 400 m² Quadrats of the
Beni Biosphere Reserve Plot 01, $N = 25$

		Individuals per Quadrat No.									
Sp. No.	Species Name	1	2	3	4	5	6	7	8	9	10
1	Scheelea princeps	0	5	8	6	9	10	17	17	13	14
2	Brosimum lactescens	15	10	4	3	7	7	2	5	6	6
3	Pouteria macrophylla	1	0	3	1	1	7	4	0	1	0
4	Calycophyllum spruceanum	0	0	0	1	1	1	0	3	0	0
5	Heisteria nitida	4	1	1	0	0	0	1	1	0	0
6	Ampelocera ruizii	0	0	0	2	2	0	1	2	2	0
7	Astrocaryum macrocalyx	0	0	0	1	4	0	0	0	2	1
8	LIANA (Bejuco)	0	0	3	3	1	0	0	0	1	0
9	Saiacia	1	2	1	1	0	1	0	0	3	1
10	Sorocea saxicola	0	0	0	1	0	0	0	0	0	0
11	Trichilia pleena	0	0	0	0	0	0	1	0	0	0
12	Aspidosperma rigidum	0	0	0	1	2	0	1	0	0	0
13	Pithecellobium latifolium	0	1	0	0	0	0	0	0	0	0
14	Swartzia jorori	0	0	0	1	0	0	0	1	1	0
15	Guapira	0	0	2	0	0	0	0	0	0	0
16	Triplaris americana	0	0	0	0	0	0	0	0	0	0
17	LEGUM (Leguminosae-Mim)	0	0	0	0	0	0	0	0	0	0
18	Acacia loretensis	0	0	0	0	0	0	0	0	0	0
19	Cupania cinerea	2	1	0	0	0	0	0	0	0	0
20	Gallesia integrifolia	0	0	1	1	0	0	0	0	0	0
21	Pera benensis	0	0	0	1	0	0	1	1	0	0
22	Spondius mombin	0	0	0	1	0	0	0	0	0	1
23	Swartzia	0	0	0	0	0	0	0	0	0	0
24	Erythroxyl (Erythroxylaceae)	0	0	0	0	0	0	0	0	0	0

Sp. No.	Species Name	Individuals per Quadrat No.									
		1	2	3	4	5	6	7	8	9	10
25	Eugenia	0	0	0	0	0	2	0	0	0	0
26	Garcinia brasiliensis	0	0	0	0	1	0	0	0	0	0
27	Inga cinnamomea	1	0	0	0	0	0	0	0	0	0
28	Inga edulis	0	0	0	0	0	0	0	0	0	0
29	Pithecellobium angustifolium	0	0	0	0	0	1	0	1	0	0
30	Simaba graicile	0	0	0	0	0	0	0	0	0	0
31	Virola sebifera	0	0	0	0	0	0	0	0	0	0
32	Xylopia ligustrifolia	2	0	0	0	0	0	0	0	0	0
33	Byrsonima spicata	0	0	1	0	0	0	0	0	0	0
34	Copaifera reticulata	0	0	0	0	0	0	0	0	0	0
35	Cordia	1	0	0	0	0	0	0	0	0	0
36	Drypetes	0	0	0	0	1	0	0	0	0	0
37	Faramea	0	0	0	0	0	0	0	0	0	0
38	Ficus trigona	1	0	0	0	0	0	0	0	0	0
39	Genipa americana	0	0	0	0	0	0	0	0	0	0
40	Guazuma ulmifolia	0	0	0	0	0	0	0	0	0	0
41	Inga	1	0	0	0	0	0	0	0	0	0
42	Luehea cymulosa	0	0	0	0	0	1	0	0	0	0
43	Maytenus ebenifolia	1	0	0	0	0	0	0	0	0	0
44	Miconia	0	0	0	0	0	0	0	0	0	0
45	Platymiscium	0	0	0	0	0	0	0	0	0	0
46	Randia armata	0	0	0	0	0	0	0	0	0	0
47	Salacia	0	0	0	0	0	0	0	1	0	0
48	Toulicia reticulata	0	0	0	0	0	0	0	0	0	0
49	Unonopsis mathewsii	0	0	0	0	0	0	0	0	0	0
50	Vochysia mapirensis	0	0	0	0	0	0	0	0	0	0
51	Unidentifiable genus	0	0	0	0	0	0	0	0	0	1
52	Unidentifiable genus	0	0	0	0	0	0	0	0	0	0
Total:		30	20	24	24	29	30	28	32	29	24
Occurrences:		11	6	9	14	10	8	8	9	8	6

Individuals per Quadrat No.

Sp. No.	11	12	13	14	15	16	17	18	19	20	21	22	23	24	25
1	12	13	8	6	12	11	3	13	14	15	5	11	7	14	9
2	3	1	5	3	4	3	6	6	8	3	7	1	5	6	5
3	5	1	1	1	2	3	0	1	8	5	5	0	2	4	6
4	0	1	0	2	1	0	1	1	2	5	0	0	2	1	3
5	0	1	0	0	3	1	0	0	0	1	0	4	1	0	0
6	1	2	2	1	0	1	0	0	0	0	0	1	0	0	0
7	0	1	0	2	0	0	1	0	0	0	2	3	0	0	0
8	1	1	1	1	1	1	2	0	0	0	0	0	0	0	1
9	0	0	0	1	0	1	0	0	0	1	1	2	0	0	1
10	0	0	0	0	1	0	0	0	1	0	0	0	0	4	2
11	0	0	2	1	0	0	1	0	1	1	0	1	0	1	0
12	2	0	0	1	0	1	0	0	0	0	0	0	0	0	0
13	0	0	0	0	0	1	0	0	1	1	1	1	0	0	0
14	1	0	1	1	0	0	0	0	0	0	0	0	0	0	0
15	1	1	0	0	0	0	0	0	0	0	0	0	0	0	0
16	1	0	0	0	1	0	0	0	0	1	1	0	0	0	0
17	0	0	0	0	0	0	0	1	1	1	0	1	0	0	0
18	1	0	0	0	0	0	0	0	1	0	0	0	1	0	0
19	0	0	0	0	0	0	0	0	0	0	0	0	0	0	0
20	0	0	0	0	0	0	0	0	0	1	0	0	0	0	0
21	0	0	0	0	0	0	0	0	0	0	0	0	0	0	0
22	0	0	0	0	0	1	0	0	0	0	0	0	0	0	0
23	0	0	0	0	1	0	0	0	0	1	0	0	0	0	1
24	0	0	0	0	0	0	0	1	0	0	0	0	0	1	0

Individuals per Quadrat No.

Sp. No.	11	12	13	14	15	16	17	18	19	20	21	22	23	24	25
25	0	0	0	0	0	0	0	0	0	0	0	0	0	0	0
26	1	0	0	0	0	0	0	0	0	0	0	0	0	0	0
27	0	0	0	0	0	1	0	0	0	0	0	0	0	0	0
28	0	1	0	0	0	1	0	0	0	0	0	0	0	0	0
29	0	0	0	0	0	0	0	0	0	0	0	0	0	0	0
30	0	1	0	0	0	0	0	0	0	0	0	1	0	0	0
31	0	0	0	0	0	0	2	0	0	0	0	0	0	0	0
32	0	0	0	0	0	0	0	0	0	0	0	0	0	0	0
33	0	0	0	0	0	0	0	0	0	0	0	0	0	0	0
34	0	0	0	0	0	0	0	0	0	0	0	0	1	0	0
35	0	0	0	0	0	0	0	0	0	0	0	0	0	0	0
36	0	0	0	0	0	0	0	0	0	0	0	0	0	0	0
37	1	0	0	0	0	0	0	0	0	0	0	0	0	0	0
38	0	0	0	0	0	0	0	0	0	0	0	0	0	0	0
39	0	0	0	0	0	0	1	0	0	0	0	0	0	0	0
40	1	0	0	0	0	0	0	0	0	0	0	0	0	0	0
41	0	0	0	0	0	0	0	0	0	0	0	0	0	0	0
42	0	0	0	0	0	0	0	0	0	0	0	0	0	0	0
43	0	0	0	0	0	0	0	0	0	0	0	0	0	0	0
44	0	0	0	0	0	0	0	0	0	1	0	0	0	0	0
45	0	0	0	0	0	0	0	0	0	0	0	0	0	1	0
46	0	0	0	0	0	1	0	0	0	0	0	0	0	0	0
47	0	0	0	0	0	0	0	0	0	0	0	0	0	0	0
48	0	0	0	0	0	0	0	0	0	0	0	1	0	0	0
49	0	0	0	0	0	0	0	0	0	0	0	1	0	0	0
50	0	0	1	0	0	0	0	0	0	0	0	0	0	0	0
51	0	0	0	0	0	0	0	0	0	0	0	0	0	0	0
52	0	0	0	0	0	1	0	0	0	0	0	0	0	0	0
Total:	31	24	21	20	26	28	17	23	37	37	22	28	19	32	28
Occurrences:	13	11	8	11	9	14	8	6	9	13	7	12	7	8	8

Sp. No.	Total	Mean (μ)	Variance (σ^2)	Standard Deviation (σ)	Proportion	Occurrences
1	252	10.08	18.3136	4.2794	0.3801	24
2	131	5.24	8.5024	2.9159	0.1976	25
3	62	2.48	5.4496	2.3344	0.0935	20
4	25	1.00	1.5200	1.2329	0.0377	14
5	19	0.76	1.3824	1.1758	0.0287	11
6	17	0.68	0.6976	0.8352	0.0256	11
7	17	0.68	1.1776	1.0852	0.0256	9
8	17	0.68	0.7776	0.8818	0.0256	12
9	17	0.68	0.6176	0.7859	0.0256	13
10	9	0.36	0.7904	0.8890	0.0136	5
11	9	0.36	0.3104	0.5571	0.0136	8
12	8	0.32	0.3776	0.6145	0.0121	6
13	6	0.24	0.1824	0.4271	0.0090	6
14	6	0.24	0.1824	0.4271	0.0090	6
15	4	0.16	0.2144	0.4630	0.0060	3
16	4	0.16	0.1344	0.3666	0.0060	4
17	4	0.16	0.1344	0.3666	0.0060	4
18	3	0.12	0.1056	0.3250	0.0045	3
19	3	0.12	0.1856	0.4308	0.0045	2
20	3	0.12	0.1056	0.3250	0.0045	3
21	3	0.12	0.1056	0.3250	0.0045	3
22	3	0.12	0.1056	0.3250	0.0045	3
23	3	0.12	0.1056	0.3250	0.0045	3
24	2	0.08	0.0736	0.2713	0.0030	2

Sp. No.	Total	Mean (μ)	Variance (σ^2)	Standard Deviation (σ)	Proportion	Occurrences
25	2	0.08	0.1536	0.3919	0.0030	1
26	2	0.08	0.0736	0.2713	0.0030	2
27	2	0.08	0.0736	0.2713	0.0030	2
28	2	0.08	0.0736	0.2713	0.0030	2
29	2	0.08	0.0736	0.2713	0.0030	2
30	2	0.08	0.0736	0.2713	0.0030	2
31	2	0.08	0.1536	0.3919	0.0030	1
32	2	0.08	0.1536	0.3919	0.0030	1
33	1	0.04	0.0384	0.1960	0.0015	1
34	1	0.04	0.0384	0.1960	0.0015	1
35	1	0.04	0.0384	0.1960	0.0015	1
36	1	0.04	0.0384	0.1960	0.0015	1
37	1	0.04	0.0384	0.1960	0.0015	1
38	1	0.04	0.0384	0.1960	0.0015	1
39	1	0.04	0.0384	0.1960	0.0015	1
40	1	0.04	0.0384	0.1960	0.0015	1
41	1	0.04	0.0384	0.1960	0.0015	1
42	1	0.04	0.0384	0.1960	0.0015	1
43	1	0.04	0.0384	0.1960	0.0015	1
44	1	0.04	0.0384	0.1960	0.0015	1
45	1	0.04	0.0384	0.1960	0.0015	1
46	1	0.04	0.0384	0.1960	0.0015	1
47	1	0.04	0.0384	0.1960	0.0015	1
48	1	0.04	0.0384	0.1960	0.0015	1
49	1	0.04	0.0384	0.1960	0.0015	1
50	1	0.04	0.0384	0.1960	0.0015	1
51	1	0.04	0.0384	0.1960	0.0015	1
52	1	0.04	0.0384	0.1960	0.0015	1
Totals:	663				1.0000	233

$\mu = 26.52$

$\sigma^2 = 26.6496$

$\sigma = 5.1623$

APPENDIX 3

Table of Random Numbers

This is a small set of random numbers to aid the reader with solving the problems at the end of chapters. A "blind" start anywhere in the table allows you to begin. Once you start, read the numbers in row or column order. Each set of four numbers can be divided into a couplet. For example, when choosing $N = 100$, the first four numbers are 18, 69, 03, and 97 starting at the upper left. For $N = 25$, read the numbers in order, choosing only the applicable ones. The first four numbers are 18, 03, 11, and 01 when reading across the row. Here 69 and 97 were not used because they are each bigger than 25. (Table prepared by Charles R. Mann Associates, Inc., Washington, D.C.)

1869	0397	1101	8532	6660	1968	7704	0187	0020	1468	7087	0951	1273
7739	8620	5251	6452	3500	3371	1091	7609	2869	9628	3931	9218	6107
3564	4871	5198	3555	6024	2068	5724	3033	9519	6310	3996	6207	1496
9833	7139	5689	1792	7888	6489	1645	4095	8903	5139	1391	5524	3302
7400	0551	7830	7991	7324	8209	6703	8085	7060	4928	2532	3583	8379
0183	1481	7791	5715	9452	0053	1460	1041	1732	3023	4717	6041	6789
5554	7259	6724	0122	1776	6145	1040	0875	3543	8031	5696	7025	5193
8632	5812	3846	9286	1914	2435	2731	0452	6465	9860	3628	7932	0580
2568	9996	7072	9599	2120	9421	8947	1768	5660	3342	8956	8542	9981
8849	9056	1616	6689	9505	2832	2136	2055	4683	1028	0424	0389	3607
2214	6827	5817	2988	6278	0377	2669	2825	6319	6686	5644	7502	4256
7921	0296	4028	4883	0405	1210	2002	5264	5552	5792	6882	1695	8796
0369	0513	9897	4571	8965	7939	7006	3123	9397	9399	7760	4203	5770
4768	7160	4203	5770	4768	7160	7186	4962	0181	4240	4821	2303	3568
9581	8740	4959	6983	3950	7849	6861	9050	4809	8746	7637	8230	4629
0172	3280	2477	6909	8720	8736	0118	3197	7607	8675	1205	0484	6084

1539	7880	6768	5301	5346	5825	8286	7460	4947	6748	7510	7052	6393
6299	5286	3835	6418	6622	2723	6440	2553	3512	0046	2346	7677	6282
4511	1008	7680	1591	6423	9665	6559	4816	5508	2358	6730	9539	6895
2203	5747	3647	8428	9966	8746	9779	4438	7800	5277	2492	0994	9150
9187	1904	2947	7762	4752	3801	2886	4948	6638	8733	7947	3688	0075
2299	2418	7281	3354	3542	3775	3556	0371	8422	6819	0433	3160	8699
4028	7823	0329	3426	2410	2150	7246	8782	7090	1965	6379	8056	2288
0533	2288	7375	0604	9915	1962	4923	9404	3056	0826	7875	8044	4780
5363	4671	5233	8408	4040	8108	2921	5395	4921	5374	6902	7430	5716
9577	4920	5816	6431	1776	1020	9865	3546	3887	9959	6507	8001	7106
4243	8347	1768	4277	8205	2858	8021	5206	8595	9428	1293	6885	1866
5638	8501	3718	9575	0486	3836	9168	4114	4361	2069	0462	8075	0652
6093	1407	7266	6292	7496	6858	5236	4045	8541	1990	7434	8133	1937
5142	5309	4239	2418	3746	8267	6023	9431	4849	3741	0897	7307	3380
4644	0119	5592	5855	4839	0049	5223	4227	8601	5837	6420	8763	1652
8798	2175	2062	0667	3699	0792	7457	5152	1828	9466	1191	1217	7070
2461	6401	3156	4081	3376	5125	3675	4911	3870	6141	6494	5825	3941
6528	4055	4986	5306	3534	9172	8617	5200	5521	5909	6943	9111	1113
8741	5865	1504	6314	7529	2149	3175	4987	6858	9941	6733	2967	0982
4980	0879	3541	0505	9043	3707	8129	3872	1263	8394	9456	0934	1968
1512	3639	8685	3897	3736	1460	7387	8164	0187	2046	8861	0435	8368
4736	7727	3035	0783	3863	2115	4462	9660	4886	8508	9435	0554	3667
1567	5984	1303	3379	5546	6515	3447	5768	0884	4848	5498	7067	9980
4055	4069	5910	3045	6498	5001	2261	8123	0259	4934	3172	4405	8395

APPENDIX 4

Values of the Log Series Parameter α for a Given Number of Individuals (*N*) and Species (*S*)

S	50	55	60	65	70	75	80	85	90	95
1	0.177	0.174	0.170	0.168	0.165	0.163	0.161	0.159	0.157	0.156
2	0.417	0.407	0.398	0.391	0.384	0.378	0.372	0.367	0.362	0.358
3	0.701	0.681	0.665	0.650	0.637	0.626	0.615	0.606	0.597	0.590
4	1.023	0.992	0.965	0.941	0.921	0.903	0.886	0.871	0.859	0.846
5	1.383	1.336	1.297	1.262	1.232	1.206	1.182	1.161	1.142	1.124
6	1.780	1.715	1.660	1.612	1.571	1.535	1.502	1.473	1.447	1.423
7	2.215	2.127	2.054	1.991	1.936	1.889	1.846	1.808	1.774	1.743
8	2.689	2.574	2.479	2.398	2.328	2.267	2.213	2.165	2.121	2.082
9	3.203	3.057	2.936	2.834	2.747	2.670	2.603	2.543	2.490	2.441
10	3.759	3.577	3.427	3.301	3.193	3.099	3.017	2.944	2.879	2.820
11	4.359	4.135	3.951	3.797	3.666	3.553	3.454	3.366	3.288	3.218
12	**5.007**	4.733	4.511	4.325	4.168	4.033	3.915	3.811	3.719	3.636
13	5.705	5.374	5.107	4.886	4.700	4.540	4.401	4.279	4.171	4.073
14	6.456	**6.060**	5.743	5.482	5.262	5.075	4.913	4.770	4.644	4.532
15	7.265	6.794	**6.419**	6.113	5.856	5.638	5.450	5.285	5.140	5.010
16	8.137	7.579	7.139	**6.781**	6.483	6.231	6.014	5.825	5.658	5.510
17	9.075	8.419	7.905	7.489	7.145	6.855	6.606	6.390	6.200	6.032
18	10.086	9.317	8.719	8.238	**7.843**	7.511	7.227	6.981	6.766	6.575
19	11.177	10.278	9.584	9.031	8.579	**8.200**	7.877	7.599	7.356	7.142
20	12.355	11.306	10.505	9.871	9.354	8.924	**8.559**	8.245	7.972	7.732
21	13.628	12.408	11.485	10.759	10.172	9.685	9.273	**8.921**	8.614	8.345
22	15.007	13.590	12.529	11.700	11.034	10.484	10.021	9.626	9.284	8.984
23	16.503	14.859	13.641	12.697	11.942	11.323	10.804	10.363	**9.982**	9.648
24	18.128	16.223	14.826	13.753	12.901	12.205	11.625	11.132	10.709	**10.339**

| | | | | | N | | | | | |
S	50	55	60	65	70	75	80	85	90	95
25	19.898	17.691	16.091	14.873	13.912	13.132	12.484	11.936	11.466	11.058
26	21.830	19.273	17.442	16.061	14.979	14.105	13.383	12.776	12.255	11.805
27	23.946	20.981	18.887	17.323	16.106	15.129	14.326	13.652	13.078	12.581
28	26.270	22.830	20.435	18.664	17.297	16.206	15.314	14.568	13.935	13.389
29	28.832	24.835	22.095	20.090	18.556	17.339	16.349	15.525	14.827	14.228
30	31.665	27.014	23.877	21.610	19.888	18.532	17.434	16.525	15.758	15.101
31		29.389	25.795	23.229	21.299	19.789	18.573	17.570	16.728	16.008
32		31.985	27.863	24.958	22.794	21.114	19.768	18.663	17.739	16.952
33		34.832	30.097	26.807	24.380	22.511	21.022	19.807	18.793	17.934
34			32.515	28.786	26.065	23.985	22.340	21.003	19.893	18.955
35			35.141	30.909	27.857	25.544	23.726	22.256	21.041	20.017
36			37.999	33.191	29.764	27.191	25.183	23.568	22.239	21.123
37				35.647	31.799	28.936	26.717	24.943	23.490	22.275
38				38.298	33.971	30.784	28.333	26.386	24.797	23.474
39				41.165	36.294	32.745	30.038	27.899	26.163	24.724
40					38.784	34.829	31.836	29.488	27.592	26.027
41					41.456	37.045	33.736	31.158	29.087	27.385
42					44.332	39.405	35.746	32.914	30.652	28.802
43						41.924	37.873	34.762	32.292	30.281
44						44.616	40.129	36.709	34.012	31.825
45						47.498	42.523	38.763	35.816	33.439
46							45.067	40.931	37.710	35.127
47							47.776	43.222	39.701	36.892
48							50.665	45.645	41.794	38.740
49								48.213	43.999	40.677
50								50.937	46.322	42.707
51								53.831	48.773	44.838
52									51.361	47.076
53									54.099	49.428
54									56.998	51.904
55										54.511
56										57.261
57										60.164

S	N 100	105	110	115	120	125	130	135	140	145
1	0.154	0.153	0.152	0.151	0.149	0.148	0.147	0.146	0.146	0.145
2	0.354	0.351	0.347	0.344	0.341	0.338	0.335	0.333	0.331	0.328
3	0.582	0.576	0.569	0.564	0.558	0.553	0.548	0.544	0.539	0.535
4	0.834	0.824	0.814	0.805	0.797	0.789	0.781	0.774	0.767	0.761
5	1.108	1.093	1.079	1.066	1.054	1.043	1.032	1.022	1.013	1.004
6	1.401	1.381	1.363	1.345	1.329	1.314	1.300	1.287	1.274	1.263
7	1.714	1.688	1.664	1.642	1.621	1.602	1.584	1.567	1.551	1.536
8	2.046	2.014	1.983	1.955	1.929	1.905	1.883	1.862	1.842	1.823
9	2.397	2.357	2.320	2.285	2.254	2.224	2.197	2.171	2.146	2.124
10	2.766	2.718	2.673	2.632	2.593	2.558	2.525	2.494	2.465	2.438
11	3.154	3.096	3.043	2.994	2.949	2.907	2.868	2.831	2.797	2.765
12	3.561	3.492	3.430	3.373	3.320	3.270	3.225	3.182	3.142	3.105
13	3.986	3.906	3.834	3.767	3.706	3.649	3.596	3.547	3.501	3.458
14	4.430	4.339	4.255	4.178	4.108	4.042	3.982	3.925	3.873	3.823
15	4.894	4.789	4.693	4.606	4.525	4.451	4.382	4.318	4.258	4.202
16	5.378	5.258	5.149	5.050	4.958	4.874	4.796	4.724	4.656	4.593
17	5.881	5.746	5.623	5.551	5.408	5.313	5.225	5.144	5.068	4.998
18	6.405	6.253	6.114	5.988	5.873	5.767	5.669	5.578	5.494	5.415
19	6.951	6.779	6.624	6.483	6.355	6.236	6.127	6.026	5.932	5.845
20	7.518	7.326	7.153	6.996	6.853	6.722	6.601	6.489	6.385	6.288
21	8.107	7.894	7.702	7.527	7.369	7.224	7.090	6.966	6.852	6.745
22	8.719	8.482	8.269	8.077	7.902	7.742	7.594	7.458	7.332	7.215
23	9.354	9.092	8.858	8.645	8.452	8.276	8.115	7.965	7.827	7.699
24	10.014	9.725	9.466	9.233	9.021	8.828	8.651	8.488	8.337	8.197
25	**10.699**	10.381	10.096	9.840	9.608	9.397	9.204	9.025	8.861	8.709
26	11.410	**11.060**	10.748	10.468	10.215	9.984	9.773	9.579	9.400	9.230
27	12.147	11.764	11.423	11.117	10.840	10.589	10.360	10.149	9.955	9.780
28	12.913	12.493	**12.121**	11.787	11.486	11.213	10.963	10.735	10.525	10.330
29	13.707	13.249	12.842	**12.479**	12.152	11.855	11.585	11.338	11.111	10.900
30	14.531	14.031	13.589	13.194	**12.839**	12.518	12.225	11.958	11.713	11.490
31	15.386	14.841	14.360	13.932	13.547	12.200	12.884	12.596	12.331	12.090
32	16.273	15.681	15.159	14.694	14.278	**13.903**	13.562	13.252	12.967	12.700
33	17.194	16.550	15.984	15.481	15.032	14.627	**14.260**	13.926	13.620	13.340

					N					
S	100	105	110	115	120	125	130	135	140	145
34	18.150	17.451	16.838	16.294	15.809	15.373	14.978	**14.619**	14.290	13.990
35	19.142	18.384	17.720	17.134	16.611	16.141	15.717	15.331	**14.979**	14.660
36	20.173	19.351	18.633	18.000	17.437	16.932	16.477	16.063	15.686	15.340
37	21.243	20.353	19.578	18.896	18.289	17.747	17.258	16.815	16.412	**16.040**
38	22.354	21.392	20.555	19.820	19.169	18.586	18.063	17.589	17.157	16.760
39	23.509	22.469	21.566	20.775	20.075	19.451	18.890	18.383	17.922	17.500
40	24.710	23.585	22.613	21.762	21.010	20.342	19.742	19.200	18.708	18.259
41	25.958	24.744	23.696	22.781	21.975	21.259	20.617	20.039	19.515	19.040
42	27.257	25.945	24.817	23.835	22.971	22.204	21.519	20.902	20.343	19.830
43	28.608	27.193	25.978	24.923	23.998	23.178	22.446	21.789	21.194	20.650
44	30.014	28.487	27.181	26.049	25.058	24.181	23.401	22.700	22.067	21.490
45	31.479	29.832	28.427	27.213	26.152	25.216	24.383	23.637	22.964	22.350
46	33.005	31.229	29.718	28.416	27.281	26.282	25.394	24.600	23.885	23.240
47	34.596	32.681	31.057	29.661	28.447	27.381	26.435	25.590	24.830	24.140
48	36.255	34.191	32.446	30.950	29.652	28.514	27.506	26.608	25.801	25.070
49	37.987	35.761	33.886	32.284	30.896	29.682	28.610	27.655	26.799	26.030
50	39.795	37.396	35.381	33.664	32.182	30.887	29.746	28.732	27.824	27.010
51	41.685	39.097	36.933	35.094	33.511	32.131	30.917	29.840	28.877	28.010
52	43.660	40.871	38.546	36.576	34.884	33.414	32.123	30.979	29.958	29.040
53	45.728	42.719	40.221	38.112	36.305	34.738	33.365	32.152	31.070	30.099
54	47.892	44.647	41.963	39.704	37.775	36.106	34.646	33.358	32.212	31.185
55	50.161	46.659	43.775	41.356	39.296	37.518	35.967	34.600	33.387	32.300
56	52.540	48.760	45.660	43.070	40.870	38.976	37.328	35.879	34.594	33.445
57	55.038	50.956	47.624	44.849	42.500	40.484	38.732	37.196	35.835	34.621
58	57.663	53.253	49.670	46.697	44.189	42.042	40.181	38.552	37.112	35.829
59	60.424	55.657	51.803	48.618	45.939	43.653	41.676	39.949	38.425	37.070
60	63.331	58.175	54.028	50.615	47.754	45.319	43.219	41.388	39.776	38.345
61		60.816	56.351	52.693	49.637	47.043	44.813	42.872	41.166	39.655
62		63.587	58.778	54.855	51.590	48.828	46.458	44.401	42.597	41.001
63			61.315	57.107	53.619	50.676	48.159	45.979	44.070	42.385
64			63.970	59.455	55.726	52.591	49.916	47.606	45.588	43.808
65			66.750	61.903	57.916	54.575	51.734	49.285	47.150	45.272
66			69.664	64.457	60.193	56.633	53.614	51.018	48.760	46.778
67				67.125	62.563	58.768	55.559	52.807	50.419	48.327

					N					
S	100	105	110	115	120	125	130	135	140	145
68				69.914	65.030	60.984	57.572	54.655	52.130	49.921
69					67.601	63.285	59.658	56.565	53.894	51.562
70					70.281	65.675	61.819	58.539	55.713	53.252
71					73.078	68.161	64.058	60.580	57.591	54.993
72						70.747	66.381	62.691	59.529	56.786
73						73.438	68.792	64.877	61.530	58.634
74						76.242	71.294	67.139	63.597	60.540
75							73.894	69.483	65.733	62.404
76							76.596	71.911	67.941	64.531
77							79.406	74.430	70.225	66.623
78								77.042	72.588	68.782
79								79.754	75.034	71.012
80								82.570	77.567	73.316
81									80.192	75.698
82									82.913	78.160
83									85.735	80.707
84										83.343
85										86.072
86										88.899

					N					
S	150	155	160	165	170	175	180	185	190	195
1	0.144	0.143	0.142	0.142	0.141	0.140	0.140	0.139	0.138	0.138
2	0.326	0.324	0.322	0.320	0.318	0.317	0.315	0.313	0.312	0.310
3	0.531	0.528	0.524	0.521	0.517	0.514	0.511	0.509	0.506	0.503
4	0.755	0.750	0.744	0.739	0.734	0.729	0.725	0.720	0.716	0.712
5	0.996	0.988	0.980	0.973	0.966	0.959	0.953	0.947	0.941	0.936
6	1.251	1.241	1.231	1.221	1.212	1.203	1.195	1.187	1.179	1.172
7	1.521	1.508	1.495	1.483	1.471	1.460	1.449	1.439	1.429	1.420
8	1.805	1.788	1.772	1.757	1.743	1.729	1.716	1.703	1.691	1.680
9	2.102	2.082	2.062	2.044	2.026	2.010	1.994	1.979	1.964	1.950
10	2.412	2.388	2.364	2.343	2.322	2.302	2.283	2.265	2.248	2.231
11	2.734	2.706	2.679	2.653	2.629	2.605	2.583	2.562	2.542	2.523
12	3.070	3.036	3.005	2.975	2.947	2.920	2.894	2.870	2.846	2.824

	N									
S	150	155	160	165	170	175	180	185	190	195
13	3.417	3.379	3.343	3.308	3.276	3.245	3.216	3.188	3.161	3.135
14	3.777	3.733	3.692	3.653	3.616	3.581	3.548	3.516	3.486	3.457
15	4.149	4.100	4.053	4.009	3.967	3.928	3.890	3.854	3.820	3.787
16	4.534	4.479	4.426	4.377	4.330	4.285	4.243	4.203	4.164	4.128
17	4.931	4.869	4.810	4.755	4.703	4.653	4.606	4.561	4.518	4.478
18	5.341	5.272	5.206	5.145	5.087	5.032	4.979	4.930	4.882	4.837
19	5.763	5.686	5.614	5.546	5.482	5.421	5.363	5.308	5.256	5.206
20	6.198	6.113	6.033	5.958	5.888	5.821	5.757	5.697	5.639	5.585
21	6.645	6.552	6.465	6.382	6.305	6.231	6.161	6.095	6.033	5.973
22	7.106	7.004	6.903	6.818	6.733	6.652	6.576	6.504	6.436	6.370
23	7.580	7.468	7.363	7.265	7.172	7.084	7.002	6.923	6.848	6.778
24	8.066	7.945	7.831	7.723	7.623	7.527	7.437	7.352	7.271	7.194
25	8.567	8.434	8.310	8.194	8.085	7.981	7.884	7.792	7.704	7.621
26	9.080	8.937	8.803	8.677	8.558	8.447	8.341	8.242	8.147	8.057
27	9.608	9.453	9.307	9.171	9.043	8.923	8.809	8.702	8.600	8.503
28	10.150	9.982	9.825	9.678	9.541	9.411	9.288	9.173	9.063	8.959
29	10.706	10.525	10.356	10.198	10.050	9.910	9.779	9.654	9.537	9.425
30	11.277	11.082	10.900	10.730	10.571	10.421	10.280	10.147	10.021	9.902
31	11.862	11.652	11.457	11.275	11.104	10.944	10.793	10.650	10.515	10.388
32	12.462	12.238	12.028	11.833	11.650	11.478	11.317	11.165	11.021	10.884
33	13.078	12.837	12.613	12.404	12.209	12.025	11.853	11.690	11.537	11.391
34	13.710	13.452	13.212	12.989	12.780	12.584	12.400	12.227	12.064	11.909
35	14.357	14.082	13.826	13.587	13.365	13.156	12.960	12.776	12.602	12.437
36	15.021	14.727	14.454	14.200	13.962	13.740	13.532	13.336	13.151	12.976
37	15.702	15.388	15.097	14.826	14.574	14.337	14.116	13.908	13.711	13.526
38	**16.400**	16.065	15.755	15.467	15.199	14.948	14.713	14.492	14.284	14.087
39	17.115	**16.758**	16.429	16.123	15.838	15.571	15.322	15.088	14.867	14.659
40	17.848	17.469	17.118	16.793	16.491	16.209	15.944	15.696	15.463	15.243
41	18.599	18.196	**17.824**	17.479	17.159	16.860	16.580	16.317	16.071	15.838
42	19.369	18.941	18.547	**18.181**	17.841	17.525	17.229	16.951	16.691	16.445
43	20.158	19.704	19.286	18.899	**18.539**	18.204	17.891	17.598	17.323	17.064
44	20.967	20.486	20.042	19.633	19.252	**18.898**	18.568	18.258	17.968	17.695
45	21.796	21.286	20.817	20.383	19.981	19.607	19.258	18.932	18.626	18.338

					N					
S	150	155	160	165	170	175	180	185	190	195
46	22.646	22.106	21.609	21.150	20.726	20.331	**19.963**	19.619	19.297	18.994
47	23.517	22.945	22.420	21.935	21.487	21.071	20.683	**20.320**	19.981	19.662
48	24.410	23.805	23.249	22.738	22.265	21.826	21.417	21.036	**20.679**	20.344
49	25.326	24.685	24.099	23.559	23.060	22.597	22.167	21.766	21.390	21.038
50	26.263	25.587	24.967	24.398	23.872	23.385	22.932	22.510	22.116	**21.746**
51	27.225	26.510	25.857	25.256	24.703	24.190	23.714	23.270	22.855	22.467
52	28.211	27.456	26.767	26.134	25.551	25.012	24.511	24.045	23.610	23.202
53	29.222	28.426	27.699	27.032	26.418	25.851	25.325	24.835	24.379	23.951
54	30.259	29.418	28.652	27.950	27.304	26.708	26.156	25.642	25.163	24.715
55	31.322	30.436	29.628	28.889	28.210	27.584	27.004	26.465	25.962	25.493
56	32.413	31.478	30.627	29.850	29.136	28.478	27.869	27.304	26.777	26.286
57	33.531	32.546	31.650	30.833	30.083	29.392	28.753	28.160	27.609	27.094
58	34.679	33.640	32.698	31.838	31.050	30.325	29.655	29.034	28.456	27.908
59	35.856	34.762	33.771	32.867	32.039	31.278	30.576	29.925	29.320	28.757
60	37.065	35.912	34.869	33.919	33.050	32.252	31.516	30.834	30.202	29.612
61	38.305	37.091	35.994	34.996	34.084	33.247	32.475	31.762	31.100	30.484
62	39.578	38.300	37.146	36.098	35.141	34.263	33.455	32.709	32.017	31.373
63	40.885	39.540	38.326	37.225	36.222	35.302	34.456	33.675	32.951	32.278
64	42.227	40.811	39.536	38.380	37.327	36.363	35.478	34.660	33.904	33.201
65	43.605	42.116	40.775	39.561	38.457	37.448	36.521	35.666	34.876	34.142
66	45.021	43.454	42.045	40.771	39.614	38.556	37.586	36.693	35.867	35.101
67	46.476	44.827	43.346	42.010	40.797	39.689	38.675	37.741	36.878	36.079
68	47.971	46.236	44.681	43.278	42.007	40.848	39.786	38.810	37.909	37.075
69	49.508	47.682	46.049	44.578	43.245	42.032	40.922	39.902	38.961	38.091
70	51.087	49.167	47.452	45.908	44.512	43.242	42.082	41.016	40.035	39.127
71	52.712	50.692	48.890	47.272	45.809	44.480	43.267	42.154	41.130	40.183
72	54.383	52.259	50.366	48.669	47.137	45.746	44.478	43.316	42.247	41.260
73	56.102	53.868	51.388	50.100	48.495	47.041	45.715	44.502	43.387	42.358
74	57.871	55.521	53.434	51.568	49.887	48.365	46.980	45.713	44.550	43.477
75	59.693	57.221	55.029	53.072	51.312	49.720	48.273	46.950	45.737	44.619
76	61.568	58.968	56.667	54.614	52.772	51.106	49.594	48.214	46.949	45.784
77	63.500	60.765	58.348	56.196	54.266	52.525	50.945	49.505	48.185	46.972
78	65.491	62.613	60.075	57.819	55.798	53.977	52.326	50.823	49.448	48.184

						N				
S	150	155	160	165	170	175	180	185	190	195
79	67.542	64.515	61.849	59.483	57.367	55.463	53.739	52.171	50.737	49.421
80	69.657	66.472	63.672	61.191	58.976	56.984	55.184	53.548	52.053	50.682
81	71.839	68.487	65.546	62.944	60.625	58.542	56.662	54.955	53.397	51.969
82	74.089	70.561	67.473	64.744	62.315	60.138	58.721	56.393	54.770	53.283
83	76.412	72.699	69.454	66.592	64.049	61.772	59.721	57.863	56.171	54.624
84	78.810	74.901	71.492	68.491	65.827	63.447	61.305	59.367	57.603	55.992
85	81.288	77.171	73.588	70.441	67.652	65.162	62.926	60.904	59.066	57.389
86	83.848	79.512	75.747	72.445	69.524	66.921	64.585	62.476	60.561	58.815
87	86.495	81.926	77.969	74.505	71.446	68.723	66.284	64.084	62.089	60.271
88	89.232	84.418	80.257	76.623	73.419	70.572	68.024	65.729	63.650	61.759
89	92.064	86.991	82.615	78.801	75.445	72.467	69.806	67.412	65.247	63.277
90		89.647	85.045	81.042	77.526	74.411	71.631	69.135	66.878	64.829
91		92.392	87.551	83.348	79.664	76.406	73.502	70.898	68.547	66.414
92		95.229	90.135	85.722	81.861	78.453	75.420	72.703	70.253	68.033
93			92.801	88.167	84.120	80.554	77.385	74.551	71.999	69.688
94			95.553	90.685	86.443	82.711	79.401	76.444	73.784	71.379
95			98.394	93.280	88.832	84.927	81.468	78.382	75.611	73.107
96				95.955	91.291	87.203	83.589	80.369	77.481	74.875
97				98.714	93.821	89.542	85.764	82.404	79.394	76.682
98				101.559	96.427	91.946	87.997	84.490	81.354	78.530
99					99.110	94.417	90.290	86.629	83.360	80.420
100					101.875	96.959	92.644	88.823	85.415	82.354
101					104.724	99.574	95.061	91.072	87.519	84.333
102					107.663	102.266	97.545	93.380	89.676	86.359
103						105.036	100.098	95.749	91.886	88.432
104						107.890	102.723	98.180	94.152	90.554
105							105.422	100.676	96.474	92.728
106							108.198	103.239	98.857	94.954
107							111.055	105.873	101.300	97.235
108								108.579	103.808	99.572
109								111.360	106.381	101.967
110								114.220	109.023	104.423
111									111.736	106.941
112									114.523	109.524
113									117.386	112.174

					N				
S	150	155	160	165	170	175	180	185	190	195
114										114.894
115										117.685
116										120.551

					N				
S	200	210	220	230	240	250	260	270	280	290
1	0.137	0.136	0.135	0.134	0.133	0.133	0.132	0.131	0.130	0.130
2	0.309	0.306	0.304	0.301	0.299	0.297	0.295	0.293	0.291	0.289
3	0.501	0.496	0.491	0.487	0.483	0.479	0.476	0.472	0.469	0.466
4	0.708	0.701	0.694	0.688	0.682	0.676	0.671	0.666	0.661	0.656
5	0.930	0.920	0.910	0.902	0.893	0.885	0.878	0.871	0.865	0.858
6	1.165	1.151	1.139	1.127	1.116	1.106	1.096	1.087	1.079	1.070
7	1.411	1.394	1.378	1.363	1.350	1.337	1.325	1.313	1.302	1.292
8	1.669	1.648	1.628	1.610	1.593	1.577	1.562	1.548	1.535	1.522
9	1.937	1.912	1.888	1.866	1.846	1.827	1.809	1.792	1.777	1.761
10	2.215	2.185	2.158	2.132	2.108	2.086	2.065	2.045	2.026	2.008
11	2.504	2.469	2.437	2.407	2.379	2.353	2.328	2.305	2.283	2.263
12	2.803	2.762	2.725	2.691	2.658	2.628	2.600	2.574	2.545	2.525
13	3.111	3.065	3.022	2.983	2.946	2.912	2.880	2.850	2.821	2.795
14	3.429	3.377	3.328	3.284	3.242	3.204	3.168	3.134	3.102	3.071
15	3.756	3.697	3.643	3.593	3.547	3.503	3.463	3.425	3.389	3.355
16	4.093	4.027	3.967	3.911	3.859	3.811	3.765	3.723	3.683	3.656
17	4.439	4.366	4.299	4.237	4.179	4.126	4.076	4.029	3.985	3.943
18	4.794	4.713	4.639	4.571	4.507	4.448	4.393	4.341	4.293	4.247
19	5.159	5.070	4.988	4.913	4.843	4.778	4.718	4.661	4.608	4.558
20	5.533	5.435	5.346	5.263	5.187	5.116	5.050	4.988	4.930	4.875
21	5.916	5.809	5.712	5.622	5.538	5.461	5.389	5.322	5.258	5.199
22	6.308	6.192	6.086	5.988	5.898	5.814	5.735	5.662	5.594	5.529
23	6.710	6.584	6.469	6.362	6.264	6.174	6.089	6.010	5.936	5.866
24	7.121	6.985	6.860	6.745	6.639	6.541	6.450	6.364	6.285	6.210
25	7.542	7.394	7.259	7.136	7.021	6.916	6.817	6.726	6.640	6.559
26	7.972	7.813	7.668	7.534	7.412	7.298	7.192	7.094	7.002	6.915
27	8.411	8.240	8.084	7.941	7.810	7.688	7.574	7.469	7.370	7.278
28	8.861	8.677	8.510	8.356	8.215	8.085	7.964	7.851	7.746	7.647
29	9.320	9.123	8.944	8.780	8.629	8.489	8.360	8.240	8.127	8.022

					N					
S	200	210	220	230	240	250	260	270	280	290
30	9.788	9.578	9.386	9.211	9.050	8.901	8.764	8.636	8.516	8.404
31	10.267	10.042	9.838	9.651	9.479	9.321	9.174	9.038	8.911	8.792
32	10.755	10.516	10.298	10.099	9.917	9.748	9.593	9.448	9.313	9.187
33	11.254	10.999	10.767	10.556	10.362	10.183	10.018	9.864	9.721	9.588
34	11.762	11.491	11.245	11.021	10.815	10.626	10.450	10.288	10.137	9.995
35	12.281	11.993	11.732	11.494	11.276	11.076	10.890	10.719	10.558	10.409
36	12.811	12.505	12.228	11.976	11.746	11.534	11.338	11.156	10.987	10.829
37	13.351	13.027	12.734	12.467	12.224	11.999	11.793	11.601	11.423	11.256
38	13.901	13.559	13.249	12.967	12.709	12.473	12.255	12.053	11.865	11.690
39	14.463	14.100	13.773	13.475	13.204	12.954	12.725	12.512	12.314	12.130
40	15.035	14.652	14.307	13.993	13.707	13.444	13.202	12.978	12.770	12.577
41	15.619	15.214	14.850	14.519	14.218	13.942	13.687	13.452	13.233	13.030
42	16.214	15.787	15.403	15.055	14.738	14.447	14.180	13.933	13.703	13.490
43	16.820	16.370	15.966	15.600	15.266	14.961	14.680	14.421	14.181	13.957
44	17.438	16.964	16.539	16.154	15.804	15.483	15.189	14.917	14.665	14.430
45	18.067	17.569	17.122	16.718	16.350	16.014	15.705	15.520	15.156	14.911
46	18.709	18.185	17.715	17.291	16.905	16.553	16.229	15.931	15.655	15.398
47	19.362	18.812	18.319	17.873	17.469	17.100	16.761	16.449	16.160	15.892
48	20.028	19.450	18.933	18.466	18.042	17.656	17.302	16.975	16.674	16.393
49	20.707	20.100	19.557	19.068	18.625	18.221	17.850	17.509	17.194	16.902
50	21.398	20.762	20.193	19.681	19.217	18.794	18.407	18.051	17.722	17.417
51	**22.102**	21.435	20.839	20.304	19.818	19.377	18.972	18.601	18.257	17.939
52	22.820	22.121	21.497	20.936	20.429	19.968	19.546	19.158	18.800	18.469
53	23.550	**22.818**	22.166	21.580	21.050	20.568	20.128	19.724	19.351	19.006
54	24.295	23.528	22.846	22.234	21.681	21.178	20.719	20.298	19.909	19.550
55	25.053	24.251	23.538	22.898	22.321	21.797	21.319	20.880	20.475	20.102
56	25.825	24.987	**24.241**	23.574	22.972	22.425	21.927	21.470	21.049	20.661
57	26.612	25.736	24.957	24.260	23.632	23.063	22.544	22.069	21.631	21.227
58	27.414	26.497	25.685	24.958	24.304	23.711	23.171	22.676	22.221	21.801
59	28.230	27.273	26.425	**25.667**	24.985	24.368	23.806	23.292	22.819	22.383
60	29.062	28.062	27.177	26.388	25.678	25.035	24.451	23.917	23.426	22.973
61	29.909	28.865	27.943	27.120	**26.381**	25.713	25.105	24.550	24.040	23.570
62	30.772	29.683	28.721	27.864	27.095	26.400	25.769	25.192	24.663	24.175
63	31.651	30.515	29.512	28.620	27.820	27.098	26.442	25.843	25.294	24.788

					N					
S	200	210	220	230	240	250	260	270	280	290
64	32.546	31.362	30.317	29.388	28.556	**27.806**	27.125	26.503	25.934	25.409
65	33.459	32.223	31.136	30.169	29.304	28.525	27.817	27.173	26.582	26.039
66	34.388	33.101	31.968	30.963	30.064	29.254	**28.520**	27.852	27.239	26.676
67	35.335	33.993	32.814	31.769	30.835	29.994	29.233	28.540	27.905	27.322
68	36.300	34.902	33.675	32.588	31.618	30.746	29.956	29.237	28.580	27.976
69	37.283	35.827	34.550	33.421	32.413	31.508	30.689	**29.945**	29.264	28.939
70	38.284	36.768	35.441	34.267	33.221	32.282	31.433	30.662	29.957	29.311
71	39.305	37.727	36.346	35.127	34.041	33.067	32.188	31.389	**30.660**	29.991
72	40.345	38.702	37.267	36.001	34.874	33.864	32.953	32.126	31.371	30.679
73	41.405	39.695	38.203	36.889	35.720	34.673	33.729	32.873	32.092	31.377
74	42.485	40.706	39.156	37.791	36.579	35.494	34.517	33.631	32.823	**32.084**
75	43.586	41.736	40.125	38.708	37.451	36.327	35.315	34.399	33.564	32.800
76	44.708	42.784	41.110	39.640	38.337	37.173	36.125	35.177	34.314	33.525
77	45.852	43.851	42.113	40.588	39.237	38.031	36.947	35.966	35.074	34.259
78	47.018	44.937	43.133	41.550	40.150	38.902	37.780	36.766	35.845	35.002
79	48.207	46.044	44.170	42.529	41.078	39.786	38.626	37.578	36.625	35.756
80	49.420	47.171	45.225	43.524	42.021	40.683	39.483	38.400	37.416	36.519
81	50.656	48.318	46.299	44.535	42.978	41.594	40.353	39.233	38.218	37.291
82	51.916	49.487	47.391	45.562	43.950	42.518	41.235	40.078	39.030	38.074
83	53.202	50.678	48.503	46.607	44.938	43.456	42.130	40.935	39.853	38.866
84	54.513	51.891	49.634	47.669	45.941	44.408	43.038	41.804	40.686	39.669
85	55.851	53.126	50.785	48.749	46.960	45.375	43.958	42.684	41.531	40.482
86	57.215	54.385	51.956	49.847	47.995	46.356	44.892	43.577	42.387	41.306
87	58.608	55.668	53.148	50.963	49.047	47.352	45.840	44.482	43.255	42.140
88	60.028	56.975	54.362	52.098	50.115	48.363	46.801	45.400	44.134	42.984
89	61.478	58.307	55.597	53.252	51.201	49.389	47.777	46.330	45.025	43.840
90	62.958	59.664	56.854	54.425	52.303	50.431	48.766	47.274	45.927	44.706
91	64.468	61.048	58.134	55.619	53.424	51.489	49.770	48.230	46.842	45.584
92	66.010	62.458	59.437	56.833	54.562	52.563	50.788	49.200	47.769	46.473
93	67.584	63.896	60.764	58.067	55.719	53.654	51.821	50.183	48.708	47.374
94	69.192	65.362	62.115	59.323	56.895	54.761	52.870	51.180	49.660	48.285
95	70.834	66.857	63.490	60.600	58.089	55.885	53.933	52.191	50.625	49.209
96	72.511	68.381	64.892	61.900	59.304	57.027	55.013	53.216	51.603	50.145
97	74.224	69.936	66.319	63.222	60.538	58.187	56.108	54.256	52.594	51.093

S	\multicolumn{10}{c}{N}
	200	210	220	230	240	250	260	270	280	290
98	75.974	71.522	67.773	64.567	61.792	59.364	57.220	55.310	53.598	52.053
99	77.763	73.140	69.254	65.936	63.067	60.560	58.348	56.380	54.616	53.025
100	79.591	74.791	70.763	67.329	64.363	61.775	59.493	57.464	55.648	54.010
101	81.459	76.476	72.300	68.746	65.681	63.008	60.654	58.564	56.693	55.008
102	83.369	78.195	73.867	70.189	67.021	64.261	61.834	59.679	57.753	56.019
103	85.323	79.950	75.464	71.657	68.383	65.534	63.031	60.811	58.828	57.044
104	87.321	81.741	77.091	73.152	69.769	66.828	64.246	61.958	59.917	58.081
105	89.364	83.570	78.751	74.674	71.177	68.142	65.479	63.122	61.021	59.133
106	91.455	85.438	80.442	76.224	72.610	69.477	66.731	64.303	62.140	60.198
107	93.595	87.345	82.167	77.802	74.067	70.833	68.002	65.501	63.274	61.277
108	95.785	89.293	83.926	79.408	75.549	72.211	69.293	66.717	64.424	62.370
109	98.026	91.284	85.720	81.045	77.057	73.612	70.603	67.950	65.591	63.478
110	100.322	93.317	87.550	82.712	78.591	75.035	71.933	69.201	66.773	64.601
111	102.673	95.396	89.416	84.410	80.152	76.482	73.284	70.470	67.972	65.738
112	105.081	97.520	91.321	86.140	81.740	77.953	74.656	71.758	69.188	66.891
113	107.548	99.692	93.264	87.902	83.356	79.448	76.050	73.065	70.420	68.059
114	110.076	101.912	95.248	89.698	85.000	80.967	77.465	74.391	71.670	69.243
115	112.668	104.183	97.273	91.529	86.674	82.513	78.902	75.737	72.938	70.443
116	115.326	106.506	99.340	93.395	88.378	84.084	80.362	77.103	74.224	71.659
117	118.052	108.882	101.450	95.297	90.113	85.681	81.845	78.490	75.527	72.891
118	120.848	111.314	103.606	97.236	91.879	87.306	83.352	79.897	76.850	74.140
119		113.803	105.807	99.214	93.677	88.958	84.883	81.326	78.191	75.406
120		116.351	108.056	101.230	95.509	90.638	86.438	82.776	79.552	76.690
121		118.960	110.354	103.287	97.374	92.348	88.019	84.248	80.932	77.991
122		121.632	112.702	105.385	99.274	94.087	89.625	85.743	82.333	79.310
123		124.369	115.102	107.526	101.209	95.856	91.258	87.261	83.753	80.647
124		127.174	117.556	109.710	103.181	97.656	92.917	88.802	85.194	82.003
125			120.065	111.939	105.190	99.488	94.603	90.368	86.657	83.377
126			122.631	114.215	107.237	101.353	96.318	91.957	88.141	84.770
127			125.255	116.538	109.324	103.250	98.061	93.572	89.647	86.184
128			127.940	118.909	111.451	105.182	99.833	95.211	91.175	87.617
129			130.688	121.331	113.620	107.149	101.635	96.877	92.726	89.070
130			133.500	123.805	115.831	109.151	103.468	98.569	94.300	90.544
131				126.332	118.086	111.190	105.331	100.289	95.898	92.039

						N				
S	200	210	220	230	240	250	260	270	280	290
132				128.914	120.386	113.266	107.227	102.035	97.520	93.555
133				131.553	122.732	115.381	109.156	103.810	99.167	95.093
134				134.250	125.126	117.536	111.118	105.614	100.839	96.653
135				137.008	127.568	119.731	113.114	107.447	102.536	98.236
136				139.827	130.061	121.967	115.144	109.310	104.260	99.842
137					132.605	124.246	117.212	111.204	106.010	101.471
138					135.202	126.569	119.316	113.130	107.788	103.124
139					137.854	128.937	121.457	115.087	109.593	104.802
140					140.563	131.351	123.637	117.078	111.427	106.504
141					143.329	133.812	125.857	119.102	113.289	108.232
142						136.322	128.117	121.160	115.182	109.985
143						138.882	130.418	123.253	117.104	111.765
144						141.493	132.763	125.383	119.058	113.572
145						144.158	135.151	127.549	121.043	115.406
146						146.876	137.583	129.753	123.060	117.268
147						149.651	140.062	131.996	125.110	119.159
148							142.588	134.278	127.194	121.079
149							145.163	136.601	129.313	123.029
150							147.788	138.965	131.466	125.009
151							150.463	141.372	133.656	127.020
152							153.191	143.823	135.883	129.063
153							155.974	146.318	138.147	131.138
154								148.859	140.451	133.246
155								151.448	142.793	135.388
156								154.084	145.176	137.565
157								156.771	147.601	139.777
158								159.508	150.068	142.025
159								162.297	152.579	144.310
160									155.135	146.633
161									157.736	148.995
162									160.384	151.396
163									163.080	153.837
164									165.825	156.320
165									168.621	158.845

					N					
S	200	210	220	230	240	250	260	270	280	290
166										161.414
167										164.027
168										166.685
169										169.391
170										172.144
171										174.946

					N					
S	300	310	320	330	340	350	360	370	380	390
1	0.129	0.128	0.128	0.127	0.127	0.126	0.126	0.125	0.125	0.124
2	0.288	0.286	0.285	0.283	0.282	0.281	0.279	0.278	0.277	0.276
3	0.463	0.461	0.458	0.455	0.453	0.451	0.448	0.446	0.444	0.442
4	0.652	0.648	0.644	0.640	0.637	0.633	0.630	0.627	0.624	0.621
5	0.852	0.847	0.841	0.836	0.831	0.826	0.822	0.817	0.813	0.809
6	1.063	1.055	1.048	1.041	1.035	1.029	1.023	1.017	1.012	1.006
7	1.282	1.273	1.264	1.256	1.247	1.240	1.232	1.225	1.218	1.212
8	1.510	1.499	1.488	1.478	1.468	1.459	1.450	1.441	1.433	1.425
9	1.747	1.733	1.720	1.708	1.696	1.685	1.674	1.664	1.654	1.644
10	1.991	1.975	1.960	1.946	1.932	1.919	1.906	1.894	1.882	1.871
11	2.243	2.225	2.207	2.191	2.175	2.159	2.145	2.131	2.117	2.104
12	2.503	2.482	2.461	2.442	2.424	2.406	2.390	2.374	2.358	2.344
13	2.769	2.745	2.722	2.701	2.680	2.660	2.641	2.623	2.606	2.589
14	3.043	3.016	2.990	2.965	2.942	2.920	2.899	2.878	2.859	2.840
15	3.323	3.293	3.264	3.237	3.211	3.186	3.162	3.139	3.118	3.097
16	3.610	3.576	3.545	3.514	3.485	3.458	3.432	3.406	3.382	3.359
17	3.904	3.867	3.831	3.798	3.766	3.736	3.707	3.679	3.653	3.627
18	4.204	4.163	4.124	4.088	4.053	4.019	3.988	3.957	3.928	3.900
19	4.511	4.466	4.424	4.383	4.345	4.309	4.274	4.241	4.209	4.179
20	4.824	4.775	4.729	4.685	4.644	4.604	4.566	4.530	4.496	4.462
21	5.143	5.090	5.040	4.993	4.948	4.905	4.864	4.825	4.787	4.751
22	5.469	5.412	5.357	5.306	5.257	5.211	5.167	5.124	5.084	5.045
23	5.801	5.739	5.681	5.625	5.573	5.523	5.475	5.429	5.386	5.344
24	6.139	6.073	6.010	5.950	5.894	5.840	5.788	5.740	5.693	5.648
25	6.484	6.412	6.345	6.281	6.220	6.162	6.107	6.055	6.005	5.957

					N					
S	300	310	320	330	340	350	360	370	380	390
26	6.834	6.758	6.685	6.617	6.552	6.490	6.431	6.375	6.322	6.271
27	7.191	7.109	7.032	6.959	6.889	6.824	6.761	6.701	6.644	6.590
28	7.554	7.467	7.384	7.306	7.233	7.162	7.096	7.032	6.971	6.913
29	7.923	7.830	7.743	7.660	7.581	7.506	7.435	7.368	7.303	7.241
30	8.299	8.200	8.107	8.018	7.935	7.856	7.780	7.709	7.640	7.575
31	8.680	8.575	8.477	8.383	8.294	8.210	8.130	8.054	7.982	7.913
32	9.068	8.957	8.852	8.753	8.659	8.570	8.486	8.405	8.329	8.255
33	9.462	9.345	9.234	9.129	9.030	8.936	8.846	8.761	8.680	8.603
34	9.863	9.738	9.621	9.510	9.405	9.306	9.212	9.122	9.037	8.955
35	10.269	10.138	10.014	9.897	9.787	9.682	9.583	9.488	9.398	9.313
36	10.682	10.543	10.413	10.290	10.173	10.063	9.959	9.859	9.765	9.674
37	11.101	10.955	10.817	10.688	10.566	10.450	10.340	10.235	10.136	10.041
38	11.526	11.372	11.228	11.092	10.963	10.841	10.726	10.616	10.512	10.412
39	11.958	11.796	11.644	11.501	11.366	11.239	11.117	11.002	10.893	10.789
40	12.396	12.226	12.067	11.917	11.775	11.641	11.514	11.393	11.279	11.169
41	12.840	12.662	12.495	12.338	12.189	12.049	11.916	11.790	11.669	11.555
42	13.291	13.104	12.929	12.764	12.609	12.462	12.323	12.191	12.065	11.946
43	13.748	13.553	13.370	13.197	13.034	12.881	12.735	12.597	12.466	12.341
44	14.212	14.008	13.816	13.635	13.465	13.305	13.152	13.008	12.871	12.741
45	14.682	14.468	14.268	14.079	13.902	13.734	13.575	13.425	13.282	13.146
46	15.159	14.936	14.726	14.529	14.344	14.169	14.003	13.846	13.697	13.555
47	15.643	15.409	15.191	14.985	14.792	14.609	14.436	14.273	14.117	13.970
48	16.133	15.889	15.661	15.447	15.245	15.055	14.875	14.704	14.543	14.389
49	16.630	16.376	16.138	15.914	15.704	15.506	15.319	15.141	14.973	14.813
50	17.133	16.868	16.621	16.388	16.169	15.963	15.768	15.583	15.408	15.242
51	17.644	17.368	17.110	16.868	16.640	16.425	16.223	16.031	15.849	15.676
52	18.161	17.874	17.605	17.353	17.116	16.893	16.682	16.483	16.294	16.114
53	18.685	18.386	18.107	17.845	17.599	17.367	17.148	16.941	16.744	16.558
54	19.216	18.905	18.615	18.343	18.087	17.846	17.619	17.404	17.200	17.007
55	19.755	19.431	19.129	18.847	18.581	18.331	18.095	17.872	17.661	17.460
56	20.300	19.996	19.650	19.357	19.081	18.822	18.577	18.346	18.127	17.919
57	20.852	20.503	20.178	19.873	19.587	19.318	19.064	18.824	18.598	18.382
58	21.412	21.050	20.712	20.396	20.099	19.820	19.557	19.309	19.074	18.851
59	21.979	21.603	21.252	20.925	20.617	20.328	20.056	19.798	19.555	19.324

					N					
S	300	310	320	330	340	350	360	370	380	390
60	22.553	22.163	21.800	21.460	21.141	20.842	20.560	20.294	20.042	19.803
61	23.135	22.730	22.354	22.002	21.672	21.362	21.070	20.794	20.534	20.287
62	23.724	23.305	22.914	22.550	22.208	21.887	21.585	21.300	21.031	20.776
63	24.320	23.886	23.482	23.105	22.751	22.419	22.107	21.812	21.533	21.270
64	24.925	24.475	24.057	23.666	23.300	22.957	22.634	22.329	22.041	21.769
65	25.537	25.071	24.638	24.234	23.855	23.501	23.167	22.852	22.555	22.273
66	26.156	25.675	25.226	24.808	24.417	24.050	23.705	23.380	23.073	22.783
67	26.784	26.285	25.822	25.390	24.985	24.606	24.250	23.914	23.598	23.298
68	27.419	26.904	26.424	25.978	25.560	25.169	24.801	24.454	24.127	23.818
69	28.063	27.530	27.034	26.573	26.141	25.737	25.357	25.000	24.663	24.344
70	28.715	28.163	27.651	27.174	26.729	26.312	25.920	25.551	25.203	24.874
71	29.374	28.805	28.276	27.783	27.324	26.893	26.489	26.108	25.750	25.411
72	30.042	29.454	28.907	28.399	27.925	27.480	27.064	26.671	26.302	25.952
73	30.719	30.111	29.547	29.022	28.533	28.074	27.645	27.240	26.859	26.500
74	31.404	30.776	30.193	29.652	29.147	28.675	28.232	27.815	27.423	27.052
75	32.097	31.449	30.848	30.289	29.769	29.282	28.825	28.396	27.992	27.611
76	32.799	32.130	31.510	30.934	30.397	29.896	29.425	28.983	28.567	28.174
77	**33.510**	32.819	32.180	31.586	31.033	30.516	30.031	29.576	29.148	28.744
78	34.229	33.517	32.858	32.245	31.675	31.143	30.644	30.176	29.735	29.319
79	34.958	34.223	33.543	32.912	32.325	31.777	31.263	30.781	30.327	29.900
80	35.695	**34.938**	34.237	33.587	32.982	32.417	31.889	31.393	30.926	30.486
81	36.442	35.661	34.939	34.269	33.646	33.065	32.521	32.011	31.531	31.079
82	37.198	36.393	**35.649**	34.959	34.318	33.719	33.160	32.635	32.142	31.677
83	37.963	37.133	36.367	35.657	34.996	34.381	33.805	33.266	32.758	32.281
84	38.738	37.883	37.093	**36.362**	35.683	35.049	34.458	33.903	33.381	32.890
85	39.523	38.641	37.828	37.076	36.377	35.725	35.117	34.546	34.011	33.506
86	40.317	39.409	38.572	37.797	37.078	36.408	35.783	35.196	34.646	34.128
87	41.121	40.186	39.324	38.527	**37.787**	37.099	36.455	35.853	35.288	34.756
88	41.935	40.972	40.085	39.265	38.504	37.796	37.135	36.517	35.936	35.390
89	42.758	41.767	40.855	40.011	39.229	**38.501**	37.822	37.187	36.591	36.030
90	43.593	42.572	41.633	40.766	39.962	39.934	38.516	37.864	37.252	36.676
91	44.437	43.387	42.421	41.529	40.703	40.662	39.218	38.548	37.919	37.329
92	45.292	44.211	43.218	42.301	41.452	41.398	**39.926**	39.238	38.594	37.988
93	46.158	45.046	44.024	43.081	42.209	42.141	40.642	39.936	39.274	38.653

					N					
S	300	310	320	330	340	350	360	370	380	390
94	47.034	45.890	44.839	43.871	42.974	42.893	41.365	**40.641**	39.962	39.325
95	47.921	46.745	45.664	44.669	43.748	43.652	42.096	41.353	40.656	40.003
96	48.820	47.609	46.499	45.476	44.530	44.419	42.835	42.072	41.357	40.687
97	49.729	48.484	47.343	46.292	45.320	45.195	43.580	42.798	**42.065**	41.378
98	50.650	49.370	48.197	47.117	46.120	45.978	44.334	43.531	42.780	42.076
99	51.582	50.266	49.061	47.952	46.928	46.770	45.096	44.272	43.502	42.780
100	52.526	51.173	49.935	48.796	47.744	47.571	45.865	45.021	44.231	**43.491**
101	53.482	52.092	50.819	49.650	48.570	48.380	46.642	45.776	44.967	44.209
102	54.450	53.021	51.714	50.513	49.405	49.197	47.427	46.540	45.711	44.934
103	55.430	53.961	52.618	51.386	50.249	50.023	48.220	47.311	46.461	45.666
104	56.422	54.913	53.534	52.268	51.102	50.858	49.022	48.090	47.219	46.404
105	57.427	55.876	54.460	53.161	51.964	51.702	49.832	48.876	47.984	47.150
106	58.444	56.851	55.397	54.064	52.836	52.555	50.650	49.671	48.757	47.902
107	59.474	57.838	56.345	54.977	53.718	53.416	51.476	50.473	49.538	48.662
108	60.517	58.837	57.304	55.900	54.609	54.287	52.311	51.284	50.325	49.430
109	61.574	59.848	58.275	56.834	55.510	55.168	53.155	52.102	51.121	50.204
110	62.644	60.871	59.256	57.779	56.421	56.057	54.007	52.929	51.925	50.986
111	63.728	61.907	60.250	58.734	57.341	56.956	54.868	53.764	52.736	51.775
112	64.825	62.956	61.255	59.700	58.272	57.865	55.738	54.608	53.555	52.572
113	65.937	64.017	62.272	60.677	59.214	58.793	56.617	55.460	54.382	53.376
114	67.062	65.092	63.301	61.666	60.165	59.711	57.506	56.320	55.217	54.188
115	68.203	66.180	64.342	62.665	61.127	60.650	58.403	57.190	56.061	55.008
116	69.358	67.281	65.396	63.676	62.100	61.598	59.310	58.068	56.912	55.835
117	70.528	68.396	66.462	64.699	63.084	62.556	60.226	58.954	57.772	56.670
118	71.713	69.525	67.541	65.733	64.078	63.525	61.151	59.850	58.641	57.514
119	72.913	70.668	68.633	66.779	65.083	64.504	62.086	60.755	59.518	58.365
120	74.130	71.825	69.738	67.838	66.100	65.493	63.031	61.669	60.403	59.225
121	75.362	72.996	70.856	68.908	67.128	66.493	63.986	62.592	61.297	60.092
122	76.610	74.183	71.988	69.991	68.167	67.504	64.951	63.524	62.200	60.968
123	77.875	75.384	73.133	71.087	69.219	68.526	65.926	64.466	63.112	61.852
124	79.156	76.601	74.292	72.195	70.281	69.559	66.911	65.418	64.033	62.745
125	80.454	77.832	75.465	73.317	71.356	70.604	67.906	66.379	64.963	63.646
126	81.770	79.080	76.653	74.451	72.443	71.659	68.912	67.350	65.902	64.556
127	83.103	80.343	77.855	75.599	73.542	72.726	69.928	68.330	66.850	65.475

					N					
S	300	310	320	330	340	350	360	370	380	390
128	84.454	81.623	79.071	76.760	74.654	73.805	70.955	69.321	67.808	66.402
129	85.823	82.919	80.303	77.934	75.778	74.896	71.993	70.322	68.775	67.339
130	87.211	84.231	81.550	79.123	76.915	75.998	73.042	71.333	69.752	68.284
131	88.617	85.561	82.812	80.326	78.064	77.113	74.102	72.354	70.738	69.238
132	90.043	86.907	84.090	81.542	79.227	78.240	75.173	73.386	71.734	70.202
133	91.487	88.271	85.383	82.774	80.403	79.379	76.255	74.428	72.740	71.175
134	92.952	89.653	86.693	84.020	81.593	80.531	77.349	75.482	73.753	72.157
135	94.437	91.053	88.019	85.281	82.796	81.695	78.455	76.546	74.782	73.149
136	95.942	92.471	89.361	86.557	84.013	82.873	79.572	77.621	75.819	74.150
137	97.468	93.908	90.721	87.848	85.245	84.063	80.702	78.707	76.866	75.161
138	99.015	95.364	92.097	89.155	86.490	85.267	81.743	79.804	77.923	76.182
139	100.584	96.840	93.491	90.478	87.750	86.484	82.997	80.913	78.991	77.213
140	102.175	98.335	94.903	91.817	89.024	87.715	84.163	82.033	80.069	78.254
141	103.788	99.849	96.333	93.172	90.314	88.960	85.342	83.165	81.159	79.305
142	105.424	101.385	97.781	94.543	91.618	90.219	86.534	84.308	82.259	80.366
143	107.083	102.940	99.247	95.932	92.938	91.492	87.738	85.464	83.371	81.437
144	108.766	104.517	100.732	97.337	94.273	92.779	88.955	86.631	84.494	82.519
145	110.473	106.116	102.237	98.760	95.624	94.081	90.186	87.811	85.628	83.612
146	112.204	107.736	103.761	100.201	96.991	95.398	91.430	89.004	86.773	84.715
147	113.961	109.378	105.305	101.659	98.374	96.730	92.688	90.209	87.931	85.830
148	115.743	111.043	106.869	103.135	99.774	98.077	93.960	91.426	89.100	86.955
149	117.551	112.731	108.453	104.630	101.190	99.440	95.245	92.657	90.281	88.091
150	119.386	114.442	110.059	106.144	102.624	100.818	96.545	93.901	91.474	89.239
151	121.247	116.177	111.686	107.677	104.074	102.212	97.859	95.158	92.680	90.398
152	123.137	117.937	113.334	109.229	105.542	103.623	99.188	96.428	93.897	91.568
153	125.054	119.721	115.004	110.800	107.028	105.050	100.532	97.712	95.128	92.750
154	127.000	121.530	116.697	112.392	108.532	106.493	101.890	99.009	96.371	93.944
155	128.976	123.365	118.412	114.004	110.055	107.953	103.264	100.321	97.627	95.150
156	130.981	125.227	120.151	115.637	111.595	109.431	104.653	101.647	98.896	96.368
157	133.017	127.115	121.913	117.291	113.155	110.926	106.058	102.987	100.178	97.599
158	135.084	129.030	123.699	118.966	114.734	112.438	107.478	104.342	101.474	98.841
159	137.183	130.973	125.509	120.663	116.333	113.969	108.915	105.711	102.783	100.096
160	139.314	132.944	127.345	122.382	117.951	115.518	110.368	107.095	104.106	101.364
161	141.479	134.944	129.206	124.124	119.590	117.085	111.838	108.495	105.443	102.645

					N					
S	300	310	320	330	340	350	360	370	380	390
162	143.677	136.973	131.092	125.889	121.249	118.671	113.324	109.909	106.794	103.939
163	145.910	139.033	133.005	127.677	122.929	120.276	114.828	111.340	108.159	105.246
164	148.178	141.123	134.945	129.488	124.631	121.900	116.348	112.786	109.539	106.566
165	150.483	143.244	136.912	131.324	126.353	123.545	117.887	114.248	110.933	107.900
166	152.824	145.397	138.907	133.185	128.098	125.209	119.443	115.727	112.343	109.247
167	155.203	147.583	140.931	135.070	129.865	126.893	121.017	117.221	113.767	110.609
168	157.621	149.802	142.983	136.981	131.655	128.599	122.610	118.733	115.207	111.984
169	160.078	152.054	145.065	138.918	133.468	130.325	124.221	120.262	116.662	113.374
170	162.576	154.342	147.177	140.882	135.304	132.072	125.851	121.807	118.133	114.778
171	165.115	156.665	149.319	142.872	137.164	133.842	127.501	123.371	119.620	116.197
172	167.696	159.024	151.493	144.890	139.048	135.633	129.170	124.951	121.123	117.631
173	170.320	161.419	153.699	146.935	140.957	137.447	130.858	126.550	122.642	119.079
174	172.989	163.853	155.938	149.010	142.892	139.283	132.567	128.168	124.178	120.543
175	175.703	166.325	158.209	151.113	144.852	141.143	134.297	129.803	125.731	122.022
176	178.464	168.837	160.515	153.246	146.838	143.026	136.047	131.458	127.301	123.517
177		171.388	162.855	155.409	148.850	144.934	137.818	133.131	128.888	125.028
178		173.981	165.231	157.602	150.889	146.866	139.611	134.824	130.493	126.555
179		176.616	167.642	159.827	152.956	148.822	141.426	136.537	132.116	128.098
180		179.295	170.091	162.084	155.052	150.804	143.263	138.269	133.757	129.657
181		182.017	172.577	164.374	157.175	152.811	145.122	140.022	135.416	131.234
182		184.784	175.102	166.697	159.328	154.845	147.005	141.795	137.094	132.827
183			177.665	169.053	161.510	156.905	148.910	143.590	138.790	134.438
184			180.270	171.445	163.723	158.993	150.840	145.405	140.506	136.066
185			182.915	173.871	165.966	161.108	152.793	147.242	142.242	137.711
186			185.602	176.334	168.240	163.251	154.771	149.102	143.997	139.375
187			188.331	178.833	170.547	165.422	156.774	150.983	145.772	141.057
188			191.105	181.370	172.886	167.623	158.802	152.887	147.568	142.757
189				183.946	175.259	169.854	160.856	154.814	149.385	144.477
190				186.560	177.665	172.114	162.936	156.765	151.222	146.215
191				189.215	180.106	174.406	165.043	158.739	153.081	147.972
192				191.910	182.582	176.729	167.177	160.737	154.961	149.749
193				194.647	185.094	179.083	169.339	162.760	156.864	151.547
194				197.427	187.643	181.471	171.528	164.808	158.789	153.364
195					190.229	183.891	173.747	166.882	160.736	155.201

S	300	310	320	330	340	350	360	370	380	390
196					192.853	186.345	175.995	168.981	162.707	157.060
197					195.517	188.834	178.272	171.106	164.702	158.940
198					198.220	191.358	180.579	173.259	166.720	160.841
199					200.964	193.918	182.918	175.438	168.762	162.764
200						196.515	185.287	177.645	170.829	164.709
201						199.149	187.689	179.881	172.921	166.676
202						201.821	190.123	182.145	175.039	168.667
203						204.531	192.590	184.438	177.182	170.680
204						207.282	195.091	186.760	179.352	172.717
205							197.626	189.113	181.548	174.778
206							200.197	191.497	183.772	176.863
207							202.803	193.912	186.024	178.973
208							205.446	196.359	188.303	181.108
209							208.126	198.839	190.611	183.269
210							210.844	201.351	192.949	185.456
211							213.600	203.898	195.316	187.668
212							216.396	206.478	197.713	189.908
213								209.094	200.141	192.175
214								211.745	202.601	194.470
215								214.433	205.092	196.792
216								217.157	207.615	199.144
217								219.920	210.172	201.525
218									212.762	203.935
219									215.387	206.375
220									218.046	208.846
221									220.741	211.349
222									223.472	213.883
223									226.240	216.449
224										219.048
225										221.681
226										224.348
227										227.050
228										229.787

					N					
S	400	410	420	430	440	450	460	470	480	490
1	0.124	0.123	0.123	0.122	0.122	0.122	0.121	0.121	0.121	0.120
2	0.275	0.273	0.272	0.271	0.270	0.270	0.269	0.268	0.267	0.266
3	0.440	0.438	0.437	0.435	0.433	0.432	0.430	0.429	0.427	0.426
4	0.618	0.615	0.612	0.610	0.607	0.605	0.602	0.600	0.598	0.596
5	0.805	0.801	0.798	0.794	0.791	0.787	0.784	0.781	0.778	0.775
6	1.001	0.996	0.992	0.987	0.983	0.978	0.974	0.970	0.966	0.962
7	1.205	1.199	1.193	1.188	1.182	1.177	1.171	1.166	1.162	1.157
8	1.417	1.409	1.402	1.395	1.388	1.382	1.376	1.370	1.364	1.358
9	1.635	1.626	1.618	1.610	1.602	1.594	1.586	1.579	1.572	1.565
10	1.860	1.850	1.840	1.830	1.821	1.812	1.803	1.795	1.787	1.779
11	2.092	2.080	2.068	2.057	2.046	2.036	2.026	2.016	2.007	1.998
12	2.329	2.316	2.302	2.290	2.278	2.266	2.254	2.243	2.232	2.222
13	2.573	2.557	2.542	2.528	2.514	2.501	2.488	2.475	2.463	2.452
14	2.822	2.805	2.788	2.772	2.756	2.741	2.727	2.713	2.699	2.686
15	3.077	3.057	3.039	3.021	3.004	2.987	2.971	2.955	2.940	2.926
16	3.337	3.316	3.295	3.275	3.256	2.238	3.220	3.203	3.186	3.170
17	3.603	3.579	3.557	3.535	3.514	3.494	3.474	3.455	3.437	3.419
18	3.874	3.848	3.823	3.799	3.776	3.754	3.733	3.712	3.692	3.673
19	4.150	4.122	4.095	4.069	4.044	4.019	3.996	3.974	3.952	3.931
20	4.431	4.400	4.371	4.343	4.316	4.289	4.264	4.240	4.216	4.193
21	4.717	4.684	4.652	4.622	4.592	4.564	4.537	4.510	4.485	4.460
22	5.008	4.973	4.938	4.905	4.874	4.843	4.814	4.785	4.758	4.731
23	5.304	5.266	5.229	5.194	5.160	5.127	5.095	5.065	5.035	5.007
24	5.605	5.564	5.525	5.487	5.450	5.415	5.381	5.349	5.317	5.286
25	5.911	5.867	5.825	5.784	5.745	5.708	5.671	5.636	5.603	5.570
26	6.222	6.175	6.130	6.086	6.045	6.005	5.966	5.929	5.893	5.858
27	6.537	6.487	6.439	6.393	6.349	6.306	6.265	6.225	6.187	6.150
28	6.858	6.804	6.753	6.704	6.657	6.612	6.568	6.526	6.485	6.446
29	7.182	7.126	7.072	7.020	6.970	6.921	6.875	6.830	6.787	6.745
30	7.512	7.452	7.395	7.340	7.287	7.236	7.186	7.139	7.093	7.049
31	7.846	7.783	7.722	7.664	7.608	7.554	7.502	7.452	7.404	7.357
32	8.185	8.118	8.054	7.993	7.933	7.877	7.822	7.769	7.718	7.669
33	8.529	8.458	8.391	8.326	8.263	8.203	8.146	8.090	8.036	7.984
34	8.878	8.803	8.732	8.663	8.597	8.534	8.473	8.415	8.358	8.304

	N									
S	400	410	420	430	440	450	460	470	480	490
35	9.231	9.152	9.077	9.005	8.936	8.869	8.805	8.744	8.684	8.627
36	9.588	9.506	9.427	9.351	9.278	9.209	9.142	9.077	9.015	8.954
37	9.950	9.864	9.781	9.702	9.625	9.552	9.482	9.414	9.348	9.285
38	10.317	10.227	10.140	10.056	9.976	9.900	9.826	9.755	9.686	9.620
39	10.689	10.594	10.503	10.415	10.332	10.251	10.174	10.100	10.028	9.959
40	11.065	10.966	10.870	10.779	10.691	10.607	10.526	10.449	10.374	10.302
41	11.446	11.342	11.242	11.147	11.055	10.967	10.883	10.801	10.723	10.648
42	11.831	11.723	11.618	11.519	11.423	11.331	11.243	11.158	11.077	10.998
43	12.222	12.108	11.999	11.895	11.795	11.700	11.608	11.519	11.434	11.352
44	12.616	12.498	12.384	12.276	12.172	12.072	11.976	11.884	11.795	11.710
45	13.016	12.892	12.774	12.661	12.552	12.448	12.348	12.252	12.160	12.071
46	13.420	13.291	13.168	13.050	12.937	12.829	12.725	12.625	12.529	12.436
47	13.829	13.695	13.566	13.444	13.326	13.214	13.105	13.002	12.902	12.805
48	14.242	14.103	13.969	13.842	13.720	13.602	13.490	13.382	13.278	13.178
49	14.661	14.515	14.377	14.244	14.117	13.995	13.879	13.767	13.659	13.555
50	15.084	14.933	14.788	14.651	14.519	14.393	14.271	14.155	14.043	13.935
51	15.511	15.354	15.205	15.062	14.925	14.794	14.668	14.547	14.431	14.319
52	15.944	15.781	15.626	15.477	15.335	15.199	15.069	14.944	14.823	14.707
53	16.381	16.212	16.051	15.897	15.750	15.609	15.474	15.344	15.219	15.099
54	16.823	16.648	16.481	16.321	16.169	16.023	15.883	15.748	15.619	15.495
55	17.270	17.088	16.915	16.750	16.592	16.441	16.296	16.156	16.023	15.894
56	17.721	17.533	17.354	17.183	17.019	16.863	16.713	16.569	16.430	16.297
57	18.178	17.983	17.797	17.620	17.451	17.289	17.134	16.985	16.842	16.705
58	18.639	18.438	18.246	18.062	17.887	17.720	17.559	17.405	17.257	17.115
59	19.105	18.897	18.698	18.509	18.328	18.155	17.989	17.830	17.677	17.530
60	19.576	19.361	19.155	18.960	18.773	18.594	18.422	18.258	18.100	17.949
61	20.052	19.830	19.617	19.415	19.222	19.037	18.860	18.690	18.528	18.371
62	20.533	20.303	20.084	19.875	19.675	19.485	19.302	19.127	18.959	18.797
63	21.019	20.782	20.555	20.339	20.133	19.936	19.748	19.567	19.394	19.228
64	21.510	21.265	21.031	20.808	20.596	20.393	20.198	20.012	19.833	19.662
65	22.006	21.753	21.512	21.282	21.063	20.853	20.653	20.461	20.276	20.100
66	22.507	22.246	21.997	21.760	21.534	21.318	21.111	20.913	20.724	20.541
67	23.014	22.744	22.487	22.243	22.010	21.787	21.574	21.370	21.175	20.987
68	23.525	23.247	22.982	22.730	22.490	22.261	22.041	21.831	21.630	21.437

					N					
S	400	410	420	430	440	450	460	470	480	490
69	24.041	23.755	23.482	23.222	22.975	22.739	22.513	22.297	22.089	21.890
70	24.563	24.267	23.987	23.719	23.464	23.221	22.989	22.766	22.553	22.348
71	25.090	24.785	24.496	24.221	23.958	23.708	23.469	23.240	23.020	22.810
72	25.622	25.308	25.010	24.727	24.457	24.199	23.953	23.717	23.492	23.275
73	26.159	25.836	25.530	25.238	24.960	24.695	24.442	24.199	23.967	23.745
74	26.702	26.369	26.054	25.754	25.468	25.195	24.935	24.686	24.447	24.218
75	27.250	26.908	26.583	26.274	25.980	25.700	25.432	25.176	24.931	24.696
76	27.803	27.451	27.117	26.800	26.497	26.209	25.934	25.671	25.419	25.178
77	28.362	28.000	27.656	27.330	27.019	26.723	26.440	26.170	25.911	25.663
78	28.926	28.554	28.201	27.865	27.546	27.241	26.951	26.673	26.408	26.153
79	29.496	29.113	28.750	28.405	28.077	27.765	27.466	27.181	26.908	26.647
80	30.071	29.677	29.304	28.950	28.613	28.292	27.986	27.693	27.413	27.145
81	30.651	30.247	29.864	29.500	29.154	28.825	28.510	28.210	27.922	27.647
82	31.238	30.822	30.429	30.055	29.700	29.362	29.039	28.731	28.436	28.154
83	31.830	31.403	30.999	30.615	30.251	29.903	29.572	29.256	28.953	28.664
84	32.427	31.989	31.574	31.180	30.806	30.450	30.110	29.786	29.476	29.179
85	33.031	32.581	32.155	31.751	31.367	31.001	30.652	30.320	30.002	29.698
86	33.640	33.178	32.741	32.326	31.932	31.557	31.200	30.859	30.533	30.221
87	34.254	33.781	33.332	32.906	32.502	32.118	31.751	31.402	31.068	30.748
88	34.875	34.389	33.929	33.492	33.078	32.683	32.308	31.950	31.607	31.280
89	35.502	35.003	34.531	34.083	33.658	33.254	32.869	32.502	32.151	31.816
90	36.134	35.622	35.138	34.679	34.244	33.830	33.435	33.059	32.700	32.356
91	36.773	36.248	35.751	35.281	34.834	34.410	34.006	33.620	33.252	32.901
92	37.417	36.879	36.370	35.888	35.430	34.995	34.581	34.187	33.810	33.450
93	38.068	37.516	36.994	36.500	36.031	35.586	35.162	34.757	34.372	34.003
94	38.725	38.159	37.624	37.118	36.637	36.181	35.747	35.333	34.938	34.561
95	39.387	38.807	38.259	37.741	37.249	36.781	36.337	35.913	35.509	35.123
96	40.057	39.462	38.901	38.369	37.866	37.387	36.932	36.498	36.085	35.690
97	40.732	40.123	39.548	39.003	38.488	37.998	37.532	37.088	36.665	36.261
98	41.414	40.790	40.200	39.643	39.115	38.614	38.137	37.693	37.250	36.837
99	42.102	41.462	40.859	40.288	39.748	39.235	38.747	38.282	37.840	37.417
100	42.796	42.141	41.524	40.939	40.386	39.861	39.362	38.887	38.434	38.002
101	43.497	42.827	42.194	41.596	41.030	40.492	39.982	39.496	39.033	38.591
102	**44.204**	43.518	42.870	42.258	41.679	41.129	40.607	40.110	39.637	39.185

					N					
S	400	410	420	430	440	450	460	470	480	490
103	44.919	44.216	43.553	42.926	42.334	41.771	41.237	40.729	40.245	39.783
104	45.639	44.920	44.241	43.600	42.994	42.419	41.873	41.353	40.859	40.387
105	46.367	**45.630**	44.936	44.280	43.660	43.072	42.514	41.983	41.477	40.995
106	47.101	46.347	45.637	44.966	44.332	43.730	43.160	42.617	42.100	41.607
107	47.842	47.070	**46.344**	45.658	45.009	44.394	43.811	43.256	42.728	42.225
108	48.590	47.800	47.057	46.355	45.692	45.064	44.467	43.901	43.361	42.847
109	49.344	48.537	47.776	47.059	46.381	45.739	45.129	44.550	43.999	43.474
110	50.106	49.280	48.502	**47.769**	47.076	46.419	45.797	45.205	44.642	44.106
111	50.875	50.030	49.235	48.485	47.776	47.106	46.469	45.865	45.290	44.743
112	51.651	50.787	49.974	49.207	**48.483**	47.797	47.148	46.530	45.943	45.384
113	52.434	51.550	50.719	49.935	49.195	48.495	47.831	47.201	46.602	46.031
114	53.225	52.321	51.471	50.670	49.914	49.199	48.521	47.877	47.265	46.682
115	54.022	53.098	52.230	51.411	50.639	**49.908**	49.215	48.558	47.934	47.339
116	54.828	53.883	52.995	52.159	51.369	50.623	49.916	49.245	48.607	48.000
117	55.640	54.674	53.767	52.913	52.106	51.344	50.622	49.937	49.286	48.667
118	56.460	55.473	54.546	53.673	52.849	52.071	51.334	50.635	49.971	49.339
119	57.288	56.279	55.332	54.440	53.599	52.804	52.052	51.338	50.660	50.015
120	58.124	57.092	56.124	55.213	54.355	53.543	52.775	**52.047**	51.355	50.697
121	58.967	57.913	56.924	55.994	55.117	54.288	53.504	52.761	**52.056**	51.385
122	59.818	58.741	57.731	56.781	55.885	55.040	54.240	53.481	52.762	52.077
123	60.677	59.577	58.545	57.574	56.660	55.797	54.981	54.207	53.473	**52.775**
124	61.544	60.420	59.366	58.375	57.442	56.561	55.728	54.939	54.190	53.478
125	62.419	61.270	60.194	59.183	58.230	57.331	56.481	55.676	54.912	54.186
126	63.302	62.129	61.030	59.997	59.025	58.107	57.240	56.419	55.640	54.900
127	64.193	62.995	61.873	60.819	59.826	58.890	58.005	57.168	56.374	55.619
128	65.093	63.869	62.723	61.647	60.634	59.679	58.777	57.923	57.113	56.343
129	66.001	64.751	63.581	62.483	61.449	60.475	59.555	58.684	57.858	57.074
130	66.917	65.641	64.447	63.326	62.271	61.277	60.339	59.450	58.608	57.809
131	67.843	66.540	65.230	64.176	63.100	62.086	61.129	60.223	59.365	58.550
132	68.776	67.446	66.201	65.034	63.936	62.902	61.926	61.002	60.127	59.297
133	69.719	68.360	67.090	65.898	64.779	63.724	62.729	61.787	60.896	60.050
134	70.670	69.283	67.987	66.771	65.629	64.553	63.538	62.579	61.670	60.808
135	71.630	70.215	68.891	67.651	66.486	65.389	64.354	63.376	62.450	61.572
136	72.600	71.154	69.804	68.539	67.350	66.232	65.177	64.180	63.236	62.341

					N					
S	400	410	420	430	440	450	460	470	480	490
137	73.578	72.103	70.725	69.434	68.222	67.081	66.006	64.990	64.028	63.117
138	74.566	73.060	71.654	70.337	69.101	67.938	66.842	65.807	64.827	63.898
139	75.562	74.026	72.591	71.248	69.987	68.802	67.685	66.630	65.631	64.685
140	76.569	75.000	73.537	72.167	70.881	69.673	68.534	67.459	66.442	65.478
141	77.584	75.984	74.491	73.093	71.783	70.551	69.391	68.295	67.259	66.278
142	76.610	76.977	75.453	74.028	72.692	71.437	70.254	69.138	68.082	67.083
143	79.645	77.979	76.424	74.971	73.609	72.329	71.124	69.987	68.912	67.894
144	80.690	78.990	77.404	75.923	74.534	73.229	72.001	70.843	69.748	68.712
145	81.745	80.010	78.393	76.882	75.466	74.137	72.886	71.706	70.591	69.535
146	82.810	81.040	79.391	77.850	76.407	75.052	73.777	72.575	71.440	70.365
147	83.885	82.079	80.397	78.826	77.355	75.975	74.676	73.452	72.296	71.201
148	84.970	83.128	81.413	79.811	78.312	76.905	75.582	74.335	73.158	72.044
149	86.066	84.187	82.437	80.805	79.277	77.843	76.496	75.226	74.027	72.893
150	87.172	85.255	83.471	81.807	80.250	78.789	77.416	76.123	74.902	73.748
151	88.289	86.334	84.515	82.818	81.231	79.743	78.345	77.028	75.785	74.610
152	89.417	87.422	85.567	83.838	82.221	80.705	79.280	77.939	76.674	75.478
153	90.555	88.521	86.630	84.867	83.219	81.675	80.224	78.858	77.570	76.353
154	91.705	89.630	87.702	85.905	84.226	82.652	81.175	79.785	78.473	77.235
155	92.865	90.749	88.783	86.952	85.241	83.638	82.134	80.718	79.384	78.123
156	94.037	91.879	89.875	88.008	86.265	84.633	83.101	81.659	80.301	79.018
157	95.220	93.019	90.976	89.074	87.298	85.635	84.075	82.608	81.225	79.920
158	96.415	94.171	92.088	90.149	88.340	86.646	85.058	83.564	82.157	80.828
159	97.621	95.333	93.210	91.234	89.391	87.666	86.048	84.528	83.096	81.744
160	98.839	96.506	94.341	92.328	90.450	88.694	87.047	85.499	84.042	82.666
161	100.069	97.690	95.484	93.433	91.520	89.731	88.054	86.479	84.995	83.596
162	101.312	98.885	96.637	94.547	92.598	90.776	89.069	87.466	85.956	84.533
163	102.566	100.092	97.800	95.670	93.686	91.831	90.093	88.461	86.925	85.476
164	103.833	101.310	98.974	96.805	94.783	92.894	91.125	89.464	87.901	86.428
165	105.112	102.540	100.160	97.949	95.889	93.966	92.165	90.475	88.885	87.386
166	106.404	103.782	101.356	99.103	97.006	95.047	93.214	91.494	89.876	88.352
167	107.709	105.036	102.563	100.268	98.132	96.138	94.272	92.521	90.875	89.325
168	109.027	106.301	103.781	101.443	99.268	97.238	95.338	93.557	91.882	90.305
169	110.358	107.579	105.011	102.629	100.414	98.347	96.414	94.601	92.897	91.293
170	111.702	108.870	106.252	103.826	101.570	99.466	97.498	95.653	93.920	92.289

S	400	410	420	430	440	450	460	470	480	490
171	113.060	110.172	107.505	105.034	102.736	100.594	98.591	96.714	94.951	93.292
172	114.431	111.488	108.770	106.252	103.912	101.732	99.693	97.784	95.991	94.303
173	115.816	112.816	110.047	107.482	105.099	102.879	100.805	98.862	97.038	95.322
174	117.215	114.157	111.335	108.723	106.297	104.037	101.926	99.949	98.094	96.349
175	118.629	115.511	112.636	109.975	107.505	105.204	103.056	101.045	99.158	97.383
176	120.057	116.879	113.949	111.239	108.723	106.382	104.196	102.150	100.230	98.426
177	121.499	118.260	115.275	112.514	109.953	107.569	105.345	103.264	101.312	99.477
178	122.956	119.654	116.613	113.801	111.193	108.767	106.504	104.386	102.401	100.536
179	124.428	121.063	117.964	115.100	112.445	109.975	107.672	105.518	103.500	101.603
180	125.916	122.485	119.328	116.411	113.708	111.194	108.851	106.660	104.607	102.678
181	127.418	123.922	120.705	117.734	114.982	112.424	110.039	107.811	105.723	103.762
182	128.936	125.373	122.095	119.070	116.268	113.664	111.238	108.971	106.847	104.854
183	130.470	126.838	123.499	120.418	117.565	114.915	112.446	110.141	107.981	105.955
184	132.020	128.318	124.916	121.778	118.874	116.177	113.665	111.320	109.124	107.064
185	133.586	129.813	126.347	123.151	120.194	117.450	114.895	112.509	110.277	108.182
186	135.169	131.323	127.792	124.537	121.527	118.734	116.134	113.708	111.438	109.309
187	136.768	132.848	129.251	125.936	122.872	120.029	117.385	114.917	112.609	110.445
188	138.384	134.389	130.724	127.349	124.229	121.336	118.646	116.136	113.789	111.589
189	140.017	135.945	132.211	128.774	125.599	122.655	119.918	117.365	114.979	112.743
190	141.667	137.517	133.714	130.213	126.981	123.985	121.201	118.605	116.179	113.906
191	143.335	139.106	135.231	131.666	128.376	125.327	122.495	119.855	117.388	115.078
192	145.021	140.710	136.763	133.133	129.783	126.681	123.800	121.115	118.607	116.259
193	146.725	142.331	138.310	134.614	131.204	128.047	125.116	122.386	119.836	117.449
194	148.447	143.969	139.872	136.109	132.638	129.426	126.444	123.667	121.075	118.649
195	150.188	145.624	141.451	137.618	134.085	130.816	127.783	124.960	122.325	119.859
196	151.948	147.296	143.045	139.142	135.545	132.219	129.134	126.263	123.584	121.078
197	153.727	148.986	144.654	140.680	137.019	133.635	130.497	127.577	124.854	122.307
198	155.525	150.693	146.281	142.234	138.507	135.064	131.872	128.903	126.134	123.546
199	157.343	152.418	147.923	143.802	140.009	136.505	133.258	130.240	127.425	124.795
200	159.181	154.162	149.582	145.386	141.525	137.960	134.657	131.588	128.727	126.054
201	161.039	155.924	151.258	146.985	143.055	139.428	136.068	132.947	130.039	127.323
202	162.918	157.704	152.951	148.600	144.600	140.909	137.492	134.319	131.363	128.602
203	164.818	159.504	154.662	150.231	146.160	142.404	138.929	135.702	132.697	129.891
204	166.739	161.322	156.390	151.878	147.734	143.913	140.378	137.097	134.042	131.191

S	400	410	420	430	440	450	460	470	480	490
205	168.681	163.161	158.136	153.541	149.323	145.435	141.840	138.503	135.399	132.502
206	170.646	165.019	159.899	155.221	150.927	146.972	143.315	139.923	135.767	133.823
207	172.632	166.897	161.682	156.918	152.547	148.523	144.803	141.354	138.146	135.155
208	174.642	168.795	163.482	158.631	154.183	150.088	146.305	142.798	139.537	136.497
209	176.674	170.714	165.302	160.362	155.834	151.668	147.820	144.254	140.940	137.851
210	178.729	172.655	167.140	162.110	157.501	153.262	149.348	145.723	142.355	139.216
211	180.808	174.616	168.998	163.876	159.185	154.871	150.891	147.205	143.781	140.592
212	182.911	176.599	170.875	165.659	160.884	156.496	152.447	148.700	145.220	141.979
213	185.038	178.604	172.773	167.461	162.601	158.136	154.018	150.208	146.671	143.378
214	187.190	180.631	174.690	169.281	164.334	159.791	155.603	151.729	148.134	144.789
215	189.367	182.681	176.628	171.120	166.084	161.462	157.203	153.264	149.610	146.211
216	191.569	184.754	178.586	172.977	167.852	163.149	158.817	154.812	151.099	147.645
217	193.798	186.850	180.566	174.854	169.637	164.852	160.446	156.374	152.600	149.090
218	196.053	188.970	182.567	176.750	171.440	166.571	162.090	157.950	154.114	150.548
219	198.335	191.113	184.590	178.666	173.260	168.306	163.749	159.540	155.642	152.018
220	200.644	193.281	186.635	180.601	175.099	170.059	165.423	161.145	157.182	163.501
221	202.980	195.474	188.702	182.557	176.956	171.828	167.114	162.763	158.736	154.996
222	205.345	197.692	190.791	184.534	178.833	173.614	168.819	164.397	160.303	156.503
223	207.739	199.936	192.904	186.531	180.728	175.418	170.541	166.045	161.885	158.023
224	210.161	202.206	195.040	188.550	182.642	177.240	172.279	167.708	163.479	159.556
225	212.614	204.502	197.200	190.589	184.575	179.079	174.034	169.386	165.088	161.103
226	215.096	206.824	199.383	192.651	186.529	180.936	175.805	171.079	166.711	162.662
227	217.609	209.175	201.591	194.735	188.502	182.811	177.592	172.788	168.349	164.234
228	220.153	211.552	203.824	196.840	190.496	184.705	179.397	174.512	170.001	165.820
229	222.729	213.958	206.082	198.969	192.510	186.618	181.219	176.252	171.667	167.420
230	225.337	216.393	208.366	201.121	194.545	188.550	183.058	178.009	173.349	169.033
231	227.978	218.856	210.676	203.296	196.602	190.501	184.915	179.781	175.045	170.661
232	230.652	221.349	213.012	205.494	198.680	192.471	186.790	181.570	176.756	172.302
233	233.360	223.872	215.375	207.717	200.779	194.461	188.683	183.376	178.483	173.958
234	236.103	226.426	217.765	209.964	202.901	196.472	190.594	185.198	180.226	175.628
235		229.011	220.182	212.236	205.045	198.503	192.524	187.038	181.984	177.312
236		231.627	222.628	214.534	207.211	200.554	194.473	188.894	183.758	179.011
237		234.276	225.103	216.856	209.401	202.626	196.440	190.769	185.548	180.726
238		236.957	227.606	219.205	211.614	204.720	198.427	192.661	187.355	182.455

						N				
S	400	410	420	430	440	450	460	470	480	490
239		239.672	230.139	221.580	213.112	206.835	200.434	194.571	189.178	184.199
240		242.420	232.702	223.982	216.112	208.971	202.461	196.499	191.017	185.959
241			235.295	226.411	218.398	211.130	204.507	198.445	192.874	187.735
242			237.920	228.868	220.708	213.311	206.574	200.410	194.747	189.526
243			240.576	231.353	223.043	215.516	208.662	202.394	196.639	191.333
244			243.264	233.866	225.404	217.743	210.771	204.397	198.547	193.157
245			245.984	236.409	227.791	219.993	212.900	206.420	200.473	194.997
246			248.738	238.980	230.205	222.267	215.052	208.462	202.418	196.853
247				241.582	232.645	224.566	217.225	210.524	204.380	198.726
248				244.214	235.112	226.889	219.420	212.606	206.361	200.616
249				246.877	237.607	229.236	221.638	214.708	208.361	202.524
250				249.571	240.130	231.609	223.879	216.832	210.380	204.448
251				252.298	242.681	234.007	226.142	218.976	212.417	206.391
252				255.057	245.261	236.431	228.429	221.142	214.475	208.351
253					247.871	238.882	230.740	223.329	216.551	210.329
254					250.510	241.359	233.075	225.538	218.648	212.325
255					253.180	243.864	235.434	227.769	220.765	214.340
256					255.880	246.396	237.818	230.022	222.903	216.374
257					258.612	248.956	240.228	232.298	225.061	218.427
258					261.376	251.544	242.662	234.598	227.240	220.499
259						254.161	245.123	236.921	229.441	222.590
260						256.807	247.610	239.267	231.663	224.701
261						259.484	250.124	241.638	233.906	226.832
262						262.190	252.664	244.033	236.173	228.983
263						264.927	255.233	246.453	238.461	231.155
264						267.695	257.829	248.897	240.772	233.347
265							260.453	251.368	243.107	235.560
266							263.106	253.864	245.464	237.795
267							265.789	256.387	247.846	240.051
268							268.501	258.936	250.252	242.330
269							271.243	261.512	252.682	244.630
270								264.116	255.136	246.953
271								266.747	257.616	249.298
272								269.407	260.121	251.667

					N					
S	400	410	420	430	440	450	460	470	480	490
273								272.095	262.652	254.059
274								274.812	265.210	256.475
275								277.559	267.793	258.914
276								280.336	270.404	261.378
277									273.042	263.867
278									275.708	266.381
279									278.402	268.921
280									281.125	271.486
281									283.877	274.077
282									286.658	276.695
283										279.339
284										282.011
285										284.711
286										287.438
287										290.195

					N					
S	500	510	520	530	540	550	560	570	580	590
1	0.120	0.120	0.120	0.119	0.119	0.118	0.118	0.118	0.118	0.117
2	0.265	0.264	0.264	0.263	0.262	0.261	0.261	0.260	0.259	0.259
3	0.424	0.423	0.421	0.420	0.419	0.418	0.416	0.415	0.414	0.413
4	0.594	0.592	0.590	0.588	0.586	0.584	0.582	0.581	0.579	0.577
5	0.772	0.769	0.767	0.764	0.762	0.759	0.757	0.754	0.752	0.750
6	0.959	0.955	0.952	0.948	0.945	0.942	0.939	0.935	0.932	0.930
7	1.152	1.148	1.143	1.139	1.135	1.131	1.127	1.123	1.120	1.116
8	1.352	1.347	1.342	1.337	1.332	1.327	1.322	1.317	1.313	1.309
9	1.559	1.552	1.546	1.540	1.534	1.528	1.523	1.517	1.512	1.507
10	1.771	1.763	1.756	1.749	1.742	1.736	1.729	1.723	1.717	1.711
11	1.989	1.980	1.972	1.964	1.956	1.948	1.941	1.933	1.926	1.919
12	2.212	2.202	2.193	2.183	2.174	2.166	2.157	2.149	2.141	2.133
13	2.440	2.429	2.418	2.408	2.398	2.388	2.378	2.369	2.360	2.351
14	2.674	2.661	2.649	2.638	2.626	2.615	2.604	2.594	2.584	2.574
15	2.912	2.898	2.885	2.872	2.859	2.847	2.835	2.823	2.812	2.801
16	3.155	3.139	3.125	3.110	3.097	3.083	3.070	3.057	3.045	3.033

					N					
S	500	510	520	530	540	550	560	570	580	590
17	3.402	3.385	3.369	3.354	3.338	3.324	3.309	3.295	3.282	3.268
18	3.654	3.636	3.618	3.601	3.585	3.568	3.553	3.537	3.523	3.508
19	3.910	3.891	3.872	3.853	3.835	3.817	3.800	3.784	3.768	3.752
20	4.171	4.150	4.129	4.109	4.089	4.070	4.052	4.034	4.016	3.999
21	4.436	4.413	4.391	4.369	4.348	4.327	4.307	4.288	4.269	4.251
22	4.706	4.681	4.657	4.633	4.610	4.588	4.567	4.546	4.526	4.506
23	4.979	4.952	4.926	4.901	4.877	4.853	4.830	4.808	4.786	4.765
24	5.257	5.228	5.200	5.173	5.147	5.122	5.097	5.073	5.050	5.028
25	5.538	5.508	5.478	5.449	5.422	5.395	5.368	5.343	5.318	5.294
26	5.824	5.792	5.760	5.729	5.700	5.671	5.643	5.616	5.589	5.564
27	6.114	6.079	6.046	6.013	5.982	5.951	5.921	5.892	5.864	5.837
28	6.408	6.371	6.335	6.301	6.267	6.235	6.203	6.172	6.143	6.114
29	6.705	6.666	6.628	6.592	6.556	6.522	6.489	6.456	6.425	6.394
30	7.007	6.965	6.925	6.887	6.849	6.813	6.778	6.743	6.710	6.678
31	7.312	7.268	7.226	7.185	7.146	7.107	7.070	7.034	6.999	6.965
32	7.621	7.575	7.531	7.488	7.446	7.406	7.366	7.328	7.291	7.255
33	7.934	7.886	7.839	7.794	7.750	7.707	7.666	7.626	7.587	7.549
34	8.251	8.200	8.151	8.103	8.057	8.012	7.969	7.927	7.886	7.846
35	8.572	8.518	8.467	8.417	8.368	8.321	8.276	8.231	8.188	8.147
36	8.896	8.840	8.786	8.734	8.683	8.633	8.586	8.539	8.494	8.450
37	9.225	9.166	9.109	9.054	9.001	8.949	8.899	8.850	8.803	8.758
38	9.557	9.495	9.436	9.378	9.322	9.268	9.216	9.165	9.116	9.068
39	9.892	9.828	9.766	9.706	9.647	9.591	9.536	9.483	9.431	9.381
40	10.232	10.165	10.100	10.037	9.976	9.917	9.860	9.804	9.750	9.698
41	10.575	10.505	10.437	10.371	10.308	10.246	10.187	10.129	10.073	10.018
42	10.922	10.849	10.778	10.710	10.643	10.579	10.517	10.457	10.398	10.341
43	11.273	11.196	11.123	11.051	10.982	10.915	10.851	10.788	10.727	10.668
44	11.627	11.548	11.471	11.396	11.325	11.255	11.188	11.122	11.059	10.997
45	11.985	11.902	11.822	11.745	11.670	11.598	11.528	11.460	11.394	11.330
46	12.347	12.261	12.178	12.097	12.020	11.945	11.872	11.801	11.733	11.666
47	12.713	12.623	12.537	12.453	12.373	12.294	12.219	12.145	12.074	12.005
48	13.082	12.989	12.899	12.812	12.729	12.648	12.569	12.493	12.419	12.348
49	13.455	13.358	13.265	13.175	13.088	13.004	12.923	12.844	12.767	12.693
50	13.831	13.731	13.635	13.542	13.451	13.364	13.280	13.198	13.119	13.042

	N									
S	500	510	520	530	540	550	560	570	580	590
51	14.212	14.108	14.008	13.911	13.818	13.728	13.640	13.555	13.473	13.394
52	14.596	14.488	14.385	14.285	14.188	14.094	14.004	13.916	13.831	13.749
53	14.984	14.872	14.765	14.661	14.561	14.464	14.371	14.280	14.192	14.107
54	15.375	15.260	15.149	15.042	14.938	14.838	14.741	14.647	14.556	14.468
55	15.771	15.651	15.536	15.425	15.318	15.215	15.114	15.017	14.923	14.832
56	16.170	16.046	15.927	15.813	15.702	15.595	15.491	15.391	15.294	15.200
57	16.572	16.445	16.322	16.203	16.089	15.978	15.871	15.768	15.668	15.570
58	16.979	16.847	16.720	16.598	16.480	16.365	16.255	16.148	16.044	15.944
59	17.389	17.253	17.122	16.996	16.874	16.756	16.642	16.531	16.425	16.321
60	17.803	17.663	17.527	17.397	17.271	17.149	17.032	16.918	16.808	16.701
61	18.221	18.076	17.936	17.802	17.672	17.547	17.425	17.308	17.194	17.084
62	18.642	18.493	18.349	18.210	18.076	17.947	17.822	17.701	17.584	17.471
63	19.068	18.914	18.765	18.622	18.484	18.351	18.222	18.098	17.977	17.860
64	19.497	19.338	19.185	19.038	18.896	18.758	18.626	18.497	18.373	18.253
65	19.930	19.766	19.609	19.457	19.310	19.169	19.033	18.900	18.773	18.649
66	20.366	20.198	20.036	19.879	19.729	19.583	19.443	19.307	19.175	19.048
67	20.807	20.633	20.466	20.306	20.151	20.001	19.856	19.716	19.581	19.450
68	21.251	21.073	20.901	20.735	20.576	20.422	20.273	20.129	19.990	19.856
69	21.699	21.516	21.339	21.169	21.005	20.846	20.693	20.545	20.402	20.264
70	22.152	21.963	21.781	21.606	21.437	21.274	21.117	20.965	20.818	20.676
71	22.608	22.413	22.226	22.046	21.873	21.706	21.544	21.388	21.237	21.091
72	23.067	22.868	22.676	22.491	22.312	22.140	21.974	21.814	21.659	21.509
73	23.531	23.326	23.128	22.938	22.755	22.579	22.408	22.243	22.084	21.930
74	23.999	23.788	23.585	23.390	23.202	23.020	22.845	22.676	22.513	22.355
75	24.470	24.254	24.046	23.845	23.652	23.466	23.286	23.112	22.945	22.782
76	24.946	24.724	24.510	24.304	24.106	23.914	23.730	23.552	23.380	23.213
77	25.425	25.197	24.978	24.766	24.563	24.367	24.177	23.995	23.818	23.647
78	25.909	25.675	25.449	25.232	25.024	24.822	24.628	24.441	24.260	24.085
79	26.396	26.156	25.925	25.702	25.488	25.282	25.083	24.891	24.705	24.526
80	26.888	26.641	26.404	26.176	25.956	25.745	25.541	25.344	25.153	24.969
81	27.383	27.130	26.887	26.653	26.428	26.211	26.002	25.800	25.605	25.417
82	27.883	27.623	27.374	27.134	26.903	26.681	26.467	26.260	26.060	25.867
83	28.387	28.120	27.865	27.619	27.383	27.155	26.935	26.723	26.518	26.321
84	28.894	28.622	28.360	28.108	27.865	27.632	27.407	27.190	26.980	26.778

S	N									
	500	510	520	530	540	550	560	570	580	590
85	29.406	29.127	28.858	28.600	28.352	28.113	27.882	27.660	27.445	27.238
86	29.922	29.636	29.361	29.096	28.842	28.597	28.361	28.134	27.914	27.702
87	30.442	30.149	29.867	29.596	29.336	29.085	28.844	28.611	28.386	28.168
88	30.966	30.666	30.377	30.100	29.833	29.577	29.330	29.091	28.861	28.639
89	31.495	31.187	30.891	30.608	30.335	30.072	29.819	29.575	29.340	29.112
90	32.027	31.712	31.410	31.119	30.840	30.571	30.312	30.063	29.822	29.589
91	32.564	32.241	31.932	31.635	31.349	31.074	30.809	30.554	30.307	30.070
92	33.105	32.775	32.458	32.154	31.861	31.580	31.309	31.048	30.796	30.553
93	33.650	33.312	32.988	32.677	32.378	32.090	31.813	31.546	31.289	31.040
94	34.200	33.854	33.523	33.204	32.898	32.604	32.321	32.048	31.785	31.531
95	34.754	34.400	34.061	33.735	33.423	33.122	32.832	32.553	32.284	32.025
96	35.312	34.950	34.603	34.270	33.951	33.643	33.347	33.062	32.787	32.522
97	35.874	35.504	35.150	34.809	34.483	34.168	33.866	33.574	33.294	33.022
98	36.441	36.063	35.700	35.353	35.018	34.697	34.388	34.090	33.803	33.527
99	37.013	36.626	36.255	35.900	35.558	35.230	34.914	34.610	34.317	34.034
100	37.588	37.193	36.814	36.451	36.102	35.767	35.444	35.133	34.834	34.545
101	38.168	37.764	37.377	37.006	36.650	36.307	35.978	35.660	35.355	35.060
102	38.753	38.340	37.945	37.565	37.201	36.851	36.515	36.191	35.879	35.578
103	39.342	38.920	38.516	38.129	37.757	37.399	37.056	36.725	36.407	36.099
104	39.936	39.505	39.092	38.696	38.316	37.952	37.601	37.263	36.938	36.624
105	40.534	40.094	39.672	39.268	38.800	38.508	38.150	37.805	37.473	37.153
106	41.137	40.687	40.256	39.844	39.448	39.068	38.702	38.350	38.012	37.685
107	41.744	41.285	40.845	40.424	40.020	39.631	39.258	38.900	38.554	38.221
108	42.356	41.887	41.438	41.008	40.595	40.199	39.819	39.453	39.100	38.760
109	42.973	42.494	42.035	41.596	41.175	40.771	40.383	40.009	39.650	39.303
110	43.594	43.105	42.637	42.189	41.760	41.347	40.951	40.570	40.203	39.850
111	44.220	43.721	43.243	42.786	42.348	41.927	41.523	41.134	40.760	40.400
112	44.851	44.341	43.854	43.387	42.940	42.511	42.099	41.703	41.321	40.954
113	45.486	44.966	44.469	43.993	43.537	43.099	42.679	42.275	41.886	41.51.1
114	46.127	45.596	45.089	44.603	44.138	43.691	43.263	42.851	42.454	42.072
115	46.772	46.230	45.713	45.217	44.743	44.288	43.851	43.431	43.026	42.637
116	47.422	46.869	46.341	45.836	45.352	44.888	44.443	44.014	43.602	43.206
117	48.077	47.513	46.975	46.459	45.966	45.493	45.039	44.602	44.182	43.778
118	48.736	48.162	47.612	47.087	46.584	46.102	45.639	45.194	44.766	44.354

					N					
S	500	510	520	530	540	550	560	570	580	590
119	49.401	48.815	48.255	47.719	47.206	46.715	46.243	45.789	45.354	44.934
120	50.071	49.473	48.902	48.356	47.833	47.332	46.851	46.389	45.945	45.518
121	50.745	50.136	49.554	48.997	48.464	47.953	47.463	46.993	46.540	46.105
122	51.425	50.804	50.210	49.643	49.100	48.579	48.080	47.600	47.140	46.696
123	52.110	51.476	50.871	50.293	49.740	49.209	48.701	48.212	47.743	47.291
124	52.800	52.154	51.537	50.948	50.384	49.844	49.326	48.828	48.350	47.890
125	53.495	52.837	52.208	51.607	51.033	50.482	49.955	49.448	48.961	48.493
126	**54.195**	53.524	52.884	52.272	51.686	51.126	50.588	50.072	49.576	49.100
127	54.901	54.217	53.564	52.941	52.344	51.773	51.226	50.700	50.196	49.710
128	55.612	54.914	54.249	53.614	53.007	52.425	51.686	51.333	50.819	50.325
129	56.328	55.617	54.940	54.293	53.674	53.082	52.514	51.969	51.446	50.943
130	57.049	**56.325**	55.635	54.976	54.346	53.742	53.164	52.610	52.078	51.566
131	57.776	57.038	56.335	55.664	55.022	54.408	53.819	53.255	52.713	52.192
132	58.508	57.757	57.040	56.357	55.703	55.078	54.479	53.904	53.353	52.823
133	59.245	58.480	**57.751**	57.054	56.389	55.752	55.142	54.558	53.996	53.457
134	59.988	59.209	58.466	57.757	57.079	56.431	55.811	55.216	54.644	54.096
135	60.737	59.943	59.186	58.464	57.775	57.115	56.483	55.878	55.297	54.738
136	61.491	60.682	59.912	**59.177**	58.475	57.803	57.160	56.544	55.953	55.385
137	62.251	61.427	60.643	59.894	59.180	58.496	57.842	57.215	56.614	56.036
138	63.016	62.177	61.379	60.617	**59.889**	59.194	58.528	57.890	57.278	56.690
139	63.787	62.933	62.120	61.344	60.604	59.896	59.219	58.570	57.948	57.350
140	64.564	63.694	62.866	62.077	61.318	**60.604**	59.911	59.252	58.619	58.012
141	65.346	64.461	63.618	62.815	62.048	61.316	60.614	59.943	59.299	58.681
142	66.134	65.233	64.375	63.558	62.778	62.032	61.315	60.633	59.979	59.353
143	66.929	66.011	65.138	64.306	63.512	62.754	**62.028**	61.334	60.668	60.029
144	67.728	66.795	65.906	65.059	64.252	63.480	62.738	62.032	61.359	60.709
145	68.534	67.584	66.668	65.818	64.997	64.212	63.461	**62.743**	62.054	61.394
146	69.346	68.379	67.459	66.582	65.746	64.948	64.185	63.454	62.754	62.083
147	70.164	69.179	68.243	67.351	66.501	65.689	64.913	64.170	**63.459**	62.776
148	70.988	69.986	69.033	68.126	67.261	66.436	65.646	64.891	64.168	63.474
149	71.818	70.798	69.829	68.906	68.027	67.187	66.384	65.617	64.881	**64.176**
150	72.654	71.617	70.630	69.692	68.797	67.943	67.127	66.347	65.599	64.882
151	73.497	72.441	71.438	70.483	69.573	68.705	67.875	67.082	66.322	65.593
152	74.346	73.271	72.251	71.279	70.354	69.471	68.628	67.821	67.049	66.309

					N					
S	500	510	520	530	540	550	560	570	580	590
153	75.201	74.108	73.069	72.082	71.141	70.243	69.386	68.566	67.781	67.029
154	76.062	74.950	73.894	72.889	71.933	71.020	70.149	69.315	68.518	67.753
155	76.930	75.798	74.724	73.703	72.730	71.802	70.916	70.070	69.259	68.482
156	77.804	76.653	75.561	74.522	73.533	72.590	71.689	70.829	70.005	69.216
157	78.684	77.514	76.403	75.347	74.341	73.382	72.467	71.593	70.756	69.954
158	79.572	78.381	77.251	76.177	75.155	74.181	73.251	72.362	71.511	70.697
159	80.466	79.255	78.106	77.014	75.974	74.984	74.039	73.136	72.272	71.444
160	81.366	80.134	78.966	77.856	76.800	75.793	74.832	73.915	73.037	72.197
161	82.273	81.021	79.833	78.704	77.630	76.607	75.631	74.699	73.807	72.954
162	83.187	81.913	80.706	79.558	78.467	77.427	76.435	75.488	74.582	73.715
163	84.108	82.813	81.585	80.418	79.309	78.252	77.245	76.283	75.363	74.482
164	85.036	83.719	82.470	81.284	80.157	79.083	78.060	77.082	76.148	75.253
165	85.970	84.631	83.362	82.157	81.011	79.920	78.880	77.887	76.938	76.029
166	86.912	85.550	84.260	83.035	81.870	80.762	79.705	78.697	77.733	76.810
167	87.861	86.476	85.164	83.919	82.736	81.610	80.536	79.512	78.533	77.596
168	88.816	87.409	86.075	84.810	83.608	82.463	81.373	80.332	79.338	78.387
169	89.779	88.348	86.993	85.707	84.485	83.323	82.215	81.158	80.149	79.183
170	90.749	89.294	87.917	86.610	85.369	84.188	83.063	81.989	80.964	79.984
171	91.727	90.248	88.847	87.520	86.258	85.059	83.916	82.826	81.785	80.790
172	92.712	91.208	89.785	88.436	87.154	85.936	84.775	83.668	82.611	81.601
173	93.704	92.175	90.729	89.358	88.056	86.818	85.640	84.516	83.442	82.417
174	94.704	93.150	91.680	90.287	88.964	87.707	86.510	85.369	84.279	83.238
175	95.711	94.132	92.638	91.222	89.879	88.602	87.386	86.227	85.121	84.064
176	96.726	95.121	93.603	92.164	90.800	89.503	88.268	87.091	85.969	84.896
177	97.748	96.117	94.574	93.113	91.727	90.410	89.156	87.961	86.821	85.732
178	98.778	97.120	95.553	94.069	92.660	91.323	90.050	88.837	87.680	86.574
179	99.817	98.132	96.539	95.031	93.601	92.242	90.949	89.718	88.543	87.421
180	100.862	99.150	97.532	96.000	94.547	93.167	91.855	90.605	89.413	88.274
181	101.916	100.176	98.532	96.976	95.500	94.099	92.767	91.498	90.287	89.132
182	102.978	101.210	99.539	97.958	96.460	95.037	93.684	92.396	91.168	89.995
183	104.048	102.251	100.554	98.948	97.426	95.982	94.608	93.301	92.054	90.864
184	105.126	103.300	101.576	99.945	98.399	96.933	95.538	94.211	92.946	91.738
185	106.213	104.357	102.606	100.949	99.379	97.890	96.475	95.127	93.843	92.618
186	107.307	105.422	103.643	101.960	100.366	98.854	97.417	96.050	94.747	93.503

S	N									
	500	510	520	530	540	550	560	570	580	590
187	108.411	106.495	104.687	102.978	101.360	99.824	98.366	96.978	95.656	94.394
188	109.522	107.576	105.739	104.004	102.360	100.802	99.321	97.912	96.571	95.291
189	110.642	108.665	106.799	105.036	103.368	101.785	100.282	98.853	97.491	96.193
190	111.771	109.762	107.867	106.077	104.382	102.776	101.250	99.800	98.418	97.101
191	112.908	110.867	108.942	107.124	105.404	103.773	102.225	100.753	99.351	98.014
192	114.054	111.981	110.026	108.180	106.433	104.777	103.206	101.712	100.290	98.934
193	115.209	113.103	111.117	109.242	107.469	105.788	104.193	102.677	101.234	99.859
194	116.373	114.233	112.216	110.313	108.512	106.806	105.188	103.649	102.185	100.790
195	117.546	115.372	113.324	111.391	109.563	107.831	106.189	104.628	103.142	101.727
196	118.728	116.520	114.439	112.477	110.621	108.863	107.196	105.612	104.105	102.670
197	119.919	117.676	115.563	113.570	111.686	109.902	108.211	106.604	105.075	103.618
198	121.120	118.841	116.695	114.672	112.759	110.949	109.232	107.601	106.050	104.573
199	122.330	120.015	117.836	115.781	113.840	112.002	110.260	108.606	107.032	105.534
200	123.549	121.198	118.985	116.899	114.928	113.063	111.295	109.617	108.021	106.501
201	124.778	122.390	120.143	118.024	116.024	114.131	112.337	110.634	109.016	107.474
202	126.017	123.591	121.309	119.158	117.128	115.207	113.387	111.659	110.017	108.454
203	127.265	124.801	122.484	120.300	118.239	116.290	114.443	112.690	111.025	109.439
204	128.523	126.020	123.667	121.451	119.359	117.380	115.506	113.728	112.039	110.431
205	129.791	127.249	124.860	122.610	120.486	118.478	116.577	114.773	113.060	111.429
206	131.069	128.487	126.061	123.777	121.621	119.584	117.655	115.825	114.087	112.434
207	132.357	129.735	127.272	124.953	122.765	120.698	118.740	116.884	115.122	113.445
208	133.655	130.992	128.491	126.137	123.917	121.819	119.833	117.950	116.163	114.463
209	134.964	132.260	129.720	127.330	125.077	122.948	120.933	119.024	117.211	115.487
210	136.283	133.536	130.958	128.532	126.245	124.085	122.041	120.104	118.265	116.517
211	137.613	134.823	132.205	129.743	127.422	125.230	123.156	121.192	119.327	117.554
212	138.953	136.120	133.462	130.962	128.607	126.383	124.279	122.287	120.396	118.598
213	140.304	137.427	134.728	132.191	129.800	127.544	125.410	123.389	121.471	119.649
214	141.666	138.744	136.004	133.428	131.003	128.713	126.549	124.498	122.554	120.706
215	143.039	140.072	137.290	134.675	132.213	129.891	127.695	125.616	123.644	121.770
216	144.423	141.410	138.585	135.931	133.433	131.076	128.849	126.740	124.741	122.841
217	145.818	142.758	139.890	137.197	134.662	132.271	130.011	127.873	125.845	123.919
218	147.224	144.117	141.206	138.472	135.899	133.473	131.181	129.012	126.956	125.004
219	148.642	145.486	142.531	139.756	137.146	134.684	132.360	130.160	128.075	126.096
220	150.071	146.867	143.867	141.050	138.401	135.904	133.546	131.315	129.202	127.195

					N					
S	500	510	520	530	540	550	560	570	580	590
221	151.512	148.258	145.212	142.354	139.666	137.133	134.741	132.479	130.335	128.301
222	152.965	149.661	146.569	143.667	140.940	138.370	135.944	133.650	131.477	129.415
223	154.429	151.074	147.935	144.991	142.223	139.616	137.155	134.829	132.625	130.535
224	155.906	152.499	149.312	146.324	143.516	140.871	138.375	136.016	133.782	131.663
225	157.395	153.935	150.700	147.667	144.818	142.135	139.603	137.211	134.946	132.798
226	158.896	155.383	152.099	149.021	146.130	143.408	140.840	138.415	136.118	133.941
227	160.409	156.843	153.509	150.385	147.451	144.690	142.086	139.626	137.298	135.091
228	161.935	158.314	154.929	151.759	148.782	145.981	143.340	140.846	138.486	136.249
229	163.474	159.796	156.361	153.143	150.123	147.282	144.604	142.074	139.682	137.414
230	165.025	161.291	157.804	154.539	151.474	148.592	145.876	143.311	140.885	138.587
231	166.590	162.798	159.258	155.944	152.835	149.912	147.157	144.556	142.097	139.767
232	168.167	164.318	160.724	157.361	154.206	151.241	148.447	145.810	143.317	140.956
233	169.758	165.849	162.201	158.788	155.588	152.579	149.746	147.073	144.545	142.152
234	171.362	167.393	163.690	160.227	156.979	153.928	151.055	148.344	145.782	143.356
235	172.980	168.950	165.191	161.676	158.382	155.286	152.373	149.624	147.027	144.568
236	174.611	170.519	166.704	163.137	159.794	156.655	153.700	150.913	148.280	145.788
237	176.256	172.102	168.229	164.609	161.218	158.033	155.036	152.211	149.542	147.016
238	177.915	173.697	169.766	166.092	162.652	159.421	156.383	153.518	150.812	148.253
239	179.589	175.305	171.315	167.587	164.096	160.820	157.738	154.834	152.091	149.497
240	181.276	176.927	172.876	169.094	165.552	162.229	159.104	156.159	153.379	150.750
241	182.978	178.563	174.451	170.612	167.019	163.648	160.479	157.494	154.676	152.011
242	184.695	180.211	176.038	172.142	168.497	165.078	161.864	158.838	155.981	153.281
243	186.427	181.874	177.637	173.684	169.986	166.518	163.260	160.191	157.296	154.559
244	188.173	183.551	179.250	175.238	171.487	167.970	164.665	161.554	158.619	155.845
245	189.935	185.241	180.876	176.805	172.999	169.431	166.080	162.926	159.952	157.141
246	191.712	186.946	182.515	178.384	174.522	170.904	167.506	164.309	161.293	158.445
247	193.504	188.665	184.167	179.975	176.058	172.388	168.942	165.701	162.644	159.758
248	195.312	190.399	185.833	181.579	177.605	173.883	170.389	167.102	164.005	161.079
249	197.136	192.147	187.513	183.196	179.164	175.389	171.846	168.514	165.374	162.410
250	198.976	193.911	189.206	184.826	180.735	176.906	173.314	169.936	166.754	163.749
251	200.833	195.689	190.914	186.468	182.319	178.435	174.792	171.368	168.142	165.098
252	202.705	197.482	192.635	188.124	183.914	179.976	176.282	172.810	169.541	166.456
253	204.594	199.291	194.371	189.794	185.523	181.528	177.782	174.263	170.949	167.823
254	206.501	201.116	196.122	191.476	187.143	183.092	179.294	175.726	172.367	169.200

					N					
S	500	510	520	530	540	550	560	570	580	590
255	208.424	202.956	197.886	193.172	188.777	184.667	180.816	177.199	173.795	170.585
256	210.364	204.812	199.666	194.882	190.423	186.255	182.350	178.684	175.234	171.981
257	212.322	206.684	201.461	196.606	192.082	187.855	183.895	180.179	176.682	173.386
258	214.297	208.573	203.270	198.344	193.754	189.467	185.452	181.684	178.140	174.800
259	216.291	210.477	205.095	200.096	195.440	191.091	187.021	183.201	179.609	176.224
260	218.302	212.399	206.935	201.862	197.139	192.728	188.601	184.729	181.088	177.659
261	220.332	214.338	208.791	203.643	198.851	194.378	190.193	186.267	182.578	179.103
262	222.380	216.293	210.663	205.439	200.577	196.040	191.797	187.817	184.078	180.557
263	224.447	218.266	212.551	207.249	202.317	197.716	193.413	189.379	185.589	182.021
264	226.533	220.256	214.454	209.074	204.071	199.404	195.041	190.951	187.110	183.495
265	228.638	222.264	216.375	210.915	205.838	201.105	196.681	192.536	188.643	184.979
266	230.763	224.290	218.311	212.771	207.621	202.820	198.334	194.132	190.186	186.474
267	232.907	226.334	220.265	214.642	209.417	204.548	200.000	195.740	191.741	187.980
268	235.072	228.397	222.235	216.529	211.228	206.290	201.678	197.359	193.307	189.495
269	237.256	230.478	224.223	218.432	213.054	208.045	203.369	198.991	194.884	191.022
270	239.462	232.577	226.227	220.350	214.894	209.815	205.073	200.635	196.472	192.559
271	241.688	234.696	228.250	222.285	216.750	211.598	206.790	202.291	198.072	194.107
272	243.935	236.834	230.290	224.237	218.621	213.395	208.520	203.959	199.684	195.666
273	246.203	238.992	232.348	226.205	220.507	215.207	210.263	205.640	201.307	197.236
274	248.493	241.169	234.424	228.189	222.409	217.033	212.020	207.334	202.942	198.818
275	250.805	243.366	236.518	230.191	224.326	218.874	213.791	209.040	204.589	200.410
276	253.138	245.584	238.631	232.210	226.259	220.729	215.575	210.760	206.249	202.014
277	255.495	247.822	240.763	234.246	228.209	222.600	217.374	212.492	207.920	203.629
278	257.874	250.081	242.914	236.300	230.174	224.485	219.186	214.237	209.604	205.256
279	260.276	252.361	245.084	238.371	232.156	226.386	221.013	215.996	211.300	206.895
280	262.702	254.662	247.274	240.460	234.155	228.302	222.853	217.767	213.008	208.545
281	265.151	256.985	249.484	242.568	236.171	230.234	224.709	219.553	214.730	210.207
282	267.624	259.329	251.713	244.694	238.203	232.181	226.579	221.352	216.464	211.881
283	270.122	261.696	253.963	246.839	240.253	234.145	228.464	223.165	218.211	213.567
284	272.644	264.085	256.233	249.002	242.320	236.124	230.363	224.992	219.971	215.266
285	275.191	266.497	258.525	251.185	244.404	238.120	232.278	226.833	221.744	216.977
286	277.764	268.932	260.837	253.387	246.507	240.132	234.208	228.688	223.531	218.700
287	280.362	271.391	263.170	255.608	248.627	242.161	236.154	230.558	225.331	220.436
288	282.987	273.873	265.525	257.849	250.766	244.207	238.115	232.442	227.144	222.185

					N					
S	500	510	520	530	540	550	560	570	580	590
289	285.637	276.379	267.902	260.110	252.923	246.269	240.092	234.341	228.971	223.946
290	288.315	278.909	270.301	262.392	255.098	248.349	242.085	236.255	230.812	225.721
291	291.020	281.464	272.723	264.694	257.293	250.447	244.094	238.183	232.668	227.508
292	293.752	284.044	275.167	267.017	259.506	252.562	246.120	240.127	234.537	229.309
293	296.513	286.649	277.634	269.361	261.739	254.694	248.162	242.086	236.420	231.123
294		289.280	280.125	271.726	263.992	256.845	250.220	244.061	238.318	232.951
295		291.937	282.639	274.112	266.264	259.014	252.296	246.051	240.231	234.792
296		294.620	285.177	276.521	268.556	261.202	254.388	248.057	242.158	236.647
297		297.330	287.739	278.952	270.869	263.408	256.498	250.079	244.100	238.516
298		300.067	290.326	281.405	273.202	265.633	258.625	252.118	246.057	240.399
299		302.832	292.938	283.880	275.556	267.877	260.770	254.172	248.030	242.296
300			295.575	286.379	277.931	270.140	262.933	256.244	250.018	244.207
301			298.238	288.901	280.327	272.423	265.113	258.331	252.021	246.133
302			300.926	291.447	282.745	274.726	267.312	260.436	254.040	248.074
303			303.641	294.016	285.184	277.049	269.530	262.558	256.075	250.029
304			306.383	296.610	287.646	279.392	271.766	264.697	258.125	251.999
305				299.228	290.130	281.755	274.021	266.854	260.192	253.984
306				301.871	292.636	284.140	276.295	269.028	262.276	255.985
307				304.540	295.166	286.545	278.588	271.220	264.376	258.001
308				307.233	297.719	288.972	280.901	273.430	266.492	260.032
309				309.953	300.295	291.420	283.234	275.658	268.626	262.079
310				312.699	302.896	293.890	285.586	277.905	270.776	264.142
311					305.520	296.382	287.959	280.170	272.944	266.221
312					308.169	298.896	290.353	282.455	275.130	268.317
313					310.843	301.433	292.767	284.758	277.333	270.428
314					313.541	303.993	295.202	287.081	279.554	272.556
315					316.266	306.576	297.659	289.423	281.793	274.701
316					319.016	309.183	300.137	291.785	284.050	276.863
317						311.813	302.637	294.168	286.326	279.042
318						314.468	305.158	296.570	288.620	281.239
319						317.147	307.703	298.993	290.933	283.452
320						319.850	310.269	301.436	293.266	285.684
321						322.579	312.859	303.901	295.617	287.933
322						325.333	315.472	306.387	297.989	290.201

S						N			
	500	510	520	530	540	550	560	570	580	590
323							318.108	308.894	300.380	292.486
324							320.768	311.423	302.791	294.790
325							323.452	313.974	305.222	297.113
326							326.160	316.547	307.673	299.455
327							328.893	319.143	310.146	301.815
328								321.762	312.639	304.196
329								324.404	315.154	306.595
330								327.069	317.690	309.015
331								329.758	320.248	311.454
332								332.470	322.827	313.913
333								335.208	325.429	316.394
334									328.054	318.894
335									330.701	321.416
336									333.371	323.959
337									336.065	326.523
338									338.782	329.109
339									341.523	331.716
340										334.346
341										336.999
342										339.674
343										342.372
344										345.094
345										347.839

S				N					
	600	650	700	750	800	850	900	950	1000
1	0.117	0.116	0.115	0.114	0.113	0.112	0.111	0.110	0.110
2	0.258	0.255	0.252	0.250	0.247	0.245	0.243	0.242	0.240
3	0.412	0.407	0.402	0.398	0.394	0.390	0.387	0.384	0.381
4	0.576	0.568	0.561	0.555	0.549	0.544	0.539	0.535	0.530
5	0.747	0.737	0.728	0.719	0.712	0.705	0.698	0.692	0.686
6	0.927	0.913	0.901	0.891	0.881	0.872	0.863	0.856	0.848
7	1.112	1.096	1.081	1.068	1.056	1.044	1.034	1.024	1.015
8	1.304	1.284	1.266	1.250	1.236	1.222	1.210	1.198	1.187

	N								
S	600	650	700	750	800	850	900	950	1000
9	1.502	1.478	1.457	1.438	1.421	1.405	1.390	1.376	1.364
10	1.705	1.677	1.653	1.630	1.610	1.592	1.575	1.559	1.544
11	1.912	1.881	1.853	1.827	1.804	1.783	1.764	1.746	1.729
12	2.125	2.089	2.057	2.029	2.002	1.978	1.956	1.936	1.917
13	2.343	2.302	2.266	2.234	2.205	2.178	2.153	2.130	2.109
14	2.564	2.519	2.479	2.443	2.411	2.381	2.353	2.328	2.304
15	2.791	2.741	2.696	2.656	2.620	2.587	2.557	2.529	2.503
16	3.021	2.966	2.917	2.873	2.834	2.797	2.764	2.733	2.705
17	3.255	3.195	3.142	3.094	3.050	3.011	2.974	2.941	2.910
18	3.494	3.428	3.370	3.318	3.270	3.227	3.188	3.151	3.118
19	3.736	3.665	3.602	3.545	3.494	3.447	3.404	3.365	3.328
20	3.983	3.906	3.837	3.776	3.721	3.670	3.624	3.581	3.542
21	4.233	4.150	4.076	4.010	3.951	3.896	3.847	3.801	3.758
22	4.487	4.398	4.319	4.248	4.184	4.125	4.072	4.023	3.978
23	4.744	4.649	4.564	4.488	4.420	4.357	4.300	4.248	4.199
24	5.006	4.903	4.813	4.732	4.659	4.592	4.531	4.475	4.424
25	5.270	5.161	5.065	4.979	4.901	4.830	4.765	4.706	4.651
26	5.539	5.423	5.320	5.228	5.146	5.070	5.002	4.939	4.880
27	5.810	5.687	5.578	5.481	5.393	5.314	5.241	5.174	5.113
28	6.085	5.955	5.840	5.737	5.644	5.560	5.483	5.412	5.347
29	6.364	6.226	6.104	5.995	5.897	5.808	5.727	5.653	5.584
30	6.646	6.500	6.372	6.257	6.154	6.060	5.974	5.896	5.823
31	6.931	6.778	6.642	6.521	6.412	6.314	6.224	6.141	6.065
32	7.220	7.059	6.916	6.789	6.674	6.570	6.476	6.389	6.309
33	7.512	7.342	7.192	7.059	6.938	6.829	6.730	6.639	6.555
34	7.808	7.629	7.472	7.331	7.205	7.091	6.987	6.892	6.804
35	8.106	7.919	7.754	7.607	7.475	7.355	7.247	7.147	7.055
36	8.408	8.212	8.039	7.885	7.747	7.622	7.508	7.404	7.308
37	8.713	8.508	8.327	8.166	8.022	7.891	7.772	7.664	7.563
38	9.021	8.807	8.618	8.450	8.299	8.163	8.039	7.925	7.821
39	9.333	9.109	8.911	8.736	8.579	8.437	8.308	8.190	8.081
40	9.647	9.414	9.208	9.025	8.861	8.714	8.579	8.456	8.343
41	9.965	9.721	9.507	9.317	9.146	8.992	8.852	8.724	8.607
42	10.286	10.032	9.809	9.611	9.434	9.274	9.128	8.995	8.873

	N								
S	600	650	700	750	800	850	900	950	1000
43	10.610	10.346	10.114	9.908	9.724	9.557	9.406	9.268	9.141
44	10.938	10.663	10.422	10.208	10.016	9.844	9.687	9.543	9.411
45	11.268	10.983	10.732	10.510	10.311	10.132	9.969	9.820	9.684
46	11.602	11.305	11.045	10.815	10.609	10.423	10.254	10.100	9.958
47	11.938	11.631	11.361	11.122	10.908	10.716	10.541	10.381	10.235
48	12.278	11.959	11.679	11.432	11.211	11.011	10.830	10.665	10.514
49	12.621	12.290	12.001	11.744	11.515	11.309	11.122	10.951	10.794
50	12.967	12.624	12.325	12.059	11.822	11.609	11.415	11.239	11.077
51	13.316	12.961	12.651	12.377	12.132	11.911	11.711	11.529	11.362
52	13.669	13.301	12.981	12.697	12.444	12.216	12.009	11.821	11.648
53	14.024	13.644	13.313	13.019	12.758	12.523	12.310	12.115	11.937
54	14.382	13.990	13.647	13.345	13.075	12.832	12.612	12.411	12.228
55	14.744	14.338	13.985	13.672	13.394	13.143	12.916	12.710	12.520
56	15.108	14.690	14.325	14.002	13.715	13.457	13.223	13.010	12.815
57	15.476	15.044	14.667	14.335	14.039	13.773	13.532	13.313	13.112
58	15.847	15.401	15.013	14.670	14.365	14.091	13.843	13.617	13.410
59	16.221	15.761	15.361	15.008	14.693	14.411	14.156	13.924	13.711
60	16.598	16.124	15.711	15.348	15.024	14.734	14.471	14.232	14.013
61	16.978	16.490	16.065	15.690	15.357	15.059	14.788	14.543	14.318
62	17.361	16.858	16.421	16.035	15.693	15.386	15.108	14.855	14.624
63	17.747	17.230	16.779	16.383	16.031	15.715	15.429	15.170	14.932
64	18.137	17.604	17.141	16.733	16.371	16.046	15.753	15.486	15.242
65	18.529	17.981	17.505	17.086	16.714	16.380	16.079	15.805	15.555
66	18.925	18.361	17.871	17.441	17.058	16.716	16.407	16.126	15.869
67	19.323	18.744	18.240	17.798	17.406	17.054	16.736	16.448	16.185
68	19.725	19.129	18.612	18.158	17.755	17.394	17.068	16.773	16.502
69	20.130	19.518	18.987	18.521	18.107	17.737	17.403	17.099	16.822
70	20.538	19.909	19.364	18.885	18.461	18.081	17.739	17.428	17.144
71	20.949	20.303	19.744	19.253	18.818	18.428	18.077	17.758	17.467
72	21.364	20.701	20.126	19.623	19.176	18.777	18.417	18.091	17.793
73	21.781	21.101	20.512	19.995	19.537	19.128	18.760	18.425	18.120
74	22.202	21.503	20.899	20.370	19.901	19.482	19.104	18.762	18.449
75	22.625	21.909	21.290	20.747	20.267	19.837	19.451	19.100	18.780
76	23.052	22.318	21.683	21.127	20.635	20.195	19.799	19.441	19.113

					N				
S	600	650	700	750	800	850	900	950	1000
77	23.482	22.729	22.079	21.509	21.005	20.555	20.150	19.783	19.448
78	23.915	23.144	22.477	21.894	21.378	20.917	20.503	20.127	19.785
79	24.352	23.561	22.878	22.281	21.753	21.282	20.858	20.473	20.123
80	24.791	23.981	23.282	22.671	22.130	21.648	21.214	20.822	20.464
81	25.234	24.404	23.689	23.063	22.510	22.017	21.573	21.172	20.806
82	25.680	24.831	24.098	23.458	22.892	22.388	21.934	21.524	21.150
83	26.129	25.259	24.510	23.855	23.276	22.761	22.297	21.878	21.496
84	26.582	25.691	24.924	24.254	23.663	23.136	22.663	22.234	21.844
85	27.038	26.126	25.341	24.656	24.052	23.513	23.030	22.592	22.194
86	27.496	26.564	25.761	25.061	24.443	23.893	23.399	22.952	22.545
87	27.959	27.005	26.184	25.468	24.837	24.275	23.770	23.314	22.899
88	28.424	27.448	26.609	25.878	25.233	24.659	24.144	23.678	23.254
89	28.893	27.895	27.037	26.290	25.631	25.045	24.519	24.044	23.611
90	29.365	28.344	27.468	26.704	26.032	25.433	24.897	24.412	23.970
91	29.840	28.797	27.901	27.122	26.435	25.824	25.276	24.781	24.331
92	30.318	29.253	28.338	27.541	26.840	26.217	25.658	25.153	24.694
93	30.800	29.711	28.776	27.963	27.248	26.612	26.041	25.527	25.059
94	31.285	30.173	29.218	28.388	27.658	27.009	26.427	25.902	25.425
95	31.774	30.637	29.662	28.815	28.070	27.408	26.815	26.280	25.793
96	32.266	31.105	30.110	29.245	28.485	27.810	27.205	26.659	26.164
97	32.761	31.575	30.560	29.677	28.902	28.213	27.597	27.041	26.536
98	33.259	32.049	31.012	30.112	29.321	28.619	27.991	27.424	26.910
99	33.761	32.525	31.468	30.549	29.743	29.027	28.387	27.809	27.285
100	34.267	33.005	31.926	30.989	30.167	29.438	28.785	28.197	27.663
101	34.775	33.488	32.387	31.432	30.594	29.850	29.185	28.586	28.042
102	35.287	33.974	32.851	31.877	31.022	30.265	29.588	28.977	28.424
103	35.803	34.462	33.317	32.324	31.454	30.682	29.992	29.371	28.807
104	36.322	34.954	33.786	32.775	31.887	31.101	30.399	29.766	29.192
105	36.844	35.449	34.259	33.227	32.323	31.523	30.807	30.163	29.579
106	37.370	35.948	34.734	33.683	32.762	31.946	31.218	30.562	29.967
107	37.900	36.449	35.211	34.141	33.202	32.372	31.630	30.963	30.358
108	38.433	36.953	35.692	34.601	33.646	32.800	32.045	31.366	30.751
109	38.969	37.461	36.176	35.064	34.091	33.231	32.462	31.771	31.145
110	39.509	37.971	36.662	35.530	34.539	33.663	32.881	32.178	31.541

S	N								
	600	650	700	750	800	850	900	950	1000
111	40.053	38.485	37.151	35.998	34.990	34.098	33.302	32.587	31.939
112	40.600	39.002	37.643	36.469	35.442	34.535	33.726	32.998	32.339
113	41.150	39.523	38.138	36.943	35.898	34.974	34.151	33.411	32.741
114	41.704	40.046	38.636	37.419	36.355	35.416	34.578	33.826	33.144
115	42.262	40.573	39.137	37.898	36.815	35.860	35.008	34.242	33.550
116	42.824	41.102	39.640	38.379	37.278	36.306	35.439	34.661	33.957
117	43.389	41.635	40.147	38.863	37.743	36.754	35.873	35.082	34.367
118	43.958	42.172	40.656	39.350	38.210	37.205	36.309	35.505	34.778
119	44.530	42.711	41.169	39.840	38.680	37.658	36.747	35.930	35.191
120	45.106	43.254	41.684	40.332	39.153	38.113	37.187	36.357	35.606
121	45.686	43.800	42.202	40.827	39.628	38.570	37.629	36.785	36.023
122	46.269	44.350	42.723	41.324	40.105	39.030	38.074	37.216	36.441
123	46.857	44.902	43.248	41.825	40.585	39.492	38.520	37.649	36.862
124	47.448	45.458	43.775	42.327	41.067	39.956	38.969	38.084	37.284
125	48.042	46.018	44.305	42.833	41.552	40.423	39.420	38.521	37.709
126	48.641	46.580	44.838	43.342	42.039	40.892	39.873	38.959	38.135
127	49.243	47.146	45.374	43.853	42.528	41.363	40.328	39.400	38.563
128	49.849	47.716	45.914	44.366	43.021	41.837	40.785	39.288	38.993
129	50.459	48.289	46.456	44.883	43.515	42.313	41.245	40.288	39.425
130	51.073	48.865	47.001	45.402	44.013	42.791	41.706	40.735	39.859
131	51.691	49.444	47.549	45.925	44.513	43.272	42.170	41.184	40.295
132	52.313	50.028	48.101	46.449	45.015	43.754	42.636	41.635	40.733
133	52.938	50.614	48.655	46.977	45.520	44.240	43.104	42.088	41.172
134	53.568	51.204	49.213	47.508	46.027	44.727	43.574	42.543	41.614
135	54.201	51.798	49.773	48.041	46.537	45.217	44.047	43.000	42.057
136	54.839	52.394	50.337	48.577	47.050	45.710	44.521	43.459	42.502
137	55.480	52.995	50.904	49.116	47.565	46.204	44.998	43.920	42.950
138	56.126	53.599	51.474	49.658	48.083	46.701	45.477	44.384	43.399
139	56.776	54.206	52.047	50.202	48.602	47.201	45.969	44.849	43.850
140	57.429	54.817	52.624	50.749	49.126	47.703	46.442	45.316	44.303
141	58.087	55.432	53.203	51.300	49.651	48.207	46.928	45.786	44.758
142	58.749	56.050	53.786	51.853	50.179	48.713	47.416	46.257	45.215
143	59.415	56.672	54.372	52.409	50.710	49.222	47.906	46.731	45.674
144	60.085	57.298	54.961	52.968	51.244	49.734	48.398	47.207	46.135

S	N								
	600	650	700	750	800	850	900	950	1000
145	60.759	57.927	55.553	53.530	51.780	50.248	48.893	47.684	46.598
146	61.438	58.559	56.149	54.094	52.318	50.764	49.390	48.164	47.062
147	62.120	59.196	56.748	54.662	52.859	51.283	49.889	48.646	47.529
148	62.807	59.836	57.350	55.233	53.403	51.804	50.390	49.130	47.998
149	63.499	60.480	57.955	55.806	53.950	52.327	50.894	49.616	48.468
150	64.194	61.127	58.564	56.383	54.499	52.853	51.400	50.104	48.941
151	64.894	61.779	59.176	56.962	55.051	53.382	51.908	50.595	49.415
152	65.598	62.434	59.791	57.544	55.606	53.913	52.418	51.087	49.892
153	**66.307**	63.093	60.410	58.130	56.163	54.446	52.931	51.582	50.370
154	67.020	63.755	61.032	58.718	56.723	54.982	53.446	52.078	50.851
155	67.737	64.422	61.657	59.309	57.286	55.521	53.963	52.577	51.333
156	68.459	65.092	62.286	59.904	57.852	56.062	54.483	53.078	51.818
157	69.185	65.766	62.918	60.501	58.420	56.605	55.005	53.581	52.304
158	69.916	66.444	63.553	61.102	58.991	57.151	55.529	54.086	52.793
159	70.651	67.126	64.192	61.705	59.565	57.699	56.056	54.594	53.283
160	71.391	67.812	64.835	62.312	60.141	58.250	56.584	55.103	53.776
161	72.135	68.502	65.480	62.921	60.721	58.804	57.116	55.615	54.270
162	72.884	69.195	66.130	63.534	61.303	59.360	57.649	56.129	54.767
163	73.638	69.893	66.782	64.150	61.888	59.918	58.185	56.645	55.265
164	74.396	70.595	67.439	64.769	62.475	60.479	58.723	57.163	55.766
165	75.159	**71.300**	68.098	65.391	63.066	61.043	59.264	57.683	56.269
166	75.927	72.010	68.762	66.016	63.659	61.609	59.807	58.206	56.773
167	76.699	72.724	69.429	66.645	64.255	62.178	60.352	58.731	57.280
168	77.476	73.442	70.099	67.276	64.854	62.750	60.899	59.258	57.789
169	78.258	74.164	70.773	67.911	65.456	63.324	61.449	59.787	58.299
170	79.045	74.890	71.451	68.549	66.061	63.900	62.002	60.318	58.812
171	79.837	75.620	72.132	69.190	66.669	64.479	62.557	60.852	59.327
172	80.633	76.354	72.817	69.834	67.279	65.061	63.114	61.388	59.844
173	81.435	77.093	73.505	70.482	67.893	65.646	63.674	61.926	60.363
174	82.241	77.836	74.197	71.132	68.509	66.233	64.236	62.466	60.884
175	83.053	78.583	74.893	71.787	69.128	66.823	64.800	63.008	61.407
176	83.869	79.334	75.593	72.444	69.750	67.415	65.367	63.553	61.933
177	84.690	80.090	76.296	73.104	70.376	68.010	65.936	64.100	62.460
178	85.517	80.850	**77.003**	73.768	71.004	68.608	66.508	64.649	62.989

					N				
S	600	650	700	750	800	850	900	950	1000
179	86.349	81.614	77.714	74.436	71.635	69.208	67.082	65.201	63.521
180	87.185	82.383	78.428	75.106	72.269	69.812	67.659	65.754	64.054
181	88.027	83.155	79.146	75.780	72.609	70.417	68.238	66.310	64.590
182	88.874	83.933	79.869	76.457	73.546	71.026	68.820	66.869	65.128
183	89.727	84.715	80.595	77.138	74.189	71.637	69.404	67.529	65.668
184	90.584	85.501	81.324	77.822	74.835	72.251	69.991	67.992	66.210
185	91.447	86.292	82.058	78.509	75.484	72.868	70.580	68.557	66.754
186	92.315	87.087	82.796	79.200	76.136	73.488	71.171	69.125	67.301
187	93.189	87.887	83.537	79.894	76.791	74.110	71.765	69.695	67.849
188	94.068	88.691	84.282	80.592	77.449	74.735	72.362	70.267	68.400
189	94.953	89.500	85.032	81.293	78.111	75.363	72.961	70.841	68.953
190	95.843	90.313	85.785	**81.998**	78.775	75.993	73.563	71.418	69.507
191	96.738	91.131	86.542	82.706	79.442	76.627	74.167	71.997	70.065
192	97.639	91.954	87.304	83.417	80.113	77.263	74.774	72.578	70.624
193	98.546	92.782	88.069	84.133	80.787	77.902	75.383	73.162	71.185
194	99.458	93.614	88.838	84.851	81.464	78.544	75.995	73.748	71.749
195	100.376	94.451	89.612	85.574	82.144	79.188	76.610	74.337	72.315
196	101.300	95.292	90.389	86.299	82.827	79.836	77.227	74.928	72.883
197	102.229	96.139	91.171	87.029	83.514	80.486	77.847	75.521	73.453
198	103.164	96.990	91.957	87.762	84.203	81.140	78.469	76.117	74.025
199	104.105	97.847	92.746	88.498	84.896	81.796	79.094	76.715	74.600
200	105.052	98.708	93.541	89.239	85.592	82.455	79.722	77.315	75.177
201	106.005	99.574	94.339	89.983	86.291	83.117	80.352	77.918	75.756
202	106.964	100.445	95.141	90.730	**86.994**	83.782	80.985	78.523	76.337
203	107.929	101.320	95.948	91.482	87.700	84.449	81.620	79.131	76.920
204	108.899	102.201	96.759	92.237	88.409	85.120	82.258	79.741	77.506
205	109.876	103.087	97.574	92.995	89.121	85.794	82.899	80.354	78.094
206	110.859	103.978	98.394	93.758	89.837	86.471	83.543	80.969	78.684
207	111.848	104.875	99.218	94.524	90.556	87.150	84.189	81.586	79.277
208	112.844	105.776	100.046	95.294	91.278	87.833	84.838	82.206	79.872
209	113.845	106.682	100.879	96.068	92.004	88.518	85.489	82.829	80.469
210	114.853	107.594	101.716	96.846	92.733	89.207	86.144	83.453	81.068
211	115.867	108.511	102.558	97.627	93.466	89.899	86.801	84.081	81.670
212	116.888	109.433	103.404	98.412	94.201	90.593	87.461	84.711	82.274

					N				
S	600	650	700	750	800	850	900	950	1000
213	117.915	110.360	104.254	99.202	94.941	91.291	88.123	85.343	82.880
214	118.948	111.293	105.109	99.995	95.683	91.992	88.788	85.978	83.488
215	119.988	112.231	105.969	100.792	96.430	**92.695**	89.456	86.615	84.099
216	121.035	113.175	106.833	101.593	97.179	93.402	90.127	87.255	84.712
217	122.088	114.124	107.701	102.398	97.932	94.112	90.801	87.898	85.328
218	123.148	115.078	108.575	103.207	98.689	94.825	91.477	88.543	85.945
219	124.215	116.038	109.453	104.019	99.449	95.542	92.156	89.190	86.566
220	125.288	117.003	110.335	104.836	100.212	96.261	92.838	89.840	87.188
221	126.368	117.974	111.222	105.657	100.979	96.983	93.523	90.493	87.813
222	127.455	118.951	112.114	106.482	101.750	97.709	94.211	91.148	88.440
223	128.549	119.933	113.011	107.311	102.524	98.438	94.902	91.806	89.070
224	129.650	120.921	113.912	108.144	103.302	99.170	95.595	92.467	89.701
225	130.758	121.915	114.819	108.982	104.083	99.905	96.291	93.130	90.336
226	131.873	122.914	115.730	109.823	104.868	100.643	96.990	93.795	90.972
227	132.996	123.919	116.646	110.669	105.657	101.385	**97.692**	94.463	91.612
228	134.125	124.930	117.566	111.518	106.449	102.130	98.397	95.134	92.253
229	135.262	125.947	118.492	112.372	107.245	102.878	99.105	95.808	92.897
230	136.406	126.970	119.423	113.230	108.045	103.629	99.816	96.484	93.543
231	137.557	127.999	120.358	114.093	108.848	104.383	100.529	97.163	94.192
232	138.716	129.034	121.299	114.959	109.655	105.141	101.246	97.844	94.843
233	139.882	130.074	122.245	115.830	110.466	105.902	101.966	98.529	95.497
234	141.055	131.121	123.195	116.706	111.280	106.667	102.688	99.215	96.153
235	142.237	132.174	124.151	117.585	112.099	107.435	103.414	99.905	96.812
236	143.426	133.233	125.112	118.469	112.921	108.206	104.142	100.597	97.473
237	144.622	134.299	126.078	119.358	113.746	108.980	104.874	101.292	98.136
238	145.827	135.370	127.049	120.250	114.576	109.758	105.608	101.990	98.802
239	147.039	136.448	128.026	121.147	115.410	110.540	106.346	102.690	99.470
240	148.259	137.532	129.007	122.049	116.247	111.324	107.086	**103.393**	100.141
241	149.487	138.623	129.994	122.955	117.088	112.112	107.830	104.099	100.815
242	150.723	139.720	130.986	123.866	117.933	112.904	108.577	104.808	101.491
243	151.967	140.823	131.984	124.781	118.783	113.699	109.327	105.519	102.169
244	153.220	141.933	132.987	125.700	119.636	114.497	110.079	106.234	102.850
245	154.480	143.049	133.995	126.625	120.493	115.299	110.835	106.951	103.534
246	155.749	144.172	135.009	127.554	121.353	116.104	111.594	107.670	104.220

S	600	650	700	750	800	850	900	950	1000
247	157.026	145.302	136.028	128.487	122.218	116.913	112.357	108.393	104.909
248	158.312	146.438	137.053	129.425	123.087	117.726	113.122	109.119	105.600
249	159.606	147.581	138.083	130.368	123.960	118.542	113.890	109.847	106.294
250	160.909	148.731	139.119	131.315	124.837	119.361	114.662	110.578	106.990
251	162.220	149.888	140.160	132.268	125.718	120.184	115.437	111.312	**107.689**
252	163.540	151.052	141.207	133.225	126.604	121.011	116.215	112.049	108.391
253	164.869	152.222	142.260	134.186	127.493	121.841	116.996	112.788	109.095
254	166.206	153.400	143.319	135.153	128.386	122.675	117.780	113.531	109.802
255	167.553	154.584	144.383	136.124	129.284	123.513	118.568	114.277	110.511
256	168.908	155.776	145.453	137.100	130.186	124.354	119.359	115.025	111.223
257	170.273	156.974	146.529	138.082	131.092	125.198	120.153	115.776	111.938
258	171.646	158.180	147.610	139.068	132.002	126.047	120.950	116.530	112.655
259	173.029	159.394	148.698	140.059	132.916	126.899	121.751	117.288	113.375
260	174.422	160.614	149.791	141.054	133.835	127.755	122.555	118.048	114.098
261	175.823	161.842	150.891	142.055	134.758	128.615	123.362	118.811	114.823
262	177.234	163.077	151.996	143.061	135.685	129.478	124.172	119.577	115.551
263	178.655	164.320	153.108	144.072	136.617	130.345	124.986	120.346	116.282
264	180.085	165.570	154.225	145.089	137.553	131.216	125.804	121.118	117.016
265	181.525	166.827	155.349	146.110	138.493	132.091	126.624	121.893	117.752
266	182.975	168.093	156.479	147.136	139.437	132.970	127.448	122.671	118.491
267	184.434	169.366	157.615	148.168	140.387	133.852	128.275	123.452	119.232
268	185.904	170.646	158.758	149.204	141.340	134.738	129.106	124.236	119.977
269	187.384	171.935	159.906	150.246	142.298	135.628	129.940	125.023	120.724
270	188.873	173.231	161.061	151.294	143.261	136.522	130.778	125.814	121.474
271	190.373	174.536	162.223	152.346	144.228	137.420	131.619	126.607	122.227
272	191.884	175.848	163.390	153.404	145.199	138.322	132.464	127.403	122.982
273	193.405	177.168	164.565	154.467	146.175	139.228	133.311	128.203	123.740
274	194.936	178.497	165.745	155.536	147.156	140.138	134.163	129.006	124.501
275	196.478	179.833	166.933	156.610	148.141	141.052	135.018	129.811	125.265
276	198.030	181.178	168.127	157.690	149.131	141.969	135.876	130.620	126.032
277	199.594	182.531	169.327	158.775	150.126	142.891	136.738	131.432	126.802
278	201.168	183.892	170.534	159.865	151.125	143.817	137.604	132.247	127.574
279	202.753	185.262	171.748	160.961	152.129	144.747	138.473	133.066	128.349
280	204.349	186.641	172.969	162.063	153.137	145.681	139.346	133.887	129.127

N

S	N 600	650	700	750	800	850	900	950	1000
281	205.957	188.027	174.196	163.170	154.151	146.619	140.222	134.712	129.908
282	207.576	189.423	175.431	164.283	155.169	147.561	141.102	135.540	130.692
283	209.206	190.827	176.672	165.402	156.192	148.508	141.986	136.371	131.479
284	210.848	192.240	177.920	166.526	157.220	149.458	142.873	137.206	132.269
285	212.501	193.662	179.176	167.656	158.252	150.413	143.764	138.043	133.061
286	214.166	195.092	180.438	168.792	159.290	151.372	144.659	138.884	133.857
287	215.843	196.532	181.707	169.934	160.333	152.335	145.557	139.729	134.656
288	217.532	197.980	182.984	171.081	161.380	153.302	146.459	140.576	135.457
289	219.233	199.438	184.268	172.235	162.432	154.274	147.365	141.427	136.261
290	220.946	200.905	185.559	173.394	163.490	155.250	148.274	142.281	137.069
291	222.671	202.381	186.857	174.560	164.552	156.230	149.187	143.139	137.879
292	224.409	203.867	188.163	175.731	165.620	157.215	150.104	143.999	138.693
293	226.159	205.361	189.476	176.909	166.692	158.204	151.025	144.864	139.509
294	227.922	206.866	190.797	178.092	167.770	159.197	151.950	145.731	140.328
295	229.697	208.380	192.125	179.282	168.853	160.195	152.878	146.602	141.151
296	231.486	209.903	193.460	180.478	169.940	161.197	153.810	147.476	141.976
297	233.288	211.437	194.804	181.680	171.034	162.204	154.747	148.354	142.805
298	235.102	212.980	196.155	182.888	172.132	163.215	155.687	149.235	143.636
299	236.930	214.533	197.513	184.103	173.236	164.230	156.631	150.120	144.471
300	238.772	216.096	198.880	185.324	174.344	165.250	157.578	151.008	145.309
301	240.626	217.669	200.254	186.551	175.459	166.275	158.530	151.900	146.150
302	242.495	219.252	201.636	187.785	176.578	167.304	159.486	152.795	146.994
303	244.377	220.845	203.027	189.025	177.703	168.337	160.446	153.693	147.841
304	246.274	222.449	204.425	190.272	178.833	169.376	161.409	154.595	148.691
305	248.184	224.063	205.831	191.525	179.969	170.419	162.377	155.501	149.544
306	250.108	225.687	207.246	192.784	181.110	171.466	163.349	156.410	150.401
307	252.047	227.322	208.668	194.051	182.257	172.518	164.325	157.323	151.261
308	254.001	228.968	210.099	195.324	183.409	173.575	165.305	158.239	152.124
309	255.969	230.624	211.539	196.603	184.567	174.637	166.289	159.159	152.990
310	257.952	232.292	212.986	197.890	185.730	175.703	167.277	160.083	153.859
311	259.949	233.970	214.442	199.183	186.899	176.775	168.269	161.010	154.732
312	261.962	235.659	215.907	200.483	188.074	177.851	169.265	161.941	155.608
313	263.991	237.360	217.380	201.790	189.254	178.931	170.266	162.875	156.487
314	266.034	239.071	218.862	203.104	190.440	180.017	171.271	163.813	157.369

					N				
S	600	650	700	750	800	850	900	950	1000
315	268.093	240.794	220.352	204.425	191.632	181.108	172.280	164.755	158.254
316	270.168	242.529	221.852	205.753	192.830	182.203	173.293	165.700	159.143
317	272.259	244.275	223.360	207.088	194.033	183.303	174.310	166.650	160.036
318	274.366	246.032	224.877	208.430	195.242	184.409	175.332	167.603	160.931
319	276.489	247.801	226.403	209.779	196.458	185.519	176.358	168.559	161.830
320	278.628	249.582	227.938	211.135	197.679	186.634	177.388	169.520	162.732
321	280.784	251.375	229.482	212.499	198.906	187.755	178.423	170.484	163.638
322	282.957	253.181	231.035	213.870	200.139	188.880	179.461	171.452	164.546
323	285.147	254.998	232.598	215.248	201.378	190.010	180.505	172.424	165.459
324	287.354	256.827	234.170	216.634	202.623	191.146	181.552	173.400	166.374
325	289.578	258.669	235.751	218.027	203.875	192.287	182.605	174.379	167.294
326	291.820	260.524	237.342	219.428	205.132	193.433	183.661	175.363	168.216
327	294.079	262.390	238.942	220.836	206.396	194.584	184.722	176.350	169.142
328	296.357	264.270	240.552	222.252	207.666	195.740	185.788	177.341	170.071
329	298.652	266.163	242.172	223.675	208.942	196.901	186.858	178.336	171.004
330	300.965	268.068	243.801	225.106	210.224	198.068	187.932	179.336	171.941
331	303.297	269.987	245.440	226.545	211.513	199.240	189.011	180.339	172.880
332	305.648	271.918	247.089	227.992	212.808	200.418	190.095	181.346	173.824
333	308.018	273.863	248.749	229.447	214.110	201.601	191.183	182.357	174.771
334	310.406	275.822	250.418	230.909	215.418	202.789	192.276	183.372	175.721
335	312.814	277.794	252.097	232.380	216.732	203.983	193.373	184.391	176.675
336	315.242	279.780	253.787	233.859	218.053	205.182	194.475	185.414	177.633
337	317.689	281.779	255.487	235.345	219.381	206.386	195.582	186.441	178.594
338	320.156	283.793	257.197	236.840	220.715	207.597	196.694	187.472	179.559
339	322.644	285.820	258.918	238.343	222.056	208.812	197.810	188.508	180.527
340	325.152	287.862	260.650	239.855	223.404	210.034	198.931	189.547	181.499
341	327.680	289.918	262.392	241.374	224.758	211.261	200.057	190.591	182.475
342	330.229	291.989	264.145	242.902	226.119	212.493	201.187	191.639	183.454
343	332.800	294.074	265.909	244.439	227.487	213.731	202.323	192.691	184.437
344	335.392	296.174	267.683	245.984	228.862	214.975	203.463	193.747	185.423
345	338.005	298.289	269.469	247.538	230.244	216.225	204.608	194.807	186.414
346	340.641	300.419	271.266	249.100	231.632	217.480	205.758	195.872	187.408
347	343.298	302.564	273.074	250.671	233.028	218.742	206.913	196.941	188.406
348	345.978	304.725	274.894	252.251	234.431	220.009	208.073	198.014	189.407

					N				
S	600	650	700	750	800	850	900	950	1000
349	348.681	306.901	276.724	253.839	235.841	221.282	209.238	199.092	190.413
350	351.406	309.093	278.567	255.437	237.258	222.561	210.408	200.173	191.422
351	354.155	311.300	280.421	257.043	238.682	223.846	211.583	201.260	192.435
352		313.524	282.286	258.658	240.113	225.136	212.763	202.350	193.451
353		315.763	284.164	260.283	241.552	226.433	213.948	203.445	194.472
354		318.019	286.053	261.917	242.998	227.736	215.138	204.544	195.496
355		320.292	287.954	263.559	244.452	229.045	216.334	205.648	196.525
356		322.581	289.868	265.212	245.913	230.360	217.534	206.756	197.557
357		324.887	291.793	266.873	247.381	231.682	218.740	207.869	198.593
358		327.209	292.731	268.544	248.857	233.009	219.951	208.986	199.633
359		329.549	295.681	270.225	250.340	234.343	221.167	210.108	200.677
360		331.907	297.644	271.915	251.831	235.683	222.389	211.234	201.725
361		334.281	299.620	273.614	253.330	237.029	223.615	212.365	202.777
362		336.674	301.608	275.324	254.836	238.381	224.847	213.500	203.833
363		339.084	303.609	277.043	256.351	239.740	226.085	214.640	204.893
364		341.512	305.623	278.772	257.873	241.106	227.328	215.784	205.957
365		343.959	307.650	280.511	259.403	242.477	228.576	216.933	207.024
366		346.423	309.690	282.260	260.940	243.856	229.830	218.087	208.097
367		348.907	311.744	284.019	262.486	245.241	231.089	219.246	209.173
368		351.409	313.811	285.788	264.040	246.632	232.354	220.409	210.253
369		353.931	315.891	287.568	265.602	248.030	233.624	221.577	211.337
370		356.471	317.985	289.357	267.172	249.435	234.900	222.750	212.426
371		359.031	320.093	291.158	268.751	250.846	236.181	223.927	213.518
372		361.610	322.215	292.968	270.337	252.264	237.468	225.109	214.615
373		364.210	324.351	294.790	271.932	253.689	238.761	226.296	215.716
374		366.829	326.501	296.622	273.536	255.121	240.059	227.488	216.821
375		369.469	328.666	298.464	275.147	256.559	241.363	228.685	217.930
376		372.129	330.845	300.318	276.768	258.005	242.673	229.887	219.044
377		374.810	333.038	302.182	278.396	259.457	243.988	231.094	220.162
378		377.511	335.246	304.058	280.034	260.916	245.310	232.305	221.284
379		380.234	337.469	305.944	281.680	262.383	246.637	233.522	222.411
380		382.978	339.707	307.842	283.335	263.856	247.970	234.744	223.542
381			341.960	309.751	284.998	265.336	249.309	235.970	224.677
382			344.229	311.671	286.671	266.824	250.654	237.202	225.816

S	600	650	700	750	800	850	900	950	1000
383			346.512	313.603	288.352	268.319	252.005	238.439	226.960
384			348.811	315.546	290.042	269.821	253.362	239.680	228.109
385			351.126	317.501	291.742	271.331	254.725	240.927	229.261
386			353.457	319.468	293.450	272.847	256.094	242.179	230.419
387			355.804	321.446	295.168	274.371	257.470	243.437	231.580
388			358.167	323.436	296.895	275.903	258.851	244.699	232.746
389			360.546	325.439	298.631	277.442	260.239	245.967	233.917
390			362.941	327.453	300.376	278.989	261.632	247.240	235.092
391			365.353	329.480	302.131	280.543	263.032	248.518	236.272
392			367.782	331.519	303.896	282.104	264.439	249.802	237.456
393			370.228	333.570	305.670	283.674	265.851	251.091	238.645
394			372.691	335.634	307.453	285.251	267.270	252.385	239.839
395			375.171	337.710	309.247	286.836	268.696	253.685	241.037
396			377.669	339.799	311.050	288.429	270.128	254.990	242.240
397			380.184	341.901	312.863	290.029	271.566	256.301	243.448
398			382.717	344.016	314.686	291.638	273.011	257.617	244.660
399			385.268	346.144	316.519	293.254	274.462	258.938	245.877
400			387.837	348.286	318.362	294.879	275.920	260.265	247.098
401			390.424	350.440	320.215	296.511	277.385	261.598	248.325
402			393.030	352.608	322.079	298.152	278.856	262.936	249.556
403			395.654	354.789	323.952	299.801	280.334	264.280	250.792
404			398.297	356.984	325.836	301.458	281.819	265.630	252.033
405			400.960	359.193	327.731	303.123	283.310	266.985	253.279
406			403.641	361.415	329.636	304.797	284.808	268.346	254.530
407			406.342	363.651	331.552	306.479	286.314	269.713	255.785
408			409.063	365.902	333.478	308.170	287.826	271.085	257.046
409			411.803	368.167	335.415	309.869	289.345	272.463	258.311
410				370.446	337.363	311.577	290.871	273.848	259.582
411				372.739	339.322	313.293	292.404	275.238	260.857
412				375.047	341.292	315.018	293.944	276.633	262.138
413				377.370	343.273	316.752	295.491	278.035	263.424
414				379.708	345.265	318.494	297.045	279.443	264.714
415				382.060	347.268	320.245	298.607	280.857	266.010
416				384.428	349.283	322.006	300.176	282.277	267.311

					N				
S	600	650	700	750	800	850	900	950	1000
417				386.811	351.309	323.775	301.752	283.703	268.617
418				389.209	353.347	325.553	303.335	285.135	269.928
419				391.623	355.397	327.341	304.926	286.573	271.245
420				394.052	357.458	329.137	306.524	288.017	272.566
421				396.498	359.531	330.943	308.130	289.468	273.893
422				398.959	361.615	332.758	309.743	290.925	275.226
423				401.436	363.712	334.582	311.363	292.388	276.563
424				403.930	365.821	336.416	312.992	293.857	277.906
425				406.440	367.942	338.260	314.628	295.332	279.255
426				408.966	370.075	340.113	316.271	296.815	280.608
427				411.509	372.221	341.975	317.923	298.303	281.967
428				414.069	374.379	343.847	319.582	299.798	283.332
429				416.646	376.550	345.729	321.249	301.299	284.702
430				419.240	378.733	347.621	322.924	302.807	286.077
431				421.851	380.930	349.522	324.607	304.322	287.458
432				424.480	383.139	351.434	326.297	305.843	288.845
433				427.126	385.361	353.356	327.996	307.370	290.237
434				429.791	387.596	355.287	329.703	308.904	291.635
435				432.473	389.844	357.229	331.418	310.445	293.039
436				435.173	392.106	359.181	333.142	311.993	294.448
437				437.892	394.381	361.144	334.873	313.548	295.863
438				440.629	396.669	363.117	336.613	315.109	297.283
439					398.971	365.100	338.361	316.677	298.710
440					401.287	367.094	340.118	318.252	300.142
441					403.617	369.098	341.883	319.834	301.580
442					405.961	371.113	343.656	321.423	303.024
443					408.318	373.139	345.438	323.020	304.473
444					410.690	375.176	347.229	324.623	305.929
445					413.077	377.224	349.028	326.233	307.391
446					415.477	379.282	350.836	327.850	308.858
447					417.893	381.352	352.653	329.475	310.332
448					420.323	383.433	354.479	331.106	311.812
449					422.767	385.525	356.314	332.745	313.297
450					425.227	387.629	358.157	334.392	314.789

S					N				
	600	650	700	750	800	850	900	950	1000
451					427.702	389.744	360.010	336.045	316.287
452					430.192	391.870	361.872	337.706	317.791
453					432.697	394.008	363.742	339.375	319.301
454					435.218	396.158	365.622	341.051	320.818
455					437.754	398.320	367.512	342.734	322.341
456					440.306	400.493	369.410	344.425	323.870
457					442.874	402.678	371.318	346.124	325.405
458					445.458	404.875	373.236	347.830	326.947
459					448.058	407.085	375.163	349.544	328.496
460					450.674	409.306	377.099	351.266	330.050
461					453.306	411.540	379.045	352.995	331.611
462					455.956	413.787	381.001	354.733	333.179
463					458.621	416.046	382.967	356.478	334.753
464					461.304	418.317	384.942	358.231	336.334
465					464.004	420.601	386.927	359.993	337.922
466					466.721	422.898	388.923	361.762	339.516
467					469.455	425.208	390.928	363.539	341.117
468						427.531	392.944	365.325	342.724
469						429.867	394.969	367.118	344.338
470						432.216	397.005	368.920	345.959
471						434.578	399.051	370.730	347.587
472						436.954	401.108	372.549	349.222
473						439.343	403.175	374.376	350.864
474						441.746	405.252	376.211	352.512
475						444.162	407.340	378.055	354.168
476						446.593	409.439	379.907	355.831
477						449.037	411.549	381.768	357.501
478						451.495	413.669	383.638	359.178
479						453.968	415.800	385.516	360.862
480						456.455	417.943	387.403	362.553
481						458.956	420.096	389.299	364.251
482						461.472	422.260	391.204	365.957
483						464.002	424.436	393.118	367.670
484						466.547	426.623	395.040	369.390

S	600	650	700	750	800	850	900	950	1000
485						469.107	428.821	396.972	371.118
486						471.682	431.031	398.913	372.853
487						474.272	433.252	400.863	374.596
488						476.877	435.485	402.822	376.346
489						479.497	437.730	404.790	378.104
490						482.133	439.986	406.768	379.870
491						484.785	442.255	408.755	381.643
492						487.452	444.535	410.752	383.424
493						490.136	446.827	412.758	385.212
494						492.835	449.131	414.774	387.008
495						495.550	451.448	416.799	388.813
496						498.282	453.777	418.834	390.625
497						501.030	456.118	420.879	392.445
498							458.472	422.934	394.273
499							460.839	424.998	396.109
500							463.218	427.073	397.953
501							465.610	429.157	399.805
502							468.014	431.252	401.665
503							470.432	433.356	403.534
504							472.863	435.471	405.410
505							475.307	437.597	407.295
506							477.764	439.732	409.189
507							480.235	441.878	411.091
508							482.719	444.035	413.001
509							485.216	446.202	414.920
510							487.727	448.379	416.847
511							490.252	450.568	418.793
512							492.791	452.767	420.728
513							495.355	454.977	422.681
514							497.911	457.198	424.644
515							500.492	459.430	426.615
516							503.088	461.673	428.595
517							505.698	463.927	430.583
518							508.322	466.193	432.581

	N								
S	600	650	700	750	800	850	900	950	1000
519							510.961	468.470	434.588
520							513.615	470.758	436.604
521							516.284	473.057	438.629
522							518.967	475.369	440.663
523							521.666	477.691	442.707
524							524.380	480.026	444.760
525							527.109	482.372	446.822
526							529.854	484.730	448.894
527								487.100	450.975
528								489.483	453.066
529								491.877	455.166
530								494.283	457.276
531								496.702	459.396
532								499.133	461.526
533								501.577	463.665
534								504.033	465.815
535								506.502	467.974
536								508.983	470.144
537								511.477	472.323
538								513.985	474.513
539								516.505	476.713
540								519.038	478.923
541								521.585	481.144
542								524.145	483.375
543								526.718	485.617
544								529.305	487.869
545								531.905	490.132
546								534.519	492.406
547								537.147	494.690
548								539.789	496.986
549								542.445	499.292
550								545.115	501.609
551								547.799	503.937
552								550.497	506.277

S	N								
	600	650	700	750	800	850	900	950	1000
553								553.210	508.628
554								555.937	510.990
555								558.679	513.363
556									515.748
557									518.144
558									520.552
559									522.972
560									525.403
561									527.846
562									530.302
563									532.769
564									535.248
565									537.740
566									540.243
567									542.759
568									545.288
569									547.829
570									550.382
571									552.949
572									555.528
573									558.120
574									560.724
575									563.342
576									565.973
577									568.617
578									571.275
579									573.946
580									576.630
581									579.328
582									582.040
583									584.766
584									587.505

S					N				
	1100	1200	1300	1400	1500	2000	3000	4000	5000
1	0.108	0.107	0.106	0.105	0.104	0.101	0.097	0.094	0.092
2	0.237	0.234	0.232	0.229	0.227	0.219	0.209	0.202	0.197
3	0.376	0.371	0.367	0.363	0.360	0.346	0.329	0.318	0.310
4	0.523	0.516	0.510	0.504	0.500	0.480	0.455	0.439	0.427
5	0.676	0.667	0.659	0.652	0.645	0.619	0.585	0.564	0.548
6	0.835	0.824	0.813	0.804	0.795	0.762	0.720	0.693	0.673
7	0.999	0.985	0.972	0.961	0.950	0.910	0.858	0.825	0.801
8	1.168	1.151	1.136	1.122	1.110	1.061	0.999	0.960	0.931
9	1.341	1.321	1.303	1.287	1.272	1.215	1.143	1.097	1.064
10	1.518	1.495	1.474	1.456	1.439	1.373	1.290	1.237	1.200
11	1.699	1.672	1.649	1.628	1.608	1.533	1.439	1.380	1.337
12	1.883	1.853	1.827	1.803	1.781	1.397	1.591	1.524	1.476
13	2.071	2.038	2.008	1.981	1.957	1.862	1.745	1.671	1.618
14	2.262	2.225	2.192	2.162	2.135	2.031	1.901	1.819	1.761
15	2.456	2.415	2.379	2.346	2.317	2.202	2.059	1.969	1.905
16	2.654	2.609	2.569	2.533	2.501	2.375	2.219	2.121	2.052
17	2.854	2.805	2.761	2.722	2.687	2.550	2.381	2.275	2.199
18	3.057	3.004	2.956	2.914	2.876	2.728	2.545	2.430	2.349
19	3.263	3.205	3.154	3.108	3.067	2.907	2.710	2.587	2.499
20	3.471	3.409	3.354	3.305	3.261	3.089	2.878	2.745	2.652
21	3.682	3.616	3.557	3.504	3.457	3.273	3.046	2.905	2.805
22	3.896	3.825	3.762	3.706	3.655	3.458	3.217	3.067	2.960
23	4.112	4.036	3.969	3.909	3.855	3.645	3.389	3.229	3.116
24	4.331	4.250	4.179	4.115	4.057	3.835	3.562	3.393	3.273
25	4.552	4.467	4.391	4.323	4.262	4.026	3.737	3.558	3.432
26	4.776	4.685	4.605	4.533	4.468	4.218	3.914	3.725	3.591
27	5.002	4.906	4.821	4.745	4.676	4.413	4.092	3.893	3.752
28	5.230	5.129	5.039	4.959	4.887	4.609	4.271	4.062	3.914
29	5.461	5.354	5.259	5.175	5.099	4.807	4.452	4.232	4.077
30	5.694	5.581	5.482	5.393	5.313	5.006	4.633	4.404	4.241
31	5.929	5.810	5.706	5.613	5.529	5.207	4.817	4.576	4.407
32	6.166	6.042	5.932	5.835	5.747	5.410	5.001	4.750	4.573
33	6.406	6.275	6.161	6.058	5.966	5.614	5.187	4.925	4.740
34	6.647	6.511	6.391	6.284	6.188	5.819	5.374	5.101	4.908

	N								
S	1100	1200	1300	1400	1500	2000	3000	4000	5000
35	6.891	6.748	6.623	6.511	6.411	6.026	5.562	5.278	5.077
36	7.137	6.988	6.857	6.740	6.636	6.235	5.752	5.456	5.247
37	7.385	7.230	7.093	6.971	6.862	6.445	5.943	5.635	5.419
38	7.635	7.473	7.331	7.204	7.091	6.657	6.134	5.815	5.591
39	7.887	7.718	7.570	7.439	7.321	6.870	6.328	5.996	5.763
40	8.141	7.966	7.812	7.675	7.552	7.084	6.522	6.178	5.937
41	8.397	8.215	8.055	7.913	7.786	7.300	6.717	6.361	6.112
42	8.655	8.466	8.300	8.153	8.021	7.517	6.913	6.545	6.288
43	8.915	8.719	8.547	8.394	8.257	7.735	7.111	6.730	6.464
44	9.177	8.974	8.795	8.637	8.495	7.955	7.309	6.916	6.641
45	9.441	9.230	9.046	8.882	8.735	8.176	7.509	7.103	6.819
46	9.707	9.489	9.298	9.128	8.976	8.399	7.710	7.291	6.998
47	9.974	9.749	9.551	9.376	9.219	8.623	7.912	7.480	7.178
48	10.244	10.011	9.807	9.626	9.464	8.848	8.114	7.669	7.359
49	10.516	10.275	10.064	9.877	9.710	9.074	8.318	7.860	7.540
50	10.789	10.541	10.323	10.130	9.957	9.302	8.523	8.051	7.722
51	11.065	10.808	10.583	10.384	10.206	9.531	8.729	8.244	7.905
52	11.342	11.077	10.845	10.640	10.457	9.761	8.936	8.437	8.089
53	11.621	11.348	11.109	10.898	10.709	9.992	9.144	8.631	8.274
54	11.902	11.621	11.375	11.157	10.962	10.225	9.353	8.826	8.459
55	12.185	11.895	11.642	11.417	11.217	10.459	9.562	9.021	8.645
56	12.469	12.171	11.910	11.680	11.474	10.694	9.773	9.218	8.832
57	12.756	12.449	12.181	11.943	11.731	10.930	9.985	9.415	9.020
58	13.044	12.728	12.452	12.209	11.991	11.168	10.198	9.613	9.208
59	13.334	13.009	12.726	12.475	12.252	11.407	10.411	9.812	9.397
60	13.626	13.292	13.001	12.744	12.514	11.647	10.626	10.012	9.587
61	13.919	13.577	13.278	13.013	12.777	11.888	10.841	10.213	9.777
62	14.215	13.863	13.556	13.285	13.043	12.130	11.058	10.414	9.968
63	14.512	14.151	13.836	13.557	13.309	12.374	11.275	10.616	10.160
64	14.811	14.440	14.117	13.832	13.577	12.618	11.493	10.819	10.353
65	15.112	14.732	14.400	14.107	13.846	12.864	11.713	11.023	10.546
66	15.415	15.024	14.684	14.384	14.117	13.111	11.933	11.227	10.740
67	15.719	15.319	14.970	14.663	14.389	13.359	12.154	11.433	10.934
68	16.025	15.615	15.258	14.943	14.663	13.608	12.375	11.639	11.130

					N				
S	1100	1200	1300	1400	1500	2000	3000	4000	5000
69	16.333	15.913	15.547	15.225	14.937	13.858	12.598	11.845	11.326
70	16.642	16.212	15.838	15.508	15.214	14.110	12.822	12.053	11.522
71	16.954	16.513	16.130	15.792	15.491	14.362	13.046	12.261	11.719
72	17.267	16.816	16.424	16.078	15.770	14.616	13.271	12.470	11.917
73	17.582	17.120	16.719	16.365	16.051	14.871	13.498	12.680	12.116
74	17.898	17.426	17.015	16.654	16.332	15.127	13.725	12.890	12.315
75	18.216	17.733	17.314	16.944	16.615	15.384	13.953	13.101	12.515
76	18.536	18.043	17.613	17.236	16.900	15.642	14.181	13.313	12.716
77	18.858	18.353	17.914	17.529	17.185	15.901	14.411	13.526	12.917
78	19.182	18.665	18.217	17.823	17.472	16.161	14.641	13.739	13.119
79	19.507	18.979	18.521	18.119	17.761	16.423	14.873	13.953	13.321
80	19.834	19.295	18.827	18.416	18.051	16.685	15.105	14.168	13.524
81	20.162	19.612	19.134	18.714	18.342	16.949	15.338	14.383	13.728
82	20.492	19.930	19.443	19.014	18.634	17.213	15.571	14.599	13.932
83	20.825	20.251	19.753	19.316	18.928	17.479	15.806	14.816	14.137
84	21.158	20.573	20.065	19.619	19.223	17.745	16.041	15.034	14.342
85	21.494	20.896	20.378	19.923	19.519	18.013	16.277	15.252	14.548
86	21.831	21.221	20.692	20.228	19.816	18.282	16.514	15.470	14.755
87	22.170	21.547	21.008	20.535	20.115	18.552	16.752	15.690	14.962
88	22.510	21.875	21.326	20.843	20.416	18.823	16.991	15.910	15.170
89	22.852	22.205	21.645	21.153	20.717	19.095	17.230	16.131	15.379
90	23.196	22.536	21.965	21.464	21.020	19.368	17.470	16.352	15.588
91	23.542	22.869	22.287	21.776	21.324	19.642	17.711	16.575	15.797
92	23.889	23.203	22.610	22.090	21.629	19.917	17.953	16.797	16.007
93	24.238	23.539	22.935	22.405	21.936	20.193	18.195	17.021	16.218
94	24.589	23.877	23.261	22.722	22.244	20.470	18.438	17.245	16.430
95	24.941	24.216	23.589	23.040	22.553	20.748	18.682	17.470	16.642
96	25.295	24.556	23.918	23.359	22.864	21.027	18.927	17.695	16.854
97	25.651	24.899	24.249	23.679	23.175	21.307	19.173	17.921	17.067
98	26.008	25.242	24.581	24.001	23.489	21.588	19.419	18.148	17.281
99	26.368	25.588	24.914	24.325	23.803	21.871	19.666	18.375	17.495
100	26.728	25.934	25.249	24.649	24.119	22.154	19.914	18.603	17.710
101	27.091	26.283	25.585	24.975	24.435	22.438	20.163	18.832	17.925
102	27.455	26.633	25.923	25.302	24.754	22.723	20.412	19.061	18.141

					N				
S	1100	1200	1300	1400	1500	2000	3000	4000	5000
103	27.821	26.984	26.262	25.631	25.073	23.009	20.662	19.291	18.357
104	28.188	27.337	26.603	25.961	25.394	23.297	20.913	19.522	18.574
105	28.558	27.692	26.945	26.292	25.716	23.585	21.165	19.753	18.792
106	28.929	28.048	27.288	26.625	26.039	23.874	21.417	19.985	19.010
107	29.301	28.405	27.633	26.959	26.363	24.164	21.670	20.217	19.229
108	29.676	28.765	27.980	27.294	26.689	24.455	21.924	20.450	19.448
109	30.052	29.125	28.328	27.631	27.016	24.748	22.179	20.684	19.667
110	30.429	29.488	28.677	27.969	27.344	25.041	22.434	20.918	19.888
111	30.809	29.851	29.028	28.309	27.674	25.335	22.690	21.153	20.108
112	31.190	30.217	29.380	28.649	28.005	25.630	22.947	21.388	20.330
113	31.572	30.584	29.733	28.991	28.337	25.926	23.205	21.624	20.552
114	31.957	30.952	30.088	29.335	28.670	26.223	23.463	21.861	20.774
115	32.343	31.322	30.445	29.679	29.004	26.521	23.722	22.098	20.997
116	32.731	31.694	30.802	30.025	29.340	26.820	23.982	22.336	21.220
117	33.120	32.067	31.162	30.373	29.677	27.120	24.242	22.574	21.444
118	33.511	32.442	31.522	30.721	30.015	27.422	24.503	22.813	21.669
119	33.904	32.818	31.884	31.071	30.355	27.723	24.765	23.053	21.894
120	34.299	33.195	32.248	31.423	30.696	28.026	25.028	23.293	22.119
121	34.695	33.575	32.613	31.775	31.038	28.330	25.291	23.534	22.345
122	35.093	33.956	32.979	32.129	31.381	28.635	25.555	23.776	22.572
123	35.493	34.338	33.347	32.485	31.725	28.941	25.820	24.018	22.799
124	35.894	34.772	33.716	32.841	32.071	29.248	26.085	24.260	23.026
125	36.297	35.107	34.087	33.199	32.418	29.556	26.351	24.503	23.254
126	36.702	35.494	34.459	33.559	32.766	29.865	26.618	24.747	23.483
127	37.108	35.883	34.833	33.919	33.116	30.174	26.886	24.991	23.712
128	37.517	36.273	35.208	34.281	33.466	30.485	27.154	25.236	23.942
129	37.927	36.665	35.584	34.645	33.818	30.797	27.423	25.482	24.172
130	38.338	37.058	35.962	35.009	34.171	31.109	27.693	25.728	24.402
131	38.751	37.453	36.341	35.375	34.526	31.423	27.963	25.975	24.633
132	39.166	37.849	36.722	35.742	34.881	31.737	28.234	26.222	24.865
133	39.583	38.247	37.104	36.111	35.238	32.053	28.506	26.469	25.097
134	40.002	38.647	37.487	36.481	35.596	32.369	28.778	26.718	25.329
135	40.422	39.048	37.872	36.852	35.956	32.686	29.051	26.967	25.563
136	40.844	39.450	38.259	37.225	36.316	33.005	29.325	27.216	25.796

					N				
S	1100	1200	1300	1400	1500	2000	3000	4000	5000
137	41.267	39.854	38.646	37.598	36.678	33.324	29.600	27.466	26.030
138	41.693	40.260	39.036	37.974	37.041	33.644	29.875	27.717	26.265
139	42.120	40.667	39.426	38.350	37.405	33.965	30.151	27.968	26.500
140	42.548	41.076	39.818	38.728	37.771	34.288	30.427	28.220	26.735
141	42.979	41.486	40.212	39.107	38.138	34.611	30.705	28.472	26.971
142	43.411	41.898	40.607	39.488	38.506	34.935	30.983	28.725	27.208
143	43.845	42.312	41.003	39.870	38.875	35.260	31.261	28.978	27.445
144	44.281	42.727	41.401	40.253	39.245	35.585	31.541	29.232	27.682
145	44.718	43.144	41.800	40.637	39.617	35.912	31.821	29.487	27.920
146	45.157	43.562	42.201	41.023	39.990	36.240	32.101	29.742	28.158
147	45.598	43.982	42.603	41.410	40.364	36.569	32.383	29.998	28.397
148	46.041	44.403	43.007	41.799	40.740	36.898	32.665	30.254	28.637
149	46.485	44.826	43.412	42.188	41.116	37.229	32.947	30.510	28.876
150	46.931	45.251	43.818	42.580	41.494	37.561	33.231	30.768	29.117
151	47.379	45.677	44.226	42.972	41.873	37.893	33.515	31.025	29.357
152	47.829	46.104	44.636	43.366	42.254	38.226	33.799	31.284	29.599
153	48.280	46.534	45.047	43.761	42.635	38.561	34.085	31.543	29.840
154	48.733	46.964	45.459	44.157	43.018	38.896	34.371	31.802	30.083
155	49.188	47.397	45.872	44.555	43.402	39.232	34.657	32.062	30.325
156	49.645	47.831	46.288	44.954	43.788	39.569	34.945	32.323	30.568
157	50.103	48.267	46.704	45.355	44.174	39.908	35.233	32.584	30.812
158	50.564	48.704	47.122	45.757	44.562	40.247	35.522	32.845	31.056
159	51.026	49.143	47.542	46.160	44.951	40.587	35.811	33.108	31.300
160	51.489	49.583	47.963	46.564	45.341	40.927	36.101	33.370	31.545
161	51.955	50.025	48.385	46.970	45.733	41.269	36.392	33.633	31.791
162	52.422	50.468	48.809	47.377	46.126	41.612	36.683	33.897	32.037
163	52.891	50.914	49.234	47.786	46.520	41.956	36.975	34.161	32.283
164	53.362	51.360	49.661	48.195	46.915	42.300	37.268	34.426	32.530
165	53.835	51.809	50.089	48.607	47.312	42.646	37.561	34.692	32.777
166	54.309	52.259	50.519	49.019	47.709	42.992	37.855	34.958	33.025
167	54.786	52.710	50.950	49.433	48.108	43.340	38.150	35.224	33.273
168	55.264	53.164	51.383	49.848	48.509	43.688	38.445	35.491	33.522
169	55.744	53.619	51.817	50.265	48.910	44.037	38.741	35.758	33.771
170	56.225	54.075	52.252	50.683	49.313	44.388	39.038	36.026	34.020

					N				
S	1100	1200	1300	1400	1500	2000	3000	4000	5000
171	56.709	54.533	52.689	51.102	49.717	44.739	39.335	36.295	34.270
172	57.194	54.993	53.128	51.522	50.122	45.091	39.633	36.564	34.521
173	57.681	55.454	53.568	51.944	50.528	45.444	39.931	36.834	34.772
174	58.170	55.917	54.009	52.368	50.936	45.798	40.231	37.104	35.023
175	58.661	56.382	54.452	52.792	51.345	46.153	40.531	37.374	35.275
176	59.154	56.848	54.897	53.218	51.755	46.508	40.831	37.645	35.527
177	59.648	57.316	55.342	53.646	52.167	46.865	41.132	37.917	35.780
178	60.144	57.785	55.790	54.074	52.580	47.223	41.434	38.189	36.033
179	60.642	58.256	56.239	54.504	52.994	47.581	41.737	38.462	36.286
180	61.142	58.729	56.689	54.936	53.409	47.941	42.040	38.735	36.540
181	61.644	59.203	57.141	55.369	53.825	48.301	42.343	39.009	36.795
182	62.148	59.679	57.594	55.803	54.243	48.663	42.648	39.283	37.050
183	62.653	60.157	58.049	56.238	54.662	49.025	42.953	39.558	37.305
184	63.160	60.637	58.505	56.675	55.082	49.388	43.259	39.833	37.561
185	63.670	61.118	58.963	57.113	55.504	49.752	43.565	40.109	37.817
186	64.181	61.600	59.422	57.553	55.927	50.117	43.872	40.385	38.074
187	64.694	62.085	59.883	57.994	56.351	50.483	44.180	40.662	38.331
188	65.208	62.570	60.345	58.536	56.776	50.850	44.488	40.940	38.588
189	65.725	63.058	60.809	58.880	57.203	51.218	44.797	41.217	38.846
190	66.244	63.547	61.274	59.325	57.630	51.587	45.106	41.496	39.105
191	66.764	64.038	61.741	59.771	58.060	51.957	45.417	41.775	39.364
192	67.286	64.531	62.209	60.219	58.490	52.327	45.727	42.054	39.623
193	67.810	65.025	62.679	60.668	58.922	52.699	46.039	42.334	39.883
194	68.337	65.521	63.150	61.119	59.354	53.071	46.351	42.614	40.143
195	68.865	66.019	63.623	61.571	59.789	53.445	46.664	42.895	40.403
196	69.394	66.519	64.097	62.024	60.224	53.819	46.977	43.177	40.664
197	69.926	67.020	64.573	62.479	60.661	54.194	47.291	43.459	40.926
198	70.460	67.522	65.051	62.935	61.099	54.571	47.606	43.741	41.188
199	70.996	68.027	65.530	63.393	61.538	54.948	47.921	44.024	41.450
200	71.533	68.533	66.010	63.852	61.979	55.326	48.237	44.307	41.713
201	72.073	69.041	66.492	64.312	62.420	55.705	48.554	44.591	41.976
202	72.614	69.550	66.975	64.773	62.864	56.085	48.871	44.876	42.240
203	73.157	70.062	67.460	65.237	63.308	56.466	49.189	45.161	42.504
204	73.703	70.575	67.947	65.701	63.754	56.847	49.507	45.446	42.768

					N				
S	1100	1200	1300	1400	1500	2000	3000	4000	5000
205	74.250	71.089	68.435	66.167	64.201	57.230	49.826	45.732	43.033
206	74.799	71.606	68.925	66.634	64.649	57.614	50.146	46.019	43.298
207	75.350	72.124	69.416	67.103	65.098	57.998	50.466	46.306	43.564
208	75.903	72.644	69.909	67.573	65.549	58.384	50.787	46.593	43.830
209	76.458	73.165	70.403	68.044	66.001	58.770	51.109	46.881	44.097
210	77.015	73.689	70.899	68.517	66.455	59.157	51.431	47.170	44.364
211	77.574	74.214	71.396	68.991	66.909	59.546	51.754	47.459	44.631
212	78.135	74.741	71.895	69.467	67.365	59.935	52.078	47.748	44.899
213	78.698	75.269	72.396	69.944	67.823	60.325	52.402	48.038	45.167
214	79.263	75.800	72.898	70.423	68.281	60.716	52.727	48.328	45.436
215	79.830	76.332	73.401	70.903	68.741	61.108	53.052	48.619	45.705
216	80.399	76.865	73.906	71.384	69.202	61.501	53.378	48.911	45.974
217	80.970	77.401	74.413	71.867	69.665	61.895	53.705	49.203	46.244
218	81.543	77.938	74.922	72.351	70.128	62.290	54.032	49.495	46.515
219	82.117	78.477	75.431	72.837	70.593	62.685	54.360	49.788	46.786
220	82.694	79.018	75.943	73.324	71.060	63.082	54.689	50.081	47.057
221	83.273	79.561	76.456	73.S12	71.527	63.480	55.018	50.375	47.328
222	83.854	80.105	76.971	74.302	71.996	63.878	55.348	50.670	47.600
223	84.437	80.651	77.487	74.794	72.467	64.278	55.678	50.965	47.873
224	85.022	81.199	78.005	75.286	72.938	64.678	56.009	51.260	48.146
225	85.609	81.749	78.524	75.781	73.411	65.079	56.341	51.556	48.419
226	86.198	82.301	79.045	76.276	73.886	65.482	56.673	51.852	48.693
227	86.790	82.854	79.568	76.773	74.361	65.885	57.006	52.149	48.967
228	87.383	83.409	80.092	77.272	74.838	66.289	57.340	52.447	49.241
229	87.978	83.966	80.618	77.772	75.316	66.694	57.674	52.744	49.516
230	88.575	84.525	81.145	78.273	75.796	67.100	58.009	53.043	49.792
231	89.175	85.085	81.674	78.776	76.276	67.507	58.344	53.342	50.067
232	89.776	85.647	82.205	79.280	76.759	67.915	58.680	53.641	50.343
233	90.380	86.212	82.737	79.786	77.242	68.324	59.017	53.941	50.620
234	90.986	86.778	83.271	80.294	77.727	68.733	59.354	54.241	50.897
235	91.593	87.345	83.806	80.802	78.213	69.144	59.692	54.542	51.174
236	92.203	87.915	84.343	81.312	78.701	69.556	60.031	54.843	51.452
237	92.815	88.487	84.882	81.824	79.189	69.968	60.370	55.145	51.730
238	93.429	89.060	85.423	82.337	79.680	70.382	60.710	55.447	52.009

					N				
S	1100	1200	1300	1400	1500	2000	3000	4000	5000
239	94.045	89.635	85.965	82.852	80.171	70.796	61.050	55.750	52.288
240	94.664	90.212	86.508	83.368	80.664	71.212	61.391	56.053	52.568
241	95.284	90.791	87.053	83.885	81.158	71.628	61.733	56.357	52.847
242	95.907	91.372	87.600	84.404	81.654	72.045	62.075	56.661	53.128
243	96.532	91.954	88.149	84.925	82.150	72.464	62.418	56.965	53.408
244	97.158	92.539	88.699	85.447	82.649	72.883	62.762	57.271	53.689
245	97.787	93.125	89.251	85.970	83.148	73.303	63.106	57.576	53.971
246	98.419	93.713	89.805	86.495	83.649	73.724	63.451	57.882	54.253
247	99.052	94.303	90.360	87.022	84.151	74.146	63.796	58.189	54.535
248	99.687	94.895	90.917	87.550	84.655	74.569	64.142	58.496	54.818
249	100.325	95.489	91.475	88.079	85.160	74.993	64.489	58.803	55.101
250	100.965	96.085	92.035	88.610	85.666	75.418	64.836	59.112	55.384
251	101.607	96.682	92.597	89.142	86.174	75.843	65.184	59.420	55.668
252	102.251	97.282	93.161	89.676	86.683	76.270	65.532	59.729	55.953
253	102.898	97.883	93.726	90.212	87.193	76.698	65.881	60.039	56.237
254	103.546	98.486	94.293	90.749	87.705	77.126	66.231	60.348	56.522
255	104.197	99.092	94.861	91.287	88.218	77.556	66.582	60.659	56.808
256	104.850	99.699	95.432	91.827	88.733	77.987	66.933	60.970	57.094
257	105.506	100.308	96.004	92.369	89.249	78.418	67.284	61.281	57.380
258	106.163	100.919	96.577	92.911	89.766	78.850	67.636	61.593	57.667
259	106.823	101.532	97.153	93.456	90.285	79.284	67.989	61.905	57.954
260	107.485	102.147	97.730	94.002	90.805	79.718	68.343	62.218	58.241
261	108.149	102.763	98.308	94.550	91.326	80.153	68.697	62.532	58.429
262	108.816	103.382	98.889	95.099	91.849	80.590	69.051	62.845	58.818
263	109.484	104.003	99.471	95.649	92.373	81.027	69.407	63.160	59.106
264	110.156	104.626	100.055	96.202	92.899	81.465	69.763	63.474	59.395
265	110.829	105.250	100.641	96.755	93.426	81.904	70.119	63.790	59.685
266	111.505	105.877	101.228	97.310	93.954	82.344	70.476	64.105	59.975
267	112.182	106.505	101.817	97.867	94.484	82.785	70.834	64.421	60.265
268	112.863	107.136	102.408	98.426	95.015	83.227	71.193	64.738	60.556
269	113.545	107.768	103.001	98.985	95.548	83.670	71.552	65.055	60.847
270	114.230	108.403	103.595	99.547	96.082	84.114	71.911	65.373	61.138
271	114.917	109.039	104.191	100.110	96.617	84.559	72.272	65.691	61.430
272	115.607	109.678	104.789	100.674	97.154	85.005	72.633	66.010	61.723

					N				
S	1100	1200	1300	1400	1500	2000	3000	4000	5000
273	116.299	110.318	105.388	101.241	97.692	85.452	72.994	66.329	62.015
274	116.993	110.961	105.990	101.808	98.232	85.899	73.356	66.648	62.308
275	117.689	111.605	106.593	102.377	98.773	86.348	73.719	66.968	62.602
276	118.388	112.252	107.198	102.948	99.315	86.798	74.082	67.288	62.896
277	119.089	112.901	107.804	103.521	99.859	87.248	74.446	67.609	63.190
278	119.793	113.551	108.413	104.095	100.404	87.700	74.811	67.931	63.485
279	120.499	114.204	109.023	104.670	100.951	88.153	75.176	68.253	63.780
280	121.207	114.858	109.635	105.247	101.499	88.606	75.542	68.575	64.075
281	121.918	115.515	110.249	105.826	102.049	89.061	75.908	68.898	64.371
282	<u>122.631</u>	116.174	110.864	106.406	102.600	89.516	76.276	69.221	64.667
283	123.347	116.835	111.482	106.988	103.152	89.973	76.643	69.545	64.964
284	124.065	117.498	112.101	107.572	103.706	90.430	77.012	69.869	65.261
285	124.785	118.162	112.722	108.157	104.261	90.888	77.380	70.194	65.558
286	125.508	118.829	113.344	108.744	104.818	91.348	77.750	70.519	65.856
287	126.233	119.499	113.969	109.332	105.376	91.808	78.120	70.845	66.154
288	126.961	120.170	114.595	109.922	105.936	92.270	78.491	71.171	66.453
289	127.691	120.843	115.223	110.513	106.497	92.732	78.862	71.497	66.752
290	128.424	121.518	115.853	111.107	107.059	93.195	79.234	71.824	67.051
291	129.159	122.196	116.485	111.701	107.623	93.659	79.607	72.152	67.351
292	129.896	122.875	117.119	112.298	108.189	94.125	79.980	72.480	67.651
293	130.636	123.557	117.755	112.896	108.756	94.591	80.354	72.809	67.952
294	131.379	124.241	118.392	113.495	109.324	95.058	80.729	73.137	68.252
295	132.124	124.927	119.031	114.097	109.894	95.526	81.104	73.467	68.554
296	132.871	125.615	119.672	114.699	110.465	95.995	81.480	73.797	68.855
297	133.621	126.305	120.315	115.304	111.038	96.465	81.856	74.127	69.158
298	134.374	126.997	120.960	115.910	111.612	96.937	82.233	74.458	69.460
299	135.129	127.692	121.606	116.518	112.188	97.409	82.610	74.789	69.763
300	135.887	128.388	122.255	117.128	112.765	97.882	82.989	75.121	70.066
301	136.647	129.087	122.905	117.739	113.344	98.356	83.368	75.453	70.370
302	137.410	129.788	123.557	118.351	113.924	98.831	83.747	75.786	70.674
303	138.175	130.491	124.211	118.966	114.505	99.307	84.127	76.119	70.978
304	138.943	131.196	124.867	119.582	115.089	99.784	84.508	76.453	71.283
305	139.713	131.904	125.525	120.200	115.673	100.262	84.889	76.787	71.588
306	140.486	132.613	126.185	120.819	116.259	100.741	85.271	77.122	71.894

	N								
S	1100	1200	1300	1400	1500	2000	3000	4000	5000
307	141.262	133.325	126.847	121.440	116.847	101.221	85.653	77.457	72.200
308	142.040	**134.039**	127.510	122.063	117.436	101.702	86.037	77.792	72.506
309	142.821	134.756	128.176	122.688	118.027	102.184	86.420	78.128	72.813
310	143.605	135.474	128.843	123.314	118.619	102.667	86.805	78.465	73.120
311	144.391	136.195	129.513	123.942	119.212	103.151	87.190	78.801	73.427
312	145.179	136.918	130.184	124.571	119.808	103.636	87.876	79.139	73.735
313	145.971	137.643	130.857	125.203	120.404	104.122	87.962	79.477	74.044
314	146.765	138.370	131.532	125.835	121.002	104.609	88.349	79.815	74.352
315	147.562	139.100	132.209	126.470	121.602	105.097	88.736	80.154	74.661
316	148.361	139.832	132.888	127.106	122.203	105.586	89.124	80.493	74.971
317	149.163	140.566	133.569	127.745	122.806	106.076	89.513	80.833	75.281
318	149.968	141.302	134.252	128.384	123.410	106.567	89.902	81.173	75.591
319	150.776	142.041	134.937	129.026	124.016	107.059	90.292	81.514	75.901
320	151.586	142.782	135.624	129.669	124.623	107.552	90.683	81.855	76.212
321	152.399	143.525	136.312	130.314	125.232	108.046	91.074	82.196	76.524
322	153.215	144.271	137.003	130.961	125.843	108.541	91.466	82.539	76.835
323	154.033	145.019	137.696	131.609	126.455	109.037	91.858	82.881	77.147
324	154.854	145.769	138.391	132.259	127.068	109.534	92.252	83.224	77.460
325	155.678	146.521	139.087	132.911	127.683	110.032	92.645	83.568	77.773
326	156.505	147.276	139.786	133.565	128.300	110.531	93.040	83.911	78.086
327	157.334	148.033	140.487	134.220	128.918	111.031	93.435	84.256	78.400
328	158.167	148.792	141.189	134.877	129.538	111.532	93.830	84.601	78.714
329	159.002	149.554	141.894	135.536	130.159	112.034	94.226	84.946	79.028
330	159.840	150.318	142.601	136.197	130.782	112.537	94.623	85.292	79.343
331	160.680	151.085	143.309	136.859	131.406	113.041	95.021	85.638	79.658
332	161.524	151.854	144.020	137.523	132.032	113.547	95.419	85.985	79.974
333	162.370	152.625	**144.733**	138.189	132.660	114.053	95.817	86.332	80.290
334	163.220	153.399	145.448	138.857	133.289	114.560	96.217	86.680	80.606
335	164.072	154.175	146.165	139.527	133.920	115.068	96.617	87.028	80.923
336	164.927	154.953	146.883	140.198	134.552	115.577	97.017	87.376	81.240
337	165.785	155.734	147.604	140.871	135.186	116.088	97.418	87.725	81.557
338	166.645	156.517	148.327	141.546	135.822	116.599	97.820	88.075	81.875
339	167.509	157.302	149.052	142.223	136.459	117.111	98.222	88.425	82.193
340	168.376	158.090	149.780	142.901	137.097	117.625	98.626	88.775	82.512

S	N								
	1100	1200	1300	1400	1500	2000	3000	4000	5000
341	169.245	158.881	150.509	143.581	137.738	118.139	99.029	89.126	82.831
342	170.117	159.674	151.240	144.264	138.380	118.654	99.433	89.478	83.150
343	170.993	160.469	151.973	144.948	139.023	119.171	99.838	89.829	83.470
344	171.871	161.267	152.709	145.633	139.668	119.688	100.244	90.182	83.790
345	172.752	162.067	153.446	146.321	140.315	120.207	100.650	90.535	84.111
346	173.637	162.870	154.186	147.010	140.963	120.726	101.057	90.888	84.431
347	174.524	163.675	154.928	147.702	141.613	121.247	101.464	91.242	84.753
348	175.414	164.483	155.672	148.395	142.265	121.768	101.872	91.596	85.074
349	176.307	165.293	156.418	149.089	142.918	122.291	102.281	91.950	85.396
350	177.204	166.105	157.166	149.786	143.573	122.815	102.690	92.305	85.719
351	178.103	166.920	157.916	150.485	144.230	123.339	103.100	92.661	86.042
352	179.005	167.738	158.668	151.185	144.888	123.865	103.511	93.017	86.365
353	179.911	168.558	159.423	151.888	145.547	124.392	103.922	93.373	86.688
354	180.819	169.381	160.180	152.592	146.209	124.920	104.334	93.730	87.012
355	181.731	170.206	160.938	153.298	146.872	125.449	104.746	94.088	87.336
356	182.645	171.034	161.699	154.006	147.537	125.979	105.159	94.446	87.661
357	183.563	171.864	162.463	154.716	148.203	126.510	105.573	94.804	87.986
358	184.484	172.697	163.228	155.427	148.871	127.042	105.987	95.163	88.312
359	185.408	173.533	163.995	**156.141**	149.541	127.575	106.402	95.522	88.637
360	186.335	174.371	164.765	156.856	150.212	128.109	106.818	95.882	88.964
361	187.265	175.211	165.537	157.574	150.885	128.644	107.234	96.242	89.290
362	188.198	176.054	166.311	158.293	151.560	129.180	107.651	96.603	89.617
363	189.135	176.900	167.087	159.014	152.236	129.718	108.068	96.964	89.944
364	190.074	177.749	167.866	159.737	152.914	130.256	108.486	97.325	90.272
365	191.017	178.600	168.646	160.462	153.594	130.796	108.905	97.687	90.600
366	191.963	179.453	169.429	161.189	154.276	131.336	109.324	98.050	90.928
367	192.913	180.310	170.214	161.918	154.959	131.878	109.744	98.413	91.257
368	193.865	181.169	171.002	162.649	155.644	132.420	110.165	98.776	91.586
369	194.821	182.030	171.791	163.381	156.330	132.964	110.586	99.140	91.916
370	195.780	182.894	172.583	164.116	157.018	133.509	111.008	99.505	92.246
371	196.742	183.761	173.377	164.853	157.708	134.054	111.430	99.870	92.476
372	197.708	184.631	174.174	165.591	158.400	134.601	111.853	100.235	92.907
373	198.677	185.503	174.972	166.332	159.093	135.149	112.277	100.601	93.238
374	199.649	186.378	175.773	167.074	159.788	135.698	112.701	100.967	93.569

					N				
S	1100	1200	1300	1400	1500	2000	3000	4000	5000
375	200.624	187.256	176.576	167.818	160.485	136.248	113.126	101.334	93.901
376	201.603	188.136	177.382	168.565	161.184	136.799	113.552	101.701	94.233
377	202.585	189.020	178.189	169.313	161.884	137.352	113.978	102.068	94.566
378	203.571	189.905	178.999	170.063	162.586	137.905	114.405	102.436	94.899
379	204.559	190.794	179.812	170.816	163.290	138.459	114.833	102.805	95.232
380	205.552	191.685	180.626	171.570	163.995	139.015	115.261	103.174	95.566
381	206.547	192.579	181.443	172.326	164.703	139.571	115.690	103.543	95.900
382	207.546	193.476	182.263	173.085	165.412	140.129	116.119	103.913	96.235
383	208.549	194.376	183.084	173.845	166.122	140.688	116.549	104.284	96.569
384	209.555	195.278	183.908	174.607	166.835	141.248	116.980	104.655	96.905
385	210.564	196.184	184.735	175.371	**167.549**	141.809	117.411	105.026	97.240
386	211.577	197.092	185.563	176.138	168.265	142.371	117.843	105.398	97.576
387	212.593	198.003	186.394	176.906	168.983	142.934	118.276	105.770	97.912
388	213.613	198.916	187.228	177.677	169.703	143.498	118.709	106.143	98.249
389	214.636	199.833	188.063	178.449	170.424	144.063	119.143	106.516	98.586
390	215.663	200.752	188.901	179.223	171.147	144.629	119.577	106.890	98.924
391	216.693	201.674	189.742	180.000	171.872	145.197	120.012	107.264	99.261
392	217.727	202.600	190.585	180.778	172.599	145.765	120.448	107.638	99.600
393	218.764	203.528	191.430	181.559	173.327	146.335	120.885	108.013	99.938
394	219.805	204.458	192.278	182.342	174.057	146.906	121.322	108.389	100.277
395	220.849	205.392	193.128	183.126	174.790	147.478	121.759	108.765	100.616
396	221.897	206.329	193.981	183.913	175.524	148.051	122.198	109.141	100.956
397	222.949	207.268	194.836	184.702	176.259	148.625	122.637	109.518	101.296
398	224.005	208.211	195.693	185.493	176.997	149.200	123.076	109.895	101.637
399	225.063	209.156	196.553	186.286	177.736	149.776	123.516	110.273	101.977
400	226.126	210.105	197.415	187.081	178.477	150.353	123.957	110.652	102.319
401	227.192	211.056	198.280	187.878	179.220	150.932	124.399	111.030	102.660
402	228.262	212.010	199.147	188.678	179.965	151.511	124.841	111.410	103.002
403	229.336	212.967	200.017	189.479	180.712	152.092	125.284	111.789	103.344
404	230.413	213.928	200.889	190.283	181.460	152.674	125.727	112.169	103.687
405	231.495	214.891	201.764	191.088	182.211	153.257	126.171	112.550	104.030
406	232.579	215.857	202.641	191.896	182.963	153.841	126.616	112.931	104.373
407	233.668	216.826	203.521	192.706	183.717	154.426	127.061	113.313	104.717
408	234.761	217.799	204.403	193.518	184.473	155.012	127.507	113.695	105.061

S	N 1100	1200	1300	1400	1500	2000	3000	4000	5000
409	235.857	218.774	205.288	194.333	185.231	155.600	127.954	114.077	105.406
410	236.957	219.752	206.175	195.149	185.991	156.188	128.401	114.460	105.751
411	238.061	220.734	207.065	195.967	186.752	156.778	128.849	114.843	106.096
412	239.168	221.718	207.957	196.788	187.516	157.369	129.298	115.227	106.442
413	240.280	222.706	208.852	197.611	188.281	157.961	129.747	115.612	106.788
414	241.395	223.697	209.749	198.436	189.048	158.554	130.197	115.996	107.134
415	242.514	224.690	210.649	199.263	189.817	159.148	130.647	116.382	107.481
416	243.638	225.687	211.552	200.093	190.588	159.743	131.098	116.767	107.828
417	244.765	226.687	212.457	200.924	191.361	160.340	131.550	117.154	108.176
418	245.896	227.690	213.365	201.758	192.136	160.938	132.002	117.540	108.524
419	247.031	228.697	214.275	202.594	192.913	161.536	132.456	117.927	108.872
420	248.170	229.706	215.188	203.432	193.691	162.136	132.909	118.315	109.220
421	249.313	230.719	216.103	204.273	194.472	162.737	133.364	118.703	109.569
422	250.460	231.734	217.022	205.115	195.254	163.339	133.819	119.091	109.919
423	251.611	232.753	217.942	205.960	196.038	163.943	134.274	119.480	110.269
424	252.766	233.776	218.866	206.807	196.825	164.547	134.731	119.870	110.619
425	253.925	234.801	219.792	207.657	197.613	165.153	135.188	120.260	110.969
426	255.088	235.830	220.721	208.508	198.403	165.760	135.645	120.650	111.320
427	256.255	236.862	221.652	209.362	199.195	166.367	136.103	121.041	111.671
428	257.426	237.897	222.586	210.218	199.989	166.976	136.562	121.432	112.023
429	258.602	238.935	223.523	211.077	200.786	167.587	137.022	121.824	112.375
430	259.781	239.977	224.462	211.937	201.584	168.198	137.482	122.216	112.727
431	260.965	241.022	225.404	212.800	202.384	168.811	137.943	122.609	113.080
432	262.153	242.070	226.349	213.666	203.185	169.424	138.404	123.002	113.433
433	263.345	243.122	227.297	214.533	203.989	170.039	138.866	123.396	113.787
434	264.541	244.176	228.247	215.403	204.795	170.655	139.329	123.790	114.140
435	265.742	245.235	229.200	216.275	205.603	171.272	139.793	124.184	114.495
436	266.946	246.296	230.156	217.149	206.413	171.891	140.257	124.579	114.849
437	268.155	247.361	231.114	218.026	207.225	172.510	140.721	124.975	115.204
438	269.369	248.429	232.076	218.905	208.039	173.131	141.187	125.371	115.559
439	270.586	249.501	233.040	219.786	208.855	173.753	141.653	125.767	115.915
440	271.808	250.576	234.007	220.670	209.673	174.376	142.120	126.164	116.271
441	273.035	251.655	234.976	221.556	210.492	175.000	142.587	126.562	116.628
442	274.265	252.737	235.949	222.445	211.314	175.626	143.055	126.959	116.984

S	N								
	1100	1200	1300	1400	1500	2000	3000	4000	5000
443	275.500	253.822	236.924	223.335	212.138	176.252	143.524	127.358	117.342
444	276.740	254.911	237.902	224.229	212.964	176.880	143.993	127.756	117.699
445	277.983	256.033	238.883	225.124	213.792	177.509	144.463	128.156	118.057
446	279.232	257.099	239.866	226.022	214.622	178.139	144.934	128.555	118.415
447	280.484	258.198	240.853	226.922	215.455	178.770	145.405	128.955	118.774
448	281.742	259.301	241.842	227.825	216.289	179.403	145.877	129.356	119.133
449	283.003	260.407	242.834	228.730	217.125	180.037	146.349	129,757	119.493
450	284.269	261.517	243.829	229.637	217.963	180.672	146.823	130.159	119.852
451	285.540	262.630	244.827	230.547	218.804	181.308	147.296	130.561	120.212
452	286.815	263.747	245.828	231.460	219.646	181.945	147.771	130.963	120.573
453	288.095	264.867	246.832	232.374	220.491	182.583	148.246	131.366	120.934
454	289.380	265.991	247.839	233.291	221.337	183.223	148.722	131.770	121.295
455	290.669	267.119	248.848	234.211	222.186	183.864	149.199	132.174	121.657
456	291.962	268.250	249.861	235.133	223.037	184.506	149.676	132.578	122.019
457	293.261	269.385	250.876	236.057	223.889	185.149	150.154	132.983	122.381
458	294.564	270.523	251.895	236.984	224.744	185.794	150.632	133.388	122.744
459	295.872	271.665	252.916	237.914	225.602	186.440	151.111	133.794	123.107
460	297.184	272.811	253.940	238.846	226.461	187.087	151.591	134.200	123.471
461	298.501	273.960	254.967	239.780	227.322	187.735	152.072	134.607	123.834
462	299.823	275.114	255.998	240.717	228.186	188.384	152.553	135.014	124.199
463	301.150	276.270	257.031	241.656	229.051	189.035	153.035	135.422	124.563
464	302.481	277.431	258.067	242.598	229.919	189.687	153.517	135.830	124.928
465	303.818	278.595	259.106	243.542	230.789	190.340	154.000	136.238	125.294
466	305.159	279.763	260.149	244.489	231.661	190.994	154.484	136.647	125.659
467	306.505	280.935	261.194	245.439	232.535	191.649	154.969	137.057	126.025
468	307.856	282.111	262.242	246.391	233.411	192.306	155.454	137.467	126.392
469	309.212	283.290	263.294	247.345	234.290	192.964	155.940	137.877	126.759
470	310.573	284.473	264.348	248.302	235.170	193.623	156.426	138.288	127.126
471	311.939	285.660	265.406	249.262	236.053	194.283	156.913	138.700	127.493
472	313.309	286.851	266.467	250.224	236.938	194.945	157.401	139.112	127.861
473	314.685	288.046	267.530	251.189	237.826	195.608	157.890	139.524	128.230
474	316.066	289.245	268.597	252.156	238.715	196.272	158.379	139.937	128.598
475	317.452	290.447	269.667	253.126	239.607	196.937	158.869	140.350	128.967

					N				
S	1100	1200	1300	1400	1500	2000	3000	4000	5000
476	318.843	291.653	270.740	254.098	240.501	197.604	159.359	140.764	129.337
477	320.239	292.864	271.817	255.074	241.397	198.272	159.850	141.178	129.707
478	321.640	294.078	272.896	256.051	242.295	198.941	160.342	141.593	130.077
479	323.046	295.296	273.979	257.032	243.195	199.611	160.835	142.008	130.447
480	324.458	296.518	275.064	258.015	244.098	200.283	161.328	142.423	130.818
481	325.875	297.744	276.153	259.000	245.003	200.956	161.822	142.840	131.190
482	327.296	298.974	277.245	259.988	245.910	201.630	162.316	143.256	131.561
483	328.724	300.208	278.341	260.979	246.820	202.305	162.811	143.673	131.933
484	330.156	301.447	279.439	261.973	247.731	202.982	163.307	144.091	132.306
485	331.594	302.689	280.541	262.969	248.645	203.660	163.804	144.509	132.678
486	333.037	303.935	281.646	263.968	249.562	204.339	164.301	144.927	133.052
487	334.485	305.185	282.754	264.970	250.480	205.019	164.799	145.346	133.425
488	335.939	306.440	283.866	265.974	251.401	205.701	165.297	145.765	133.799
489	337.398	307.698	284.981	266.981	252.324	206.384	165.796	146.185	134.173
490	338.863	308.961	286.099	267.990	253.249	207.068	166.296	146.606	134.548
491	340.333	310.227	287.220	269.003	254.177	207.753	166.797	147.027	134.923
492	341.808	311.498	288.345	270.018	255.107	208.440	167.298	147.448	135.298
493	343.289	312.773	289.473	271.036	256.039	209.128	167.800	147.870	135.674
494	344.776	314.053	290.604	272.056	256.974	209.817	168.303	148.292	136.050
495	346.268	315.336	291.739	273.080	257.911	210.508	168.806	148.715	136.427
496	347.766	316.624	292.877	274.106	258.850	211.200	169.310	149.138	136.803
497	349.269	317.916	294.018	275.135	259.792	211.893	169.815	149.561	137.181
498	350.778	319.212	295.163	276.166	260.736	212.588	170.320	149.985	137.558
499	352.292	320.513	296.311	277.201	261.682	213.283	170.826	150.410	137.936
500	353.813	321.817	297.463	278.238	262.631	213.980	171.333	150.835	138.314
501	355.339	323.126	298.618	279.278	263.582	214.679	171.840	151.261	138.693
502	356.870	324.440	299.776	280.321	264.535	215.378	172348	151.687	139.072
503	358.408	325.758	300.938	281.366	265.491	216.079	172.857	152.113	139.452
504	359.951	327.080	302.103	282.415	266.449	216.781	173.366	152.540	139.832
505	361.500	328.406	303.272	283.466	267.410	217.485	173.876	152.968	140.212
506	363.055	329.737	304.444	284.520	268.373	218.190	174.387	153.396	140.592
507	364.616	331.073	305.620	285.577	269.338	218.896	174.898	153.824	140.973
508	366.183	332.412	306.799	286.637	270.306	219.603	175.411	154.253	141.355

S	N								
	1100	1200	1300	1400	1500	2000	3000	4000	5000
509	367.756	333.757	307.982	287.700	271.276	220.312	175.923	154.682	141.736
510	369.334	335.105	309.168	288.766	272.248	221.022	176.437	155.112	142.118
511	370.919	336.459	310.358	289.834	273.223	221.734	176.951	155.542	142.501
512	372.510	337.816	311.552	290.906	274.201	222.446	177.466	155.973	142.883
513	374.107	339.179	312.748	291.980	275.181	**223.161**	177.982	156.404	143.267
514	375.710	340.545	313.949	293.057	276.163	223.876	178.498	156.836	143.650
515	377.319	341.917	315.153	294.138	277.148	224.593	179.015	157.268	144.034
516	378.934	343.293	316.361	295.221	278.135	225.311	179.532	157.701	144.418
517	380.555	344.674	317.572	296.307	279.125	226.030	180.051	158.134	144.803
518	382.183	346.059	318.787	297.396	280.117	226.751	180.570	158.568	145.188
519	383.817	347.449	320.006	298.488	281.112	227.473	181.089	159.002	145.573
520	385.457	348.843	321.228	299.583	282.109	228.196	181.610	159.436	145.959
521	387.103	350.242	322.454	300.681	283.109	228.921	182.131	159.871	146.345
522	388.756	351.646	323.684	301.782	284.111	229.647	182.653	160.307	146.732
523	390.415	353.055	324.917	302.886	285.116	230.374	183.175	160.743	147.119
524	392.081	354.469	326.154	303.993	286.123	231.103	183.698	161.179	147.506
525	393.753	355.887	327.395	305.103	287.133	231.833	184.222	161.616	147.894
526	395.431	357.310	328.639	306.216	288.145	232.564	184.747	162.054	148.282
527	397.116	358.738	329.887	307.332	289.160	233.297	185.272	162.492	148.670
528	398.808	360.170	331.139	308.451	290.177	234.031	185.798	162.930	149.059
529	400.506	361.608	332.395	309.573	291.197	234.767	186.324	163.369	149.448
530	402.211	363.050	333.655	310.698	292.220	235.504	186.852	163.808	149.837
531	403.922	364.497	334.918	311.827	293.245	236.242	187.380	164.248	150.227
532	405.640	365.950	336.185	312.958	294.272	236.982	187.908	164.688	150.617
533	407.365	367.407	337.457	314.093	295.303	237.723	188.438	165.129	151.008
534	409.097	368.869	338.731	315.230	296.335	238.465	188.968	165.571	151.399
535	410.835	370.336	340.010	316.371	297.371	239.209	189.499	166.012	151.790
536	412.580	371.808	341.293	317.515	298.409	239.954	190.030	166.455	152.182
537	414.332	373.285	342.579	318.662	299.450	240.700	190.563	166.897	152.574
538	416.091	374.767	343.870	319.813	300.493	241.448	191.096	167.340	152.967
539	417.857	376.255	345.164	320.966	301.539	242.197	191.629	167.784	153.359
540	419.629	377.747	346.463	322.122	302.587	242.948	192.163	168.228	153.753
541	421.409	379.244	347.765	323.282	303.639	243.700	192.698	168.673	154.146

					N				
S	1100	1200	1300	1400	1500	2000	3000	4000	5000
542	423.196	380.747	349.071	324.445	304.693	244.453	193.234	169.118	154.540
543	424.990	382.255	350.382	325.611	305.749	245.208	193.771	169.564	154.934
544	426.791	383.768	351.696	326.781	306.808	245.964	194.308	170.010	155.329
545	428.599	385.286	353.014	327.953	307.870	246.722	194.846	170.456	155.724
546	430.414	386.809	354.336	329.129	308.935	247.481	195.384	170.903	156.120
547	432.237	388.338	355.663	330.308	310.002	248.241	195.923	171.351	156.516
548	434.066	389.872	356.993	331.491	311.072	249.003	196.463	171.799	156.912
549	435.903	391.411	358.328	332.676	312.145	249.766	197.004	172.247	157.308
550	437.748	392.955	359.666	333.865	313.220	250.531	197.545	172.696	157.705
551	439.600	394.505	361.009	335.058	314.298	251.297	198.087	173.146	158.103
552	441.459	396.060	362.356	336.253	315.379	252.064	198.630	173.596	158.500
553	443.325	397.621	363.707	337.452	316.463	252.833	199.174	174.046	158.898
554	445.200	399.187	365.062	338.654	317.549	253.603	199.718	174.497	159.297
555	447.081	400.759	366.421	339.860	318.638	254.375	200.263	174.948	159.696
556	448.971	402.336	367.785	341.069	319.730	255.148	200.809	175.400	160.095
557	450.867	403.918	369.153	342.281	320.825	255.923	201.355	175.852	160.494
558	452.772	405.506	370.524	343.496	321.922	256.699	201.902	176.305	160.894
559	454.684	407.100	371.901	344.715	323.023	257.476	202.450	176.759	161.294
560	456.604	408.699	373.281	345.938	324.126	258.255	202.998	177.212	161.695
561	458.532	410.303	374.666	347.164	325.232	259.035	203.547	177.667	162.096
562	460.468	411.914	376.055	348.393	326.340	259.817	204.097	178.121	162.498
563	462.411	413.530	377.448	349.625	327.452	260.600	204.648	178.577	162.899
564	464.363	415.151	378.846	350.862	328.566	261.385	205.199	179.032	163.302
565	466.322	416.779	380.248	352.101	329.683	262.171	205.751	179.489	163.704
566	468.289	418.412	381.654	353.344	330.803	262.958	206.304	179.945	164.107
567	470.264	420.051	383.065	354.591	331.926	263.747	206.858	180.402	164.510
568	472.248	421.695	384.480	355.841	333.052	264.538	207.412	180.860	164.914
569	474.240	423.346	385.899	357.094	334.181	265.330	207.967	181.318	165.318
570	476.240	425.002	387.323	358.351	335.312	266.123	208.522	181.777	165.722
571	478.248	426.664	388.752	359.612	336.447	266.918	209.079	182.236	166.127
572	480.264	428.332	390.184	360.876	337.584	267.714	209.636	182.696	166.532
573	482.289	430.006	391.622	362.144	338.724	268.512	210.194	183.156	166.938
574	484.322	431.686	393.064	363.415	339.868	269.311	210.752	183.616	167.344

S					N				
	1100	1200	1300	1400	1500	2000	3000	4000	5000
575	486.364	433.371	394.510	364.690	341.014	270.112	211.311	184.077	167.750
576	488.413	435.063	395.961	365.968	342.163	270.914	211.871	184.539	168.157
577	490.472	436.761	397.416	367.250	343.315	271.718	212.432	185.001	168.564
578	492.539	438.465	398.876	368.535	344.470	272.523	212.994	185.464	168.971
579	494.615	440.175	400.341	369.825	345.628	273.329	213.556	185.927	169.379
580	496.699	441.891	401.810	371.118	346.789	274.138	214.119	186.390	169.787
581	498.792	443.613	403.284	372.414	347.953	274.947	214.682	186.854	170.196
582	500.893	445.342	404.762	373.714	349.120	275.758	215.247	187.319	170.605
583	503.004	447.076	406.245	375.018	350.290	276.571	215.812	187.784	171.014
584	505.123	448.817	407.733	376.326	351.463	277.385	216.378	188.249	171.424
585	507.252	450.564	409.226	377.637	352.638	278.201	216.944	188.715	171.834
586	509.389	452.318	410.723	378.952	353.818	279.018	217.511	189.182	172.244
587	511.535	454.078	412.225	380.271	355.000	279.836	218.079	189.649	172.655
588	513.690	455.844	413.732	381.593	356.185	280.657	218.648	190.116	173.066
589	515.855	457.616	415.243	382.919	357.373	281.478	219.218	190.584	173.478
590	518.029	459.395	416.760	384.249	358.564	282.302	219.788	191.052	173.889
591	520.211	461.180	418.281	385.583	359.758	283.126	220.359	191.521	174.302
592	522.403	462.972	419.807	386.921	360.956	283.952	220.930	191.991	174.714
593	524.605	464.770	421.338	388.262	362.156	284.780	221.503	192.460	175.127
594	526.816	466.575	422.873	389.607	363.360	285.610	222.076	192.931	175.541
595	529.036	468.386	424.414	390.956	364.567	286.440	222.650	193.402	175.955
596	531.266	470.204	425.959	392.309	365.777	287.273	223.224	193.873	176.369
597	533.505	472.029	427.510	393.666	366.990	288.107	223.800	194.345	176.783
598	535.754	473.860	429.065	395.027	368.206	288.942	224.376	194.817	177.198
599	538.012	475.698	430.626	396.391	369.425	289.779	224.953	195.290	177.614
600	540.281	477.543	432.191	397.760	370.648	290.618	225.530	195.763	178.029
601	542.559	479.395	433.762	399.132	371.873	291.458	226.108	196.237	178.445
602	544.847	481.253	435.337	400.508	373.102	292.299	226.687	196.712	178.862
603	547.144	483.118	436.918	401.889	374.334	293.143	227.267	197.186	179.279
604	549.452	484.990	438.503	403.273	375.570	293.987	227.848	197.662	179.696
605	551.770	486.869	440.094	404.661	376.808	294.834	228.429	198.138	180.113
606	554.098	488.755	441.690	406.053	378.050	295.682	229.011	198.614	180.531
607	556.436	490.647	443.291	407.450	379.295	296.531	229.594	199.091	180.950
608	558.784	492.547	444.897	408.850	380.543	297.382	230.177	199.568	181.368

					N				
S	1100	1200	1300	1400	1500	2000	3000	4000	5000
609	561.142	494.454	446.509	410.254	381.795	298.235	230.761	200.046	181.787
610	563.511	496.368	448.125	411.663	383.049	299.089	231.346	200.524	182.207
611	565.890	498.289	449.747	413.075	384.308	299.945	231.932	201.003	182.627
612	568.280	500.217	451.374	414.491	385.569	300.802	232.519	201.482	183.047
613	570.680	502.152	453.007	415.912	386.834	301.661	233.106	201.962	183.467
614	573.090	504.095	454.644	417.337	388.102	302.522	233.694	202.442	183.888
615	575.512	506.044	456.288	418.766	389.373	303.384	234.283	202.923	184.310
616	577.944	508.001	457.936	420.199	390.648	304.247	234.872	203.404	184.731
617	580.386	509.966	459.590	421.636	391.926	305.113	235.462	203.886	185.154
618	582.840	511.938	461.249	423.077	393.207	305.980	236.053	204.368	185.576
619	585.304	513.917	462.913	424.523	394.492	306.848	236.645	204.851	185.999
620	587.780	515.903	464.584	425.972	395.780	307.718	237.238	205.334	186.422
621	590.266	517.897	466.259	427.426	397.071	308.590	237.831	205.818	186.846
622	592.764	519.899	467.940	428.885	398.366	309.463	238.425	206.302	187.270
623	595.273	521.908	469.627	430.347	399.665	310.338	239.020	206.787	187.694
624	597.793	523.925	471.319	431.814	400.966	311.215	239.615	207.272	188.119
625	600.324	525.949	473.016	433.285	402.272	312.093	240.211	207.758	188.544
626	602.867	527.981	474.719	434.760	403.580	312.973	240.808	208.244	188.969
627	605.421	530.021	476.428	436.240	404.892	313.855	241.406	208.731	189.395
628	607.986	532.068	478.143	437.724	406.208	314.738	242.005	209.218	189.822
629	610.564	534.123	479.863	439.212	407.527	315.623	242.604	209.706	190.248
630	613.153	536.187	481.589	440.705	408.850	316.509	243.204	210.194	190.675
631	615.753	538.257	483.320	442.202	410.176	317.397	243.805	210.683	191.103
632	618.366	540.336	485.057	443.703	411.505	318.287	244.407	211.172	191.530
633	620.990	542.423	486.800	445.209	412.839	319.178	245.009	211.661	191.959
634	623.627	544.518	488.549	446.719	414.175	320.071	245.612	212.152	192.387
635	626.275	546.621	490.304	448.234	415.516	320.966	246.216	212.642	192.816
636	628.935	548.732	492.064	449.753	416.859	321.862	246.821	213.134	193.245
637	631.608	550.851	493.831	451.277	418.207	322.760	247.426	213.625	193.675
638	634.293	552.978	495.603	452.805	419.558	323.660	248.032	214.118	194.105
639	636.991	555.113	497.381	454.338	420.912	324.561	248.639	214.610	194.535
640	639.701	557.257	499.165	455.875	422.270	325.464	249.247	215.104	194.966
641	642.423	559.409	500.955	457.417	423.632	326.369	249.855	215.597	195.397
642	645.158	561.569	502.751	458.964	424.998	327.275	250.465	216.092	195.829

S	N								
	1100	1200	1300	1400	1500	2000	3000	4000	5000
643		563.738	504.553	460.515	426.367	328.183	251.075	216.586	196.261
644		565.915	506.361	462.070	427.740	329.093	251.686	217.082	196.693
645		568.100	508.175	463.631	429.116	330.004	252.297	217.577	197.126
646		570.294	509.996	465.196	430.496	330.918	252.909	218.074	197.559
647		572.497	511.822	466.765	431.880	331.832	253.523	218.571	197.993
648		574.708	513.655	468.339	433.268	332.749	254.137	219.068	198.427
649		576.928	515.493	469.918	434.659	333.667	254.751	219.566	198.861
650		579.156	517.338	471.502	436.054	334.587	255.367	220.064	199.295
651		581.394	519.190	473.090	437.453	335.509	255.983	220.563	199.730
652		583.640	521.047	474.683	438.855	336.432	256.600	221.062	200.166
653		585.895	522.911	476.281	440.262	337.357	257.218	221.562	200.602
654		588.159	524.781	477.884	441.672	338.284	257.836	222.062	201.038
655		590.431	526.657	479.492	443.086	339.213	258.456	222.563	201.474
656		592.713	528.540	481.104	444.503	340.143	259.076	223.064	201.911
657		595.004	530.430	482.721	445.925	341.075	259.697	223.566	202.348
658		597.304	532.325	484.343	447.350	342.009	260.318	224.068	202.786
659		599.613	534.227	485.970	448.780	342.944	260.941	224.571	203.224
660		601.931	536.136	487.602	450.213	343.881	261.564	225.075	203.663
661		604.258	538.051	489.239	451.650	344.820	262.188	225.579	204.101
662		606.595	539.973	490.880	453.091	345.761	262.813	226.083	204.541
663		608.941	541.902	492.527	454.536	346.703	263.438	226.588	204.980
664		611.296	543.837	494.179	455.984	347.648	264.065	227.093	205.420
665		613.661	545.778	495.835	457.437	348.594	264.692	227.599	205.861
666		616.036	547.727	497.497	458.894	349.541	265.320	228.106	206.301
667		618.419	549.682	499.164	460.354	350.491	265.949	228.612	206.742
668		620.813	551.644	500.836	461.819	351.442	266.578	229.120	207.184
669		623.216	553.612	502.513	463.287	352.395	267.209	229.628	207.626
670		625.629	555.588	504.195	464.760	353.350	267.840	230.136	208.068
671		628.051	557.570	505.882	466.237	354.307	268.472	230.645	208.511
672		630.484	559.559	507.574	467.717	355.265	269.104	231.155	208.954
673		632.926	561.555	509.271	469.202	356.226	269.738	231.665	209.397
674		635.378	563.558	510.974	470.691	357.188	270.372	232.175	209.841
675		637.841	565.568	512.682	472.184	358.152	271.007	232.686	210.285

S	1100	1200	1300	1400	1500	2000	3000	4000	5000
676		640.313	567.585	514.395	473.680	359.117	271.643	233.198	210.730
677		642.795	569.609	516.113	475.182	360.085	272.280	233.710	211.175
678		645.288	571.640	517.837	476.687	361.054	272.917	234.222	211.620
679		647.790	573.679	519.565	478.196	362.025	273.556	234.736	212.066
680		650.303	575.724	521.300	479.709	362.998	274.195	235.249	212.512
681		652.826	577.776	523.039	481.227	363.973	274.835	235.763	212.959
682		655.360	579.836	524.784	482.749	364.949	275.475	236.278	213.405
683		657.904	581.903	526.534	484.275	365.927	276.117	236.793	213.853
684		660.459	583.977	528.290	485.805	366.908	276.759	237.309	214.300
685		663.024	586.059	530.051	487.339	367.890	277.402	237.825	214.748
686		665.600	588.148	531.818	488.878	368.874	278.046	238.342	215.197
687		668.186	590.244	533.590	490.421	369.859	278.691	238.859	215.646
688		670.784	592.348	535.367	491.968	370.847	279.336	239.376	216.095
689		673.392	594.459	537.150	493.520	371.836	279.983	239.895	216.544
690		676.011	596.578	538.939	495.075	372.828	280.630	240.413	216.994
691		678.641	598.704	540.733	496.636	373.821	281.278	240.933	217.445
692		681.281	600.838	542.532	498.200	374.816	281.927	241.452	217.896
693		683.933	602.979	544.338	499.769	375.813	282.576	241.973	218.347
694		686.596	605.128	546.149	501.342	376.811	283.227	242.494	218.798
695		689.271	607.285	547.965	502.919	377.812	283.878	243.015	219.250
696		691.956	609.450	549.787	504.501	378.814	284.530	243.537	219.702
697		694.653	611.622	551.615	506.088	379.819	285.183	244.059	220.155
698		697.362	613.802	553.449	507.678	380.825	285.836	244.582	220.608
699		700.081	615.990	555.288	509.274	381.833	286.491	245.105	221.062
700		702.813	618.185	557.134	510.873	382.843	287.146	245.629	221.515
701			620.389	558.985	512.477	383.855	287.802	246.154	221.970
702			622.600	560.841	514.086	384.869	288.459	246.679	222.424
703			624.820	562.704	515.699	385.885	289.117	247.204	222.879
704			627.047	564.573	517.317	386.903	289.775	247.730	223.335
705			629.283	566.447	518.939	387.922	290.435	248.257	223.790
706			631.527	568.327	520.566	388.944	291.095	248.784	224.247
707			633.779	570.214	522.197	389.967	291.756	249.312	224.703
708			636.039	572.106	523.833	390.993	292.418	249.840	225.160

					N				
S	1100	1200	1300	1400	1500	2000	3000	4000	5000
709			638.307	574.004	525.474	392.020	293.080	250.368	225.617
710			640.583	575.908	527.119	393.050	293.744	250.897	226.075
711			642.868	577.819	528.769	394.081	294.408	251.427	226.533
712			645.162	579.735	530.423	395.114	295.073	251.957	226.992
713			647.463	581.658	532.082	396.149	295.739	252.488	227.451
714			649.773	583.586	533.746	397.186	296.406	253.019	227.910
715			652.092	585.521	535.415	398.225	297.074	253.551	228.370
716			654.419	587.462	537.088	399.266	297.742	254.083	228.830
717			656.754	589.409	538.766	400.309	298.411	254.616	229.290
718			659.099	591.363	540.449	401.354	299.082	255.150	229.751
719			661.451	593.322	542.137	402.401	299.752	255.684	230.212
720			663.813	595.288	543.829	403.450	300.424	256.218	230.674
721			666.183	597.261	545.526	404.501	301.097	256.753	231.136
722			668.562	599.239	547.228	405.554	301.770	257.288	231.598
723			670.950	601.224	548.935	406.609	302.445	257.824	232.061
724			673.347	603.216	550.647	407.665	303.120	258.361	232.524
725			675.753	605.213	552.364	408.724	303.796	258.898	232.988
726			678.168	607.218	554.086	409.785	304.473	259.435	233.452
727			680.591	609.228	555.812	410.848	305.150	259.974	233.916
728			683.024	611.246	557.544	411.613	305.829	260.512	234.381
729			685.466	613.269	559.280	412.980	306.508	261.051	234.846
730			687.917	615.300	561.022	414.049	307.188	261.591	235.311
731			690.377	617.337	562.768	415.120	307.869	262.131	235.777
732			692.847	619.380	564.520	416.193	308.551	262.672	236.244
733			695.326	621.430	566.276	417.268	309.234	263.213	236.710
734			697.814	623.487	568.038	418.345	309.917	263.755	237.177
735			700.311	625.551	569.804	419.424	310.602	264.297	237.645
736			702.819	627.621	571.576	420.506	311.287	264.840	238.113
737			705.335	629.698	573.353	421.589	311.973	265.384	238.581
738			707.861	631.782	575.135	422.674	312.660	265.928	239.050
739			710.397	633.873	576.923	423.762	313.348	266.472	239.519
740			712.942	635.971	578.715	424.851	314.037	267.017	239.988
741			715.497	638.075	580.513	425.943	314.726	267.563	240.458

S	1100	1200	N 1300	1400	1500	2000	3000	4000	5000
742			718.062	640.187	582.315	427.036	315.417	268.109	240.928
743			720.636	642.305	584.124	428.132	316.108	268.655	241.399
744			723.221	644.431	585.937	429.230	316.800	269.203	241.870
745			725.815	646.563	587.756	430.330	317.493	269.750	242.341
746			728.420	648.703	589.580	431.432	318.187	270.298	242.813
747			731.034	650.849	591.409	432.536	318.881	270.847	243.285
748			733.658	653.003	593.243	433.642	319.577	271.396	243.758
749			736.293	655.164	595.083	434.750	320.273	271.946	244.231
750			738.938	657.332	596.929	435.861	320.971	272.497	244.704
751			741.593	659.507	598.780	436.974	321.669	273.048	245.178
752			744.258	661.690	600.636	438.088	322.368	273.599	245.652
753			746.933	663.879	602.498	439.205	323.067	274.151	246.127
754			749.619	666.076	604.365	440.324	323.768	274.703	246.602
755			752.316	668.281	606.237	441.445	324.470	275.257	247.077
756			755.023	670.493	608.116	442.568	325.172	275.810	247.553
757			757.740	672.712	609.999	443.694	325.876	276.364	248.029
758			760.469	674.939	611.889	444.822	326.580	276.919	248.506
759			763.207	677.173	613.783	445.951	327.285	277.474	248.983
760				679.414	615.684	447.083	327.991	278.030	249.460
761				681.664	617.590	448.217	328.698	278.586	249.938
762				683.921	619.502	449.354	329.405	279.143	250.416
763				686.185	621.419	450.492	330.114	279.700	250.894
764				688.457	623.342	451.633	330.823	280.258	251.373
765				690.737	625.271	452.775	331.533	280.817	251.852
766				693.024	627.206	453.921	332.245	281.376	252.332
767				695.320	629.146	455.068	332.957	281.935	252.812
768				697.623	631.092	456.217	333.670	282.495	253.293
769				699.934	633.044	457.369	334.384	283.056	253.774
770				702.253	635.002	458.523	**335.098**	283.617	254.255
771				704.579	636.966	459.679	335.814	284.179	254.737
772				706.914	638.935	460.837	336.530	284.741	255.219
773				709.257	640.910	461.998	337.248	285.304	255.701
774				711.607	642.892	463.161	337.966	285.867	256.184

				N					
S	1100	1200	1300	1400	1500	2000	3000	4000	5000
775				713.966	644.879	464.326	338.685	286.431	256.667
776				716.333	646.872	465.493	339.405	286.996	257.151
777				718.708	648.872	466.662	340.126	287.561	257.635
778				721.091	650.877	467.834	340.848	288.126	258.120
779				723.483	652.888	469.008	341.570	288.692	258.605
780				725.882	654.906	470.185	342.294	289.259	259.090
781				728.290	656.930	471.363	343.019	289.826	259.576
782				730.707	658.959	472.544	343.744	290.394	260.062
783				733.131	660.995	473.727	344.470	290.962	260.548
784				735.565	663.037	474.913	345.197	291.531	261.035
785				738.006	665.085	476.101	345.925	292.100	261.522
786				740.456	667.140	477.291	346.654	292.670	262.010
787				742.915	669.200	478.483	347.384	293.240	262.498
788				745.382	671.267	479.678	348.115	293.812	262.987
789				747.858	673.341	480.875	348.847	294.383	263.475
790				750.342	675.420	482.074	349.579	294.955	263.965
791				752.840	677.506	483.276	350.313	295.528	264.454
792				755.340	679.599	484.480	351.047	296.101	264.944
793				757.850	681.698	485.686	351.782	296.675	265.435
794				760.370	683.803	486.895	352.519	297.249	265.926
795				762.900	685.915	488.106	353.256	297.824	266.417
796				765.430	688.033	489.319	353.994	298.399	266.909
797				767.980	690.157	490.535	354.733	298.975	267.401
798				770.540	692.289	491.753	355.472	299.552	267.893
799				773.100	694.427	492.974	356.213	300.129	268.386
800				775.670	696.571	494.197	356.955	300.707	268.880
801				778.256	698.722	495.422	357.697	301.285	269.373
802				780.848	700.880	496.650	358.441	301.864	269.867
803				783.449	703.044	497.880	359.185	302.443	270.263
804				786.059	705.215	499.112	359.930	303.023	270.857
805				788.679	707.393	500.347	360.677	303.603	271.352
806				791.308	709.578	501.585	361.424	304.184	271.848
807				793.947	711.769	502.824	362.172	304.766	272.344

S	1100	1200	1300	1400	1500	2000	3000	4000	5000
					N				
808				796.595	713.968	504.067	362.921	305.348	272.841
809				799.252	716.173	505.311	363.671	305.931	273.337
810				801.919	718.385	506.558	364.422	306.514	273.835
811				804.596	720.604	507.808	365.173	307.098	274.333
812				807.282	722.830	509.060	365.926	307.682	274.831
813				809.978	725.063	510.314	366.680	308.267	275.329
814				812.684	727.303	511.571	367.434	308.852	275.828
815				815.400	729.550	512.830	368.190	309.438	276.327
816				818.125	731.804	514.092	368.946	310.025	276.827
817				820.861	734.065	515.356	369.704	310.612	277.327
818					736.333	516.523	370.462	311.200	277.828
819					738.609	517.892	371.221	311.788	278.329
820					740.891	519.164	371.981	312.377	278.830
821					743.181	520.438	372.742	312.966	279.332
822					745.478	521.715	373.504	313.556	279.834
823					747.783	522.994	374.267	314.147	280.337
824					750.094	524.276	375.031	314.738	280.840
825					752.414	525.560	375.796	315.330	281.343
826					754.740	526.847	376.562	315.922	281.847
827					757.074	528.137	377.329	316.515	282.351
828					759.415	529.429	378.096	317.108	282.856
829					761.764	530.723	378.865	317.702	283.361
830					764.120	532.020	379.634	318.296	283.867
831					766.484	533.320	380.405	318.891	284.372
832					768.856	534.622	381.176	319.487	284.879
833					771.235	535.927	381.949	320.083	285.385
834					773.622	537.234	382.722	320.680	285.892
835					776.016	538.544	383.496	321.277	286.400
836					778.418	539.857	384.272	321.875	286.908
837					780.828	541.172	385.048	322.473	287.416
838					783.246	542.490	385.825	323.073	287.925
839					785.671	543.810	386.603	323.672	288.434
840					788.105	545.133	387.382	324.272	288.944

S	N 1500	2000	3000	4000	5000	S	N 1500	2000	3000	4000	5000
841	790.546	546.459	388.162	324.873	289.453	871	864.945	587.485	412.016	343.163	304.943
842	792.995	547.787	388.943	325.474	289.964	872	867.641	588.896	412.826	343.782	305.466
843	795.453	549.118	389.725	326.076	290.475	873	870.346	590.309	413.637	344.401	305.989
844	797.918	550.451	390.508	326.679	290.986	874	873.060	591.725	414.450	345.021	306.513
845	800.391	551.788	391.292	327.282	291.497	875	875.783	593.145	415.263	345.641	307.037
846	802.872	553.127	392.077	327.885	292.009	876	878.516	594.567	416.078	346.262	307.561
847	805.362	554.468	392.863	328.489	292.522	877		595.992	416.893	346.884	308.086
848	807.859	555.812	393.650	329.094	293.035	878		597.419	417.709	347.506	308.611
849	810.365	557.159	394.437	329.700	293.548	879		598.850	418.527	348.129	309.137
850	812.879	558.509	395.226	330.305	294.061	880		600.284	419.345	348.752	309.663
851	815.401	559.861	396.016	330.912	294.576	881		601.720	420.165	349.376	310.189
852	817.932	561.216	396.806	331.519	295.090	882		603.160	420.985	350.000	310.716
853	820.471	562.574	397.598	332.127	295.605	883		604.602	421.806	350.626	311.244
854	823.018	563.935	398.391	332.735	296.120	884		606.047	422.629	351.251	311.771
855	825.574	565.298	399.184	333.344	296.636	885		607.496	423.452	351.878	312.300
856	828.138	566.664	399.979	333.953	297.152	886		608.947	424.277	352.504	312.828
857	830.138	568.032	400.775	334.563	297.668	887		610.401	425.102	353.132	313.357
858	830.710	569.404	401.571	335.173	298.185	888		611.858	425.929	353.760	313.887
859	833.291	570.778	402.369	335.785	298.703	889		613.318	426.756	354.389	314.416
860	835.881	572.155	403.167	336.396	299.221	890		614.781	427.585	355.018	314.947
861	838.480	573.535	403.967	337.008	299.739	891		616.248	428.414	355.648	315.477
862	841.087	574.917	404.767	337.621	300.257	892		617.717	429.245	356.278	316.008
863	843.702	576.302	405.568	338.235	300.776	893		619.189	430.077	356.909	316.540
864	846.327	577.690	406.371	338.849	301.296	894		620.664	430.909	357.541	317.072
865	848.960	579.081	407.174	339.463	301.816	895		622.142	431.743	358.173	317.604
866	851.602	580.475	407.979	340.078	302.336	896		623.623	432.577	358.806	318.137
867	854.253	581.871	408.784	340.694	302.857	897		625.107	433.413	359.439	318.670
868	856.912	583.270	409.591	341.310	303.378	898		626.594	434.250	360.073	319.204
869	859.581	584.672	410.398	341.927	303.899	899		628.085	435.088	360.708	319.738
870	862.259	586.077	411.206	342.545	304.421	900		629.578	435.926	361.343	320.272

S	N				S	N			
	2000	3000	4000	5000		2000	3000	4000	5000
901	631.074	436.766	361.979	320.807	931	677.436	462.449	381.333	337.051
902	632.574	437.607	362.615	321.343	932	679.031	463.321	381.988	337.599
903	634.077	438.449	363.252	321.878	933	680.631	464.195	382.643	338.147
904	635.582	439.292	363.890	322.414	934	682.233	465.070	383.299	338.696
905	637.091	440.136	364.528	322.951	935	683.839	465.946	383.955	339.246
906	638.603	440.981	365.167	323.488	936	685.448	466.823	384.612	339.795
907	640.118	441.827	365.806	324.025	937	687.061	467.701	385.270	340.346
908	641.636	442.674	366.446	324.563	938	688.677	468.580	385.928	340.896
909	643.157	443.522	367.087	325.102	939	690.296	469.460	386.587	341.447
910	644.682	444.372	367.728	325.640	940	691.919	470.341	387.246	341.999
911	646.209	445.222	368.370	326.179	941	693.545	471.223	387.906	342.551
912	647.740	446.073	369.012	326.719	942	695.175	472.107	388.567	343.103
913	649.274	446.926	369.655	327.259	943	696.808	472.991	389.228	343.656
914	650.811	447.779	370.299	327.799	944	698.444	473.877	389.890	344.209
915	652.351	448.633	370.943	328.340	945	700.084	474.763	390.553	344.763
916	653.895	449.489	371.588	328.881	946	701.728	475.651	391.216	345.317
917	655.441	450.345	372.233	329.423	947	703.375	476.540	391.880	345.871
918	656.991	451.203	372.879	329.965	948	705.025	477.430	392.544	346.426
919	658.544	452.062	373.526	330.507	949	706.679	478.321	393.209	346.982
920	660.101	452.921	374.173	331.050	950	708.336	479.213	393.875	347.537
921	661.660	453.782	374.821	331.594	951	709.997	480.107	394.541	348.094
922	663.223	454.644	375.470	332.138	952	711.662	481.001	395.208	348.650
923	664.789	455.507	376.119	332.682	953	713.330	481.897	395.876	349.207
924	666.358	456.371	376.768	333.226	954	715.001	482.793	396.544	349.765
925	667.931	457.236	377.419	333.771	955	716.676	483.691	397.212	350.323
926	669.507	458.102	378.070	334.317	956	718.355	484.590	397.882	350.881
927	671.086	458.969	378.721	334.863	957	720.037	485.490	398.552	351.440
928	672.668	459.838	379.373	335.409	958	721.723	486.391	399.223	351.999
929	674.254	460.707	380.026	335.956	959	723.412	487.293	399.894	352.559
930	675.843	461.577	380.679	336.503	960	725.106	488.196	400.566	353.119

		N					N		
S	2000	3000	4000	5000	S	2000	3000	4000	5000
961	726.802	489.101	401.238	353.680	991	779.435	516.761	421.708	370.700
962	728.502	490.006	401.911	354.241	992	781.249	517.700	422.400	371.274
963	730.206	490.913	402.585	354.802	993	783.067	518.641	423.093	371.849
964	731.914	491.820	403.260	355.364	994	784.889	519.584	423.786	372.424
965	733.625	492.729	403.935	355.926	995	786.715	520.527	424.480	372.999
966	735.340	493.639	404.610	356.489	996	788.545	521.472	425.175	373.575
967	737.059	494.551	405.287	357.052	997	790.379	522.417	425.871	374.152
968	738.781	495.463	405.963	357.616	998	792.217	523.364	426.567	374.728
969	740.507	496.376	406.641	358.180	999	794.059	524.312	427.263	375.306
970	742.236	497.291	407.319	358.744	1000	795.905	525.261	427.961	375.883
971	743.970	498.206	407.998	359.309	1001	797.755	526.212	428.659	376.462
972	745.707	499.123	408.677	359.875	1002	799.609	527.164	429.357	377.040
973	747.448	500.041	409.358	360.441	1003	801.468	528.116	430.057	377.619
974	749.192	500.960	410.038	361.007	1004	803.330	529.070	430.757	378.199
975	750.941	501.881	410.720	361.573	1005	805.197	530.025	431.457	378.779
976	752.693	502.802	411.402	362.141	1006	807.067	530.982	432.159	379.359
977	754.449	503.724	412.084	362.708	1007	808.942	531.939	432.860	379.940
978	756.208	504.648	412.767	363.276	1008	810.821	532.898	433.563	380.521
979	757.972	505.573	413.451	363.845	1009	812.704	533.858	434.266	381.103
980	759.739	506.499	414.136	364.413	1010	814.591	534.819	434.970	381.685
981	761.510	507.426	414.821	364.983	1011	816.482	535.782	435.675	382.268
982	763.285	508.354	415.507	365.552	1012	818.378	536.745	436.380	382.851
983	765.064	509.284	416.193	366.123	1013	820.278	537.710	437.086	383.434
984	766.847	510.214	416.880	366.693	1014	822.182	538.676	437.792	384.018
985	768.634	511.146	417.568	367.264	1015	824.090	539.643	438.499	384.602
986	770.424	512.079	418.256	367.836	1016	826.002	540.611	439.207	385.187
987	772.219	513.013	418.945	368.408	1017	827.919	541.581	439.915	385.773
988	774.017	513.948	419.635	368.980	1018	829.840	542.552	440.624	386.358
989	775.819	514.884	420.325	369.553	1019	831.765	543.524	441.334	386.945
990	777.625	515.822	421.016	370.126	1020	833.695	544.497	442.045	387.531

	N						N			
S	2000	3000	4000	5000	S	2000	3000	4000	5000
1021	835.629	545.471	442.756	388.118	1051	895.714	575.277	464.397	405.941
1022	837.567	546.447	443.467	388.706	1052	897.788	576.290	465.129	406.542
1023	839.509	547.424	444.180	389.294	1053	899.867	577.304	465.861	407.144
1024	841.456	548.402	444.893	389.882	1054	901.950	578.319	466.594	407.746
1025	843.407	549.381	445.607	390.471	1055	904.039	579.336	467.328	408.348
1026	845.363	550.361	446.321	391.060	1056	906.132	580.354	468.063	408.951
1027	847.323	551.343	447.036	391.650	1057	908.230	581.374	468.798	409.555
1028	849.287	552.326	447.752	392.240	1058	910.333	582.394	469.534	410.159
1029	851.256	553.310	448.468	392.831	1059	912.440	583.416	470.270	410.763
1030	853.229	554.296	449.185	393.422	1060	914.553	584.439	471.007	411.368
1031	855.207	555.282	449.903	394.014	1061	916.670	585.464	471.745	411.973
1032	857.189	556.270	450.621	394.606	1062	918.792	586.489	472.484	412.579
1033	859.176	557.259	451.340	395.198	1063	920.920	587.516	473.223	413.185
1034	861.167	558.250	452.060	395.791	1064	923.052	588.545	473.963	413.792
1035	863.162	559.241	452.780	396.385	1065	925.189	589.574	474.704	414.399
1036	865.162	560.234	453.501	396.979	1066	927.331	590.605	475.445	415.006
1037	867.167	561.228	454.223	397.573	1067	929.478	591.637	476.187	415.614
1038	869.176	562.224	454.945	398.168	1068	931.629	592.671	476.930	416.223
1039	871.190	563.220	455.668	398.763	1069	933.786	593.706	477.673	416.832
1040	873.208	564.218	456.392	399.359	1070	935.948	594.742	478.417	417.441
1041	875.231	565.217	457.116	399.955	1071	938.115	595.779	479.162	418.051
1042	877.258	566.217	457.841	400.551	1072	940.287	596.818	479.907	418.661
1043	879.290	567.219	458.567	401.148	1073	942.464	597.858	480.653	419.272
1044	881.327	568.222	459.294	401.746	1074	944.646	598.899	481.400	419.884
1045	883.368	569.226	460.021	402.344	1075	946.834	599.942	482.148	420.495
1046	885.414	570.231	460.748	402.942	1076	949.026	600.986	482.896	421.107
1047	887.464	571.238	461.477	403.541	1077	951.223	602.031	483.645	421.720
1048	889.520	572.246	462.206	404.140	1078	953.426	603.078	484.394	422.333
1049	891.580	573.255	462.936	404.740	1079	955.634	604.126	485.145	422.947
1050	893.644	574.265	463.666	405.340	1080	957.847	605.175	485.896	423.561

		N					N		
S	2000	3000	4000	5000	S	2000	3000	4000	5000
1081	960.065	606.225	486.647	424.175	1111	1029.000	638.368	509.523	442.829
1082	962.288	607.277	487.400	424.790	1112	1031.000	639.461	510.296	443.458
1083	964.517	608.331	488.153	425.406	1113	1034.000	640.554	511.070	444.087
1084	966.750	609.385	488.906	426.022	1114	1036.000	641.650	511.845	444.717
1085	968.990	610.441	489.661	426.638	1115	1039.000	642.747	512.621	445.348
1086	971.234	611.498	490.416	427.255	1116	1041.000	643.845	513.397	445.979
1087	973.483	612.557	491.172	427.872	1117	1044.000	644.944	514.174	446.610
1088	975.738	613.617	491.928	428.490	1118	1046.000	646.045	514.952	447.242
1089	977.999	614.678	492.686	429.108	1119	1048.000	647.148	515.730	447.875
1090	980.264	615.740	493.443	429.727	1120	1051.000	648.251	516.510	448.507
1091	982.535	616.804	494.202	430.346	1121	1053.000	649.357	517.290	449.141
1092	984.812	617.870	494.961	430.966	1122	1056.000	650.463	518.070	449.775
1093	987.093	618.936	495.721	431.586	1123	1058.000	651.571	518.852	450.409
1094	989.380	620.004	496.482	432.207	1124	1061.000	652.681	519.634	451.044
1095	991.673	621.073	497.244	432.828	1125	1063.000	653.791	520.417	451.679
1096	993.971	622.144	498.006	433.449	1126	1066.000	654.904	521.200	452.315
1097	996.275	623.216	498.769	434.071	1127	1068.000	656.017	521.984	452.951
1098	998.584	624.290	499.532	434.694	1128	1070.000	657.132	522.769	453.588
1099	1001.000	625.364	500.296	435.317	1129	1073.000	658.249	523.555	454.225
1100	1003.000	626.440	501.061	435.940	1130	1075.000	659.367	524.341	454.862
1101	1006.000	627.518	501.827	436.564	1131	1078.000	660.486	525.129	455.501
1102	1008.000	628.597	502.593	437.188	1132	1080.000	661.607	525.917	456.139
1103	1011.000	629.677	503.360	437.813	1133	1083.000	662.729	526.705	456.778
1104	1013.000	630.758	504.128	438.438	1134	1086.000	663.853	527.494	457.418
1105	1015.000	631.841	504.897	439.064	1135	1088.000	664.978	528.285	458.058
1106	1017.000	632.926	505.666	439.690	1136	1091.000	666.104	529.075	458.698
1107	1020.000	634.011	506.436	440.317	1137	1093.000	667.232	529.867	459.339
1108	1022.000	635.098	507.206	440.944	1138	1096.000	668.362	530.659	459.981
1109	1024.000	636.187	507.978	441.572	1139	1098.000	669.492	531.452	460.623
1110	1027.000	637.277	508.750	442.200	1140	1101.000	670.625	532.246	461.265

S	N 2000	3000	4000	5000	S	N 2000	3000	4000	5000
1141	1103.000	671.758	533.040	461.908	1172	707.635	558.036	482.081	
1142	1106.000	672.894	533.835	462.552	1173	708.816	558.854	482.739	
1143	1108.000	674.030	534.631	463.196	1174	709.999	559.673	483.398	
1144	1111.000	675.168	535.428	463.840	1175	711.183	560.493	484.058	
1145	1114.000	676.308	536.225	464.485	1176	712.369	561.313	484.718	
1146	1116.000	677.449	537.024	465.130	1177	713.557	562.135	485.378	
1147	1119.000	678.591	537.822	465.776	1178	714.746	562.957	486.039	
1148	1121.000	679.735	538.622	466.422	1179	715.936	563.779	486.701	
1149	1124.000	680.881	539.422	467.069	1180	717.128	564.603	487.363	
1150	1127.000	682.028	540.223	467.717	1181	718.322	565.427	488.025	
1151	1129.000	683.176	541.025	468.364	1182	719.517	566.252	488.688	
1152	1132.000	684.326	541.828	469.013	1183	720.713	567.078	489.352	
1153	1135.000	685.477	542.631	469.661	1184	721.912	567.905	490.016	
1154	1137.000	686.630	543.435	470.311	1185	723.111	568.732	490.680	
1155	1140.000	687.784	544.240	470.960	1186	724.313	569.561	491.345	
1156	1143.000	688.940	545.046	471.611	1187	725.515	570.389	492.011	
1157	1145.000	690.097	545.852	472.261	1188	726.720	571.219	492.677	
1158	1148.000	691.256	546.659	472.913	1189	727.926	572.050	493.343	
1159	1151.000	692.416	547.467	473.564	1190	729.133	572.881	494.010	
1160	1153.000	693.578	548.275	474.216	1191	730.343	573.713	494.678	
1161	1156.000	694.741	549.084	474.869	1192	731.553	574.546	495.345	
1162	1159.000	695.905	549.894	475.522	1193	732.765	575.379	496.345	
1163	1161.000	697.072	550.705	476.176	1194	733.979	576.213	496.683	
1164	1164.000	698.239	551.517	476.830	1195	735.195	577.048	497.352	
1165	1167.000	699.408	552.329	477.485	1196	736.412	577.884	498.022	
1166		700.579	553.142	478.140	1197	737.630	578.721	498.693	
1167		701.751	553.956	478.795	1198	738.850	579.558	499.364	
1168		702.925	554.770	479.451	1199	740.072	580.396	500.035	
1169		704.100	555.585	480.108	1200	741.295	581.235	500.707	
1170		705.277	556.401	480.765					
1171		706.455	557.218	481.423					

	N					N		
S	3000	4000	5000		S	3000	4000	5000
1201	742.520	582.075	501.380		1231	780.020	607.631	521.788
1202	743.747	582.916	502.053		1232	781.295	608.495	522.476
1203	744.975	583.757	502.726		1233	782.572	609.360	523.165
1204	746.204	584.599	503.400		1234	783.851	610.225	523.854
1205	747.436	585.442	504.074		1235	785.132	611.092	524.544
1206	748.669	586.285	504.749		1236	786.414	611.959	525.234
1207	749.903	587.130	505.425		1237	787.698	612.827	525.925
1208	751.139	587.975	506.101		1238	788.983	613.696	526.616
1209	752.377	588.821	506.777		1239	790.271	614.566	527.308
1210	753.616	589.667	507.454		1240	791.560	615.437	528.000
1211	754.857	590.515	508.132		1241	792.850	616.308	528.693
1212	756.100	591.363	508.810		1242	794.143	617.180	529.386
1213	757.344	592.212	509.488		1243	795.437	618.053	530.080
1214	758.590	593.062	510.167		1244	796.733	618.927	530.774
1215	759.837	593.913	510.847		1245	798.030	619.801	531.469
1216	761.086	594.764	511.527		1246	799.329	620.677	532.164
1217	762.337	595.616	512.207		1247	800.630	621.553	532.860
1218	763.589	596.469	512.888		1248	801.933	622.430	533.557
1219	764.843	597.323	513.570		1249	803.237	623.308	534.253
1220	766.099	598.178	514.252		1250	804.543	624.187	534.951
1221	767.356	599.033	514.935		1251	805.851	625.066	535.649
1222	768.615	599.889	515.618		1252	807.161	625.946	**536.347**
1223	769.876	600.746	516.301		1253	808.472	626.827	537.046
1224	771.138	601.604	516.985		1254	809.785	627.709	537.746
1225	772.402	602.463	517.670		1255	811.100	628.592	538.446
1226	773.667	603.322	518.355		1256	812.416	629.476	539.146
1227	774.934	604.182	519.040		1257	813.734	630.360	539.847
1228	776.203	605.043	519.727		1258	815.054	631.245	540.549
1229	777.474	605.905	520.413		1259	816.376	632.131	541.251
1230	778.746	606.767	521.100		1260	817.699	633.018	541.954

S	N				S	N		
	3000	4000	5000			3000	4000	5000
1261	819.025	633.906	542.657		1291	859.611	660.921	563.995
1262	820.352	634.794	543.361		1292	860.993	661.835	564.715
1263	821.680	635.683	544.065		1293	862.375	662.749	565.435
1264	823.011	636.573	544.769		1294	863.760	663.665	566.156
1265	824.343	637.464	545.475		1295	865.147	664.581	566.877
1266	825.677	638.356	546.180		1296	866.535	665.498	567.598
1267	827.013	639.249	546.887		1297	867.926	666.416	568.320
1268	828.351	640.142	547.301		1298	869.318	667.334	569.043
1269	829.690	641.036	548.301		1299	870.712	668.254	569.766
1270	831.031	641.932	549.009		1300	872.108	669.174	570.490
1271	832.374	642.827	549.717		1301	873.506	670.095	571.214
1272	833.719	643.724	550.426		1302	874.906	671.017	571.939
1273	835.065	644.622	551.135		1303	876.307	671.940	572.665
1274	836.413	645.520	551.845		1304	877.711	672.864	573.390
1275	837.764	646.419	552.556		1305	879.116	673.789	574.117
1276	839.115	647.319	553.267		1306	880.523	674.714	574.844
1277	840.469	648.220	553.978		1307	881.932	675.641	575.571
1278	841.825	649.122	554.690		1308	883.344	676.568	576.300
1279	843.182	650.025	555.403		1309	884.756	677.496	577.028
1280	844.541	650.928	556.116		1310	886.171	678.425	577.757
1281	845.902	651.832	556.830		1311	887.588	679.355	578.487
1282	847.265	652.738	557.544		1312	889.007	680.286	579.217
1283	848.629	653.644	558.259		1313	890.427	681.217	579.948
1284	849.996	654.550	558.974		1314	891.850	682.150	580.679
1285	851.364	655.458	559.690		1315	893.274	683.083	581.411
1286	852.734	656.366	560.406		1316	894.701	684.017	582.143
1287	854.106	657.276	561.123		1317	896.129	684.952	582.876
1288	855.479	658.186	561.840		1318	897.559	685.888	583.610
1289	856.855	659.097	562.558		1319	898.992	686.825	584.344
1290	858.232	660.009	563.276		1320	900.426	687.763	585.078

S	N 3000	4000	5000	S	N 3000	4000	5000
1321	901.862	688.701	585.814	1351	945.863	717.268	608.121
1322	903.300	689.641	586.549	1352	947.361	718.234	608.873
1323	904.740	690.581	587.285	1353	948.861	719.201	609.626
1324	906.181	691.522	588.022	1354	950.363	720.169	610.379
1325	907.625	692.464	588.759	1355	951.867	721.138	611.133
1326	909.071	693.407	589.497	1356	953.373	722.108	611.887
1327	910.519	694.351	590.236	1357	954.882	723.078	612.642
1328	911.969	695.295	590.974	1358	956.392	724.050	613.398
1329	913.420	696.241	591.714	1359	957.904	725.022	614.154
1330	914.874	697.187	592.454	1360	959.419	725.996	614.910
1331	916.330	698.135	593.194	1361	960.935	726.970	615.668
1332	917.787	699.083	593.936	1362	962.454	727.945	616.425
1333	919.247	700.032	594.677	1363	963.975	728.921	617.184
1334	920.709	700.982	595.419	1364	965.498	729.898	617.942
1335	922.172	701.933	596.162	1365	967.023	730.876	618.702
1336	923.638	702.885	596.905	1366	968.550	731.855	619.462
1337	925.105	703.837	597.649	1367	970.079	732.835	620.222
1338	926.575	704.791	598.394	1368	971.610	733.815	620.983
1339	928.046	705.745	599.139	1369	973.143	734.797	621.745
1340	929.520	706.701	599.884	1370	974.679	735.779	622.507
1341	930.996	707.657	600.630	1371	976.216	736.763	623.270
1342	932.473	708.614	601.377	1372	977.756	737.747	624.033
1343	933.953	709.572	602.124	1373	979.298	738.732	624.797
1344	935.435	710.531	602.872	1374	980.842	739.719	625.562
1345	936.918	711.491	603.620	1375	982.388	740.706	626.327
1346	938.404	712.451	604.369	1376	983.936	741.694	627.092
1347	939.892	713.413	605.118	1377	985.487	742.683	627.858
1348	941.381	714.375	605.868	1378	987.039	743.673	628.625
1349	942.873	715.339	606.618	1379	988.594	744.663	629.392
1350	944.367	716.303	607.370	1380	990.151	745.655	630.160

S	N 3000	4000	5000	S	N 3000	4000	5000
1381	991.710	746.648	630.929	1411	1040.000	776.866	654.247
1382	993.271	747.641	631.698	1412	1041.000	777.888	655.034
1383	994.834	748.636	632.467	1413	1043.000	778.911	655.820
1384	996.400	749.631	633.237	1414	1044.000	779.935	656.608
1385	997.968	750.628	634.008	1415	1046.000	780.960	657.396
1386	999.537	751.625	634.779	1416	1048.000	781.986	658.184
1387	1001.000	752.623	635.551	1417	1049.000	783.013	658.973
1388	1003.000	753.623	636.324	1418	1051.000	784.040	659.763
1389	1004.000	754.623	637.097	1419	1053.000	785.069	660.553
1390	1006.000	755.624	637.870	1420	1054.000	786.099	661.344
1391	1007.000	756.626	638.644	1421	1056.000	787.130	662.136
1392	1009.000	757.629	639.419	1422	1058.000	788.161	662.928
1393	1011.000	758.633	640.194	1423	1059.000	789.194	663.720
1394	1012.000	759.638	640.970	1424	1061.000	790.228	664.514
1395	1014.000	760.644	641.746	1425	1063.000	791.262	665.307
1396	1015.000	761.650	642.523	1426	1064.000	792.298	666.102
1397	1017.000	762.658	643.301	1427	1066.000	793.335	666.897
1398	1019.000	763.667	644.079	1428	1067.000	794.372	667.692
1399	1020.000	764.676	644.858	1429	1069.000	795.411	668.488
1400	1022.000	765.687	645.637	1430	1071.000	796.450	669.285
1401	1023.000	766.698	646.417	1431	1073.000	797.491	670.082
1402	1025.000	767.711	647.197	1432	1074.000	798.532	670.880
1403	1027.000	768.724	647.978	1433	1076.000	799.575	671.679
1404	1028.000	769.739	648.760	1434	1078.000	800.618	672.478
1405	1030.000	770.754	649.542	1435	1079.000	801.663	673.278
1406	1031.000	771.770	650.325	1436	1081.000	802.708	674.079
1407	1033.000	772.787	651.108	1437	1083.000	803.755	674.879
1408	1035.000	773.806	651.892	1438	1084.000	804.802	675.680
1409	1036.000	774.825	652.677	1439	1086.000	805.850	676.482
1410	1038.000	775.845	653.462	1440	1088.000	806.900	677.285

		N					N	
S	3000	4000	5000		S	3000	4000	5000
1441	1089.000	807.950	678.088		1471	1141.000	839.929	702.463
1442	1091.000	809.002	678.892		1472	1143.000	841.011	703.285
1443	1093.000	810.054	679.696		1473	1145.000	842.094	704.107
1444	1094.000	811.107	680.000		1474	1147.000	843.178	704.930
1445	1096.000	812.162	681.307		1475	1148.000	844.263	705.754
1446	1098.000	813.217	682.113		1476	1150.000	845.348	706.578
1447	1100.000	814.274	682.920		1477	1152.000	846.435	707.403
1448	1101.000	815.331	683.727		1478	1154.000	847.523	708.228
1449	1103.000	816.389	684.535		1479	1156.000	848.612	709.055
1450	1105.000	817.449	685.344		1480	1157.000	849.702	709.881
1451	1106.000	818.509	686.153		1481	1159.000	850.793	710.709
1452	1108.000	819.570	686.963		1482	1161.000	851.885	711.536
1453	1110.000	820.633	687.773		1483	1163.000	852.978	712.365
1454	1112.000	821.696	688.584		1484	1165.000	854.073	713.194
1455	1113.000	822.761	689.396		1485	1166.000	855.168	714.024
1456	1115.000	823.826	690.208		1486	1168.000	856.264	714.854
1457	1117.000	824.893	691.021		1487	1170.000	857.361	715.685
1458	1119.000	825.960	691.834		1488	1172.000	858.460	716.517
1459	1120.000	827.029	692.648		1489	1174.000	859.559	717.349
1460	1122.000	828.098	693.463		1490	1176.000	860.660	718.182
1461	1124.000	829.168	694.278		1491	1177.000	861.761	719.015
1462	1126.000	830.240	695.094		1492	1179.000	862.864	719.849
1463	1127.000	831.313	695.910		1493	1181.000	863.967	720.684
1464	1129.000	832.386	696.727		1494	1183.000	865.072	721.519
1465	1131.000	833.461	697.545		1495	1185.000	866.178	722.355
1466	1133.000	834.536	698.363		1496	1186.000	867.284	723.192
1467	1134.000	835.613	699.182		1497	1188.000	868.392	724.029
1468	1136.000	836.690	700.001		1498	1190.000	869.501	724.867
1469	1138.000	837.769	700.821		1499	1192.000	870.611	725.705
1470	1140.000	838.849	701.642		1500	1194.000	871.722	726.544

S	N 3000	4000	5000	S	N 3000	4000	5000
1501	1196.000	872.834	727.384	1531	1252.000	906.695	752.864
1502	1198.000	873.947	728.224	1532	1254.000	907.841	753.723
1503	1199.000	875.061	729.065	1533	1256.000	908.988	754.582
1504	1201.000	876.176	729.906	1534	1258.000	910.135	755.443
1505	1203.000	877.293	730.749	1535	1260.000	911.284	756.304
1506	1205.000	878.410	731.591	1536	1262.000	912.434	757.165
1507	1207.000	879.529	732.435	1537	1264.000	913.585	758.028
1508	1209.000	880.648	733.279	1538	1266.000	914.738	758.891
1509	1211.000	881.769	734.123	1539	1268.000	915.891	759.754
1510	1212.000	882.890	734.968	1540	1270.000	917.045	760.618
1511	1214.000	884.013	735.814	1541	1272.000	918.358	761.483
1512	1216.000	885.137	736.661	1542	1274.000	919.358	762.349
1513	1218.000	886.262	737.508	1543	1276.000	920.515	763.215
1514	1220.000	887.388	738.356	1544	1278.000	921.674	764.082
1515	1222.000	888.515	739.204	1545	1280.000	922.834	764.949
1516	1224.000	889.643	740.053	1546	1282.000	923.995	765.817
1517	1226.000	890.772	740.903	1547	1284.000	925.158	766.686
1518	1228.000	891.903	741.753	1548	1286.000	926.321	767.555
1519	1229.000	893.034	742.604	1549	1288.000	927.486	768.425
1520	1231.000	894.166	743.455	1550	1290.000	928.651	769.296
1521	1233.000	895.300	744.307	1551	1292.000	929.818	770.167
1522	1235.000	896.435	745.160	1552	1294.000	930.986	771.039
1523	1237.000	897.570	746.014	1553	1296.000	932.155	771.911
1524	1239.000	898.707	746.868	1554	1298.000	933.325	772.785
1525	1241.000	899.845	747.722	1555	1300.000	934.496	773.659
1526	1243.000	900.984	748.578	1556	1302.000	935.668	774.533
1527	1245.000	902.124	749.434	1557	1304.000	936.842	775.408
1528	1247.000	903.265	750.290	1558	1306.000	938.017	776.284
1529	1249.000	904.408	751.147	1559	1308.000	939.192	777.160
1530	1251.000	905.551	752.005	1560	1310.000	940.369	778.038

	N				N		
S	3000	4000	5000	S	3000	4000	5000
1561	1312.000	941.547	778.915	1591	1374.000	977.425	805.553
1562	1314.000	942.727	779.794	1592	1376.000	978.639	806.451
1563	1316.000	943.907	780.673	1593	1378.000	979.854	807.350
1564	1318.000	945.088	781.552	1594	1380.000	981.071	808.249
1565	1320.000	946.271	782.433	1595	1382.000	982.288	809.149
1566	1322.000	947.455	783.314	1596	1385.000	983.507	810.050
1567	1324.000	948.826	784.195	1597	1387.000	984.727	810.952
1568	1326.000	949.826	785.078	1598	1390.000	985.948	811.854
1569	1328.000	951.013	785.961	1599	1391.000	987.170	812.757
1570	1330.000	952.201	786.844	1600	1393.000	988.394	813.660
1571	1332.000	953.391	787.729	1601	1395.000	989.618	814.564
1572	1334.000	954.581	788.614	1602	1397.000	990.844	815.469
1573	1336.000	955.773	789.499	1603	1400.000	992.071	816.375
1574	1338.000	956.966	790.385	1604	1402.000	993.300	817.281
1575	1340.000	958.160	791.272	1605	1404.000	994.529	818.188
1576	1343.000	959.356	792.160	1606	1406.000	995.760	819.095
1577	1345.000	960.552	793.048	1607	1408.000	996.992	820.004
1578	1347.000	961.750	793.937	1608	1410.000	998.225	820.913
1579	1349.000	962.949	794.827	1609	1413.000	999.459	821.822
1580	1351.000	964.148	795.717	1610	1415.000	1001.000	822.732
1581	1353.000	965.350	796.608	1611	1417.000	1002.000	823.643
1582	1355.000	966.552	797.499	1612	1419.000	1003.000	824.555
1583	1357.000	967.755	798.391	1613	1421.000	1004.000	825.467
1584	1359.000	968.960	799.284	1614	1424.000	1006.000	826.380
1585	1361.000	970.166	800.178	1615	1426.000	1007.000	827.294
1586	1363.000	971.373	801.072	1616	1428.000	1008.000	828.208
1587	1365.000	972.581	801.967	1617	1430.000	1009.000	829.123
1588	1368.000	973.790	802.862	1618	1432.000	1011.000	830.039
1589	1370.000	975.000	803.758	1619	1435.000	1012.000	830.955
1590	1372.000	976.212	804.655	1620	1437.000	1013.000	831.872

S	N			S	N		
	3000	4000	5000		3000	4000	5000
1621	1439.000	1014.000	832.790	1651	1507.000	1052.000	860.642
1622	1441.000	1016.000	833.708	1652	1509.000	1054.000	861.581
1623	1443.000	1017.000	834.627	1653	1512.000	1055.000	862.521
1624	1446.000	1018.000	835.547	1654	1514.000	1056.000	863.461
1625	1448.000	1019.000	836.468	1655	1516.000	1058.000	864.402
1626	1450.000	1021.000	837.389	1656	1519.000	1059.000	865.344
1627	1452.000	1022.000	838.311	1657	1521.000	1060.000	866.287
1628	1455.000	1023.000	839.233	1658	1524.000	1061.000	867.230
1629	1457.000	1024.000	840.156	1659	1526.000	1063.000	868.175
1630	1459.000	1026.000	841.080	1660	1528.000	1064.000	869.119
1631	1461.000	1027.000	842.005	1661	1531.000	1065.000	870.065
1632	1464.000	1028.000	842.930	1662	1533.000	1067.000	871.011
1633	1466.000	1029.000	843.856	1663	1535.000	1068.000	871.958
1634	1468.000	1031.000	844.783	1664	1538.000	1069.000	872.905
1635	1470.000	1032.000	845.710	1665	1540.000	1071.000	873.854
1636	1473.000	1033.000	846.638	1666	1542.000	1072.000	874.803
1637	1475.000	1035.000	847.567	1667	1545.000	1073.000	875.752
1638	1477.000	1036.000	848.496	1668	1548.000	1074.000	876.703
1639	1479.000	1037.000	849.426	1669	1550.000	1076.000	877.654
1640	1482.000	1038.000	850.357	1670	1552.000	1077.000	878.606
1641	1484.000	1040.000	851.289	1671	1554.000	1078.000	879.558
1642	1486.000	1041.000	852.221	1672	1557.000	1080.000	880.512
1643	1489.000	1042.000	853.154	1673	1559.000	1081.000	881.466
1644	1491.000	1043.000	854.087	1674	1562.000	1082.000	882.420
1645	1493.000	1045.000	855.021	1675	1564.000	1084.000	883.376
1646	1496.000	1046.000	855.956	1676	1566.000	1085.000	884.332
1647	1498.000	1047.000	856.892	1677	1569.000	1086.000	885.289
1648	1500.000	1049.000	857.828	1678	1571.000	1088.000	886.246
1649	1503.000	1050.000	858.765	1679	1574.000	1089.000	887.205
1650	1505.000	1051.000	859.701	1680	1576.000	1090.000	888.164

S	N			S	N		
	3000	4000	5000		3000	4000	5000
1681	1579.000	1092.000	889.123	1711	1654.000	1132.000	918.251
1682	1581.000	1093.000	890.084	1712	1656.000	1133.000	919.233
1683	1584.000	1094.000	891.045	1713	1659.000	1135.000	920.216
1684	1586.000	1096.000	892.007	1714	1661.000	1136.000	921.199
1685	1588.000	1097.000	892.969	1715	1664.000	1137.000	922.184
1686	1591.000	1098.000	893.933	1716	1667.000	1139.000	923.169
1687	1593.000	1100.000	894.897	1717	1669.000	1140.000	924.155
1688	1596.000	1101.000	895.861	1718	1672.000	1142.000	925.142
1689	1598.000	1102.000	896.827	1719	1674.000	1143.000	926.129
1690	1601.000	1104.000	897.793	1720	1677.000	1144.000	927.117
1691	1603.000	1105.000	898.760	1721	1680.000	1146.000	928.106
1692	1606.000	1106.000	899.727	1722	1682.000	1147.000	929.096
1693	1608.000	1108.000	900.696	1723	1685.000	1148.000	930.086
1694	1611.000	1109.000	901.665	1724	1687.000	1150.000	931.077
1695	1613.000	1111.000	902.635	1725	1690.000	1151.000	932.069
1696	1616.000	1112.000	903.605	1726	1693.000	1153.000	933.061
1697	1618.000	1113.000	904.576	1727	1695.000	1154.000	934.055
1698	1621.000	1114.000	905.548	1728	1698.000	1155.000	935.049
1699	1623.000	1116.000	906.521	1729	1701.000	1157.000	936.044
1700	1626.000	1117.000	907.495	1730	1703.000	1158.000	937.039
1701	1628.000	1118.000	908.469	1731	1706.000	1160.000	938.036
1702	1631.000	1120.000	909.444	1732	1709.000	1161.000	939.033
1703	1633.000	1121.000	910.419	1733	1711.000	1162.000	940.030
1704	1636.000	1122.000	911.396	1734	1714.000	1164.000	941.029
1705	1638.000	1124.000	912.373	1735	1716.000	1165.000	942.028
1706	1641.000	1125.000	913.350	1736	1719.000	1167.000	943.028
1707	1643.000	1127.000	914.329	1737	1722.000	1168.000	944.029
1708	1646.000	1128.000	915.308	1738	1725.000	1169.000	945.031
1709	1649.000	1129.000	916.288	1739	1727.000	1171.000	946.033
1710	1651.000	1131.000	917.269	1740	1730.000	1172.000	947.036

S	N			S	N		
	3000	4000	5000		3000	4000	5000
1741	1733.000	1174.000	948.040	1771		1216.000	978.509
1742	1735.000	1175.000	949.045	1772		1218.000	979.537
1743	1738.000	1176.000	950.050	1773		1219.000	980.565
1744	1741.000	1178.000	951.056	1774		1221.000	981.594
1745	1743.000	1179.000	952.063	1775		1222.000	982.624
1746	1746.000	1181.000	953.071	1776		1224.000	983.654
1747	1749.000	1182.000	954.079	1777		1225.000	984.686
1748		1183.000	955.088	1778		1227.000	985.718
1749		1185.000	956.098	1779		1228.000	986.751
1750		1186.000	957.109	1780		1230.000	987.785
1751		1188.000	958.120	1781		1231.000	988.855
1752		1189.000	959.132	1782		1232.000	989.855
1753		1191.000	960.145	1783		1234.000	990.891
1754		1192.000	961.159	1784		1235.000	991.928
1755		1193.000	962.173	1785		1237.000	992.965
1756		1195.000	963.189	1786		1238.000	994.004
1757		1196.000	964.204	1787		1240.000	995.043
1758		1198.000	965.221	1788		1241.000	996.083
1759		1199.000	966.239	1789		1243.000	997.124
1760		1201.000	967.257	1790		1244.000	998.165
1761		1202.000	968.276	1791		1246.000	999.208
1762		1203.000	969.296	1792		1247.000	1000.000
1763		1205.000	970.317	1793		1249.000	1001.000
1764		1206.000	971.338	1794		1250.000	1002.000
1765		1208.000	972.360	1795		1252.000	1003.000
1766		1209.000	973.383	1796		1253.000	1004.000
1767		1211.000	974.407	1797		1255.000	1005.000
1768		1212.000	975.431	1798		1256.000	1007.000
1769		1214.000	976.456	1799		1258.000	1008.000
1770		1215.000	977.482	1800		1259.000	1009.000

S	N 4000	5000	S	N 4000	5000	S	N 4000	5000
1801	1261.00	1010.00	1821	1291.00	1031.00	1841	1322.00	1052.00
1802	1262.00	1011.00	1822	1292.00	1032.00	1842	1323.00	1053.00
1803	1264.00	1012.00	1823	1294.00	1033.00	1843	1325.00	1055.00
1804	1265.00	1013.00	1824	1295.00	1034.00	1844	1326.00	1056.00
1805	1267.00	1014.00	1825	1297.00	1035.00	1845	1328.00	1057.00
1806	1268.00	1015.00	1826	1299.00	1036.00	1846	1330.00	1058.00
1807	1270.00	1016.00	1827	1300.00	1037.00	1847	1331.00	1059.00
1808	1271.00	1017.00	1828	1302.00	1038.00	1848	1333.00	1060.00
1809	1273.00	1018.00	1829	1303.00	1039.00	1849	1334.00	1061.00
1810	1274.00	1019.00	1830	1305.00	1040.00	1850	1336.00	1062.00
1811	1276.00	1020.00	1831	1306.00	1042.00	1851	1337.00	1063.00
1812	1277.00	1021.00	1832	1308.00	1043.00	1852	1339.00	1064.00
1813	1279.00	1022.00	1833	1309.00	1044.00	1853	1341.00	1065.00
1814	1280.00	1023.00	1834	1311.00	1045.00	1854	1342.00	1066.00
1815	1282.00	1024.00	1835	1312.00	1046.00	1855	1344.00	1068.00
1816	1283.00	1026.00	1836	1314.00	1047.00	1856	1345.00	1069.00
1817	1285.00	1027.00	1837	1316.00	1048.00	1857	1347.00	1070.00
1818	1286.00	1028.00	1838	1317.00	1049.00	1858	1349.00	1071.00
1819	1288.00	1029.00	1839	1319.00	1050.00	1859	1350.00	1072.00
1820	1289.00	1030.00	1840	1320.00	1051.00	1860	1352.00	1073.00

S	N 4000	5000	S	N 4000	5000	S	N 4000	5000
1861	1353.00	1074.00	1871	1369.00	1085.00	1881	1385.00	1096.00
1862	1355.00	1075.00	1872	1371.00	1086.00	1882	1387.00	1097.00
1863	1356.00	1076.00	1873	1373.00	1087.00	1883	1389.00	1099.00
1864	1358.00	1077.00	1874	1374.00	1089.00	1884	1390.00	1100.00
1865	1360.00	1079.00	1875	1376.00	1090.00	1885	1392.00	1101.00
1866	1361.00	1080.00	1876	1377.00	1091.00	1886	1394.00	1102.00
1867	1363.00	1081.00	1877	1379.00	1092.00	1887	1395.00	1103.00
1868	1364.00	1082.00	1878	1381.00	1093.00	1888	1397.00	1104.00
1869	1366.00	1083.00	1879	1382.00	1094.00	1889	1399.00	1105.00
1870	1368.00	1084.00	1880	1384.00	1095.00	1890	1400.00	1106.00

S	N 4000	5000	S	N 4000	5000	S	N 4000	5000
1891	1402.00	1108.00	1901	1418.00	1119.00	1911	1435.00	1130.00
1892	1403.00	1109.00	1902	1420.00	1120.00	1912	1437.00	1131.00
1893	1405.00	1110.00	1903	1422.00	1121.00	1913	1438.00	1133.00
1894	1407.00	1111.00	1904	1423.00	1122.00	1914	1440.00	1134.00
1895	1408.00	1112.00	1905	1425.00	1123.00	1915	1442.00	1135.00
1896	1410.00	1113.00	1906	1427.00	1125.00	1916	1443.00	1136.00
1897	1412.00	1114.00	1907	1428.00	1126.00	1917	1445.00	1137.00
1898	1413.00	1115.00	1908	1430.00	1127.00	1918	1447.00	1138.00
1899	1415.00	1117.00	1909	1432.00	1128.00	1919	1449.00	1139.00
1900	1417.00	1118.00	1910	1433.00	1129.00	1920	1450.00	1141.00

S	N 4000	5000	S	N 4000	5000	S	N 4000	5000
1921	1452.00	1142.00	1941	1486.00	1165.00	1961	1521.00	1188.00
1922	1454.00	1143.00	1942	1488.00	1166.00	1962	1523.00	1190.00
1923	1455.00	1144.00	1943	1490.00	1167.00	1963	1525.00	1191.00
1924	1457.00	1145.00	1944	1491.00	1168.00	1964	1527.00	1192.00
1925	1459.00	1146.00	1945	1493.00	1170.00	1965	1528.00	1193.00
1926	1460.00	1147.00	1946	1495.00	1171.00	1966	1530.00	1194.00
1927	1462.00	1149.00	1947	1497.00	1172.00	1967	1532.00	1196.00
1928	1464.00	1150.00	1948	1498.00	1173.00	1968	1534.00	1197.00
1929	1466.00	1151.00	1949	1500.00	1174.00	1969	1535.00	1198.00
1930	1467.00	1152.00	1950	1502.00	1175.00	1970	1537.00	1199.00
1931	1469.00	1153.00	1951	1504.00	1177.00	1971	1539.00	1200.00
1932	1471.00	1154.00	1952	1505.00	1178.00	1972	1541.00	1202.00
1933	1472.00	1156.00	1953	1507.00	1179.00	1973	1543.00	1203.00
1934	1474.00	1157.00	1954	1509.00	1180.00	1974	1544.00	1204.00
1935	1476.00	1158.00	1955	1511.00	1181.00	1975	1546.00	1205.00
1936	1478.00	1159.00	1956	1512.00	1183.00	1976	1548.00	1206.00
1937	1479.00	1160.00	1957	1514.00	1184.00	1977	1550.00	1208.00
1938	1481.00	1161.00	1958	1516.00	1185.00	1978	1552.00	1209.00
1939	1483.00	1163.00	1959	1518.00	1186.00	1979	1553.00	1210.00
1940	1484.00	1164.00	1960	1519.00	1187.00	1980	1555.00	1211.00

S	N 4000	5000	S	N 4000	5000	S	N 4000	5000
1981	1557.00	1212.00	2001	1594.00	1237.00	2021	1631.00	1261.00
1982	1559.00	1214.00	2002	1596.00	1238.00	2022	1633.00	1263.00
1983	1561.00	1215.00	2003	1597.00	1239.00	2023	1635.00	1264.00
1984	1562.00	1216.00	2004	1599.00	1240.00	2024	1637.00	1265.00
1985	1564.00	1217.00	2005	1601.00	1242.00	2025	1639.00	1266.00
1986	1566.00	1218.00	2006	1603.00	1243.00	2026	1641.00	1268.00
1987	1568.00	1220.00	2007	1605.00	1244.00	2027	1642.00	1269.00
1988	1570.00	1221.00	2008	1607.00	1245.00	2028	1644.00	1270.00
1989	1572.00	1222.00	2009	1609.00	1247.00	2029	1646.00	1271.00
1990	1573.00	1223.00	2010	1610.00	1248.00	2030	1648.00	1273.00
1991	1575.00	1225.00	2011	1612.00	1249.00	2031	1650.00	1274.00
1992	1577.00	1226.00	2012	1614.00	1250.00	2032	1652.00	1275.00
1993	1579.00	1227.00	2013	1616.00	1251.00	2033	1654.00	1276.00
1994	1581.00	1228.00	2014	1618.00	1253.00	2034	1656.00	1278.00
1995	1583.00	1229.00	2015	1620.00	1254.00	2035	1658.00	1279.00
1996	1584.00	1231.00	2016	1622.00	1255.00	2036	1660.00	1280.00
1997	1586.00	1232.00	2017	1624.00	1256.00	2037	1662.00	1281.00
1998	1588.00	1233.00	2018	1625.00	1258.00	2038	1664.00	1283.00
1999	1590.00	1234.00	2019	1627.00	1259.00	2039	1665.00	1284.00
2000	1592.00	1235.00	2020	1629.00	1260.00	2040	1667.00	1285.00

S	N 4000	5000	S	N 4000	5000	S	N 4000	5000
2041	1669.00	1286.00	2051	1689.00	1299.00	2061	1708.00	1312.00
2042	1671.00	1288.00	2052	1691.00	1300.00	2062	1710.00	1313.00
2043	1673.00	1289.00	2053	1693.00	1302.00	2063	1712.00	1315.00
2044	1675.00	1290.00	2054	1695.00	1303.00	2064	1714.00	1316.00
2045	1677.00	1292.00	2055	1697.00	1304.00	2065	1716.00	1317.00
2046	1679.00	1293.00	2056	1699.00	1306.00	2066	1718.00	1318.00
2047	1681.00	1294.00	2057	1701.00	1307.00	2067	1720.00	1320.00
2048	1683.00	1295.00	2058	1703.00	1308.00	2068	1722.00	1321.00
2049	1685.00	1297.00	2059	1704.00	1309.00	2069	1724.00	1322.00
2050	1687.00	1298.00	2060	1706.00	1311.00	2070	1726.00	1324.00

S	N 4000	5000
2071	1728.00	1325.00
2072	1730.00	1326.00
2073	1732.00	1327.00
2074	1734.00	1329.00
2075	1736.00	1330.00
2076	1738.00	1331.00
2077	1740.00	1333.00
2078	1742.00	1334.00
2079	1744.00	1335.00
2080	1746.00	1337.00

S	N 4000	5000
2081	1748.00	1338.00
2082	1750.00	1339.00
2083	1752.00	1340.00
2084	1755.00	1342.00
2085	1757.00	1343.00
2086	1759.00	1344.00
2087	1761.00	1346.00
2088	1763.00	1347.00
2089	1765.00	1348.00
2090	1767.00	1350.00

S	N 4000	5000
2091	1769.00	1351.00
2092	1771.00	1352.00
2093	1773.00	1354.00
2094	1775.00	1355.00
2095	1777.00	1356.00
2096	1779.00	1358.00
2097	1781.00	1359.00
2098	1783.00	1360.00
2099	1785.00	1362.00
2100	1787.00	1363.00

S	N 4000	5000
2101	1789.00	1364.00
2102	1791.00	1365.00
2103	1794.00	1367.00
2104	1796.00	1368.00
2105	1798.00	1369.00
2106	1800.00	1371.00
2107	1802.00	1372.00
2108	1804.00	1373.00
2109	1806.00	1375.00
2110	1808.00	1376.00
2111	1810.00	1377.00
2112	1812.00	1379.00
2113	1814.00	1380.00
2114	1816.00	1381.00
2115	1819.00	1383.00
2116	1821.00	1384.00
2117	1823.00	1385.00
2118	1825.00	1387.00
2119	1827.00	1388.00
2120	1829.00	1390.00

S	N 4000	5000
2121	1831.00	1391.00
2122	1833.00	1392.00
2123	1835.00	1394.00
2124	1838.00	1395.00
2125	1840.00	1396.00
2126	1842.00	1398.00
2127	1844.00	1399.00
2128	1846.00	1400.00
2129	1848.00	1402.00
2130	1850.00	1403.00
2131	1853.00	1404.00
2132	1855.00	1406.00
2133	1857.00	1407.00
2134	1859.00	1408.00
2135	1861.00	1410.00
2136	1863.00	1411.00
2137	1865.00	1413.00
2138	1868.00	1414.00
2139	1870.00	1415.00
2140	1872.00	1417.00

S	N 4000	5000
2141	1874.00	1418.00
2142	1876.00	1419.00
2143	1878.00	1421.00
2144	1881.00	1422.00
2145	1883.00	1424.00
2146	1885.00	1425.00
2147	1887.00	1426.00
2148	1889.00	1428.00
2149	1891.00	1429.00
2150	1894.00	1430.00
2151	1896.00	1432.00
2152	1898.00	1433.00
2153	1900.00	1435.00
2154	1902.00	1436.00
2155	1905.00	1437.00
2156	1907.00	1439.00
2157	1909.00	1440.00
2158	1911.00	1441.00
2159	1913.00	1443.00
2160	1916.00	1444.00

S	N 4000	5000	S	N 4000	5000	S	N 4000	5000
2161	1918.00	1446.00	2181	1963.00	1474.00	2201	2009.00	1502.00
2162	1920.00	1447.00	2182	1965.00	1475.00	2202	2011.00	1504.00
2163	1922.00	1448.00	2183	1967.00	1476.00	2203	2013.00	1505.00
2164	1925.00	1450.00	2184	1970.00	1478.00	2204	2016.00	1506.00
2165	1927.00	1451.00	2185	1972.00	1479.00	2205	2018.00	1508.00
2166	1929.00	1453.00	2186	1974.00	1481.00	2206	2020.00	1509.00
2167	1931.00	1454.00	2187	1976.00	1482.00	2207	2023.00	1511.00
2168	1934.00	1455.00	2188	1979.00	1484.00	2208	2025.00	1512.00
2169	1936.00	1457.00	2189	1981.00	1485.00	2209	2027.00	1514.00
2170	1938.00	1458.00	2190	1983.00	1486.00	2210	2030.00	1515.00
2171	1940.00	1460.00	2191	1986.00	1488.00	2211	2032.00	1517.00
2172	1942.00	1461.00	2192	1988.00	1489.00	2212	2035.00	1518.00
2173	1945.00	1462.00	2193	1990.00	1491.00	2213	2037.00	1519.00
2174	1947.00	1464.00	2194	1993.00	1492.00	2214	2039.00	1521.00
2175	1949.00	1465.00	2195	1995.00	1494.00	2215	2042.00	1522.00
2176	1951.00	1467.00	2196	1997.00	1495.00	2216	2044.00	1524.00
2177	1954.00	1468.00	2197	1999.00	1496.00	2217	2046.00	1525.00
2178	1956.00	1469.00	2198	2002.00	1498.00	2218	2049.00	1527.00
2179	1958.00	1471.00	2199	2004.00	1499.00	2219	2051.00	1528.00
2180	1961.00	1472.00	2200	2006.00	1501.00	2220	2053.00	1530.00

S	N 4000	5000	S	N 4000	5000	S	N 4000	5000
2221	2056.00	1531.00	2231	2080.00	1546.00	2241	2104.00	1561.00
2222	2058.00	1533.00	2232	2082.00	1547.00	2242	2109.00	1562.00
2223	2061.00	1534.00	2233	2085.00	1549.00	2243	2111.00	1564.00
2224	2063.00	1535.00	2234	2087.00	1550.00	2244	2114.00	1565.00
2225	2065.00	1537.00	2235	2089.00	1552.00	2245	2116.00	1566.00
2226	2068.00	1538.00	2236	2092.00	1553.00	2246	2119.00	1568.00
2227	2070.00	1540.00	2237	2094.00	1555.00	2247	2121.00	1569.00
2228	2073.00	1541.00	2238	2097.00	1556.00	2248	2124.00	1571.00
2229	2075.00	1543.00	2239	2099.00	1558.00	2249	2126.00	1572.00
2230	2077.00	1544.00	2240	2102.00	1559.00	2250	2129.00	1574.00

S	N 4000	5000	S	N 4000	5000	S	N 4000	5000
2251	2131.00	1575.00	2261	2156.00	1590.00	2271	2181.00	1606.00
2252	2134.00	1577.00	2262	2158.00	1592.00	2272	2184.00	1607.00
2253	2136.00	1578.00	2263	2161.00	1593.00	2273	2186.00	1609.00
2254	2139.00	1580.00	2264	2163.00	1595.00	2274	2189.00	1610.00
2255	2141.00	1581.00	2265	2166.00	1597.00	2275	2191.00	1612.00
2256	2143.00	1583.00	2266	2169.00	1598.00	2276	2194.00	1613.00
2257	2146.00	1584.00	2267	2171.00	1600.00	2277	2196.00	1615.00
2258	2148.00	1586.00	2268	2174.00	1601.00	2278	2199.00	1616.00
2259	2151.00	1587.00	2269	2176.00	1603.00	2279	2202.00	1618.00
2260	2153.00	1589.00	2270	2179.00	1604.00	2280	2204.00	1619.00

S	N 4000	5000	S	N 4000	5000	S	N 4000	5000
2281	2207.00	1621.00	2301	2259.00	1652.00	2321	2312.00	1683.00
2282	2209.00	1622.00	2302	2261.00	1653.00	2322	2315.00	1685.00
2283	2212.00	1624.00	2303	2264.00	1655.00	2323	2317.00	1686.00
2284	2214.00	1625.00	2304	2267.00	1656.00	2324	2320.00	1688.00
2285	2217.00	1627.00	2305	2269.00	1658.00	2325	2323.00	1690.00
2286	2220.00	1629.00	2306	2272.00	1660.00	2326	2325.00	1691.00
2287	2222.00	1630.00	2307	2274.00	1661.00	2327	2328.00	1693.00
2288	2225.00	1632.00	2308	2277.00	1663.00	2328	2331.00	1694.00
2289	2227.00	1633.00	2309	2280.00	1664.00	2329	2334.00	1696.00
2290	2230.00	1635.00	2310	2282.00	1666.00	2330		1698.00
2291	2232.00	1636.00	2311	2285.00	1667.00	2331		1699.00
2292	2235.00	1638.00	2312	2288.00	1669.00	2332		1701.00
2293	2238.00	1639.00	2313	2290.00	1671.00	2333		1702.00
2294	2240.00	1641.00	2314	2293.00	1672.00	2334		1704.00
2295	2243.00	1642.00	2315	2296.00	1674.00	2335		1706.00
2296	2246.00	1644.00	2316	2298.00	1675.00	2336		1707.00
2297	2248.00	1646.00	2317	2301.00	1677.00	2337		1709.00
2298	2251.00	1647.00	2318	2304.00	1679.00	2338		1710.00
2299	2253.00	1649.00	2319	2307.00	1680.00	2339		1712.00
2300	2256.00	1650.00	2320	2309.00	1682.00	2340		1714.00

S	N 5000	S	N 5000	S	N 5000	S	N 5000
2341	1715.0	2356	1740.0	2371	1764.0	2386	1789.0
2342	1717.0	2357	1741.0	2372	1766.0	2387	1791.0
2343	1718.0	2358	1743.0	2373	1768.0	2388	1793.0
2344	1720.0	2359	1744.0	2374	1769.0	2389	1794.0
2345	1722.0	2360	1746.0	2375	1771.0	2390	1796.0
2346	1723.0	2361	1748.0	2376	1773.0	2391	1798.0
2347	1725.0	2362	1749.0	2377	1774.0	2392	1799.0
2348	1727.0	2363	1751.0	2378	1776.0	2393	1801.0
2349	1728.0	2364	1753.0	2379	1777.0	2394	1803.0
2350	1730.0	2365	1754.0	2380	1779.0	2395	1804.0
2351	1731.0	2366	1756.0	2381	1781.0	2396	1806.0
2352	1733.0	2367	1758.0	2382	1782.0	2397	1808.0
2353	1735.0	2368	1759.0	2383	1784.0	2398	1809.0
2354	1736.0	2369	1761.0	2384	1786.0	2399	1811.0
2355	1738.0	2370	1763.0	2385	1788.0	2400	1813.0

S	N 5000	S	N 5000	S	N 5000	S	N 5000
2401	1814.0	2416	1840.0	2431	1866.0	2446	1892.0
2402	1816.0	2417	1842.0	2432	1868.0	2447	1894.0
2403	1818.0	2418	1844.0	2433	1869.0	2448	1896.0
2404	1820.0	2419	1845.0	2434	1871.0	2449	1898.0
2405	1821.0	2420	1847.0	2435	1873.0	2450	1899.0
2406	1823.0	2421	1849.0	2436	1875.0	2451	1901.0
2407	1825.0	2422	1850.0	2437	1876.0	2452	1903.0
2408	1826.0	2423	1852.0	2438	1878.0	2453	1905.0
2409	1828.0	2424	1854.0	2439	1880.0	2454	1906.0
2410	1830.0	2425	1856.0	2440	1882.0	2455	1908.0
2411	1831.0	2426	1857.0	2441	1883.0	2456	1910.0
2412	1833.0	2427	1859.0	2442	1885.0	2457	1912.0
2413	1835.0	2428	1861.0	2443	1887.0	2458	1914.0
2414	1837.0	2429	1863.0	2444	1889.0	2459	1915.0
2415	1838.0	2430	1864.0	2445	1891.0	2460	1917.0

S	N 5000	S	N 5000	S	N 5000	S	N 5000
2461	1919.0	2476	1946.0	2491	1973.0	2506	2001.0
2462	1921.0	2477	1948.0	2492	1975.0	2507	2003.0
2463	1922.0	2478	1949.0	2493	1977.0	2508	2005.0
2464	1924.0	2479	1951.0	2494	1979.0	2509	2006.0
2465	1926.0	2480	1953.0	2495	1981.0	2510	2008.0
2466	1928.0	2481	1955.0	2496	1982.0	2511	2010.0
2467	1930.0	2482	1957.0	2497	1984.0	2512	2012.0
2468	1931.0	2483	1959.0	2498	1986.0	2513	2014.0
2469	1933.0	2484	1960.0	2499	1988.0	2514	2016.0
2470	1935.0	2485	1962.0	2500	1990.0	2515	2018.0
2471	1937.0	2486	1964.0	2501	1992.0	2516	2020.0
2472	1939.0	2487	1966.0	2502	1993.0	2517	2021.0
2473	1940.0	2488	1968.0	2503	1995.0	2518	2023.0
2474	1942.0	2489	1970.0	2504	1997.0	2519	2025.0
2475	1944.0	2490	1971.0	2505	1999.0	2520	2027.0

S	N 5000	S	N 5000	S	N 5000	S	N 5000
2521	2029.0	2536	2057.0	2551	2086.0	2566	2115.0
2522	2031.0	2537	2059.0	2552	2088.0	2567	2117.0
2523	2033.0	2538	2061.0	2553	2090.0	2568	2119.0
2524	2035.0	2539	2063.0	2554	2092.0	2569	2121.0
2525	2036.0	2540	2065.0	2555	2094.0	2570	2123.0
2526	2038.0	2541	2067.0	2556	2096.0	2571	2125.0
2527	2040.0	2542	2069.0	2557	2098.0	2572	2127.0
2528	2042.0	2543	2071.0	2558	2100.0	2573	2129.0
2529	2044.0	2544	2073.0	2559	2102.0	2574	2131.0
2530	2046.0	2545	2075.0	2560	2104.0	2575	2133.0
2531	2048.0	2546	2077.0	2561	2106.0	2576	2135.0
2532	2050.0	2547	2078.0	2562	2108.0	2577	2137.0
2533	2052.0	2548	2080.0	2563	2109.0	2578	2139.0
2534	2054.0	2549	2082.0	2564	2111.0	2579	2141.0
2535	2055.0	2550	2084.0	2565	2113.0	2580	2143.0

S	N 5000	S	N 5000	S	N 5000	S	N 5000
2581	2145.0	2596	2175.0	2611	2205.0	2626	2236.0
2582	2147.0	2597	2177.0	2612	2207.0	2627	2238.0
2583	2149.0	2598	2179.0	2613	2209.0	2628	2240.0
2584	2151.0	2599	2181.0	2614	2211.0	2629	2242.0
2585	2153.0	2600	2183.0	2615	2214.0	2630	2244.0
2586	2155.0	2601	2185.0	2616	2216.0	2631	2247.0
2587	2157.0	2602	2187.0	2617	2218.0	2632	2249.0
2588	2159.0	2603	2189.0	2618	2220.0	2633	2251.0
2589	2161.0	2604	2191.0	2619	2222.0	2634	2253.0
2590	2163.0	2605	2193.0	2620	2224.0	2635	2255.0
2591	2165.0	2606	2195.0	2621	2226.0	2636	2257.0
2592	2167.0	2607	2197.0	2622	2228.0	2637	2259.0
2593	2169.0	2608	2199.0	2623	2230.0	2638	2261.0
2594	2171.0	2609	2201.0	2624	2232.0	2639	2263.0
2595	2173.0	2610	2203.0	2625	2234.0	2640	2265.0

S	N 5000	S	N 5000	S	N 5000	S	N 5000
2641	2267.0	2656	2299.0	2671	2331.0	2686	2364.0
2642	2270.0	2657	2301.0	2672	2333.0	2687	2366.0
2643	2272.0	2658	2303.0	2673	2336.0	2688	2368.0
2644	2274.0	2659	2305.0	2674	2338.0	2689	2370.0
2645	2276.0	2660	2308.0	2675	2340.0	2690	2373.0
2646	2278.0	2661	2310.0	2676	2342.0	2691	2375.0
2647	2280.0	2662	2312.0	2677	2344.0	2692	2377.0
2648	2282.0	2663	2314.0	2678	2346.0	2693	2379.0
2649	2284.0	2664	2316.0	2679	2349.0	2694	2381.0
2650	2286.0	2665	2318.0	2680	2351.0	2695	2384.0
2651	2288.0	2666	2320.0	2681	2353.0	2696	2386.0
2652	2291.0	2667	2323.0	2682	2355.0	2697	2388.0
2653	2293.0	2668	2325.0	2683	2357.0	2698	2390.0
2654	2295.0	2669	2327.0	2684	2359.0	2699	2392.0
2655	2297.0	2670	2329.0	2685	2362.0	2700	2395.0

S	N 5000	S	N 5000	S	N 5000	S	N 5000
2701	2397.0	2716	2430.0	2731	2464.0	2746	2499.0
2702	2399.0	2717	2433.0	2732	2467.0	2747	2501.0
2703	2401.0	2718	2435.0	2733	2469.0	2748	2503.0
2704	2403.0	2719	2437.0	2734	2471.0	2749	2506.0
2705	2406.0	2720	2439.0	2735	2473.0	2750	2508.0
2706	2408.0	2721	2442.0	2736	2476.0	2751	2510.0
2707	2410.0	2722	2444.0	2737	2478.0	2752	2513.0
2708	2412.0	2723	2446.0	2738	2480.0	2753	2515.0
2709	2415.0	2724	2448.0	2739	2483.0	2754	2517.0
2710	2417.0	2725	2451.0	2740	2485.0	2755	2520.0
2711	2419.0	2726	2453.0	2741	2487.0	2756	2522.0
2712	2421.0	2727	2455.0	2742	2490.0	2757	2524.0
2713	2424.0	2728	2457.0	2743	2492.0	2758	2527.0
2714	2426.0	2729	2460.0	2744	2494.0	2759	2529.0
2715	2428.0	2730	2462.0	2745	2496.0	2760	2531.0

S	N 5000	S	N 5000	S	N 5000	S	N 5000
2761	2534.0	2776	2569.0	2791	2605.0	2806	2642.0
2762	2536.0	2777	2572.0	2792	2608.0	2807	2644.0
2763	2538.0	2778	2574.0	2793	2610.0	2808	2647.0
2764	2541.0	2779	2576.0	2794	2612.0	2809	2649.0
2765	2543.0	2780	2579.0	2795	2615.0	2810	2652.0
2766	2545.0	2781	2581.0	2796	2617.0	2811	2654.0
2767	2548.0	2782	2584.0	2797	2620.0	2812	2656.0
2768	2550.0	2783	2586.0	2798	2622.0	2813	2659.0
2769	2553.0	2784	2588.0	2799	2625.0	2814	2661.0
2770	2555.0	2785	2591.0	2800	2627.0	2815	2664.0
2771	2557.0	2786	2593.0	2801	2629.0	2816	2666.0
2772	2560.0	2787	2596.0	2802	2632.0	2817	2669.0
2773	2562.0	2788	2598.0	2803	2634.0	2818	2671.0
2774	2564.0	2789	2600.0	2804	2637.0	2819	2674.0
2775	2567.0	2790	2603.0	2805	2639.0	2820	2676.0

S	N 5000	S	N 5000	S	N 5000	S	N 5000
2821	2679.0	2836	2716.0	2851	2754.0	2866	2793.0
2822	2681.0	2837	2719.0	2852	2757.0	2867	2796.0
2823	2684.0	2838	2721.0	2853	2760.0	2868	2798.0
2824	2686.0	2839	2724.0	2854	2762.0	2869	2801.0
2825	2689.0	2840	2726.0	2855	2764.0	2870	2804.0
2826	2691.0	2841	2729.0	2856	2767.0	2871	2806.0
2827	2694.0	2842	2732.0	2857	2770.0	2872	2809.0
2828	2696.0	2843	2734.0	2858	2772.0	2873	2811.0
2829	2699.0	2844	2737.0	2859	2775.0	2874	2814.0
2830	2701.0	2845	2739.0	2860	2778.0	2875	2817.0
2831	2704.0	2846	2742.0	2861	2780.0	2876	2819.0
2832	2706.0	2847	2744.0	2862	2783.0	2877	2822.0
2833	2709.0	2848	2747.0	2863	2785.0	2878	2825.0
2834	2711.0	2849	2749.0	2864	2788.0	2879	2827.0
2835	2714.0	2850	2752.0	2865	2791.0	2880	2830.0

APPENDIX 5

Subset of Bat Counts from Venezuela

	Quadrats														
Sp. No.	1	2	3	4	5	6	7	8	9	10	11	12	13	14	15
1															
2						4									
3															3
4						3	6				8				
5															3
6															
7															
8															
9															
10			4										4		
11	3														
12									21						
13															
14											5				
15		4													
16											6		9	12	
17									5						
18													5		
19	4	3													
20															8
21													17		5
22											5				
23									9		8				
24															

Sp. No.	Quadrats														
	1	2	3	4	5	6	7	8	9	10	11	12	13	14	15
25	3														
26	6														
27							5							10	
28															
29															
30									4						
31									4						
32															
33															
34											3				
35						19									
36												7			
37															5
38															
39															
40								11							
41															4
42															
43															
44															
45															
46					4		3								
47							4								
48															
49															6
50											5				
N	16	7	4	4	19	14	11	11	43	3	37	7	30	27	34
S	4	2	1	1	1	4	2	1	5	1	6	1	3	3	7
cum. N	16	23	27	31	50	64	75	86	129	132	169	176	206	233	267
cum. S	4	5	6	7	8	11	12	13	18	19	23	24	25	26	32

Subset of Bat Counts from Venezuela

	Quadrats														
Sp. No.	16	17	18	19	20	21	22	23	24	25	26	27	28	29	30
1						8									
2			5												
3															
4															
5															
6								12							
7															
8							4								
9											6				
10		5	12						11						
11															
12															
13			5	18	21				3						
14		28	5	3					8				4		
15															
16															
17			11	4										5	
18															
19														7	
20		7	20												3
21															
22			20	6	10					5					4
23															
24			64									5			

						Quadrats									
Sp. No.	16	17	18	19	20	21	22	23	24	25	26	27	28	29	30
25			21	11	12	4			5						
26												6			
27			13	24	37	6	16		8						
28	10	25													
29								18							
30															
31															
32			7												
33						4									
34			20												8
35							28		5						7
36															
37															
38					4		18		6						
39		3							5						
40															
41			3												
42															
43															
44															
45															
46															
47															
48						5									
49															
50															
N	7	70	235	75	53	37	50	30	51	5	6	6	9	12	22
S	1	6	14	5	5	5	3	2	8	1	1	1	2	2	4
cum. N	274	344	579	654	707	744	794	824	875	880	886	892	901	913	935
cum. S	32	34	38	38	39	41	42	44	44	44	45	45	45	45	45

Subset of Bat Counts from Venezuela

Sp. No.							Quadrats								
	31	32	33	34	35	36	37	38	39	40	41	42	43	44	45
1															
2	4					5									
3															
4															
5															
6															
7						6									
8															
9										18				6	
10		12													3
11															
12															
13							3				6				
14		11	7		3	4	7	6	9			7			15
15															
16															
17															
18															
19															
20									50						
21															
22								3				3			
23															
24															

Quadrats

Sp. No.	31	32	33	34	35	36	37	38	39	40	41	42	43	44	45
25					3	6					9	4			4
26															
27	4		5		3	13	15	14	3		3				
28															
29														37	
30															
31															
32															
33															
34					8			5							
35	10	6											7		
36															
37											5				
38															
39															
40															
41															
42					5										
43					5										
44				8	6				36						
45									35						23
46															
47															
48															
49															
50															
N	18	29	12	8	33	34	25	28	133	18	23	14	7	66	22
S	3	3	2	1	7	5	3	4	5	1	4	3	1	3	3
cum. N	953	982	994	1002	1035	1069	1094	1122	1255	1273	1296	1310	1317	1383	1405
cum. S	45	45	45	46	48	49	49	49	50	50	50	50	50	50	50

APPENDIX 5 (CONTINUED)

Subset of Bat Counts from Venezuela

Sp. No.	Total	Mean (μ)	Variance (σ^2)	Standard Deviation (σ)	Proportion	Occurrences
1	8	0.1778	0.0000	0.0000	0.0057	1
2	18	0.4000	0.2500	0.5000	0.0128	4
3	3	0.0667	0.0000	0.0000	0.0021	1
4	17	0.3778	4.2222	2.0548	0.0121	3
5	3	0.0667	0.0000	0.0000	0.0021	1
6	12	0.2667	0.0000	0.0000	0.0085	1
7	6	0.1333	0.0000	0.0000	0.0043	1
8	4	0.0889	0.0000	0.0000	0.0028	1
9	30	0.6667	32.0000	5.6569	0.0214	3
10	51	1.1333	14.7755	3.8439	0.0363	7
11	3	0.0667	0.0000	0.0000	0.0021	1
12	21	0.4667	0.0000	0.0000	0.0149	1
13	56	1.2444	53.5556	7.3182	0.0399	6
14	122	2.7111	37.7156	6.1413	0.0868	15
15	4	0.0889	0.0000	0.0000	0.0028	1
16	27	0.6000	6.0000	2.4495	0.0192	3
17	25	0.5556	7.6875	2.7726	0.0178	4
18	5	0.1111	0.0000	0.0000	0.0036	1
19	14	0.3111	2.8889	1.6997	0.0100	3
20	88	1.9556	294.6400	17.1651	0.0626	5
21	22	0.4889	36.0000	6.0000	0.0157	2
22	56	1.2444	28.5000	5.3385	0.0399	8
23	17	0.3778	0.2500	0.5000	0.0121	2
24	69	1.5333	870.2500	29.5000	0.0491	2

Sp. No.	Total	Mean (μ)	Variance (σ^2)	Standard Deviation (σ)	Proportion	Occurrences
25	82	1.8222	27.5207	5.2460	0.0584	11
26	12	0.2667	0.0000	0.0000	0.0085	2
27	179	3.9778	78.1523	8.8404	0.1274	16
28	35	0.7778	56.2500	7.5000	0.0249	2
29	55	1.2222	90.2500	9.5000	0.0391	2
30	4	0.0889	0.0000	0.0000	0.0028	1
31	4	0.0889	0.0000	0.0000	0.0028	1
32	7	0.1556	0.0000	0.0000	0.0050	1
33	4	0.0889	0.0000	0.0000	0.0028	1
34	44	0.9778	34.9600	5.9127	0.0313	5
35	82	1.8222	63.3469	7.9591	0.0584	7
36	7	0.1556	0.0000	0.0000	0.0050	1
37	10	0.2222	0.0000	0.0000	0.0071	2
38	28	0.6222	38.2222	6.1824	0.0199	3
39	8	0.1778	1.0000	1.0000	0.0057	2
40	11	0.2444	0.0000	0.0000	0.0078	1
41	7	0.1556	0.2500	0.5000	0.0050	2
42	5	0.1111	0.0000	0.0000	0.0036	1
43	5	0.1111	0.0000	0.0000	0.0036	1
44	50	1.1111	187.5556	13.6951	0.0356	3
45	58	1.2889	36.0000	6.0000	0.0413	2
46	7	0.1556	0.2500	0.5000	0.0050	2
47	4	0.0889	0.0000	0.0000	0.0028	1
48	5	0.1111	0.0000	0.0000	0.0036	1
49	6	0.1333	0.0000	0.0000	0.0043	1
50	5	0.1111	0.0000	0.0000	0.0036	1

APPENDIX 6

Answers to Chapter Problems

These calculations were made in a variety of mediums. Because of this your answers may vary from those provided. The amount of difference in answers should be in the decimal places, occasionally even in the first decimal place.

CHAPTER 2: DENSITY: MEAN AND VARIANCE

2.1 For the random quadrats we obtained using our random number table:

a. Estimated mean density and variance

For *Scheelea princeps* the counts were:

0, 3, 5, 0, which gave estimates of

$$\hat{\mu}_{Sp} = \frac{(0 + 3 + 5 + 0)}{4} = \frac{8}{4} = 2.0 \text{ trees (from Equation 2.2)}$$

$$\hat{\sigma}^2_{Sp} = \frac{(0-2)^2 + (3-2)^2 + (5-2)^2 + (0-2)^2}{4-1} = \frac{18}{3} = 6.0 \text{ (from Equation 2.5)}$$

$$\hat{\sigma}_{Sp} = 2.449.$$

For *Calycophyllum spruceanum*:

0, 0, 2, 1

$$\hat{\mu}_{Cs} = \frac{(2+1)}{4} = \frac{3}{4} = 0.75, \text{ or rounding we would estimate 1 tree}$$

$$\hat{\sigma}^2_{Cs} = \frac{(0-0.75)^2 + (0-0.75)^2 + (2-0.75)^2 + (1-0.75)^2}{4-1} = \frac{2.75}{3} = 0.9167$$

$$\hat{\sigma}_{Cs} = 0.9574.$$

For *Aracia loretensis*:

0, 0, 0, 0

The estimated density is $\hat{\mu}_{Al} = 0$, the estimated variance is $\hat{\sigma}^2_{Al} = 0$.

b. Repeat these steps for a sample of size $n = 8$.

For *Scheelea princeps* we got the counts:

0, 4, 2, 2, 2, 0, 1, 4, which gave estimates of

$$\hat{\mu}_{Sp} = \frac{15}{8} = 1.8750$$

$$\hat{\sigma}^2_{Sp} = 2.411$$

$$\hat{\sigma}_{Sp} = 1.553.$$

For *Calycophyllum spruceanum*:

0, 0, 0, 1, 0, 0, 0, 1

$$\hat{\mu}_{Cs} = \frac{2}{8} = 0.250, \text{ or rounding we would estimate 1 tree}$$

$$\hat{\sigma}^2_{Cs} = 0.2140$$

$$\hat{\sigma}_{Cs} = 0.4626.$$

For *Aracia loretensis*:

0, 0, 0, 0, 1, 0, 0, 0

$$\hat{\mu}_{Al} = \frac{1}{8} = 0.125$$

$$\hat{\sigma}^2_{Al} = 0.1250$$

$$\hat{\sigma}_{Al} = 0.3536.$$

c. Using Tchebychev's theorem, calculate the proportion of observations within 2 standard deviations of the mean when using first the statistical sample of size 4, and second, the sample of size 8. Comment.

Tchebychev's theorem (2.6) states that $p_2 = 1 - \frac{1}{2^2} = 1 - \frac{1}{4} = \frac{3}{4} = 0.75$. That is, 75% of the observations will be within 2 standard deviations of the mean. Therefore, we can expect AT LEAST 75% of the counts to fall in the intervals below.

SIZE 4 SAMPLES

SPECIES	EXPECTED INTERVAL	OBSERVED
S. princeps	$2 \pm 2(2.449) = (-2.898, 6.898) \approx (0,7)$	100% are in the interval
C. spruceanum	$0.75 \pm 2(0.9574) = (-1.1648, 2.6648) \approx (0,3)$	100% are in the interval
A. loretensis	$0 \pm 2(0) = 0$	this is not usable data

SIZE 8 SAMPLES

SPECIES	EXPECTED INTERVAL	OBSERVED
S. princeps	$1.875 \pm 2(1.553) = (-1.231, 4.981) \approx (0, 5)$	95 of the 100 (95%) are in the expected interval

C. spruceanum	$0.250 \pm 2(0.4626) = (-0.6752, 1.1752) \approx (0, 1)$	97 of 100 (97%) are in the expected interval
A. loretensis	$0.125 \pm 2(0.3536) = (-0.5822, 0.8322) \approx (0, 1)$	100% are in the expected interval

    d. Comment on the differences you obtained for an abundant, a common, and a rare species.

SPECIES	SIZE 4 SAMPLES		SIZE 8 SAMPLES		TRUE VALUES	
	$\hat{\mu}$	$\hat{\sigma}^2$	$\hat{\mu}$	$\hat{\sigma}^2$	$\hat{\mu}$	$\hat{\sigma}^2$
S. princeps	2.00	6.00	1.875	2.411	2.52	2.6496
C. spruceanum	0.750	0.9167	0.250	0.2140	0.25	0.2475
A. loretensis	0	0	0.125	0.125	0.03	0.0291

The most obvious problem is trying to estimate a rare species' density with a small number of samples. It is quite hopeless to expect to obtain reasonable density estimates. Indeed, increasing the sample size helps little; we are just unable to get a biologically meaningful picture of the rare species density and variability. Using an estimator with desirable statistical properties like unbiasedness does not help. As abundance increases we expect to do better with our estimation. For both common and the most abundant species, the smaller the number of samples or quadrats, the larger the variance usually. However, by the time we reached the 8 samples the variance estimates were considerably smaller and varied more tightly about the true value. The mean estimates are all unbiased, and we were just lucky with the $n = 8$ estimated density of *C. spruceanum*. Confidence intervals are quite reasonable by the time we have sampled 8 quadrats.

2.2

    a. Find the mean density per week over all sweeps of *A. maculatum*.

The 3 sweeps gave totals of 11, 2, and 4 individuals of this species, or a total of 17. Over the 16 weeks of the study, then, the mean density per week is $\dfrac{17}{16} = 1.06$ individuals.

    b. Find the mean density of *H. crucifer* for each microhabitat and for the total. Sweep counts for this species are 530, 46, and 58, for a total of 634. This gives $\dfrac{530}{16} = 33.125$ or a density of 33 individuals in the surface sweeps per week.

$\dfrac{46}{16} = 2.875$ or a weekly density of 3 individuals in the midwater sweeps.

$\dfrac{58}{16} = 3.625$ or a bottom density of 4 individuals per week.

$\dfrac{634}{16} = 39.625$ or a mean of 40 individuals of *H. crucifer* were found per week of the study.

c. Put 95% bounds on the microhabitat and total mean density of *H. crucifer*.

Thus far in the book, our only way to bound the density is by using Tchebychev's theorem. For *H. crucifer* we have

	$\hat{\mu}$	$\hat{\sigma}^2$
Surface	33.125	The weekly data is not given.
Midwater	2.875	We cannot calculate the needed variances
Bottom	3.625	for each of the microhabitat sweeps
Total	39.625	76197.33

For the total variance the calculation is
$(530 - 39.625)^2 + (46 - 39.625)^2 + (58 - 39.625)^2 = 76197.33$.
The square root of this, or the standard deviation, is 276.0386.

Now, our theorem says we expect at least 75% of our anuran larvae counts to be within $39.625 \pm 2(276.0386) = (-512.4522, 591.7022) \approx (0, 592)$. A look at the data shows that all observed counts are in this interval.

2.3

a. Each sample size was 0.025m². What is the density per m² for the first replicate?

Combining sieve sizes, we get $1,511 + 178 + 2,889 = 4,578$ sipunculids in the 0.025m² sample. Then dividing 0.025 into 1m² gives 40. Thus, because it takes 40 samples of size 0.025m² to make one sample of size 1m², we multiply $4,578 \cdot 40 = 183,120$, which gives the density for 1m², or per m².

b. Estimate the mean and total density for June per 0.025m² replicate.

The 4 replicate totals are 4,578, 2,356, 4,534, and 934. Adding these and dividing by 4 gives us $\dfrac{12402}{4} = 3100.5$ as the average density per replicate for June. Using these values in Equation 2.5 gives a total June variance per replicate of 3,162,466 and a standard deviation of 1,778.195.

c. What is the amount of deviation from the mean per replicate? What is the total amount of deviation from the mean?

$d_1 = 4578 - 3100.5 = 1477.5$
$d_2 = 2356 - 3100.5 = -743.5$
$d_3 = 4533 - 3100.5 = 1432.5$
$d_4 = 933 - 3100.5 = -2167.5$

The total amount of deviation, obtained by adding these 4 numbers, is, of course, 0.

d. Use Tchebychev's theorem to find what proportion of *P. cryptum* is expected to be within 2 standard deviations of the mean for each of the 4 samples.

Sample/Replicate	Mean	Variance	St. dev.
1	1,526.00	1,837,549.00	1,355.56
2	785.33	245,929.30	495.91
3	1,511.33	2,068,696.00	1,438.30
4	311.33	96,854.33	311.21

Now, our theorem says we expect at least 75% of our sipunculid counts to be within.

Sample/ Replicate	Theorem Interval	Observed
1	$1526 \pm 2(1355.56) = (-1187.12, 4237.12) \approx (0, 4237)$	all 4 observed values are in this interval
2	$785.33 \pm 2(495.91) = (-206.49, 1777.15) \approx (0, 1777)$	all 4 observed values are in this interval
3	$1511 \pm 2(1438.30) = (-1365.60, 4387.60) \approx (0, 4388)$	all 4 observed values are in this interval
4	$311 \pm 2(311.21) = (-311.42, 933.42) \approx (0, 933)$	all 4 observed values are in this interval

Alternatively, if we had ignored replicates and used the entire set of counts for June, we would have a mean of 1033.5 and a standard deviation of 1029.57. In this case the theorem would give

$$1033.5 \pm 2(1029.57) = (-1025.6, 3092.6) \approx (0, 3093)$$

Only 1 of the 12 observations, or about 8.3%, are outside this interval.

Chapter 3: Normal and Sampling Distributions for Fieldwork

3.1 Counting the numbers of trees found in each of the 100 quadrats, we obtain the following

Trees per Quadrat	Number of Quadrats	Product
0	28	$0 \cdot 28 = 0$
1	34	$1 \cdot 34 = 34$
2	24	$2 \cdot 24 = 48$
3	9	$3 \cdot 9 = 27$
4	3	$4 \cdot 3 = 12$
5	2	$5 \cdot 2 = 10$
Total	100 quadrats	131, which agrees with the total in Appendix 1

The mode (number of trees most often observed per quadrat) is 1. We know from Appendix 1 that the true mean is 2.52. Because the mean and mode are different, the distribution is skewed. There are 86 of the 100 observations between 0 and 2.

3.2

a. Find the $\hat{\mu}$, and $\hat{\sigma}^2$.

Using the random number table, we selected quadrats #2, #12, #21, and #22. These contained counts of 10, 1, 3, and 7 trees. Remember that this is the 25 observation/ quadrat data.

Using these counts

$$\hat{\mu} = \frac{10 + 1 + 3 + 7}{4} = \frac{21}{4} = 5.25$$

$$\hat{\sigma}^2 = 16.25$$

$$\hat{\sigma} = 4.03.$$

b. Based on these estimates, about what proportion of the quadrats (observations) would you expect to have fewer than 10 trees? Does this agree with the data?

If we examine the value of 10 trees in standard deviation units, we look at the standardized score of $z = \dfrac{x - \hat{\mu}}{\hat{\sigma}} = \dfrac{10 - 5.25}{4.03} = 1.1787$. This says that the value of 10 trees per quadrat is about 1.18 standard deviation units above the mean value. Comparing this to a Normal distribution, a z value of 1.18 cuts off about 0.83 of the area. So, you would expect about 83% of the observations to contain 10 or fewer trees. Reading across the row in the appendix for *B. lactescens* shows that all but 1 quadrat contains 10 or fewer trees. That is $\dfrac{24}{25} = 96\%$.

3.3

a. Calculate the standard error using the fpc for the sample of $n = 4$ of *Brosimum lactescens* calculated in Problem 3.2.

The standard errors are calculated by using the value for the standard deviation estimate and dividing by the square root of the sample size (Equation 3.7), or adding in the fpc (Equation 3.6). For this species we get

$$\text{standard error without the fpc} = \frac{4.03}{2} = 2.015$$

$$\text{standard error with fpc} = \left(\frac{4.03}{2}\right)\sqrt{\frac{(25 - 4)}{(25 - 1)}} = 1.885.$$

b. Is the true mean within $2\hat{\sigma}_\mu$ of $\hat{\mu}$? Had you ignored the fpc, would you still get the same answer?

The true mean for this species from Appendix 2 is 5.24. $2\hat{\sigma}_\mu = 4.03$, so the interval about the mean would be $5.25 \pm 4.03 = (1.22, 9.28)$, which includes the true mean.

If we redo the calculation with the fpc as a factor we obtain $2\hat{\sigma}_\mu = 1.885$, giving an interval of $5.25 \pm 3.77 = (1.48, 9.02)$, which also includes the true mean. Notice that our unbiased estimate is actually almost exactly equal to the true value. If we had obtained a more disparate (but still unbiased) estimate, we would have seen more discrepancy in the two intervals.

3.4

a. Divide the values for each replicate by the appropriate standard deviation estimate and calculate the new standard deviation of all the $\left(\dfrac{x_i}{\hat{\sigma}}\right)$ divided values.

In order to do this we use the standard deviation estimate for the entire set of all June counts, or 1,029.565. This figure was calculated by ignoring replicates and using each of the 12 June counts. The new data set in which each count is divided by this estimate is

Raw Data Replicate	Size 1	Size 2	Size 3
1	1,511	178	2,889
2	1,156	222	978
3	889	489	3,156
4	667	178	89

Counts in Standard Deviation Units

Replicate	Size 1	Size 2	Size 3
1	1.468	0.173	2.806
2	1.123	0.216	0.950
3	0.863	0.475	3.065
4	0.648	0.173	0.086

The sum of these numbers is 12.046 with a mean of 1.004, whereas the raw data's mean value is 1,033.5. The new variance and the standard deviation each are now equal to 1.000.

b. Standardize the 4 replicate observations on *P. cryptum*. Check that the standardized mean and variance are correct.

To standardize we take each of the original counts and subtract the mean of 1,033.5 from them. Then, we divide each of these by 1,029.565, the standard deviation of all the 12 counts. The standardized values are:

Replicate	Size 1	Size 2	Size 3
1	0.464	−0.831	1.802
2	0.119	−0.788	−0.054
3	−0.140	−0.529	2.062
4	−0.356	−0.831	−0.917

After doing this, we now have the same standard deviation of 1.000, but the mean is 0.000 in addition. The data is now standardized.

3.5 There are $\binom{6}{2} = 15$ possible samples of size 2 from a population of size 6. These 15 samples are:

Quadrat Counts	Mean
4, 6	5
4, 2	3
4, 8	6
4, 0	2
4, 4	4

The range of possible values for the mean density is {1, 2, 3, 4, 5, 6, 7}

Quadrat Counts	Mean
6, 2	4
6, 8	7
6, 0	3
6, 4	5
2, 8	5
2, 0	1
2, 4	3
8, 0	4
8, 4	6
0, 4	2

CHAPTER 4: CONFIDENCE LIMITS AND INTERVALS FOR DENSITY

4.1 Choose random numbers from Appendix 3. Our random numbers gave us from Appendix 1 the following numbers for *Scheelea princeps*: 2, 4, 6, 4. We have

$$\hat{\mu} = \frac{16}{4} = 4.00 \text{ from Equation 2.2}$$

$$\hat{\sigma}^2 = 2.6667 \text{ from Equation 2.5}$$

$$\hat{\sigma} = 1.6330.$$

At the 95% level, $z = 1.96$. We then have for the confidence interval

$$4.00 - 1.96 \left(\frac{1.6330}{\sqrt{4}} \right) \le \mu \le 4.00 + 1.96 \left(\frac{1.6330}{\sqrt{4}} \right) \text{ from Equation 4.4}$$

$$4.00 - 1.6003 \le \mu \le 4.00 + 1.6003$$

$$2.3997 \le \mu \le 5.6003.$$

The confidence limits are 4.00 ± 1.6003 from Equation 4.3. As a shortcut, take $z = 2$, then when $n = 4$, $\frac{z\hat{\sigma}}{\sqrt{n}} \approx \hat{\sigma}$ and confidence limit is 4.00 ± 1.6330. The true mean from Appendix 1 is 2.52. We should actually use t instead of z because of the small number of samples. From Table 4.1, t with three degrees of freedom, two-tailed is 3.18. We would then have for the interval

$$4.00 - 2.5965 < \mu < 4.00 + 2.5965$$

$$1.4035 < \mu < 6.5965.$$

The confidence limits are 4.00 ± 2.5965.

4.2 Using $t = 3.18$ from Equation 4.1, we have

$$d = \frac{3.18(0.8165)}{4.00} = 0.6491 \text{ or } 65\%.$$

For the true population parameters we have $d = \dfrac{1.96(0.8139)}{2.52} = 0.6330$ or 63%.

The results indicate that with 4 samples we cannot estimate the mean with any great precision.

4.3 After obtaining 4 random numbers from Appendix 2 ($N = 25$), we obtained for *Scheelea princeps* the counts 14, 6, 0, 11.

$$\hat{\mu} = \frac{31.00}{4} = 7.75$$

$$\hat{\sigma} = 6.1305.$$

Using $z = 1.96$, we have from Equation 4.4 for the confidence interval

$$7.75 - 1.96\left(\frac{6.1305}{\sqrt{4}}\right) \le \mu \le 7.75 + 1.96\left(\frac{6.1305}{\sqrt{4}}\right)$$

$$7.75 - 6.0079 \le \mu \le 7.75 + 6.0079$$

$$1.7421 \le \mu \le 13.7579.$$

The confidence limits are 7.75 ± 6.0079. The true mean from Appendix 2 is 10.08.

If we use $t = 3.18$ for 95% confidence level (two-tailed), the wider limits are

$$7.75 - 3.18\left(\frac{6.1305}{\sqrt{4}}\right) < \mu < 7.75 + 3.18\left(\frac{6.1305}{\sqrt{4}}\right)$$

$$7.75 - 9.7475 < \mu < 7.75 + 9.7475$$

$$0 < \mu < 17.4975.$$

The confidence limits are 7.75 ± 9.7475.

4.4 From Problem 4.3, $\hat{\mu} = 7.75$ and $\hat{\sigma} = 6.1305$ with $n = 4$. We have then $\sigma_\mu = \frac{6.1305}{\sqrt{4}} = 3.0653$. From Equation 4.1 using $t = 3.18$ as before, we have

$$d = \frac{3.18(3.0653)}{7.75} = 1.26 \text{ or } 126\%.$$

For the true populations parameters (Appendix 2) we have $\mu = 10.08$ and $\sigma = 4.2794$, so that $\sigma_\mu = \frac{4.2794}{\sqrt{4}} = 2.1397$. We have then

$$d = \frac{1.96(2.1397)}{10.08} = 0.4161 \text{ or } 42\%.$$

4.5 The estimates from Problem 4.1 are: $\hat{\mu} = 4.00$, $\hat{\sigma} = 1.6330$, and $\sigma_\mu = 0.8165$. Using Equation 4.8 we have

$$100(4.00) - 100(3.18 \cdot 0.8165) \le N\mu \le 100(4.00) + 100(3.18 \cdot 0.8165)$$

$$139.69 \le N\mu \le 660.31.$$

The estimates from Problem 4.3 are: $\hat{\mu} = 7.75$, $\hat{\sigma} = 6.1305$, and $\sigma_\mu = 3.0653$.

$$25(7.75) - 25(3.18 \cdot 3.0653) \le N\mu \le 25(7.75) + 25(3.18 \cdot 3.0653)$$

$$0 \le N\mu \le 437.44.$$

We have (Appendix 2) 252 trees belonging to *S. princeps*.

4.6 Using random numbers from Appendix 3 and sampling Appendix 2, we obtained 1 and 3 individuals of *Pouteria macrophylla*. We have then $\hat{\mu} = 2.00$ and $\hat{\sigma} = 1.4142$. We wish to find the number of quadrats with 0, so $t = \dfrac{0 - 2.00}{1.4142} = -1.4142$. Because $n = 2$, the df $= 1$, and looking at Table 4.1 we observe that for one-tail and df $= 1$ at 0.90 confidence, $t = 3.08$. We would expect at least 90% of the quadrats, or $0.9(25) = 22$, to have a value of less than 3.08. The values in the table are $+$ and our interest is in -1.4142, so that we have $25 - 22 = 3$ quadrats, at least, with 0 values. Referring to Appendix 2, we observe that 5 quadrats have 0 values.

4.7

a. In this problem the crayfish are our quadrats so that $n = 10$ or in one case $n = 15$. Because the number of crayfish, N, must be large, we will ignore the fpc for standard errors. From Table 4.1, $t_{95,9} = 2.26$ (two-tailed) and $t_{95,14} = 2.15$. Using Equation 4.7

Fall: $384.60 - 2.26 \left(\dfrac{273.06}{\sqrt{10}} \right) \leq \mu \leq 384.60 + 2.26 \left(\dfrac{273.06}{\sqrt{10}} \right)$

$384.60 - 195.15 \leq \mu \leq 384.60 + 195.15$

$189.45 \leq \mu \leq 579.75$

Winter: $243.20 - 2.26 \left(\dfrac{168.21}{\sqrt{10}} \right) \leq \mu \leq 243.20 + 2.26 \left(\dfrac{168.21}{\sqrt{10}} \right)$

$243.20 - 120.22 \leq \mu \leq 243.20 + 120.22$

$122.98 \leq \mu \leq 363.42$

Spring: $195.33 - 2.15 \left(\dfrac{180.12}{\sqrt{15}} \right) \leq \mu \leq 195.33 + 2.15 \left(\dfrac{180.12}{\sqrt{15}} \right)$

$195.33 - 99.99 \leq \mu \leq 195.33 + 99.99$

$95.34 \leq \mu \leq 295.32$

Summer: $95.50 - 2.26 \left(\dfrac{68.87}{\sqrt{10}} \right) \leq \mu \leq 95.50 + 2.26 \left(\dfrac{68.87}{\sqrt{10}} \right)$

$95.50 - 49.22 \leq \mu \leq 95.50 + 49.22$

$46.28 \leq \mu \leq 144.72.$

b. We have then

Fall: $189 \leq \mu \leq 580$

Winter: $123 \leq \mu \leq 363$

Spring: $95 \leq \mu \leq 295$

Summer: $46 \leq \mu \leq 145.$

The confidence limits for summer do not overlap with fall, and we would conclude, if this were our test, that they are different. However, if we were looking for seasonal differences, we would use another statistical method (analysis of variance), which Hart and colleagues (1985) did do.

4.8 Using the larval data of Heyer (1979) in Problem 2.2, Chapter 2, find the mean density and variance for *R. sylvatica* over all sweeps. Which type of sweep's count is the most deviant from the mean density in standard deviation units?

$$\hat{\mu} = \frac{1191}{3} = 63.6667$$

$$\hat{\sigma} = 62.0672$$

$$z_1 = \frac{(7 - 63.6667)}{62.0672} = -0.9130$$

$$z_2 = \frac{(130 - 63.6667)}{62.0672} = 1.0687$$

$$z_3 = \frac{(54 - 63.6667)}{62.0672} = -0.1557$$

Because the midwater count of 130 has the largest value (ignoring the sign, just looking at the number), 130 is the most deviant.

CHAPTER 5: HOW MANY FIELD SAMPLES?

5.1 For *Pouteria macrophylla*, from Appendix 1 we have

$$\mu = 0.62, \, 0.75\mu = 0.4650, \, \sigma^2 = 0.7756, \, \sigma = 0.8807$$

$$n = \frac{2^2(0.7756)}{0.4650^2} = 14.35 \text{ from Equation 5.5.}$$

From Appendix 2 for *P. macrophylla*, we have

$$\mu = 2.48, \, 0.75\mu = 1.86, \, \sigma^2 = 5.4496, \, \sigma = 2.3344$$

$$n = \frac{2^2(5.4496)}{1.86^2} = 6.30.$$

Fourteen random samples from Appendix 1 for *P. macrophylla* gave us 2, 1, 1, 0, 0, 0, 0, 0, 0, 0, 1, 0, 0, 1

$$\hat{\mu} = 0.4286, \, \hat{\sigma} = 0.6462, \, \hat{\sigma}^2 = 0.4176, \, \hat{\sigma}_\mu = \frac{0.6462}{\sqrt{14}} = 0.1727$$

using Equation 4.6 and two-tailed $t_{90,13} = 1.77$, we have

$$0.4286 - 1.77(0.1727) \leq \mu \leq 0.4286 + 1.77(0.1727)$$

$$0.4286 - 0.3057 \leq \mu \leq 0.4286 + 0.3057$$

$$0.1229 \leq \mu \leq 0.7343.$$

The true value is $\mu = 0.62$.

Six random samples of *P. macrophylla* from Appendix 2 are 8, 1, 4, 1, 2, 0, so that

$$\hat{\mu} = 2.6667, \hat{\sigma} = 2.9439, \hat{\sigma}^2 = 8.6665, \hat{\sigma}_\mu = \frac{2.9439}{\sqrt{6}} = 1.2019$$

$$t_{.90,5} = 2.01, 2.6667 - 2.01(1.2019) \leq \mu \leq 2.6667 + 2.01(1.2019)$$

$$0.2509 \leq \mu \leq 5.0825.$$

The true mean is 2.48.

We might ask, "Would it be more work to take 14 samples from 100 or 6 samples from 25?" It depends on the work involved in locating the quadrats and counting and identifying the trees within them. In the first instance you must on the average count 0.62(14.35) = 8.89, or 9 trees. In the second, 2.48(6.30) = 15.63, or 16 trees. Note in the first sample of $n = 14$ we actually counted 6 and in the second sample of $n = 6$, we counted 16. We should note in passing that if $\mu = \sigma^2$, then in each case the number of individuals needed to be counted would be the same.

5.2 For *Astrocaryum macrocalyx* from Appendix 1, we have

$$\mu = 0.17, 0.75\mu = 0.1275, \sigma^2 = 0.1811, \text{we will use a } z = 2$$

$$n = \frac{4(0.1811)}{0.1275^2} = 44.56.$$

For *A. macrocalyx* from Appendix 2, we have

$$\mu = 0.68, 0.75\mu = 0.51, \sigma^2 = 0.6976$$

$$n = \frac{4(0.6796)}{0.51^2} = 10.73.$$

Because of the small mean, the number of samples required for $\pm 75\%$ of the mean is ridiculous and we must accept a greater margin.

5.3 For *Buliminella elegantissima* we have

$$\hat{\mu} = 182.200, 0.5\hat{\mu} = 91.100, \hat{\sigma} = 116.540, \hat{\sigma}^2 = 13581.700, z = 2$$

$$n = \frac{4(13,581.700)}{91.1^2} = 6.5460 \approx 6.$$

For *Valvulineria floridana* we have

$$\hat{\mu} = 20.2000, 0.5\hat{\mu} = 10.1000, \hat{\sigma} = 11.9666, \hat{\sigma}^2 = 143.2000, z = 2$$

$$n = \frac{4(143.2)}{10.1^2} = 5.6151 \approx 6.$$

For *Epistominella pontoni* we have

$$\hat{\mu} = 4.4000, \ 0.5\hat{\mu} = 2.2000, \ \hat{\sigma} = 5.3198, \ \hat{\sigma}^2 = 28.3000, \ z = 2$$

$$n = \frac{4(28.3)}{2.2^2} = 23.3884 \approx 23.$$

The CV's (Equation 4.9) are similar for the first two and twice as large for the third. Because of time and cost, go with the 6 samples and accept the larger confidence limits for *E. pontoni*.

CHAPTER 6: SPATIAL DISTRIBUTION: THE POWER CURVE

6.1 From Appendix 2 for *Pouteria macrophylla* we took 8 random samples, each consisting of $n = 4$ quadrats. We obtained

Sample Observation (n)	1	2	3	4	5	6	7	8
1	4	1	1	1	1	0	0	4
2	5	1	0	5	0	8	4	1
3	2	3	2	2	0	0	5	4
4	1	0	0	3	0	3	5	3
μ	3.00	1.25	0.75	2.75	0.25	2.75	3.50	3.00
$\hat{\sigma}^2$	3.33	1.58	0.92	2.92	0.25	14.25	5.67	2.00

For the power curve we obtained $\hat{\sigma}^2 = 1.25\mu^{1.14}$.

Estimates	μ	$\hat{\sigma}^2$	
	0.62	0.72	
	2.48	3.50	true $\mu = 2.48$, $\hat{\sigma}^2 = 5.45$
	10.00	17.08	
	100.00	233.62	

6.2 For *Pouteria macrophylla* 8 random samples of 4 observations each from Appendix 1 ($N = 100$) are

Sample Observation (n)	1	2	3	4	5	6	7	8
1	1	2	1	1	0	1	0	0
2	1	4	0	0	0	0	1	1
3	0	1	0	1	0	0	0	3
4	0	0	0	0	2	0	0	0
μ	0.50	1.75	0.25	0.50	0.50	0.25	0.25	1.00
$\hat{\sigma}^2$	0.33	2.92	0.25	0.33	1.00	0.25	0.25	2.00

For the power curve we obtained $\hat{\sigma}^2 = 1.44\mu^{1.33}$.

Estimates	μ	$\hat{\sigma}^2$	
	0.62	0.76	true $\mu = 0.62$, $\hat{\sigma}^2 = 0.78$
	2.48	4.82	
	10.00	30.61	
	100.00	648.93	

6.3 Combining the μ and $\hat{\sigma}^2$ estimates from the first two problems gives an $n = 16$. The power curve we obtained is $\hat{\sigma}^2 = 1.26\mu^{1.18}$.

Estimates		combined	$N = 25$	$N = 100$
	μ	$\hat{\sigma}^2$	$\hat{\sigma}^2$	$\hat{\sigma}^2$
	0.62	0.72	0.72	0.76
	2.48	3.67	3.50	4.82
	10.00	18.89	17.08	30.61
	100.00	283.17	233.62	648.93

We know the true mean for $N = 25$ is 2.48 and the variance 5.45, and for $N = 100$, 0.62 and 0.78, respectively. We would expect that the power curve from Problem 6.1 would give the best estimate for $N = 25$ and the power curve from 6.2 the best estimate for $N = 100$. However, in our example the best estimates came from the $N = 100$ predictions. Notice how quickly the estimates become far apart as we go beyond the observations (at 10 and 100).

6.4 For $N = 25$ quadrats we recall Equation 5.6 and Table 4.1 for two-tailed $t_{.95,7} = 2.36$ and $t^2 = 5.57$. We have then for 0.75μ,

for $\mu = 0.62$ $n = \dfrac{(5.57)(0.72)}{0.22} = 18.68 \approx 19$

for $\mu = 2.48$ $n = \dfrac{(5.57)(3.50)}{3.46} = 5.64 \approx 6$

for $\mu = 10$ $n = \dfrac{(5.57)(17.08)}{56.25} = 1.69 \approx 2$

for $\mu = 100$ $n = \dfrac{(5.57)(233.62)}{5,625} = 0.23 \approx 1.$

For $N = 100$ quadrats t remains the same and we have for 0.75μ

for $\mu = 0.62$ $n = \dfrac{(5.57)(0.76)}{0.22} = 19.73 \approx 20$

for $\mu = 2.48$ $n = \dfrac{(5.57)(4.82)}{3.46} = 7.75 \approx 8$

for $\mu = 10$ $n = \dfrac{(5.57)(30.61)}{56.25} = 3.03 \approx 3$

for $\mu = 100$ $n = \dfrac{(5.57)(648.93)}{5,625} = 0.64 \approx 1.$

The number of quadrats required for $\pm 75\%$ of the mean is about the same. Consequently, we would have to consider the amount of work involved with placing the quadrats. Notice how the glib use of a certain specified percent gets us into trouble. It is easy to be within 75% of 100 trees, but much more difficult for 1.

6.5

a. From the data, the following μ's and $\hat{\sigma}^2$'s were estimated.

	AUG	OCT	FEB	APR	MAY	JUL
μ	462.25	507.75	731.75	430.25	336.75	469.00
$\hat{\sigma}^2$	4,577.58	146,722.25	82,528.25	59,786.25	17,812.25	67,745.33

We obtained from these data the power curve $\hat{\sigma}^2 = 0.067 \mu^{2.16}$.

Estimates	μ	$\hat{\sigma}^2$
	430	31,959.88
	450	35,252.38
	500	44,246.26
	600	65,561.96
	700	91,419.83

b. The estimated variances from the power curve are not very good. In October, the variance is very large, and looking at the data we observe that one observation was only 61. Nevertheless, the high value of b in the power curve indicates considerable aggregation, which the estimates and the original data bear out.

CHAPTER 8: SPECIES PROPORTIONS: RELATIVE ABUNDANCES

8.1 We have $N = 663$ from which we choose $n = 30$ individuals at random. Five belonged to *Pouteria macrophylla*. We have then $p = \dfrac{5}{30} = 0.1667$ and $q = 0.8333$. Using the fpc and Equation 8.13,

$$\hat{\sigma}_p^2 = \frac{663 - (30 \cdot 0.1667 \cdot 0.8333)}{29(663)} = 0.0046$$

$$\hat{\sigma}_p = 0.0676.$$

The confidence interval using $z = 1.96$ for 95% is

$$0.1667 - 1.96(0.0676) \leq p \leq 0.1667 + 1.96(0.0676)$$

$$0.1667 - 0.1325 \leq p \leq 0.1667 + 0.1325$$

$$0.0342 \leq p \leq 0.2992,$$

or for the limits 0.1667 ± 0.1325.

Ignoring the fpc we have from Equation 8.14

$$\hat{\sigma}_p^2 = \frac{(0.1667)(0.8333)}{29} = 0.0048$$

$$\hat{\sigma}_p = 0.0692.$$

The confidence interval is

$$0.1667 - 1.96(0.0692) \leq p \leq 0.1667 + 1.96(0.0692)$$

$$0.1667 - 0.1356 \leq p \leq 0.1667 + 0.1356$$

$$0.0311 \leq p \leq 0.3023,$$

or for the limits 0.1667 ± 0.1356.

Clearly, using the fpc was not worth the trouble. Because $n = 30$, the confidence limits are large and the true value of 0.09 (Appendix 1) easily falls within the limits.

8.2 We chose 4 quadrats at random and recorded the number of *Pouteria macrophylla*, a_i, and the number of individuals in the quadrats, m_i. We obtained

Quadrat	a_i	m_i	$\hat{p} = \dfrac{a_i}{m_i}$	$\hat{q} = 1 - \hat{p}$
1	1	30	0.0333	0.9667
4	1	24	0.0417	0.9583
19	8	37	0.2162	0.7838
22	0	28	0.0000	1.0000

Ignoring the fpc we have using Equation 8.15

$$\hat{\sigma}_p = \sqrt{\frac{0.0322}{29}} = 0.0062$$

for quadrat 1 and the interval $0.0333 - 1.96(0.0062) \leq p \leq 0.0333 + 1.96(0.0062)$

$$0.0212 \leq p \leq 0.0454$$

$$\hat{\sigma}_p = \sqrt{\frac{0.0399}{23}} = 0.0087$$

for quadrat 4 and the interval $0.0417 - 1.96(0.0087) \leq p \leq 0.0417 + 1.96(0.0087)$

$$0.0247 \leq p \leq 0.0587$$

$$\hat{\sigma}_p = \sqrt{\frac{0.1695}{36}} = 0.0114$$

for quadrat 19 and the interval $0.2162 - 1.96(0.0114) \leq p \leq 0.2162 + 1.96(0.0114)$

$$0.1938 \leq p \leq 0.2386$$

for quadrat 22 $\qquad\qquad \hat{\sigma}_p = 0.$

For cluster estimation we use Equation 8.25:

Quadrat	a_i	m_i	m_i^2	p_i	$(p_i - p)^2$	$m_i^2(p_i - p)^2$
1	1	30	900	0.0333	0.0026	2.3134
4	1	24	576	0.0417	0.0018	1.0306

Quadrat	a_i	m_i	m_i^2	p_i	$(p_i - p)^2$	$m_i^2(p_i - p)^2$
19	8	37	1,369	0.2162	0.0175	23.9258
22	0	28	784	0.0000	0.0071	5.5319
Totals	10	119				32.8017

$$p = \frac{10}{119} = 0.0840, \; \hat{\mu} = \frac{119}{4} = 29.75, \; \hat{\mu}^2 = 885.0625,$$

$$n(n-1)\,\hat{\mu}^2 = 10620.75$$

$$\hat{\sigma}^2 = \frac{32.8017}{10620.75} = 0.0031, \; \hat{\sigma} = 0.0556$$

$$0.0840 - 1.96(0.0556) \le p \le 0.0840 + 1.96(0.0556)$$

$$0 \le p \le 0.1929.$$

The true value is $p = 0.0935$. Only the cluster estimate bracketed the true value.

8.3

a. From Problem 8.2 we have

a_i	m_i	p_i
1	30	0.0333
1	24	0.0417
8	37	0.2162
0	28	0.0000
Totals 10	119	0.2912

b. $\hat{p} = \dfrac{\Sigma p_i}{n} = \dfrac{0.2912}{4} = 0.0728$ incorrect

$\hat{p} = \dfrac{\Sigma a_i}{\Sigma m_i} = \dfrac{10}{119} = 0.0840$ correct

8.4 For the binomial confidence of each sample we use Equation 8.23.

For G-l $\hat{p} = \dfrac{179}{450} = 0.3978, \; \hat{q} = 0.6022$

$$0.3978 - 1.96(0.0011) \le p \le 0.3978 + 1.96(0.0011)$$

$$0.3957 \le p \le 0.3999.$$

For the estimated entire sample $n = \dfrac{450}{0.75} = 600,$

Bulimenella elegantissima $= \dfrac{179}{0.75} = 239$

$$\hat{p} = \dfrac{239}{600} = 0.3983, \; \hat{q} = 0.6017$$

$$0.3983 - 1.96(0.0008) \le p \le 0.3983 + 1.96(0.0008)$$

$$0.3967 \le p \le 0.3999.$$

For G-2 $\qquad \hat{p} = \dfrac{120}{318} = 0.3774, \ \hat{q} = 0.6226,$

$0.3774 - 1.96(0.0015) \le p \le 0.3774 + 1.96(0.0015)$

$0.3740 \le p \le 0.3804,$

sample was not split.

For G-3 $\qquad \hat{p} = \dfrac{81}{237} = 0.3418, \ \hat{q} = 0.6582$

$0.3418 - 1.96(0.0020) \le p \le 0.3418 + 1.96(0.0020)$

$0.3379 \le p \le 0.3457,$

sample was not split

For G-4 $\qquad \hat{p} = \dfrac{182}{412} = 0.4417, \ \hat{q} = 0.5583$

$0.4417 - 1.96(0.0012) \le p \le 0.4417 + 1.96(0.0012)$

$0.4393 \le p \le 0.4441.$

For the estimated entire sample, we have twice the observed, so

$$\hat{p} = \frac{364}{824} = 0.4417, \ \hat{q} = 0.5583$$

$$0.4417 - 1.96(0.0006) \le p \le 0.4417 + 1.96(0.0006)$$

$$0.4405 \le p \le 0.4429.$$

For G-5 $\qquad \hat{p} = \dfrac{206}{448} = 0.4598, \ \hat{q} = 0.5402$

$0.4598 - 1.96(0.0011) \le p \le 0.4598 + 1.96(0.0011)$

$0.4576 \le p \le 0.4620.$

For the estimated entire sample, we have twice the observed, so

$$\hat{p} = \frac{412}{896} = 0.4598, \ \hat{q} = 0.5402$$

$$0.4598 - 1.96(0.0006) \le p \le 0.4598 + 1.96(0.0006)$$

$$0.4587 \le p \le 0.4609.$$

b. For the entire set of observations put together we have

$n_{B. \ elegantissima}$	n
179	450
120	318
81	237
182	412
206	448
Totals 768	1,865

$$\hat{p} = \frac{768}{1865} = 0.4118, \ \hat{q} = 0.5882$$

Note that if we took the mean of the 5 calculated p's we would have $\hat{p} = 0.4037$. Using the correct estimate of $\hat{p} = 0.4118$, we have

$$0.4118 - 1.96(0.0003) \le p \le 0.4118 + 1.96(0.0003)$$

$$0.4113 \le p \le 0.4123.$$

For the estimated total numbers from the splits we have

	$n_{B. \ elegantissima}$	n
	239	600
	120	318
	81	237
	364	824
	412	896
Totals	1,216	2,875

$$\hat{p} = \frac{1216}{2875} = 0.4230, \ \hat{q} = 0.5770$$

$$0.4230 - 1.96(0.0002) \le p \le 0.4230 + 1.96(0.0002)$$

$$0.4227 \le p \le 0.4233.$$

c. For cluster estimates we use Equation 8.25. We have

Sample	a_i	m_i	m_i^2	p_i	$(p_i - p)^2$	$m_i^2 (p_i - p)^2$
G-1	179	450	202,500	0.3978	0.0002	40.5000
G-2	120	318	101,124	0.3774	0.0012	121.3488
G-3	81	237	56,169	0.3418	0.0049	275.2281
G-4	182	412	169,744	0.4417	0.0009	152.7696
G-5	206	448	200,704	0.4598	0.0023	461.6192
Totals	768	1,865				1,051.4657

$$n = 5, \ \hat{p} = \frac{768}{1865} = 0.4118, \ \hat{\mu} = \frac{1865}{5} = 373, \ \hat{\mu}^2 = 139129, \ n(n-1)\hat{\mu}^2 = 2782580$$

$$\hat{\sigma}^2_{pclus} = \frac{1051.4657}{2782580} = 0.0004, \ \hat{\sigma}_{pclus} = 0.0194$$

$$0.4118 - 1.96(0.0194) \le p \le 0.4118 + 1.96(0.0194)$$

$$0.3737 \le p \le 0.4499.$$

When we use the estimated total number of individuals from splits, we have

Sample	a_i	m_i	m_i^2	p_i	$(p_i - p)^2$	$m_i^2(p_i - p)^2$
G-1	239	600	360,000	0.3983	0.0006	219.6324
G-2	120	318	101,124	0.3774	0.0021	210.2732
G-3	81	237	56,169	0.3418	0.0066	370.3469
G-4	364	824	678,976	0.4417	0.0003	237.4311
G-5	412	896	802,816	0.4598	0.0014	1,087.2055
Totals	1,216	2,875				2,124.8891

$$n = 5, \; \hat{p} = \frac{1216}{2875} = 0.4230, \; \hat{\mu} = \frac{2875}{5} = 575, \; \hat{\mu}^2 = 330625, \; n(n-1)\hat{\mu}^2 = 6,612,500$$

$$\hat{\sigma}^2_{pclus} = \frac{2124.8891}{6612500} = 0.0003, \; \hat{\sigma}_{pclus} = 0.0179$$

$$0.4230 - 1.96(0.0179) \le p \le 0.4230 + 1.96(0.0179)$$

$$0.3879 \le p \le 0.4581$$

d. a. When we calculated binomial confidence limits on each of the cores, we obtained the following confidence intervals

Sample	Split $<p<$		All $<p<$	
G-1	0.40	0.40	0.40	0.40
G-2	no split		0.37	0.38
G-3	no split		0.34	0.34
G-4	0.44	0.44	0.44	0.44
G-5	0.46	0.46	0.46	0.46

This summary indicates that nothing is gained by reconstituting the samples to the estimated total number of individuals (if we were interested in density, this would not be true). We also see that something is dreadfully wrong about our confidence intervals. The estimates for the species proportion of B. *elegantissima* vary from 0.34 to 0.46, and the confidence limits are less than 1%.

b. When the entire data set is taken together, the confidence intervals from the Binomial method become very small because n becomes so large. A researcher would have to be very naive or inexperienced to believe that the estimate of a species proportion spanned an interval of 0.0006.

c. Using the cluster method, we once again observe that little is gained by using total estimates versus the splits. The confidence limits of about ± 0.04 are much more reasonable and in keeping with the estimates of the original samples. Even so, two values of the individual samples (0.34 and 0.46) fall outside of the interval for the splits. If we used the value of $t_{.95,4} = 2.78$ instead of 1.96, only 0.34 would fall outside of the interval.

CHAPTER 9: SPECIES DISTRIBUTIONS

9.1 When the terms in the sequence are summed we get $1 + 2 + 3 + \cdots + 24 + 25$. Using 25 terms and the formula for the sum from this chapter we have $1 + 2 + 3 + \cdots + n = 1/2 \cdot (n) \cdot (n+1)$, or for $n = 25$ the sum is $1/2 \cdot 25 \cdot 24 = 325$. Now you can actually do the addition if you don't believe it.

9.2 For $N = 1,405$ individuals, $S = 50$ species, we obtain a Log series parameter of $\alpha = 10.12066$ and $x = 0.9928482$. Using Fisher's formula, the variance of alpha is $var(\alpha) = 0.44324$. Taking the square root of this quantity we obtain st error = 0.66576. Then, using this quantity in the formula for a confidence interval, we obtain for the 95%

level (using $z = 1.96$), $\alpha \pm 1.96 \cdot$ (st error) $= 10.12066 \pm 1.30489 = (8.81577, 11.42554)$. Finally, using the series expansion formula we can find an expected quantity for each observed term. Notice of course that the first 2 expected terms are quite different from the corresponding observed terms. This is because we have no singletons (species represented by one individual) or species with only 2 individuals in this data set. Also note that there are quite obvious discrepancies between the observed and the expected while the value for x is above 0.99.

N	Observed	Expected
1	0	10.05
2	0	5.00
3	3	3.30
4	3	2.46
5	5	1.95
6	2	1.62
7	4	1.37
8	1	1.19

9.3

a. Find the mean density and variance for each species. Adding all weekly values and dividing by 16 weeks gives the weekly densities *H. crucifer:* 39.625 and *R. sylvatica:* 11.3125, with variances of 2,098.0425, and 216.0961, respectively.

b. Find the covariance of the 2 species.

When calculating the value, we used Equation 9.1 with a change for the fact that these are anuran samples, not the population. By this point you should recognize the change in the formula. The formula used is

$$\sigma_{ij} = \sum_i \sum_j \frac{(X_i - \mu_i)(X_j - \mu_j)}{n(n-1)}.$$

We replaced the population size of N with the sample size and then corrected for bias. Thus, for the covariance of the two species we followed the calculations below.

Week	H. cruc.	(x−39.625)	(x−39.625)²	R. sylv.	(x−11.3125)	(x−11.3125)²	(x−39.625) · (x−11.3125)
1	0	−39.625	1,570.141	0	−11.3125	127.9727	448.2578
2	0	−39.625	1,570.141	0	−11.3125	127.9727	448.2578
3	0	−39.625	1,570.141	0	−11.3125	127.9727	448.2578
4	0	−39.625	1,570.141	0	−11.3125	127.9727	448.2578
5	0	−39.625	1,570.141	2	−9.3125	86.723	369.008
6	22	−17.625	310.641	23	11.688	136.598	−205.992
7	81	41.375	1,711.891	30	18.688	349.223	773.195
8	34	−5.625	31.641	45	33.688	1,134.848	−189.492
9	101	61.375	3,766.891	29	17.688	312.848	1,085.570
10	77	37.375	1,396.891	26	14.688	215.723	548.945

Week	H. cruc.	(x–39.625)	(x–39.625)²	R. sylv.	(x–11.3125)	(x–11.3125)²	(x–39.625) · (x–11.3125)
11	141	101.375	10,276.89	16	4.688	21.973	475.195
12	91	51.375	2,639.391	7	–4.313	18.598	–221.555
13	56	16.375	268.141	3	–8.313	69.098	–136.117
14	31	–8.625	74.391	0	–11.3125	127.9727	448.2578
15	0	–39.625	1,570.141	0	–11.3125	127.9727	448.2578
16	0	–39.625	1,570.141	0	–11.3125	127.9727	448.2578
	Mean	Sum	Variance	Mean	Sum	Variance	Covariance
	39.625	0	2,098.0425	11.3125	0	216.0961	23.4857

c. Place your findings (values) in a V–CV matrix (Equation 9.3). From the table we have

23.4857   2,098.0425

216.09   23.4857

CHAPTER 10: REGRESSION: OCCURRENCES AND DENSITY

10.1 Use the bat data of Problem 9.2 and calculate a regression on the occurrences for predicting density.

a. Using $N = 150$, $S = 50$, and the data in the table below, we obtain the equation

Density = 9.351449 · (Occurrences) + 0.045652.

The r^2 value is 0.7974; that is, occurrences are explaining about 80% of the variability in the bat densities.

b. The standard error of estimate is 15.98327.

c. Make a table containing the observed and expected numbers from the regression.

Species No.	Observed Species Occ.	Observed Species Density	Expected Species Density	
50	1	8	9	(= 9.351449 – 1 + 0.045652)
49	4	18	37	
48	1	3	9	
47	3	17	28	
46	1	3	9	
45	1	12	9	
44	1	6	9	
43	1	4	9	
42	3	30	28	
41	7	51	66	
40	1	3	9	
39	1	21	9	
38	6	56	56	
37	5	122	140	
36	1	4	9	

Species No.	Observed Species Occ.	Observed Species Density	Expected Species Density
35	3	27	28
34	4	25	37
33	1	5	9
32	3	14	28
31	5	88	47
30	2	22	19
29	8	56	75
28	2	17	19
27	2	69	19
26	11	82	103
25	2	12	19
24	16	179	150
23	2	35	19
22	2	55	19
21	1	4	9
20	1	4	9
19	1	7	9
18	1	4	9
17	5	44	47
16	7	82	66
15	1	7	9
14	2	10	19
13	3	28	28
12	2	8	19
11	1	11	9
10	2	7	19
9	1	5	9
8	1	5	9
7	3	50	28
6	2	58	19
5	2	7	19
4	1	4	9
3	1	5	9
2	1	6	9
1	1	5	9
	50	150	1,405

CHAPTER 11: SPECIES OCCURRENCES

11.1 The true proportion of occurrences from Appendix 1 is $\frac{15}{396} = 0.379$. To be 95% confident we select $z = 1.96$ and use $d = 0.05$ in the formula (Equation 5.5), repeated in this chapter: $n = \frac{1.96^2 \cdot 0.0379 \cdot (1 - 0.0379)}{0.05^2} = \frac{0.140079}{0.0025} \approx 56$ samples.

11.2 The true relative abundance for *A. macrocalyx* is 0.0256. Using Equation 5.5 and, from this chapter, $n = \dfrac{z^2 \cdot p_o \cdot q_o}{d^2}$, we use $d = 0.05$. Then we get

$$n = \frac{1.96^2 \cdot 0.0256 \cdot (1 - 0.0256)}{0.05^2} = \frac{0.095827}{0.0025} \approx 38 \text{ individuals.}$$

11.3 Use the bat data of problem 9.2 and fit a Log series distribution to the occurrences (see Problem 10.1).

a. For $N = 150$ occurrences and $S = 50$ species, we can solve Equation 9.14 or use appendix 4 to obtain Equation 9.15 $\alpha = 26.26307$, and Equation 9.14 $x = 0.8510007$. Calculating the terms from Equation 9.10 gives:

n	Obs.	Expected from Log series
1	22	22.35
2	11	9.51
3	6	5.41
4	2	3.44
5	2	2.34
6	1	1.66
7	2	1.21
8	1	0.90
9	0	0.68
10	0	0.52
11	1	0.40
12	0	0.32
13	0	0.25
14	0	0.20
15	1	0.16
16	1	0.12

b. Find the Log series parameter and put a confidence interval about it.

The log series parameter is $\alpha = 26.26307$ for the distribution with $N = 150$ and $S = 50$. The variance of α, $V(\alpha) = 11.58478$ and its square root, 3.40364, is the standard error calculated using Equation 9.17. Then, multiplying Equation 9.17 by 1.96 we get 6.67114, and the confidence interval is $(26.26307 \pm 6.67114) = (19.59193 \leq \alpha \leq 32.93421)$.

c. Discuss the fit of the Log series to this bat occurrence data versus the fit to the bat abundance data (Problem 9.2).

Although we can see that $x = 0.99$ for abundance data and $x = 0.85$ for occurrence data, we must check how close the observed values are to those predicted from the Log series in order to discuss the fit. We have for the first 8 terms:

	Log Series Individuals			Log Series Occurrences	
N	Observed	Expected	n	Observed	Expected
1	0	10.05	1	22	22.35
2	0	5.00	2	11	9.51
3	3	3.30	3	6	5.40

	Log Series Individuals			Log Series Occurrences	
N	Observed	Expected	n	Observed	Expected
4	3	2.46	4	2	3.44
5	5	1.95	5	2	2.34
6	2	1.62	6	1	1.66
7	4	1.37	7	2	1.21
8	1	1.19	8	1	0.90

Without any statistical test, the occurrence data appear to fit better.

CHAPTER 12: SPECIES DIVERSITY: THE NUMBER OF SPECIES

12.1 Using the numbers for cum N and cum S in Appendix 5 for the bat data, we obtained for a power curve $\hat{S} = 1.1770N^{0.5366}$. For the Bolivian trees we had an estimate of $\hat{S} = 2.144N^{0.497}$ (Table 12.2). For $N = 301$ this equation yields $\hat{S} = 36.56$, which agrees well with the observation of $S = 37$ at $N = 301$. At $N = 579$ we would expect 50.61 trees in Bolivia (at $N = 603$ we observe 51, Table 12.2). For bats at $N = 301$ we expect $\hat{S} = 25.72$ and at $N = 579$, $\hat{S} = 35.84$. We would have to disagree with the colleague.

12.2 The predictive equation is $\hat{S} = 1.1770N^{0.5366}$. We obtain:

N	$\hat{S}$	S_{obs}
206	20	25
344	27	34
579	36	38
707	40	39
901	45	45
1,002	48	46
1,253	54	50

12.3 We have from Equation 12.12, $\hat{S} = \alpha\left(ln(1) + \dfrac{N}{\alpha}\right)$ and $\alpha = 10.1207$. We obtain:

N	$\hat{S}_{\alpha}$	S_{obs}
206	31	25
344	36	34
579	41	38
707	43	39
901	46	45
1,002	47	46
1,253	49	50

Both methods work pretty well. In this case one works better lower, the other higher.

12.4 Using the $\alpha = 9.3651$ we obtained for $N = 344$ and $S = 34$, abundifaction yields:

N	$\hat{S}_{\alpha}$	S_{obs}
344	34	34
579	39	38

N	$\hat{S}_\alpha$	S_{obs}
707	41	39
901	43	45
1,002	44	46
1,253	46	50
1,405	47	50

12.5 We recall Equation 12.11. In this case $r = 344$. In Appendix 5 the first term is $p = 0.0057$ so that $1 - p = 0.9943 = q$; $q^r = 0.9943^{344} = 0.1400$, $(1 - 0.1400) = 0.8600$. This is the first term. We repeat this for all 50 species and sum. Our answer was 43.02, or 43 species. We recall at $n = 344$, $S = 34$. Rarefaction using α and regression gave us a closer answer in this case.

12.6 We wish to use Sanders's method of rarefaction for $n = 344$. Now, $\dfrac{1}{344} = 0.0029$. In Appendix 5, 41 species of bats have a larger proportion, and these make up 0.9768 of the total. $1 - 0.9768 = 0.0232$, so $\dfrac{0.0232}{0.0029} = 8$, and we predict $41 + 8 = 49$ species at $n = 344$. The reason we did so poorly is that there are no singletons (species represented by one individual) in the bat set.

12.7 For our bat data in Appendix 5 at $N = 150$ occurrences and $S = 50$, $\alpha = 26.2631$. At 17 quadrats $n = 49$ and our estimate is $S = 27.65 \approx 28$. We observed $S = 34$.

CHAPTER 13: BIODIVERSITY: DIVERSITY INDICES USING N AND S

13.1 For the first set of numbers we have $N = 100$, $S = 8$.

$$d = \frac{7}{4.605} = 1.520, \quad d_{max} = \frac{99}{4.605} = 21.498, \quad \frac{d}{d_{max}} = 0.071, \quad \iota = 0.452.$$

From Appendix 4 $\alpha = 2.046$ or $S = 2.046 \cdot ln\left(1 + \dfrac{100}{2.046}\right) = 7.999$.

For the second set of numbers we have $N = 400$, $S = 11$.

$$d = \frac{10}{5.991} = 1.669, \quad d_{max} = \frac{399}{5.991} = 66.595, \quad \frac{d}{d_{max}} = 0.025, \quad \iota = \frac{2.398}{5.991} = 0.400.$$

From Appendix 4 $\alpha = 2.092$ or $S = 2.092 \cdot ln\left(1 + \dfrac{400}{2.092}\right) = 11.001$.

Comment: For d the second set has the higher diversity, for the standardized $\dfrac{d}{d_{max}}$ the first set has the higher diversity, for ι the first set has the higher diversity, for α the second set has the higher diversity. Using different indices for both N and S can give confusing results.

13.2 The numbers 34, 22, 15, 11, 8, 5, 3, 2 add up to $N = 100$ and $S = 8$. The measures used in the first problem of 13.1 are exactly the same as we would get here. So we could not recognize any difference even though the proportions from the numbers are quite different.

CHAPTER 14: BIODIVERSITY: DIVERSITY MEASURES USING RELATIVE ABUNDANCES

14.1

a.

p_i	0.90	0.10		
$p_i ln p_i$	−0.095	−0.230		$H_1 = 0.325$
p_i^2	0.810	0.010		$\dfrac{1}{0.820} = 1.220, \; ln(1.22) = H_2 = 0.198$

b.

p_i	0.90	0.09	0.01	
$p_i ln p_i$	−0.095	−0.217	−0.046	$H_1 = 0.358$
p_i^2	0.810	0.008	0.000	$\dfrac{1}{0.818} = 1.222, \; ln(1.222) = H_2 = 0.201$

c.

p_i	0.90	0.09	0.009	0.001	
$p_i ln p_i$	−0.095	−0.217	−0.042	−0.007	$H_1 = 0.361$
p_i^2	0.810	0.008	0.000081	0.000001	$\dfrac{1}{0.818} = 1.222, \; ln(1.222) = H_2 = 0.201$

d. The values of H_1 do increase a little between b and c as rare species are added, but only in the third decimal place. For H_2 we can recognize a difference between b and c only in the fourth decimal place. So for measures of information, rare species add very little to the values of the diversity indices.

14.2 For the first set of numbers we have

n_1	p_i	p_i^2	$p_i ln p_i$
50	0.500	0.250	−0.347
25	0.250	0.062	−0.347
15	0.150	0.022	−0.285
5	0.050	0.002	−0.150
2	0.020	0.0004	−0.078
1	0.010	0.0001	−0.046
1	0.010	0.0001	−0.046
1	0.010	0.0001	−0.046
Sums 100	1.000	0.338	−1.344

$$\lambda = 0.338, \; \frac{1}{\lambda} = 2.959, \; ln(2.959) = H_2 = 1.085$$

$$H_1 = 1.344$$

For the second set of numbers we have

n_1	p_i	p_i^2	$p_i ln p_i$
150	0.375	0.141	−0.368
100	0.250	0.062	−0.347

n_1	p_i	p_i^2	$p_i lnp_i$
75	0.188	0.035	−0.314
40	0.100	0.010	−0.230
20	0.050	0.002	−0.150
10	0.025	0.0006	−0.092
1	0.002	0.000006	−0.015
1	0.002	0.000006	−0.015
1	0.002	0.000006	−0.015
1	0.002	0.000006	−0.015
1	0.002	0.000006	−0.015
Sums 400	1.000	0.251	−1.576

$$\lambda = 0.251, \frac{1}{\lambda} = 3.984, \ ln(3.984) = H_2 = 1.382$$

$$H_1 = 1.576$$

In both of these examples, values of H_1 and H_2 could distinguish between the two samples. It is still difficult, however, to distinguish how much of the difference is due to differences in S versus species proportions.

CHAPTER 15: BIODIVERSITY: DOMINANCE AND EVENNESS

15.1 Using the data from Table 15.1 for a dominant distribution, recall that as S is doubled we add $ln2$ or 0.693 to the value of H_1 We have then for H_1

S	H_1	E_1	E_1	J_1
5	0.453	0.315	0.143	0.281
10	1.146	0.315	0.238	0.498
20	1.839	0.315	0.278	0.614

For H_2 recall as we double S the value of λ is divided by two

S	λ	$ln\left(\frac{1}{\lambda}\right) = H_2$	E_2	E_2'	J_2
5	0.814	0.206	0.246	0.057	0.069
10	0.407	0.899	0.246	0.162	0.390
20	0.204	1.592	0.246	0.205	0.531

The values of E repeat nicely, but E prime and J do not.

15.2 Using the data for the dominant distribution given in Table 15.1, we have for the suggested E

S	λ	$\dfrac{1-\lambda}{\left(1-\dfrac{1}{S}\right)}$
5	0.814	0.233
10	0.407	0.659

The suggested E does very badly on a repeat.

15.3 For the suggested E at the dominant distribution given in Table 15.1, we have

S	λ	$1-\left(\dfrac{\lambda-1}{S}\right)$
5	0.814	0.386
10	0.407	0.693

The suggested E does very badly on a repeat.

15.4 We will compare the above measure of E with E_2 and E_2 at λ_{min}

S	λ_{min}	$E=1-\left(\dfrac{\lambda-1}{S}\right)$	E_2	E_2'
5	0.973	0.227	0.206	0.007
10	0.941	0.159	0.106	0.007

The measure E_2 does as well at minimum values of λ. And if we desire only a value close to zero, E_2' is best, but it won't repeat.

CHAPTER 16: BIODIVERSITY: UNIFYING DIVERSITY AND EVENNESS MEASURES WITH CANONICAL EQUATIONS

16.1

n_i	p_i	p_i^2	$p_i ln p_i$
163	0.449	0.202	−0.360
144	0.397	0.158	−0.367
25	0.069	0.005	−0.184
10	0.028	0.001	−0.100
6	0.017	0.00029	−0.069
5	0.014	0.00020	−0.060
4	0.011	0.00012	−0.050
3	0.008	0.000064	−0.039
1	0.003	0.000009	−0.017
1	0.003	0.000009	−0.017
1	0.003	0.000009	−0.017
Total 363	1.001	0.367	−1.177

$S = 11$, $lnS = 2.398$, $H = 1.177$, $E = e^H/S = 0.295$, $lnE = -1.221$

Shannon decomposition: $1.177 = 2.398 - 1.221$

16.2 From the table above in 16.1, $\lambda = 0.367$, $\dfrac{1}{\lambda} = 2.725$, $ln\left(\dfrac{1}{\lambda}\right) = 1.002 = H_2$,

$\dfrac{1}{\lambda}S = 0.248 = E_2$, $lnE_2 = -1.396$.

Simpson's decomposition: $1.002 = 2.398 - 1.396$.

16.3

$$E_1' = \frac{(e^H - 1)}{(S-1)} \qquad\qquad E_2' = \frac{\left(\frac{1}{\lambda} - 1\right)}{(S-1)}$$

$$E_1' = \frac{2.245}{10} = 0.224 \qquad\qquad E_2' = \frac{1.725}{10} = 0.173$$

$$\ln E_1' = -1.494 \qquad\qquad \ln E_2' = -1.757$$

Decomposition:

$0.904 = 2.398 - 1.494 \qquad 0.641 = 2.398 - 1.757$

$H = 1.177 \qquad\qquad H_2 = 1.002$

Both measures will not decompose and recall they could also not pass the repeat.

16.4 Both E and E' will measure evenness, but the latter will not decompose or pass a repeat.

	n_i	p_i	p_i^2	$p_i \ln p_i$
	163	0.744	0.554	−0.220
	25	0.114	0.013	−0.248
	10	0.046	0.002	−0.141
	6	0.027	0.001	−0.099
	5	0.023	0.001	−0.086
	4	0.018	0.00033	−0.073
	3	0.014	0.00019	−0.059
	1	0.005	0.000021	−0.025
	1	0.005	0.000021	−0.025
	1	0.005	0.000021	−0.025
Total	219	1.001	0.572	−1.001

$S = 10$, $\ln S = 2.303$, $H = 1.001$, $E = 0.272$, $\ln E = -1.302$, $E' = 0.191$, $\ln E' = -1.654$

Shannon decomposition for E: $1.001 = 2.303 - 1.302$

For E': $0.649 = 2.303 - 1.654$

Both measures E and E' have gone down with the deletion of 144, indicating more dominance. However, only E will decompose.

$$\lambda = 0.572, \ \frac{1}{\lambda} = 1.748, \ \ln\left(\frac{1}{\lambda}\right) = 0.559, \ E_2 = 0.175, \ \ln E_2 = -1.744,$$

$E_2' = 0.083$, $\ln E_2' = -2.488$

Simpson decomposition for E_2: $0.559 = 2.303 - 1.744$

For E_2': $-0.185 = 2.303 - 2.488$

Both measures have gone down with the deletion of 144, indicating more dominance. However, only E will decompose, and E' yields a negative value, which is an impossible situation.

CHAPTER 17: BIODIVERSITY: SHE ANALYSIS AS THE ULTIMATE UNIFICATION THEORY OF BIODIVERSITY WITH THE COMPLETE BIODIVERSITYGRAM

17.1 The BDG shows lnS versus lnN is identical for all three communites. H and lnE are nearly identical for communities 1 and 3, but distinct for community 2.

17.2 At $lnN = 5.99$, all values are identical. If we had only this point, the communities would be indistinguishable. However, using the distributional fit, we can easily distinguish community 2 from 1 and 3 by either H or lnE.

CHAPTER 18: BIODIVERSITY: SHE ANALYSIS FOR COMMUNITY STRUCTURE IDENTIFICATION, SHECSI

18.1 Consider the decomposition equation $H = lnS + lnE$.

a. If $S_1 \neq S_2$, then $lnS_1 \neq lnS_2$. Consequently, for $H_1 = H_2$, E cannot be constant.

b. If $S_1 = S_2$, and $H_1 = H_2$, then $E_1 = E_2$ to maintain the relationship.

c. If $S_1 = S_2$, and $E_1 = E_2$, then $H_1 = H_2$ to maintain the relationship.

18.2 Using the data in Appendix 5 for bats, we calculated the following using a spreadsheet:

Cum N	Cum. S	H	lnS	lnE	$\dfrac{lnE}{lnS}$
206	25	3.01	3.22	−0.21	−0.06
344	34	3.23	3.53	−0.29	−0.08
579	38	3.24	3.64	−0.39	−0.11
707	39	3.19	3.66	−0.48	−0.13
901	45	3.32	3.80	−0.48	−0.13
1,002	46	3.32	3.82	−0.51	−0.14
1,255	50	3.32	3.91	−0.59	−0.15
1,405	50	3.33	3.91	−0.58	−0.15

The most constant of the measures is H, and lnE becomes more negative as S increases.

Tabulating similar N's for trees and bats, we have

N_{Tree}	N_{Bat}	S_{Tree}	S_{Bat}	H_{Tree}	H_{Bat}	E_{Tree}	E_{Bat}
217	206	32	25	2.34	3.01	0.32	0.81
301	344	37	34	2.30	3.23	0.27	0.74
603	579	51	38	2.37	3.24	0.21	0.68
603	707	52	39	2.36	3.19	0.20	0.62

In both cases the H's are the most constant. Bats have much higher evenness (no single-tons), which causes H to be higher for bats even though S is smaller.

18.3 For the New Zealand data the last observation indicates $N = 1,300$ and $S = 72$. From Appendix 4, we have $\alpha = 16.424$ or $72.002 = 16.424 \, ln\left(1 + \dfrac{1300}{16.424}\right)$. We have then for our estimate of entropy $ln(16.424) + 0.577 = 3.38 = E(H)$, forming a new table:

N	$S_{observed}$	$S_{Log\ series}$	$H_{observed}$	$H_{Log\ series}$	$E_{observed}$	$E_{Log\ series}$
200	41	42	3.22	3.38	0.62	0.70
400	51	53	3.28	3.38	0.43	0.55
1,000	67	68	3.35	3.38	0.43	0.43
1,300	72	72	3.37	3.38	0.41	0.41

Recalling the regression for lnS versus lnE is analogous to the decomposition equation. We have for pattern:

For the Log series: $lnS = 3.38 - 1.00lnE$

For the observed: $lnS = 3.04 - 1.35lnE$

The difference is 0.35

We have for value: Log series $E(H) = 3.38$

For observed: $H = 3.37$

The difference is 0.01.

18.4 For the Canonical Log normal recall $E(H) = 1.128\sqrt{lnS}$. Now $ln(72) = 4.277$ and $\sqrt{4.277} = 2.068$, so that $1.128(2.068) = 2.33 = E(H)$. If $S = 1,000$ then $E(H) = 2.96$. The entropy estimate (May 1975) is unsuitable for high-diversity situations.

REFERENCES

Alatalo, R. V. 1981. Problems in the measurement of evenness in ecology. *Oikos* 37(2): 199–204.

Alroy, J., M. Aberhan, D. J. Bottjer, M. Foote, F. T. Fursich, P. J. Harries, A. J. Hendy, S. M. Holland, L. C. Ivany, W. Kiessling, M. A. Kosnik, C. R. Marshall, A. J. McGowan, A. I. Miller, T. D. Olszewski, M. E. Patzkowsky, S. E. Peters, L. Villier, P. J. Wagner, N. Bonuso, P. S. Borkow, B. Brenneis, M. E. Clapham, L. M. Fall, C. A. Ferguson, V. L. Hanson, A. Z. Krug, K. M. Layou, E. H. Lockey, S. Nurnberg, C. M. Powers, J. A. Sessa, C. Simpson, A. Tomasovych, and C. C. Visaggi. 2008. Phanerozoic trends in the global diversity of marine invertebrates. *Science* 321: 97–100.

Anscombe, F. J. 1950. Sampling theory of the negative binomial and logarithmic series distributions. *Biometrika* 37(3 & 4): 358–382.

Aronson, R. B. 1994. Scale-dependent biological processes in the marine environment. *Oceanography and Marine Biology: An Annual Review* 32: 435–460.

Arrhenius, O. 1921. Species and area. *Journal of Ecology* 9: 95–99.

Berger, W. H. and F. L. Parker. 1970. Diversity of planktonic foraminifera in deep-sea sediments. *Science* 168: 1345–1347.

Bliss, C. I. 1965. Analysis of some insect trap records. In G. P. Patil, ed., *Classical and Contagious Discrete Distributions*, 385–397. Oxford, UK: Pergamon.

Bliss, C. I. and R. A. Fisher. 1953. Fitting the negative binomial distribution to biological data. *Biometrics* 9: 176–199.

Brian, M. V. 1953. Species frequencies in random samples from animal populations. *Journal of Animal Ecology* 22: 57–63.

Bulmer, M. G. 1974. On fitting the Poisson lognormal distribution to species abundance data. *Biometrics* 30(1): 101–110.

Buzas, M. A. 1967. An application of canonical analysis as a method for comparing faunal areas. *Journal of Animal Ecology* 36: 563–577.

Buzas, M. A. 1990. Another look at confidence limits for species proportions. *Journal of Paleontology* 64(5): 842–843.

Buzas, M. A. 2004. Community structure of foraminifera from two Miocene beds at Calvert Cliffs, Maryland. *Journal of Foraminiferal Research* 34: 208–213.

Buzas, M. A. and S. J. Culver. 1991. Species diversity and dispersal of benthic foraminifera. *Bioscience* 41(7): 483–489.

Buzas, M. A. and T. G. Gibson. 1969. Species diversity: Benthonic foraminifera in western north Atlantic. *Science* 163: 72–75.

Buzas, M. A. and T. G. Gibson. 1990. Spatial distribution of Miocene foraminifera at Calvert Cliffs, Maryland. *Smithsonian Contributions to Paleobiology* 68: 1–35.

Buzas, M. A. and L. C. Hayek. 1996. Biodiversity resolution: An integrated approach. *Biodiversity Letters* 3: 401–403.

Buzas, M. A. and L. C. Hayek. 1998. SHE analysis for biofacies identification. *Journal of Foraminiferal Research* 28: 233–239.

Buzas, M. A. and L. C. Hayek. 2005. On richness and evenness within and between communities. *Paleobiology* 31: 199–220.

Buzas, M. A., L. C. Hayek, and S. J. Culver. 2007a. Community structure of benthic foraminifera in the Gulf of Mexico. *Marine Micropaleontology* 65: 43–53.

Buzas, M. A., L. C. Hayek, B. W. Hayward, H. R. Grenfell, and A. T. Sabaa. 2007b. Biodiversity and community structure of deep-sea foraminifera around New Zealand. *Deep-Sea Research I* 54: 1641–1654.

Buzas, M. A., C. F. Koch, S. J. Culver, and N. F. Sohl. 1982. On the distribution of species occurrences. *Paleobiology* 8(2): 143–150.

Buzas, M. A. and K. P. Severin. 1993. Foraminiferal densities and pore water chemistry in the Indian River, Florida. *Smithsonian Contributions to the Marine Sciences* 36: 1–38.

Buzas, M. A., R. K. Smith, and K. A. Beem. 1977. Ecology and systematics of foraminifera in two Thalassia habitats, Jamaica, West Indies. *Smithsonian Contributions to Paleobiology* 31: 1–139.

Buzas-Stephens, P. and M. A. Buzas. 2005. Population dynamics and dissolution of Foraminifera in Nueces Bay, Texas. *Journal of Foraminiferal Research* 35(3): 248–258.

Cassie, J. R. M. 1962. Frequency distribution models in the ecology of plankton and other organisms. *Journal of Animal Ecology* 31: 65–92.

Caswell, H. 1976. Community structure: A neutral model analysis. *Ecological Monographs* 46: 327–354.

Chao, A. 1984. Nonparametric estimation of the number of classes in a population. *Scandinavian Journal of Statistics* 11: 265–270.

Chao, A. 1987. Estimating the population size for capture–recapture data with unequal catchability. *Biometrics* 43(4): 783–791.

Chao, A., R. K. Colwell, C. W. Lin, and N. J. Gotelli. 2009. Sufficient sampling for asymptotic minimum species richness estimators. *Ecology* 90: 1125–1133.

Cohen, J. 1968. Multiple regression as a general data-analytic system. *Psychological Bulletin* 70(6): 426–443.

Connor, E. F. and E. D. McCoy. 1979. The statistics and biology of the species–area relationship. *American Naturalist* 113(6): 791–832.

Corbet, A. S. 1941. The distribution of butterflies in the Malay Peninsula (Lepid.). *Proceedings of the Royal Entomological Society of London,* series A, 16(10–12): 101–116.

Cramér, H. 1946. *Mathematical Methods of Statistics.* Princeton, NJ: Princeton University Press.

Dallmeier, E., R. B. Foster, C. B. Romano, R. Rice, and M. Kabel. 1991. *User's Guide to the Beni Biosphere Reserve Biodiversity Plots 01 and 02, Bolivia.* Washington, DC: Smithsonian Institution.

Engen, S. 1974. On species frequency models. *Biometrika* 61(2): 263–270.

Erwin, T. L. 1982. Tropical forests: Their richness in Coleoptera and other arthropod species. *Coleopterists Bulletin* 36: 74–82.

Ewens, W. J. 1972. The sampling theory of selectively neutral alleles. *Theoretical Population Biology* 3: 87–112.

Fager, E. W. 1972. Diversity: A sampling study. *The American Naturalist* 106: 293–310.

Feller, W. 1957. *An Introduction to Probability Theory and Its Applications*. New York: John Wiley and Sons.

Ferrari, F. and L. C. Hayek. 1990. Monthly differences in the distributions of sex and asymmetry in a looking-glass copepod, *Pleuromamma xithias*, off Hawaii. *Journal of Crustacean Biology* 10(1): 114–127.

Fisher, R. A. 1941. The negative binomial distribution. *Annals of Eugenics* 11: 182–187.

Fisher, R. A. 1943. The relation between the number of species and the number of individuals in a random sample of an animal population. Part 3. A theoretical distribution for the apparent abundance of different species. *Journal of Animal Ecology* 12(1): 54–58.

Fisher, R. A., A. S. Corbet, and C. B. Williams. 1943. The relation between the number of species and the number of individuals in a random sample of an animal population. *Journal of Animal Ecology* 12: 42–58.

Gadagkar, R. 1989. An undesirable property of Hill's diversity index N_2. *Occologia* 80: 140–141.

Gilbert, F. S. 1980. The equilibrium theory of island biogeography: Fact or fiction. *Journal of Biogeography* 7: 209–235.

Gini, C. 1912. Variabilitita e mutabilita. Studi economico-giuridici. *Cura Facolta Giurisprudenza Universita Cagliari*, A. 3, pt. 2, Cagliari, Italy.

Gleason, H. A. 1922. On the relation between species and area. *Ecology* 3(2):158–162.

Gotelli, N. J. and R. K. Colwell. 2001. Quantifying biodiversity procedures and pitfalls in the measurement and comparison of species richness. *Ecology Letters* 4: 379–391.

Gotelli, N. J. and B. J. McGill. 2006. Null versus neutral models: What's the difference? *Ecogeography* 29: 793–800.

Grundy, P. M. 1951. The expected frequencies in a sample of an animal population in which the abundances of species are lognormally distributed. Part I. *Biometrika* 38(3 & 4): 427–434.

Handley, C. O., Jr. 1976. Mammals of the Smithsonian Venezuelan Project. *Brigham Young University Science Bulletin, Biological Series* 20(5): 1–85.

Hart, C. W., L. C. Hayek, J. Clarke, and D. Clarke. 1985. The life history and ecology of the entocytherid ostracod *Uncinocythere occidentalis* (Kosloff and Whitman) in Idafao. *Smithsonian Contributions to Zoology* 149: 1–22.

Hayek, L. C. 1994. Analysis of amphibian biodiversity data. In R. Heyer, M. Donnelley, R. McDiarmid, L. C. Hayek, and M. Foster, eds., *Measuring and Monitoring Biological Diversity: Standard Methods for Amphibians*, 207–269. Washington, DC: Smithsonian Institution Press.

Hayek, L. C. and M. A. Buzas. 1997. *Surveying Natural Populations*. New York: Columbia University Press.

Hayek, L. C. and M. A. Buzas. 1998. SHE analysis: An integrated approach to the analysis of forest biodiversity. In F. Dallmeier and J. A. Comiskey, eds., *Forest Biodiversity Research, Monitoring and Modeling*. Paris: Parthenon Publishing Group: 311–321.

Hayek, L. C. and M. A. Buzas. 2006. The martyrdom of St. Lucie: Decimation of a meiofauna. *Bulletin of Marine Science* 79(2): 341–352.

Hayek, L. C., M. A. Buzas, and L. Osterman. 2007. Community structure of foraminiferal communities within temporal biozones for the western Arctic ocean. *Journal of Foraminiferal Research I* 37(1): 33–40.

Heip, C. 1974. A new index measuring evenness. *Journal of the Marine Biological Association UK* 54: 555–557.

Heyer, W. R. 1976. Studies in larval amphibian habitat partitioning. *Smithsonian Contributions to Zoology* 242: 1–27.

Heyer, W. R. 1979. Annual variation in larval amphibian populations within a temperate pond. *Journal of the Washington Academy of Sciences* 69(2): 65–74.

Hill, M. O. 1973. Diversity and evenness: A unifying notation and its consequences. *Ecology* 54(2): 427–432.

Holme, N. A. 1950. Population-dispersion in *Tellina tenuis* Da Costa. *Journal of the Marine Biological Association UK* 29(2): 267–280.

Hubbell, S. P. 1997. A unified theory of biogeography and relative species abundance and its application to tropical rain forests and coral reefs. *Coral Reefs* 16 suppl: S9–S21.

Hubbell, S. P. 2001. *The Unified Neutral Theory of Biodiversity and Biogeography*. Princeton, NJ: Princeton University Press.

Hurlbert, S. H. 1971. The nonconcept of species diversity: A critique and alternative parameters. *Ecology* 52(4): 577–586.

Huston, M. A. 1994. *Biological Diversity*. London: Cambridge University Press.

Hutchinson, G. E. 1965. *The Ecological Theater and the Evolutionary Play*. New Haven, CT: Yale University Press.

Johnson, R. G. 1959. Spatial distribution of *Phoronopsis viridis Hilton*. *Science* 129: 1221.

Jost, L. 2006. Entropy and diversity. *Oikos* 113(2): 363–375.

Jost, L. 2007. Partitioning diversity into independent alpha and beta components. *Ecology* 88: 2427–2439.

Koch, A. L. 1969. The logarithm in biology. II. Distributions simulating the lognormal. *Journal of Theoretical Biology* 23(2): 251–268.

Krebs, C. J. 1999. *Ecological Methodology*. New York: Harper & Row.

Lamas, G., R. K. Robbins, and D. J. Harvey. 1991. A preliminary butterfly fauna of Pakitza, Parque Nacional del Manu, Peru, with an estimate of its species richness. *Publicaciones del Museo de Historia Natural, Universidad Nacional Mayor de San Marcos* (A)40: 1–19.

Lambshead, P. J. D. and H. M. Platt. 1988. Analysing disturbance with the Ewens/Caswell neutral model: Theoretical review and practical assessment. *Marine Ecology, Progress Series* 43: 31–41.

Lambshead, P. J. D., H. M. Platt, and K. M. Shaw. 1983. The detection of differences among assemblages of marine benthic species based on an assessment of dominance and diversity. *Journal of Natural History* 17: 859–874.

Latham, R. E. and R. E. Ricklefs. 1993. Continental comparisons of temperate-zone species diversity. In Ricklefs, R. E. and D. Schluter, eds., *Species Diversity in Ecological Communities*, 294–314. Chicago: University of Chicago Press.

Levin, S. A. 1992. The problem of pattern and scale in ecology. *Ecology* 73: 1943–1967.

Lloyd, M. and R. J. Ghelardi. 1964. A table for calculating the equitability component of species diversity. *Journal of Animal Ecology* 33: 217–225.

Ludwig, J. A. and J. F. Reynolds. 1988. *Statistical Ecology: A primer on Methods and Computing*. New York: John Wiley & Sons.

MacArthur, R. H. 1957. On the relative abundance of bird species. *Proceedings of the National Academy of Sciences* 43: 293–295.

MacArthur, R. H. 1965. Patterns of species diversity. *Biological Reviews* 40: 510–533.

MacArthur, R. H. and E. O. Wilson. 1967. *The Theory of Island Biogeography*. Princeton, NJ: Princeton University Press.

Magurran, A. E. 1988. *Ecological Diversity and Its Measurement*. Princeton, NJ: Princeton University Press.

Magurran, A. E. 2004. *Measuring Biological Diversity*. Oxford, UK: Blackwell.

Margalef, D. R. 1957. Information theory in ecology. *Memorias de la Real Academia de Ciencias y Artes de Barcelona* 23: 378–449.

Mathcad. Pius Version 12.0. Cambridge, MA: MathsoftInc.

May, R. M. 1975. Patterns of species abundance and diversity. In M. L. Cody and J. M. Diamond, eds., *Ecology and Evolution of Communities*, 81–120. Cambridge, MA: Belnap Press of Harvard University.

McGill, B. J. 2003. A test of the unified neutral theory of biodiversity. *Nature* 422: 881–885.

McGill, B. J. 2006. A renaissance in the study of abundance. *Science* 314: 770–772.

McGill, B. J., R. S. Etienne, J. S. Gray, D. Alonso, M. J. Anderson, H. K. Benecha, M. Dornelas, B. J. Enquist, J. L. Green, F. He, A. J. Hurlbert, A. E. Magurran, P. A. Marquet, B. A. Maurer, A. Ostling, C. U. Soykan, K. I. Ugland, and E. P. White. 2007. Species abundance distributions: Moving beyond single prediction theories to integration within an ecological framework. *Ecology Letters* 10: 1–21.

McIntyre, A. D., J. M. Elliott, and D. V. Ellis. 1984. Introduction: Design of sampling programmes. In N. A. Holme and A. D. McIntyre, eds., *Methods for the Study of Marine Benthos*, 11–26. IPB Handbook 16. London: Blackwell Scientific Publications.

Murray, J. W. 2003. Patterns in the cumulative increase in species from foraminiferal time-series. *Marine Micropaleontology* 48: 1–21.

Olszewski, T. 2004. A unified mathematical framework for the measurement of richness and evenness within and among multiple communities. *Oikos* 104: 377–387.

Patil, G. P. and C. Taillie. 1982. Diversity as a concept and its measurement. *Journal of the American Statistical Association* 77: 548–561.

Peters, S. E. 2004. Evenness of Cambrian–Ordovician benthic marine communities in North America. *Paleobiology* 30: 325–346.

Phleger, F. B. 1956. Significance of living foraminiferal populations along the central Texas coast. *Contributions from the Cushman Foundation for Foraminiferal Research* 1(4): 106–151.

Phleger, F. B. 1960. *Ecology and Distribution of Recent Foraminifera*. Baltimore: Johns Hopkins Press.

Pielou, E. C. 1966. The measurement of diversity in different types of biological collections. *Journal of Theoretical Biology* 13: 131–144.

Pielou, E. C. 1975. *Ecological Diversity*. New York: John Wiley & Sons.

Platt, H. M. and P. J. D. Lambshead. 1985. Neutral model analysis of patterns of marine benthic species diversity. *Marine Ecology Progress Series* 24: 75–81.

Platt, H. M., K. M. Shaw, and P. J. D. Lambshead. 1984. Nematode species abundance patterns and their use in the detection of environmental perturbation. *Hydrobiologia* 118: 59–66.

Podani, J. 1992. Space series analysis of vegetation: Processes reconsidered. *Abstracta Botanica* 16: 25–29.

Preston, F. W. 1948. The commonness, and rarity, of species. *Ecology* 29(3): 254–283.

Preston, F. W. 1962. The canonical distribution of commonness and rarity: Parts I & II. *Ecology* 43(2): 185–215; 43(3): 410–432.

Ralls, K., L. A. Hayek, and C. O. Handley Jr. 1982. Correlations between three possible measures of size in neotropical bats. *Säugetierkundliche Mitteilungen* 30(3): 190–198.

Renyi, A. 1965. On the foundations of information theory. *Review of International Statistical Institute* 33: 1–14.

Rice, M. R., J. Piraino, and H. F. Reichardt. 1983. Observations on the ecology and reproduction of the sipunculan *Phascolion cryptus* in the Indian River Lagoon. *Florida Scientist* 46(3/4): 382–396.

Ricotta, C. 2005. On hierarchial diversity decomposition. *Journal of Vegetative Science* 6: 223–226.

Routledge, R. D. 1983. Evenness indices: Are any admissible? *Oikos* 40: 149–151.

Samuel-Cahn, E. 1975. Remark on a formula by Fisher. *Journal of the American Statistical Association* 70(351): 720.

Sanders, H. L. 1968. Marine benthic diversity: A comparative study. *American Naturalist* 102(925): 243–282.

SAS for Personal Computers. 2003. NC: SAS Institute.

Shannon, C. E. 1948. A mathematical theory of communication. *Bell System Technical Journal* 27(1/2): 379–423; 27(3): 623–656.

Sheldon, A. L. 1969. Equitability indices: Dependence on the species count. *Ecology* 50(3): 466–467.

Shipley, B., D. Vile, and E. Garnier. 2006. From plant traits to plant communities: A statistical mechanistic approach to biodiversity. *Science 3*, November 314(5800): 812–814.

Simpson, E. H. 1949. Measurement of diversity. *Nature* 163(4148): 688.

Singleton. (Cited in Preston 1948.)

Smith, B. and J. B. Wilson. 1996. A consumer's guide to evenness indices. *Oikos* 70: 70–82.

SPSS for Windows. 2007. Version 15. Chicago: SPSS, Inc.

Systat for Windows. 2005. Version 12. Chicago: Systat, Inc.

Taillie, C. 1979. Species equitability: A comparative approach. In J. Grassle, G. Patil, W. K. Smith, and C. Taillie, eds., *Ecological Diversity in Theory and Practice,* 51–62. Fairland, MD: Cooperative Publishing House.

Taylor, L. R. 1961. Aggregation, variance and the mean. *Nature* 189(4766): 732–735.

Taylor, L. R., R. A. Kempton, and I. P. Woiwood. 1976. Diversity statistics and the log-series model. *Journal of Animal Ecology* 45(1): 255–272.

Taylor, L. R., I. P. Woiwood, and J. N. Perry. 1980. Variance and the large-scale spatial stability of aphids, moths, and birds. *Journal of Animal Ecology* 49: 831–854.

Tokeshi, M. 1993. Species abundance patterns and community structure. *Advances in Ecological Research* 24: 111–186.

Ulrich, W. 2001. Models of relative abundance distributions II: Diversity and evenness statistics. *Polish Journal of Ecology* 49: 159–175.

Vezina, A. F. 1988. Sampling variance and the design of quantitative surveys of the marine benthos. *Marine Biology* 97: 151–155.

Wagner, P. J., M. A. Kosnik, and S. Lidgard. 2006. Abundance distributions imply elevated complexity of post-Paleozoic marine ecosystems. *Science* 314: 1289–1292.

Whittaker, R. H. 1972. Evolution and measurement of species diversity. *Taxon* 21(2/3): 213–251.

Whittaker, R. H. 1975. *Communities and Ecosystems*. 2nd ed. New York: Macmillan.

Williams, C. B. 1943. Area and number of species. *Nature* 152(3853): 264–267.

Wilson, B. 2008. Using SHEBI (SHE analysis for biozone identification): To proceed from the top down or the bottom up? A discussion using two Miocene foraminiferal successions from Trinidad, West Indies. *Palaios* 23: 636–644.

Wratten, S. D. 1974. Aggregation in the birch aphid *Euceraphis punctipennis* (Zett.) in relation to food quality. *Journal of Animal Ecology* 43: 191–198.

INDEX